HANDBOOK OF EROSION MODELLING

Handbook of Erosion Modelling

Edited by

R.P.C. Morgan
National Soil Resources Institute
Cranfield University

and

M.A. Nearing
USDA-ARS
Southwest Watershed Research Center

WILEY-BLACKWELL

A John Wiley & Sons, Ltd., Publication

Library of Congress Cataloguing-in-Publication Data
Handbook of erosion modelling / edited by R.P.C. Morgan and M.A. Nearing.
p. cm.
Includes bibliographical references and index.
ISBN 978-1-4051-9010-7 (cloth)
1. Soil erosion–Simulation methods. I. Morgan, R.P.C. (Royston Philip Charles), 1942–
II. Nearing, M.A. (Mark A.)
S627.M36H36 2011
631.4′50113–dc22
2010026596

A catalogue record for this book is available from the British Library.
This book is published in the following electronic formats: eBook 9781405190107; Wiley Online Library 9781444328455

Set in 9/11.5pt Trump Mediaeval by SPi Publisher Services, Pondicherry, India
Printed and bound in Malaysia by Vivar Printing Sdn Bhd

1   2011

Contents

Contributors

J.C. BATHURST *School of Civil Engineering and Geosciences, Newcastle University, Newcastle upon Tyne NE1 7RU, United Kingdom*

K.J. BEVEN *Lancaster Environment Centre, University of Lancaster, Lancaster LA1 4YW, United Kingdom; GeoCentrum, Uppsala University, Uppsala, Sweden; ECHO/ISTE, EPFL, Lausanne, Switzerland*

G.S. BILOTTA *School of Environment and Technology, University of Brighton, Cockcroft Building, Brighton BN2 4GJ, United Kingdom*

R.E. BRAZIER *School of Geography, University of Exeter, Amory Building, Exeter EX4 4RJ, United Kingdom*

K. COUGHLAN *P O Box 596, Annerley, Queensland, Australia 4103*

S.M. DABNEY *USDA-ARS, National Sedimentation Laboratory, 598 McElroy Drive, Oxford, MS 38655, USA*

L.K. DEEKS *National Soil Resources Institute, Cranfield University, Cranfield, Bedfordshire MK43 0AL, United Kingdom*

J.H. DUZANT *National Soil Resources Institute, Cranfield University, Cranfield, Bedfordshire MK43 0AL, United Kingdom*

W.J. ELLIOT *USDA Forest Service, Rocky Mountain Research Station, 1221 South Main Street, Moscow, ID 83843, USA*

B. FENTIE *Queensland Department of Environment and Resource Management, 80 Meiers Road, Indooroopilly, Queensland, Australia 4068*

J. FREER *School of Geographical Sciences, University of Bristol, University Road, Bristol BS8 1SS, United Kingdom*

D.C. GOODRICH *USDA-ARS, Southwest Watershed Research Center, 2000 East Allen Road, Tucson, AZ 85719, USA*

G. GOVERS *Physical and Regional Geography Research Group, Department of Earth and Environmental Sciences, Katholieke Universiteit Leuven, GEO-Institute, Celestijnenlaan 200E, 3001 Heverlee, Belgium*

A.J.T. GUERRA *Department of Geography, Institute of Geosciences, Federal University of Rio de Janeiro, Avenida Jose Luiz Ferraz 250, Apto 1706, CEP.22.790-587, Rio de Janeiro, Brazil*

D.P. GUERTIN *Landscape Studies Program, School of Natural Resources, University of Arizona, Tucson, AZ 85721, USA*

P.B. HAIRSINE *CSIRO Land and Water Division, G.P.O.Box 1666, Canberra 2601 Australian Capital Territory, Australia*

G.R. HANCOCK *School of Environment and Life Sciences, Faculty of Science, The University of Newcastle, Callaghan, New South Wales 2308, Australia*

R. HESSEL *Soil Science Centre, Alterra, Wageningen University and Research Centre, PO Box 47, 6700 AA Wageningen, The Netherlands*

C.J. HUTTON *School of Geography, University of Exeter, Amory Building, Exeter EX4 4RJ, United Kingdom*

V.G. JETTEN *Department of Earth Systems Analysis, International Institute of Geoinformation Science and Earth Observation, Hengelosestraat 99, PO Box 6, 7500 AA, Enschede, The Netherlands*

T. KRUEGER *School of Environmental Sciences, University of East Anglia, Norwich NR4 7TJ, United Kingdom*

J.M. LAFLEN *USDA-ARS (retired), 5784 Highway 9, Buffalo Center, IA 5042, USA*

D.T. LIGHTLE *USDA-NRCS, National Soil Survey Center, 100 Centennial Mall North, Lincoln, NE 68508-3866, USA*

B. LIU *School of Geography, Beijing Normal University, 19 Xinwai Street, Beijing 100875, China*

M.P. MANETA *Geosciences Department, University of Montana, 32 Campus Drive #1296, Missoula, MT 59812, USA*

R.K. MISRA *Faculty of Engineering and Surveying, University of Southern Queensland, Toowoomba, Queensland, Australia 4350*

R.P.C. MORGAN *National Soil Resources Institute, Cranfield University, Cranfield, Bedfordshire MK43 0AL, United Kingdom*

M.A. NEARING *USDA-ARS, Southwest Watershed Research Center, 2000 East Allen Road, Tucson, AZ 85719, USA*

J.P. NUNES *Centre for Environmental and Marine Studies (CESAM), Department of Environment and Planning, University of Aveiro, Campus Universitário de Santiago, 3810-193 Aveiro, Portugal*

A.J. PARSONS *Department of Geography, University of Sheffield, Sheffield S10 2TN, United Kingdom*

J.W.A. POESEN *Physical and Regional Geography Research Group, Department of Earth and Environmental Sciences, Katholieke Universiteit Leuven, GEO-Institute, Celestijnenlaan 200E, B-3001 Heverlee, Belgium*

Y. QIU *School of Geography, Beijing Normal University, 19 Xinwai Street, Beijing 100875, China*

J.N. QUINTON *Lancaster Environment Centre, University of Lancaster, Lancaster LA1 4YQ, United Kingdom*

K.G. RENARD *USDA-ARS, Southwest Watershed Research Center, 2000 East Allen Road, Tucson, AZ 85719-1596, USA*

P.R. ROBICHAUD *USDA Forest Service, Rocky Mountain Research Station, 1221 South Main Street, Moscow, ID 83843, USA*

C.W. ROSE *The Griffith School of Environment, Griffith University, Nathan Campus, Brisbane, Queensland, Australia 4111*

A. SOARES DA SILVA *Federal University of Rio de Janeiro, Rua Hermengarda 151, Apto 906 – Meier. CEP.20710-010 Rio de Janeiro, Brazil*

D.B. TORRI *IRPI CNR, Via Madonna Alta 126, 06128 Perugia, Italy*

T. VANWALLEGHEM *Department of Agronomy, Institute for Sustainable Agriculture – CSIC, Finca Alameda del Obispo, Apartado Correos 4084, Córdoba 14080, Spain*

J. WAINWRIGHT *Department of Geography, University of Sheffield, Sheffield S10 2TN, United Kingdom*

G.R. WILLGOOSE *School of Engineering, Faculty of Engineering and the Built Environment, The University of Newcastle, Callaghan, New South Wales 2308, Australia*

G.A. WOOD *Integrated Environmental Systems Institute, Cranfield University, Cranfield, Bedfordshire MK43 0AL, United Kingdom*

D.C. YODER *Biosystems Engineering and Soil Science, University of Tennessee, 2506 E J Chapman Drive, Knoxville, TN 37996-4531, USA*

B. YU *School of Engineering, Griffith University, Nathan Campus, Brisbane, Queensland, Australia 4111*

1 Introduction

R.P.C. MORGAN

National Soil Resources Institute, Cranfield University, Cranfield, Bedfordshire, UK

The movement of sediment and associated pollutants over the landscape and into water bodies is of increasing concern with respect to pollution control, prevention of muddy floods and general environmental protection. This concern exists whether the sediment is derived from farmland, road banks, construction sites, recreation areas or other sources. In today's environment it is often considered of equal or even greater importance than the effects of loss of soil on-site, with its implications for declining agricultural productivity, loss of biodiversity and decreased amenity and landscape values. With the expected changes in climate over coming decades, there is a need to predict how environmental problems associated with sediment are likely to be affected so that appropriate management systems can be put in place.

Whilst it is possible to instrument a few individual farms and catchments in order to obtain the data to evaluate the current situation and propose best management practices, it is not feasible to study every location on the Earth's surface in detail. Instead, evaluation and predictive tools need to be applied to assess current problems, predict future trends and provide a scientific base for policy and management decisions. Erosion models can fulfil this function provided that they are robust and used correctly. Despite, or maybe even because of, the vast amount of research over

the last 30 years or more on erosion modelling, potential model-users are confronted with a multiplicity of models from which to choose, often with little guidance on which might be the best for particular circumstances or the steps required to apply a selected model to a given situation. Many models have been tested for only a limited range of conditions of climate, soils and land use, and little information is available to enable a user to assess in advance how well a model might perform under different conditions. Models range from empirical to physically- or process-based, and vary considerably in their complexity and the amount of data input required. Very little guidance is available on how accurate that data input has to be, or what effect different levels of accuracy can have on the accuracy of the model output. Further, sediment problems can exist at scales that range from a farmer's field or a small construction site to the effects of sediment transport and deposition in small and large catchments. Somewhat limited information exists on the range of scales over which different models can operate successfully, leaving the user uncertain on whether a particular model is the most appropriate for a given scale. In the worst case, as a result of a lack of clear guidance, the user may choose a totally inappropriate model.

Users can obtain a list of the leading soil erosion models from the Internet site http://soilerosion.net/doc/models_menu.html. Links are provided to other sites associated specifically with each model from which the software can be

Handbook of Erosion Modelling, 1ˢᵗ edition. Edited by R.P.C. Morgan and M.A. Nearing. © 2011 Blackwell Publishing Ltd.

Table 1.1 Erosion models used in the case studies.

Title	Sources	Case studies
EUROSEM	Morgan et al. (1998) http://www.es.lancs.ac.uk/people/johnq/EUROSEM.html	Chapter 5
GUEST	Misra and Rose (1996)	Chapter 11
LISEM	Jetten and de Roo (2001) http://www.itc.nl/lisem	Chapter 12
Modified MMF	Morgan and Duzant (2008)	Chapter 13
RUSLE	Renard et al. (1997) http://fargo.nserl.purdue.edu/rusle2_dataweb/RUSLE2_Index.htm	Chapter 8
SHETRAN	Ewen et al. (2000) http://www.ceg.ncl.ac.uk/shetran	Chapter 14
SIBERIA	Willgoose et al. (1991) http://www.telluricresearch.com/siberia_8.30_manual.pdf	Chapter 18
WEPP	Flanagan and Nearing (1995) http://topsoil.nserl.purdue.edu/nserlweb/weppmain/wepp.html	Chapters 9, 10, 15, 16

downloaded along with the user manual. Whilst the majority of the links are valid and the site is a useful starting point for finding out what models exist, there are some links which are out-of-date and either do not work or are no longer the most appropriate. Clearly no such site can be fully comprehensive, and there will inevitably be some models which are not included. Table 1.1 lists the models which are used in this Handbook together with details of published sources and, where they exist, relevant Internet sites. Knowing which models are available is only a starting point. As indicated above, the user needs advice on how well the models perform and the conditions to which they can be applied. Previous experience with the models is extremely valuable, particularly where the output of several models is compared for the same conditions. Boardman and Favis-Mortlock (1998) discussed the performance of various models when applied to common sets of data at a hillslope scale, and De Roo (1999) presented the results of a similar exercise carried out at a small catchment scale. More recently, Harmon and Doe (2001) provided details of a range of models, physically-based and empirical, which can be used over various spatial and temporal scales to assess the short- and long-term effects of different land management strategies. These publications, however, describe erosion models more from a research than a user perspective. Although they are a source of useful information, they do little to help potential model users to answer the questions raised earlier, or to guide them in the selection of the most appropriate model for a specific application, taking account of the objectives, the environmental conditions and the availability of data. Also, since their publication, there has been an increasing use of geographical information system (GIS) techniques in analysing data for planning and decision-making, and erosion models have been increasingly integrated into geospatial systems, particularly at large catchment and regional scales.

The *Handbook of Erosion Modelling* seeks to address these issues and provide the model user with the tools to evaluate different erosion models and select the most appropriate for a specific purpose, compatible with the type of input data that are available. The book is aimed at model users within government, non-governmental organisations, academic institutions and consultancies involved in environmental assessment, planning, policy and research. The intention is to give existing and potential model users working in the erosion control industry greater confidence in selecting and using models by providing an insight into what users can expect of models in terms of robustness, accuracy and data requirements, and by raising the questions that users need to ask when selecting a model that is appropriate to the type and scale of their problem. It is important that users understand both the advantages and limitations of erosion models.

The Handbook is arranged in two main parts. The first part introduces the user to some important generic issues associated with erosion models. Chapter 2 sets out the various stages that a user should go through when selecting and applying an erosion model, and shows that these are much the same as erosion scientists adopt when developing their models. There is much common ground between model developers and model users, probably more so than most users are aware of. The next four chapters take key issues and discuss them in detail, along with solutions which model users might adopt. Chapter 3 looks at the question of calibration. This is a controversial topic with opinions ranging from those who consider that it is impossible to calibrate the more complex, physically-based models and those who believe that calibration is essential. This chapter is broadly in favour of calibration, showing how it can improve the quality of predictions both in terms of erosion rates and the spatial distribution of erosion. Chapter 4 raises the issue of uncertainty in model predictions. After discussing why we should worry about uncertainty, various approaches are described which can be used to reduce the level of uncertainty. How successful these are depends on the causes of the uncertainty, and model users need to be encouraged to appreciate and understand these. Uncertainty is taken further in Chapter 5, which shows how one approach is used in practice with reference to the application of one specific erosion model. Chapter 6 reviews the issues posed by scale. Many problems faced by users relate to a single scale, be it field, hillslope, small catchment or large catchment, but others need to be addressed at a range of scales. This chapter looks at the problems involved when moving from one scale to another with the difficulty of modelling interconnectivity between hillslope and river systems. At present there are

few solutions to the problems that arise when modelling across a range of scales, but several ideas for further research are presented whereby model development and data collection need to become more fully integrated. Chapter 7 shows the importance of choosing the right model for a specific problem and scale, and the implications of using inappropriate models. A frequent occurrence is the misunderstanding by the user of either the problem being addressed or what specific models are able to achieve. Although a dynamic process-based model is often the best choice, there are many situations in which it will not perform better than a simpler statistical model.

Part 2 of the Handbook looks at specific applications and shows how models are used in practice. Each chapter is really a case study in which a problem commonly faced by environmental planners, consultants and managers is presented. An appropriate model is then chosen and the user is taken through the various steps involved in setting-up and applying the model and interpreting its output. Table 1.2 lists the applications under broad subject headings and for each one identifies the relevant chapter and the spatial scale (erosion plot, field, catchment, region) of the problem being considered. Additional information is provided on the temporal scale, which ranges from individual events to mean annual conditions and long-term landform evolution.

Taking each chapter in turn, Chapter 8 reviews the issues typically faced by field officers of the Natural Resources Conservation Service of the US when predicting erosion from agricultural land and planning soil protection measures. Chapter 9 takes a specific example of a small watershed in southwest Missouri and shows how modelling can assist in designing a strategy for sustainable management under both present land use and climatic change. In Chapter 10, modelling is used to predict rates of soil loss in Brazil from hillslopes on forest roads in São Paulo State and from agricultural land under different management systems in Minas Gerais State. Chapter 11 examines how a physically-based erosion model can be used to assess soil erodibility and evaluate different soil

Table 1.2 Issues covered by the erosion modelling applications in the Handbook, together with their spatial and temporal scales.

	Plot (1–100 m²)	Field/hillslope (100–10,000 m²)	Small catchment (1–500 ha)	Medium catchment (500 ha to 1500 km²)	Regional
Agriculture	10 (SP) 11 (MA)	5 (E) 8 (MA SP) 13 (MA) 15 (A MA)	9 (MA SP R) 12 (E) 15 (A MA)	15 (A MA)	15 (A MA)
Forestry Construction Mine waste	10 (SP)	16 (D A RP) 8 (MA SP) 18 (SP LE)	16 (D A RP)		
Land-use change	14 (MM MA SP)	14 (MM MA SP)	9 (MA SP R) 12 (E) 14 (MM MA SP)		
Climate change	14 (MM MA R)	14 (MM MA SP) 15 (A MA)	9 (MA SP R) 14 (MM MA SP) 15 (A MA)	15 (A MA)	15 (A MA)

Numbers in each cell refer to the chapter, and the letters indicate the temporal scale of model outputs in the applications described (E, event; D, daily; MM, mean monthly; A, annual; MA, mean annual; SP, set period of time; R, return period; LE, long-term landform evolution).

conservation practices at four different locations on tropical steeplands, one in China, one in Malaysia and two in Thailand.

The evaluation of sediment yield from a small catchment in a highly erodible area is the focus of Chapter 12, based on a case study on the loess plateau of China. Chapter 13 addresses a problem at a very different scale, namely the transfer of sediment from individual fields to watercourses in southwest England. Chapter 14 returns to the catchment scale, using a model to examine the impacts of land use and climate change on erosion and sediment yield in small river basins where hillslope erosion, river channel and bank erosion and landslides are all important components of sediment production. Chapter 15 is also concerned with assessing the impacts of climate change, but this time over a range of spatial and temporal scales from hillslope to regional and continental. There is no single model that can apply to all situations, and several models are reviewed. Chapter 16 looks at the risk of erosion in forested areas in Montana, US, following disturbance either by timber harvesting or wildfire. Chapter 17 discusses the potential of the Internet

as both a source of data and a vehicle for operating erosion models to address problems of environmental management. Chapter 18 examines the role of longer-term landscape evolution models (LEMs) for designing hillslope landscapes to encapsulate and contain mining waste. Chapter 19 reviews the question of modelling gully erosion. Although there is no specific gully erosion model that can be recommended, various approaches that a user can adopt are described.

The Handbook ends with a review of the state-of-art of erosion modelling, as illustrated by the case studies, and discusses the developments that users can expect in the near future. These include the inclusion of more models within geospatial frameworks, associated improvements to modelling across different scales, and the increasing use of web-based approaches and risk-based applications. It is hoped that, by combining a general review of the principles of erosion modelling with examples of model applications across a range of management issues, the Handbook will enable potential users to employ models in a more informed way. Hopefully, managers, decision-makers and policy-makers within the erosion control industry will be encouraged to make more use of models to evaluate present situations, the impacts of control measures and future policies. In addition, model developers may be encouraged to provide better information to model users about the suitability and limitations of their models and what levels of accuracy in prediction they are likely to achieve.

References

Boardman, J. & Davis-Mortlock, D. (1998) *Modelling Soil Erosion by Water*. NATO ASI Series: Series 1, Global Environmental Change, Vol. 55. Springer-Verlag, Berlin.

De Roo, A.P.J. (1999) Soil erosion modelling at the catchment scale. *Catena* 37 (3–4).

Ewen, J., Parkin, G. & O'Connell, P.E. (2000) SHETRAN: distributed river basin flow and transport modeling system. *Journal of Hydrologic Engineering ASCE* 5: 250–258.

Flanagan, D.C. & Nearing, M.A. (1995) *USDA Water Erosion Prediction Project: Hillslope Profile and Watershed Model Documentation*. USDA-ARS National Soil Erosion Laboratory Report No. 10.

Harman, R.S. & Doe III, W.W. (2001) *Landscape Erosion and Evolution Modeling*. Kluwer, New York.

Jetten, V. & de Roo, A.P.J. (2001) Spatial analysis of erosion conservation measures with LISEM. In Harmon, R.S. & Doe III, W.W. (eds), *Landscape Erosion and Evolution Modeling*. Kluwer, New York: 429–45.

Misra, R. & Rose, C.W. (1996) Application and sensitivity analysis of process-based erosion model GUEST. *European Journal of Soil Science* 47: 593–604.

Morgan, R.P.C. & Duzant, J.H. (2008) Modified MMF (Morgan-Morgan-Finney) model for evaluating effects of crops and vegetation cover on soil erosion. *Earth Surface Processes and Landforms* 33: 90–106.

Morgan, R.P.C., Quinton, J.N., Smith, R.E., *et al.* (1998) The European Soil Erosion Model (EUROSEM): a dynamic approach for predicting sediment transport from fields and small catchments. *Earth Surface Processes and Landforms* 23: 527–44.

Renard, K.G., Foster, G.R., Weesies, G.A., *et al.* (1997) *Predicting soil erosion by water. A guide to conservation planning with the Revised Universal Soil Loss Equation (RUSLE)*. USDA Agricultural Handbook No. 703.

Willgoose, G., Bras, R.L. & Rodriguez-Iturbe, I. (1991) A physically based coupled network growth and hillslope evolution model: 1. Theory. *Water Resources Research* 27: 1671–84.

Part 1
Model Development

2 Model Development: A User's Perspective

National Soil Resources Institute, Cranfield University, Cranfield, Bedfordshire, UK

2.1 Introduction

The last 40 years or so have witnessed the development of a very large number of erosion models operating at different scales and different levels of complexity, with huge variations in the quantity and type of input data required and, at least according to the model developers, covering a wide range of applications. A potential user of erosion models is therefore faced with a bewildering choice when attempting to select the best model for a particular purpose. All too often, the choice of a model is made more difficult because the user is unable to define the problem precisely enough to state what output is required; for example, whether knowledge of erosion rates is needed as a mean annual value or for a specific year, season, month, day or storm, and if the latter, whether it is a storm total or a value at the storm peak which is wanted. The user is sometimes uncertain whether this information is needed for a field, a particular hillslope or a catchment. Perhaps knowledge of actual erosion rates is not needed at all, and all that is required is an idea of the location of erosion within the landscape or an indication of the time of year that it is most likely to occur. Even when the requirements are clearly defined, the user is still confronted with the

difficulty that most models are not accompanied by clear statements of the purposes and conditions for which they were designed, their limitations or indicators of the accuracy of their output.

This chapter discusses how the user might deal with these issues. It does so by proposing that users should adopt the same procedures in analysing their problem as model developers adopt in constructing their models. By understanding how model developers operate and following a common methodology, users will be better equipped to decide what questions need to be asked when selecting a model to meet their specific objectives. These questions can then be formulated into a set of design requirements that a model must meet in order to be suitable. Users will also gain an appreciation of whether they will be able to operate the model software unaided, or whether they will need to seek expert advice in how to set up the model to meet their requirements and interpret the results. Table 2.1 sets out the steps followed by model developers and lists the main points that need to be considered at each stage.

2.2 Some Fundamentals

Any model is a simplification of reality and, for some users, this creates an immediate theoretical issue. How can a problem associated with erosion in a particular location be predicted by a model that describes erosion in a generic way? Surely the only way to deal effectively with a problem in

Table 2.1 Stages in model development.

Stage	Requirements
Objectives	Definition of problem
	Required temporal and spatial scales
	Required output, e.g. rates/location of erosion/deposition
	Required level of accuracy of prediction
Conceptualization	Understanding of system being modelled
	Required level of simplicity/complexity
	Experimental foundation for modelling
	Definition of system variables
	Definition of key processes
	Decisions on which variables and processes to include and exclude
	Construction of flow diagram
Process description	Decisions of best available mathematical descriptions of processes
	Match between mathematical description and process understanding
	Parameterization of system variables
	Availability of input data
Boundary conditions	Selection of appropriate time and space boundaries
	Continuity of mass and momentum when routing water and sediment across boundaries.
Sensitivity analysis	Rationality of model
	Determination of most sensitive input parameters
	Required level of accuracy of input data
Calibration	Feasibility of calibration
	Selection of key parameters for calibration
	Selection of dataset for calibration
	Calibration procedure
	Match between calibrated values and values expected in field conditions
Validation	Criteria for goodness of fit
	Selection of dataset for validation
	Validation procedure
	Required level of accuracy for acceptance of model
	Problems associated with uncertainty
	Interpretation of results
Application	Decision on whether model is appropriate
	Data requirements
	Setting-up and running of the model
	Analysis of results

Rainfall

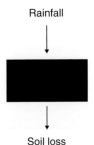

Soil loss

Fig. 2.1 An example of a simple black-box model.

a given catchment or at a given field site is to undertake detailed field observations and measurements of erosion and its controlling factors at that site and, based on an analysis of the results, to select appropriate measures to control the problem? Unfortunately, such detailed field measurements are often very costly and must be carried out over many years, probably ten or more, in order to collect representative data. In contrast, many problems must be addressed immediately and cannot wait for a solution some years later by which time considerable environmental damage may have occurred. The value of an erosion model is that it can be applied now. The question that arises, however, is how simple or complex it needs to be for it to be valid.

Broadly, simplification can be represented at three levels, resulting in what is usually termed black-box, grey-box and white-box models. In a black-box model (Fig. 2.1) a relationship exists between one or more inputs or controlling factors, such as rainfall or soil type and the output, such as soil loss. There is no understanding or modelling of the processes through which the inputs give rise to the output. Such models are usually expressed by some form of statistical relationship, like a linear regression equation or a correlation. A sediment-rating curve for a river channel whereby sediment concentration is expressed as a function of runoff is a good example of this type of model. A grey-box model (Fig. 2.2) includes some understanding of the relationship between input and output, reflecting, for example, that the effect of rainfall on erosion alters according to slope steepness and vegetation cover. The model is again operated by equations based on statistical

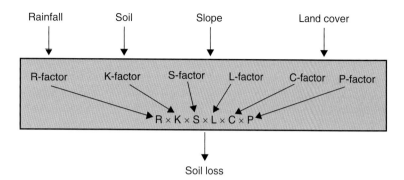

Fig. 2.2 The structure of the Universal Soil Loss Equation: an example of a grey-box model.

relationships, but these are usually more complex than those used in black-box models and involve multiple regression analysis or polynomial relationships. Good examples are the nomograph for determining the value of the soil erodibility factor (K) as a function of soil texture, structure and permeability (Wischmeier *et al.*, 1971), and the Universal Soil Loss Equation (USLE), in which the relationship between soil loss, rainfall and soil type is adjusted by coefficients representing slope steepness, slope length, crop management and physical protection measures (Wischmeier & Smith, 1978).

In white-box models an attempt is made to describe as many of the processes of the erosion, transport and deposition of sediment as possible. Mathematical equations are used to do this, or sometimes simple arithmetical calculations, but more often, differential and difference equations. Since these equations are generally based on satisfying the laws of conservation of mass and energy, the models are often described as being physically-based. The use of complex mathematics can be off-putting for many model users who do not always have the same mathematical background and expertise as the model developers. In practice, however, model users should not be frightened by this but, instead, should adopt a questioning approach. They should be more concerned about the processes being described, as to whether they are relevant to the problem and have the same level of importance within the model as in the field. The user should be sure that the equations

used are based on sound science and can be supported by underlying knowledge and measured data. Often such questioning reveals that the scientific understanding of many of the processes described in the model is limited, and that many of the equations used are actually empirical and similar to those used in grey-box models. In reality, no truly white-box model of erosion exists, but there are several which could be considered as pale grey to cream in that they are certainly process-based but only partially physically-based (Fig. 2.3). Examples include WEPP (Nearing *et al.*, 1989b), EUROSEM (Morgan *et al.*, 1998), GUEST (Rose *et al.*, 1983), LISEM (De Roo *et al.*, 1998) and SHE (Wicks & Bathurst, 1996).

2.3 Conceptual Framework

Being able to determine the questions to ask of modellers means that users need to have a good understanding of their objectives in using a model, an understanding which is enhanced if their problem is conceptualized in terms of relevant processes and outputs. Since defining objectives and conceptualizing the problem represents the first two stages in model development, there should be some common ground between model developers and model users.

The pioneering work of Meyer and Wischmeier (1969) represents a good example of setting objectives and developing a conceptual framework for erosion modelling (Fig. 2.4). Indeed, it has laid

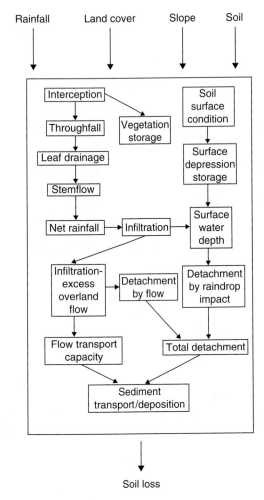

Rainfall   Land cover   Slope   Soil

Soil loss

Fig. 2.3 The structure of EUROSEM: one of many models produced as erosion modellers aspire towards a white-box model.

topographic conditions. Since the relationships on which the model is based are essentially statistical, it is an empirical model. Meyer and Wischmeier were interested in developing a more process-based approach in which the actual processes of erosion were described mathematically. Their objective was not predictive in terms of erosion rates, but rather to demonstrate that such an approach was feasible and that it could simulate the patterns of erosion and deposition observed in the landscape.

Based on research by Ellison (1947), Meyer and Wischmeier (1969) conceptualized erosion as a two-phase process comprising the detachment and transport of soil particles by rainfall and runoff. They therefore described erosion as a result of (a) the detachment of soil particles by rainfall; (b) the detachment of soil particles by runoff; (c) the transport of soil particles by rainfall; and (d) the transport of soil particles by runoff. Empirical equations were selected, derived largely from the results of laboratory experiments, to describe each of these processes. The landscape was visualized in simple terms as a single slope profile from hilltop to valley bottom which could be divided into a series of segments. Erosion was simulated by calculating for each segment the amount of soil detached and the capacity to transport it out of the segment in a downslope direction. The amount of soil supplied by detachment on each segment was that detached on the segment and that transported into the segment from upslope. A simple arithmetical calculation compared the amount of sediment available from detachment with that which could be carried in transport. If the amount of detachment was less than the capacity for transport, all the sediment was removed downslope; if the amount of detachment was greater than the capacity for transport, only the amount which could be transported was carried downslope and the rest was deposited on the segment. Since the transport capacity depended on the steepness of the slope, the model simulated a net loss of soil by erosion on the upper and mid-slope segments, and the deposition of soil on gentler concave segments at the foot of the slope.

the foundation for many subsequent erosion models. In the 1960s, the only widely-used approach to erosion prediction was the USLE developed by Wischmeier and Smith (1965) as a design tool for soil conservation workers in the US, particularly in the Corn Belt. As noted above, it is a grey-box model which predicts the mean annual rate of soil erosion at a field scale under different cropping systems and management practices for a given set of rainfall, soil and

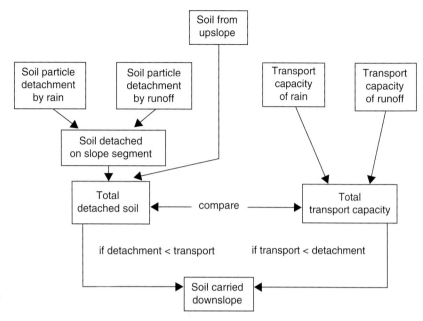

Fig. 2.4 The Meyer-Wischmeier model: a conceptual framework for erosion modelling (after Meyer & Wischmeier, 1969).

Meyer and Wischmeier (1969) carried out their simulations for relatively simple conditions in which the soil was bare, the slope planar, and there was no removal of sediment at the base of the slope, for example, by a river. There was also no addition to sediment on the slope over the long term through the breakdown of the underlying rock into soil by weathering processes. Although the authors showed their approach was feasible and that it could reproduce the patterns of erosion and deposition observed in simple landscapes, its limitations meant that further research was necessary before it could be developed into an erosion model that could satisfy the objectives of many potential users. In particular, the effects of crop or vegetation cover and soil management needed to be accounted for. Further, during the 1970s and 1980s it became increasingly clear that the USLE was no longer able to meet the demands of many users as their attention changed from one of conserving soil in a field to sustain long-term agricultural productivity, to concerns over the transfer of sediment from agricultural fields, construction sites and recreational areas to water bodies. The inability of the USLE to compute sediment yields from areas larger than single fields or for specific time periods instead of an average year meant that different models were required. In their absence, some workers used the USLE regardless of its limitations and generally with unsatisfactory results. The USLE suffered much adverse publicity because it could not deal with these new issues. This caused Wischmeier (1978) to publish a warning about its use and misuse. His paper is still relevant today because it illustrates the importance of using a model which is fit-for-purpose (see Chapter 7 for a fuller discussion of this issue) and that if, for any reason, an inappropriate model is used and does not work, it is the fault of the user and not the model. The most common reasons why an inappropriate model is chosen are that an appropriate model does not exist, or the data to operate one are not available.

The first set of erosion models that adopted the Meyer-Wischmeier approach to describe the processes of erosion and simulate the movement of sediment over the landscape operated by bolting the equations for the erosion processes on to a hydrological model, which was used to generate runoff and transport the resulting water flow over the landscape. Examples include AGNPS (Young et al., 1989), ANSWERS (Beasley et al., 1980) and GAMES (Rudra et al., 1986). Some models, such as CREAMS (Knisel, 1980), allowed a choice between daily simulations based on the USCS Curve Number, a coefficient which describes the soil, slope and land cover characteristics, and storm simulations in which runoff is calculated as the excess of the rainfall intensity over the infiltration rate of the soil. This generation of erosion models was process-based as regards the simulation of runoff and sediment, but relied on the factors of the USLE to describe soil erodibility (K), slope length (L), cropping (C) and management (P) effects.

Current research on erosion modelling is concerned with replacing the coefficients related to soil, slope and land cover with parameters that measure their properties directly and which can therefore take account of variations in both time and space. Instead of a single value to express K, soils are described by properties such as cohesion, shear strength and surface roughness, and land cover by architectural properties of the vegetation such as height, percentage cover, stem size and stem density. This means that soil, for example, can be modelled dynamically allowing for changes in cohesion as the surface crusts or seals under raindrop impact (Moussa et al., 2003) or human or animal trampling, or as surface roughness changes as a result of different tillage practices. Similarly, plant cover effects can be altered in relation to seasonal plant growth and decay. The effects of soil and plant cover are sometimes described separately for each of the four processes of erosion, namely detachment of soil particles by raindrops and runoff, and the transport of the detached material by rainfall and runoff. The outcome is that erosion models have become more complex, since they now incorporate many submodels to describe the behaviour of soil and vegetation. Future models may well account for the movement of soil over the landscape by tillage using

methodology developed by Govers et al. (1994) and van Muysen et al. (2002). WATEM (van Oost et al., 2000; Verstraeten et al., 2002) combines these equations with a modification of the Revised Universal Soil Loss Equation (RUSLE) (Renard et al., 1991) to simulate the transport of sediment by runoff and tillage on a mean annual basis. Improved descriptions of the effect of soil will take account of its aggregate structure rather than the size distribution of the primary particles of clay, silt and sand, by using parameters based on aggregate stability (Issa et al., 2006).

With greater concern environmentally about the fate of eroded sediment has come the recognition that the way in which many erosion models simulate the deposition of sediment is too simplistic. Just comparing the amount of material available for transport with the transport capacity, and dumping the sediment which cannot be transported, results in patterns of deposition over the landscape which are unrealistic. Too much material is deposited too quickly. In models such as WEPP and EUROSEM, an attempt is made to control the rate of deposition by taking account of the settling velocity of the soil particles in the flow and a coefficient expressing the efficiency of the deposition process. Although this approach produces better results, it is analogous to the use of coefficients to describe the effects of plant cover on erosion rather than simulating the physical process. Future models are likely to model deposition explicitly taking account not only of particle settling velocity, but also the velocity of the runoff and the depth of flow. The approaches developed to predict sediment deposition in filter strips (Tollner et al., 1976; Rose et al., 2003) and farm ponds (Verstraeten & Poesen, 2001) are likely to be adapted to describe deposition from runoff. Such developments will lead to even greater complexity in erosion models as they attempt to describe the erosion processes more fully. For example, the four erosion processes identified by Meyer and Wischmeier (1969) and the process of deposition will need to be simulated separately for each soil particle size. Alternatively, erosion, transport and deposition can be modelled simultaneously for different erosion/deposition domains, depending on the relative dominance of the three processes (Beuselinck *et al.*, 1999). Ideally this will not be restricted to primary particles, as in Morgan and Duzant (2008) and Fiener *et al.* (2008), but will cover the sizes of soil aggregates. In addition, since deposited material has different strengths to that of the original soil because cohesion has been lost during detachment and transport, erosion models will need to distinguish between the two when simulating the detachment and transport of the sediment (Rose *et al.*, 1983).

2.4 Operating Equations

The simplicity and number of operating equations required to run an erosion model depends on its type and level of complexity. Since this section is descriptive rather than intended for practical application, the units of the equations are not given. Readers should consult the original sources for these details. A simple grey-box model like the USLE (Wischmeier & Smith, 1978) requires only one equation which multiplies together six numbers:

$$A = R \times K \times L \times S \times C \times P \qquad (2.1)$$

where A is the mean annual soil loss, R is the rainfall erosivity factor, K is the soil erodibility factor, S is the slope steepness factor, L is the slope length factor, C is the crop management factor and P is the erosion-control practice factor. Additional equations are required to determine the values of the S and L factors and, as indicated above, a third equation can be used to estimate the value of K (Wischmeier *et al.*, 1971). Graphical solutions to these additional equations exist in the form of nomographs.

More complex process-based models use separate equations to describe the various processes of erosion and deposition, and link these together using continuity equations to ensure conservation of energy and mass. The continuity equation

for the volume or mass of sediment passing a given point on the land surface at a given time is:

$$\frac{\partial (AC)}{\partial t} + \frac{\partial (QC)}{\partial x} - e(x,t) = q_s(x,t) \qquad (2.2)$$

where A is the cross-sectional area of the flow, C is the sediment concentration in the flow, t is time, x is the horizontal distance downslope, e is the net pick-up rate or erosion of sediment on the slope segment, and q_s is the rate of input or extraction of sediment per unit length of flow from land external to the segment, for example, from the sides of a convergent slope surface. On a planar slope, $q_s = 0$, and the continuity equation can be rewritten as:

$$\frac{\partial (AC)}{\partial t} + \frac{\partial (QC)}{\partial x} = e_i + e_r \qquad (2.3)$$

where e_i is the net rate of erosion in the inter-rill area of the slope segment and e_r is the net rate of erosion by rills. This is the form of the continuity equation used in WEPP (Nearing *et al.*, 1989b), EUROSEM (Morgan *et al.*, 1998), LISEM (Jetten & de Roo, 2001) and many other process-based models. In GUEST (Rose *et al.*, 1983) the equation takes a slightly different form. In this model, the soil is described in terms of up to 50 particle-size classes, determined according to their settling velocity, and, for each particle-size class, a distinction is made between that eroded from the original soil and that eroded from previously-detached and recently deposited sediment; in addition, deposition is modelled explicitly. The continuity equation becomes:

$$\frac{\partial (AC_j)}{\partial t} + \frac{\partial (QC_j)}{\partial x} = e_{ij} + e_{idj} + e_{rj} + e_{rdj} - d_i \qquad (2.4)$$

where C_j is the concentration of sediment of particle size j in the flow, e_{ij} is the rate of erosion of particles of sediment class j in the original soil on the inter-rill area, e_{idj} is the rate of erosion of particles of sediment class j from previously detached soil on the inter-rill area, e_{rj} is the rate of erosion of particles of sediment class j from the original soil by rills, e_{rdj} is the rate of erosion of particles of sediment class j from previously detached soil by rills, and d_j is the rate of deposition of particles in sediment class j.

Even where models use the same form of the continuity equation, they differ in the operating functions used to describe erosion and deposition. The user therefore has the possibility of selecting a model according to which functions best describe the way that erosion occurs in a particular study area or which functions are theoretically more satisfying. As an illustration, the way e_i and e_r are described in WEPP and EUROSEM are compared. In WEPP (Nearing et al., 1989b), the inter-rill erosion rate (per unit rill width) is given by:

$$e_i = K_i I^2 \left(1 - F^{0.34PH}\right) e^{-2.5G} (R_s / W) \quad (2.5)$$

where K_i is the inter-rill erodibility of the soil, I is the intensity of the rainfall, F is the fraction of the soil protected by the plant canopy, PH is the height of the plant canopy, G is the fraction of the soil covered by ground vegetation or crop residue, R_s is the spacing of the rills and W is the width of the rill computed as a function of the flow discharge. The rate of rill erosion is calculated from:

$$e_r = K_r \left(\tau - \tau_c\right)\left[1 - C / \left(k_t \tau^{3/2}\right)\right] \quad (2.6)$$

where K_r is the rill erodibility of the soil, τ is the flow shear stress acting on the soil, τ_c is the critical flow shear stress for detachment to take place, C is the sediment load in the flow, and k_t is a sediment transport coefficient. This equation only operates when the sediment load in the flow is less than the sediment transport capacity of the flow. When the sediment load exceeds the transport capacity, the equation becomes:

$$e_r = K_r \left(\tau - \tau_c\right)\left\{v_s / \left[q \left(k_t \tau^{3/2} - C\right)\right]\right\} \quad (2.7)$$

where v_s is the particle settling velocity and q is the flow discharge per unit width.

In EUROSEM (Morgan et al., 1998) a single equation is used to describe the erosion rate by soil particle detachment by raindrop impact and runoff, i.e. $e_i + e_r = e$. The equation can be applied to unchannelled inter-rill flow or to runoff in rills. Where rills are present, these need to be defined in terms of their number, depth and width. The model initially places all the runoff in the rills and then uses a unified rill model to describe the hydraulic conditions of the flow as the runoff overflows the rills and spreads out over the inter-rill area. The single equation is:

$$e = \left[k \left(KE_{DT} + KE_{LD}\right)^{1.0} e^{-2h} \right] + \left\{\eta w v_s \left[\left(a(\omega - \omega_c)^b\right) - C\right]\right\} \quad (2.8)$$

where k is the detachability of the soil by raindrop impact, KE is the kinetic energy of the rainfall which is divided into direct throughfall (DT) and that falling from the plant canopy as leaf drainage (LD), h = the depth of surface water, η is an expression of the efficiency of soil particle detachment by flow which is a function of soil cohesion, w is the width of flow, v_s is the settling velocity of the particles in the flow, ω is the unit stream power of the flow (the product of slope and flow velocity), ω_c is the critical value of unit stream power for sediment transport, a and b are coefficients related to sediment particle size, and C is the sediment concentration in the flow.

The user therefore has a choice between a model that simulates the detachment of soil particles by raindrop impact as a function of rainfall intensity, and one that uses the kinetic energy of the rain. WEPP allows for the effect of the plant cover by assuming that ground-level vegetation protects the soil completely and that the plant canopy provides some protection dependent upon its height above the ground surface. In EUROSEM, the proportion of the soil surface covered by vegetation is used to split the rainfall into direct throughfall and leaf drainage. The kinetic energy of the leaf drainage is calculated as a function of plant height but in such a way that, for very tall canopies, the energy can exceed that of the direct throughfall, whereas ground-level vegetation will protect the soil completely. The user also has a choice between a model that simulates the detachment and transport of soil particles as a function of the shear stress exerted by the flow, and one that describes the same processes using unit stream power. The other fundamental difference between WEPP and EUROSEM is that the former operates over a range of time steps from individual storms to daily. Within each time step, steady-state conditions are assumed, which means that each time step is either one of erosion or deposition of sediment. In contrast, EUROSEM uses very short time steps (1–2 minutes) and therefore continuously updates the sediment concentration in the runoff and the transport capacity. The latter is assumed to be the sediment concentration at which erosion and deposition are in balance. EUROSEM is therefore a dynamic rather than a steady-state model, which implies that there is a continuous exchange of soil particles between the runoff and the soil surface which controls the sediment concentration in the flow.

There is no doubt that researchers with an interest in modelling the processes of erosion and deposition will develop even more complex models than WEPP and EUROSEM as they work towards the goal of a comprehensive description of the processes and a dynamic simulation of the factors affecting them. For practical purposes the model user may well question whether all this complexity is necessary. It is well known that in terms of the amount of sediment reaching the bottom of a hillside or discharging into a water course, some erosion processes are more important than others. Indeed, in terms of producing a relatively simple and efficient model, it is recommended that attention is focused on the most important processes and that those contributing little to the generation, transport and deposition of sediment should be ignored (Kirkby, 1980). The user therefore requires some knowledge of the most important processes that affect his or her problem so that a model which

emphasizes these can be selected. Even if a very detailed model is not chosen, the user can gain much by establishing the conceptual framework of the problem as fully as possible. By understanding the processes involved and their controlling factors, it is possible to decide which are the most relevant and which models best match what is required. Without a comprehensive framework, there is a danger that something important will go undetected.

2.5 Spatial Considerations

When selecting an erosion model it is necessary to define the area over which it should operate. This may vary from a small segment of a hillslope, to a complete hillside, a small catchment (typically 0.01–0.5 km^2 in area but sometimes as large as 10 km^2) encompassing hillslopes and a river channel, or a large catchment (typically 10–100 km^2 but sometimes as large as 100,000 km^2). A decision is then needed on whether the area can be treated as a single unit or whether it is necessary to know what is taking place at different locations within it. The first approach is suitable where only knowledge of the amount of sediment leaving the area is required. The second is essential where knowledge of the source of the sediment is needed so that the implementation of erosion protection measures can be targeted. Generally the larger the area, the greater the need for internal understanding since deposition of sediment may occur at several locations and sediment movement may be concentrated along preferred flow paths, all of which can influence the design of a system for sediment management. These two approaches are catered for respectively by the use of lumped and distributed models.

The most commonly used lumped model in erosion work is the USLE. As already noted, this predicts mean annual soil loss from a single area, in this case a field, within which rainfall, soil, slope and land cover are either considered uniform or can be represented by coefficients which express the average condition. Lumped models are more commonly used in hydrology. They are process-based in that they effectively describe the water balance of a catchment whereby a proportion of the incoming rainfall passes into runoff whilst the rest is held in various stores, such as interception by the plant cover, soil moisture and groundwater. An erosion component can be built on to such a model as illustrated by the Stanford Sediment Model (Negev, 1967), which is linked to the Stanford Watershed Model (Crawford & Linsley, 1966). Since most practical applications require knowledge of the way sediment is moved over the landscape so that protection measures can be targeted either at areas of sediment source or along pathways of movement, there is a demand for models which can describe what is happening within a catchment, and lumped models are unable to do this. Lumped models are therefore of value for predicting soil loss from relatively small areas such as an agricultural field, road embankment or construction site. They can also be used to assess erosion over large areas, as illustrated by the PESERA model (Gobin *et al.*, 2006) which estimates mean annual erosion over 1 km^2 size units. A process-based approach is used to generate infiltration-excess and saturation overland flow from daily rainfall. The calculations are integrated across the frequency distribution of daily rainfall events. Sediment transport is estimated according to the runoff, soil erodibility and slope of each cell. Both runoff generation and sediment transport are modified by land cover, surface roughness and soil crusting.

Where the need is to determine where erosion and deposition take place within a catchment, distributed models are used. These operate by dividing the catchment into discrete land units and use mathematical procedures to route water and sediment from one unit to another. Such models are necessarily process-based and, in so far as they use input data that can be measured physically in the field and use continuity equations to ensure the conservation of mass and energy as water and sediment are moved in space and over time, they are often considered as physically-based (Beven & Kirkby, 1979). Most of the recently developed erosion models, like WEPP, GUEST, LISEM and EUROSEM, are distributed models.

They are suitable for analysing the effects of changes in land use in different parts of a catchment, as well as the effects of variations in rainfall, soil type, slope and land cover within a catchment.

Some distributed models, like EUROSEM and CREAMS, require land units in a catchment to be identified in terms of similarity in soils, slope and land cover. For most practical purposes, the land units are similar in nature to the land facets identified in terrain analysis (Christian & Stewart, 1968; Webster & Beckett, 1970). These can be grouped into larger units or land systems (usually between 100 m^2 and 10,000 m^2 in size) which have been shown to be significantly different from each other in terms of both erosion status and the rate of change in erosion over time (Morgan *et al.*, 1997). The art in setting up these models is to identify the land units so that they are internally as uniform in their characteristics as possible, and then to determine the likely patterns of water flow from one unit to another (Auzet *et al.*, 1995). Although this can be done using aerial photographs, topographic maps or digital elevation models to determine the low points in the landscape along which water will concentrate, there is often the need for field observations to identify where flow paths deviate from those which would occur naturally, for example as a result of installing diversion terraces or ditches to take water across the slope to a safe outlet rather than allowing it to flow downslope.

In recent years, with the advent of geographical information systems, model users have moved away from defining land units in relation to the natural variations in the landscape, in favour of dividing the catchment into grid cells of uniform size. Although the movement of water from one cell to another is still based on the local topography, the units themselves are less likely to be internally consistent in their soils, slopes and land cover. Unless the grid cells are extremely small, say 10 m × 10 m, it is likely that they will be crossed by boundaries between soil types or slope breaks. The larger the grid cells used, the more likely that each one will need to be characterized by either its dominant soil type, slope steepness or land cover, or by some weighted average of these conditions (Wood *et al.*, 2006). As grid cells approach 1 km × 1 km in size, the model simulations are less likely to resemble the observed landscape (see also Section 6.4). A further problem with large grid cells is that the equations used in the models to calculate flow depths and flow velocities and to describe the processes of detachment and transport of soil particles are strictly valid only for virtually instantaneous conditions or extremely short time periods. It is difficult to see how these can be used to describe average conditions when the time taken for water to flow from the highest to the lowest point within a grid cell exceeds the time for which the equation may be expected to operate reliably. Smith (1979) stated that flow velocity equations can be averaged over time periods as long as 5 minutes, but only over slope lengths of 2 to 5 m. With a typical velocity for unchannelled overland flow of $0.01\,m\,s^{-1}$ averaged over 5 minutes, water will have travelled 3 m in that time. The situation is somewhat better for flow in small channels or rills; with a velocity of $2\,m\,s^{-1}$, water will have travelled 600 m over 5 minutes. The use of large grid cells, say 1 km × 1 km, may be just about workable if it is accepted that flow velocities can be averaged for periods up to 10 minutes and velocities are no greater $3\text{--}4\,m\,s^{-1}$.

The decision on whether to select a lumped or distributed approach may depend on the degree of complexity at which erosion processes need to be modelled. As noted earlier, Meyer and Wischmeier (1969) described erosion in terms of four separate processes but, in reality, the detachment and transport of sediment by runoff is more complicated because it can take place by both unchannelled flow and channelled flow, such as rills and gullies. When dealing with practical problems, it is important to decide how much information is needed on the processes by which sediment gets detached, transported and deposited. If it is sufficient to know only the quantity of material eroded over a given time, there is no need to assess the different processes and a lumped model

is likely to suffice. If, however, it is necessary to know whether conservation measures should be targeted at controlling the detachment or the transport of sediment and, if the latter, whether the transport is relatively uniform over the area or is concentrated along selected flow paths, a more detailed modelling of the separate processes is essential; a distributed model will therefore be required. At present, many erosion models separate the processes of detachment and transport, but few will separate unchannelled overland flow from rill flow. Indeed, one of the challenges for modelling research is to develop a dynamic rill erosion model which will simulate the processes by which rills develop and infill at different locations on a hillslope at different times during a storm (Favis-Mortlock *et al.*, 2000). From a geomorphological viewpoint, most erosion models are still rather incomplete in respect of processes since they are restricted to rainsplash, unchannelled overland flow and rill flow. A fully integrated model incorporating processes such as gully erosion, subsurface piping, mass movements (ranging from soil creep to mudflows and landslides) and wind erosion has yet to be developed (but see Chapter 14). Fortunately, most erosion-control problems do not require this level of complexity.

The above discussion has focused on the conditions within the area being modelled. It is also important to consider the nature of the boundary of the area and how models deal with this. In the simplest situation, the boundary is usually clearly defined, it being the edge of an erosion plot or a field, or the divide of the catchment. It is assumed that the location of the boundary is static over time and that, except for the lower boundary, there is no transfer of water or sediment across it. Most erosion models allow for the transfer of material across the lower boundary either to an area downslope or into the river system. Once the material leaves the area being modelled, no further account is taken of it. This approach is generally acceptable for relatively short time periods ranging from individual storms to a few years. Over longer time periods, the boundary condition can become more complicated. For example, the lower boundary of a hillslope represents the junction between two different process domains, the hillslope and the river channel. Through bank erosion, the river may undercut the slope and cause its boundary to retreat upslope. Alternatively, where the hillslope extends on to a flood plain, migration of the river channel away from the slope may cause the hillslope to lengthen. The need to consider changes in the location of the boundary becomes more important with long-term erosion models that simulate landscape evolution. In such models it may also be necessary to recognize changes at the upslope boundary, for example its lowering of altitude through erosion over thousands of years. Model users need to assess the importance and nature of the boundary conditions and select a model accordingly. For most applications involving hillslopes or small catchments and with short time scales, a model with a simple approach to boundary conditions will suffice.

2.6   Temporal Considerations

The USLE predicts the rate of mean annual erosion, which means that it operates on factors that express the average condition of the study area over a period of years. This is best illustrated by the way the C-factor value is calculated. The C-factor expresses the effect of crop management. It is a dimensionless coefficient based on the ratio of the mean annual soil loss under a specific management system to that from bare soil under otherwise identical conditions of rainfall, soil and slope. Since the C-factor is dynamic and varies with the percentage ground cover, percentage canopy cover, height and root density of the crop, the values change throughout a cropping season and also from one year to another within a crop rotation. In addition, the effect of these crop properties at any given time depends on how erosive the rainfall is. Erosive rain on a bare soil during the off-season is likely to result in a high rate of erosion, but if the off-season coincides with the dry season, the effect of the lack of plant cover will be minimal. Thus the C-value for a particu-

lar stage in the cropping year needs to be adjusted by the proportion of the rainfall erosivity (R-factor) occurring during that stage. The C-factor value is thus the sum of these adjusted values. A similar approach is used in the Revised USLE (Renard et al., 1991) to take account of seasonal variations in soil erodibility (K-factor).

With erosion models that operate on a storm or daily basis, it is possible to change the input parameters regularly over time to take account of changes in, for example, infiltration, soil moisture, soil erodibility and crop cover. Often calculations are made within the model for even shorter time steps, varying between one and ten minutes, which allow the effects of varying rainfall intensities within a storm to be simulated. Some models include the effect of feedback mechanisms whereby, for example, surface roughness or soil erodibility as affected by surface crusting or sealing, alter within a storm as a direct effect of the rainfall. The output from such models invariably includes a storm hydrograph and a storm sediment graph, enabling the timing and magnitude of runoff and sediment peaks to be determined. This output is essential for dealing with problems where concentrations of pollutants associated with sediment concentrations exceeding some critical value are the issue, rather than total or average sediment levels. The success of models operating over short time periods depends on how well the conditions at the start of the time period can be specified. In particular, initial soil moisture, soil erodibility, surface roughness and the nature of the vegetation or crop cover need to be stated. A distinction can be made here between single storm models, like EUROSEM, where starting conditions must be defined by the model user, and continuous simulation models, like WEPP, which calculate the starting conditions for a given time period from the simulations of previous time periods. These latter models can therefore be 'run in' because, after a certain number of simulations, their output tends to stabilize almost regardless of the initial conditions chosen. Judgment is then required on how many simulations are needed to complete the 'running in' period.

2.7   Temporal and Spatial Scale Interactions

So far it has been shown that erosion models exist for a range of spatial scales from the small plot to the large catchment, and a range of temporal scales from the single storm to thousands of years. Although the model user should be able to select an appropriate model at the scale required to address a specific problem, in reality model choice is more complex. The challenge for model developers and users is how to scale up from a storm to several years, and from a plot to a large catchment. The approach commonly adopted for temporal scaling is to take a storm or daily simulation model and run it consecutively for many storms or many days. Thus, CREAMS and WEPP can be operated by running daily simulations for more than 7000 consecutive days in order to give output for a 20-year period or more, from which average annual values of soil loss can then be calculated. Some modellers have run continuous simulation models for periods as long as 100 years (Lee, 1998). For spatial scaling, the approach is to use distributed models to simulate the movement of water and sediment over very large numbers of land units in order to accommodate large catchments. There are two problems with these approaches. Firstly, they are not very efficient. Secondly, they can become very unstable mathematically when the time step between successive calculations by the model becomes greater than the value of the ratio of the slope length between successive calculations to the speed of the flow wave. The exact value of this ratio depends on the mathematical method used to solve the equations for routing runoff and sediment over the landscape (e.g. finite difference or finite element methods). Generally, the shorter the time steps and the shorter the space steps, the more stable the model will be. Such a situation is usually achieved by undertaking calculations at shorter time steps and space steps than is given by the model output. Thus, even in a daily simulation model, calculations may be made every ten minutes and for transport distances of every 5 to 10 m within each land unit. Where time and space scales prevent such detailed calculations, the model can become very sensitive to the length of the individual spatial units. The slope length of the individual land units may then need to be restricted within a certain range, as is the case with the Revised MMF model (Morgan, 2001; Morgan & Duzant, 2008) which routes annual calculations of sediment transport over the landscape (see Chapter 13).

When it comes to assessing erosion over long time periods or large areas, the inefficiency of continuous simulation models raises the question of whether there is an alternative approach. Firstly, there is the issue that as temporal and spatial scales change, the relative importance of the factors that influence erosion also changes. Long-period and large-area models therefore need to emphasize the processes and factors that are most relevant to those scales. They may also require different equations to express those relationships compared with those used at hillslope or plot scales. At large spatial scales the important controls are climate, lithology and vegetation (Kirkby, 1998). If a large area is divided into land units, the model needs to simulate how erosion varies in relation to these three factors, whereas other factors, like slope, surface roughness and soil properties, can be represented by average values for each land unit. Averaging can take various forms depending on the nature of the relationship between each factor and erosion. Arithmetic averages may be appropriate if the relationship is linear, but where non-linear relationships apply, the root mean square value, the logarithmic mean, the geometric mean, or some value which is exceeded by a given percentage of the statistical population of values, may be more meaningful alternatives. Examples of the latter are often used in soil descriptions, which can be expressed by a grain size value at which a certain percentage of the soil (e.g. 5, 10 or 35%) is finer. At present most of the models at large spatial scales operate on a large grid cell basis to give an output of sediment yield for each cell but with no indication of its fate. Sometimes the output is simplified into classes of different levels of erosion rate, with each class being described by terms such as slight, moderate or severe. In other cases, recourse is made to a sediment delivery ratio to express what proportion of the sediment yield actually leaves the cell and discharges into the river system. The problems of obtaining a suitable relationship at this scale between time and space increments at which calculations are made precludes the possibility of routing the water and sediment physically from one grid cell to another. Nevertheless, this scale of modelling is seen as increasingly valuable because it allows the possibility of including additional components, for example, the ability to predict the growth of the plant cover in relation to long-term changes in climate and soils (Kirkby & Neale, 1987) or changes in cattle-grazing regime (Biot, 1990; Thornes, 1990). In the MEDALUS model (Kirkby et al., 2002), feedback mechanisms are included to cover long-term changes in the stoniness and roughness of the surface soil as a result of erosion.

Some care needs to be taken in using and interpreting the results of models applied at different scales. The scale at which the model operates is the key, rather than the scale at which maps are produced. For example, if the output from the USLE is mapped at a national or European scale, using 1 km or 10 km grid cells, the map shows the predicted value of mean annual soil loss at a field scale within each cell. It does not show the contribution of sediment from each cell to the river system. Nor can the values of each cell within a catchment be summed to give a value for the sediment yield from that catchment. This is because the USLE does not predict sediment delivery to rivers; nor does it predict erosion over $1–10\,km^2$ size areas.

2.8   Testing, Calibration and Validation

Once a model has been developed at the appropriate temporal and spatial scale for meeting a specific objective, it needs to be tested to show that it actually works. The first step is normally to feed in arbitrary values of the input parameters to check that the model functions mathematically and is rational. Once errors have been corrected and 'bugs' in the computer programming removed,

the model can be evaluated to see whether it yields realistic output.

2.8.1   Sensitivity analysis

An increasingly important aspect of model testing is a sensitivity analysis that is designed to determine how sensitive the output is to unit changes in the value of one or more of the input parameters. Different sensitivity indices are used (Morgan, 2005), of which the average linear sensitivity (Nearing et al., 1989a) is the most common. The results provide a check on whether the sensitivity of the different input parameters in the model accords with that observed in field situations at the appropriate temporal and spatial scales. If a parameter is found to be too sensitive or not sensitive enough, the model may need to be adjusted. Unfortunately, although conceptually simple, sensitivity analysis in reality can be extremely complex, particularly when carried out for process-based, distributed models. Since these models allow for erosion to be limited by either the detachment rate or the transport capacity, parameter sensitivity depends upon which is the limiting factor. Furthermore, many of the relationships by which erosion is related to individual parameters are non-linear, which means that the output may be sensitive when the parameter values fall within a certain range, but not when they are outside that range. Sensitivity may also be different when a model is operated for extreme conditions. A further issue is the interaction between the parameters so that a certain parameter may only be sensitive when the values of another parameter exist within a certain range. Procedures for addressing these issues are discussed more fully in Chapter 3. Understanding which parameters are most sensitive is also important because the values of those parameters need to be determined more accurately. There is then the question of whether the procedures used in field measurement are robust enough to give the required level of accuracy.

The complex nature of sensitivity means that the user cannot assume that the results of a generic sensitivity analysis for a chosen model will necessarily hold when the model is applied to a specific location. A parameter which overall is rated as moderately sensitive in the general analysis may be highly sensitive in a specific situation, either because there is little or no variation in the values of the parameters which are generically the most sensitive, or because the values of another parameter are within the range that makes the first parameter sensitive. It is recommended that the user undertakes a sensitivity analysis of the chosen model with local data to test that the parameter sensitivity matches what is known about the local situation.

2.8.2 *Data availability*

A problem frequently faced by model users is that insufficient data exist to run many of the models. This may limit their ability to select what would otherwise be the most appropriate model. In some instances there is no alternative but to choose another model which is less demanding in its data input, even though it may not meet all the user's requirements. Failure to appreciate the issues associated with data availability can lead to misconceptions about how well a model is likely to perform (see Chapter 7). Generally, models which require information on a time-averaged or spatial-averaged basis are the easiest to satisfy with respect to input data. For example, annual or mean annual rainfall and mean annual temperature data exist for most areas of the world. Even though the nearest meteorological recording station may be several kilometres away from the field area, it is possible to use the data with some degree of confidence. The devices used to record rainfall and temperature have high levels of accuracy, so that the main sources of error relate to how representative the meteorological station is of the local site. This will depend on how variable local climate is with respect to the terrain. Variability will be greatest in mountainous areas where rain-shadow effects and the differences between sunny and shady slopes are more pronounced. Such differences are most obvious with respect to individual storms and short time periods, so models that rely on annual

averages for their data are likely to be less affected. Further, it is sometimes possible to obtain additional local information from records made by farmers and other land-holders, or at schools and research stations.

The situation is less satisfactory with respect to data on soils. Some national soil surveys map soil units at a soil series level but, more often, they show only generalized categories such as soil associations or even soil orders or groups. Published data on soil properties that accompany the maps usually cover grain-size distributions, organic content and soil moisture at field capacity, but are unlikely to contain information on soil strength, erodibility, detachability or infiltration capacity. These properties need to be measured on site in the field or estimated from statistical relationships between the property concerned and already-measured properties. The nomograph (Wischmeier *et al.*, 1971), used to estimate soil erodibility (K value) for the USLE from measurements of grain-size distribution, organic content, permeability and structure, is an example of such a pedotransfer function. The confidence levels of these pedotransfer functions are often unknown. A further issue is that in most models, the properties of each soil unit are described by a single value. Yet, as already noted above, soil erodibility is not static but changes over time; similarly, whilst primary particle size may be considered static over short to medium time-scales, actual particle size in terms of soil aggregates will vary. Spatial variability in some soil properties is considerable. For example, infiltration capacity is often measured in the field at a number of points and the values averaged to give a single number for use in modelling; yet the spatial variability in that number within a soil series unit can be as high as 120% (Eyles, 1967). Furthermore, on crust-prone soils, the mean value around which this variation occurs can fall by 50 to 100% in a single rainstorm (Hoogmoed & Stroosnijder, 1984; Torri *et al.*, 1999). It is clear from all of this that input data used in models to express soil effects are subject to considerable error, the extent of which is usually not known. However, if the error is as much as 100% and the model outputs are sensitive to soil properties, it has to be recognized that the model predictions will also have an error of similar magnitude associated with them.

The situation would seem to be better with respect to data on slope and land cover because these can be obtained from several sources, ranging from field measurement to the use of aerial photography and satellite imagery. Height survey data and aerial photographs can be used to produce digital elevation models from which slope information can be derived. Unfortunately, few models can use this information direct. Crop or land cover usually needs to be expressed as percentage canopy cover or ground cover, or both, alongside parameters such as plant height. Even where the model requires a single value such as the C-factor in the USLE, this, as seen earlier, is derived from more detailed information linking crop cover and height to rainfall. Whilst slope steepness data can be used directly, slope length data cannot. Some models require a measure of the actual length of a slope segment, but others need only the length over which runoff occurs. Since the latter is highly dynamic, changing both between and within storms, an average or typical value usually needs to be chosen based on local experience.

The more detailed process-based models like WEPP and EUROSEM are more data-hungry than grey-box models like the USLE or MMF. Such models require data on, for example, inter-rill erodibility, rill erodibility, soil shear strength, soil cohesion, surface roughness, Manning's n (a roughness coefficient affecting flow velocity) and plant density. In recognition of the fact that these data do not generally exist and that many model users have neither the time nor the resources to obtain the data from measurement or experiment, model developers have produced various tables containing guide values for different conditions. Whilst using these often works reasonably well, there is little information available on the errors associated with or arising from using these guide values. Also, their use makes an assumption that the conditions at the field site fall within the range of the values given, and are

not abnormal or extreme. Experience with the MMF model indicates that using locally-measured values of the input parameters rather than the guide values yields better results (Morgan et al., 1984).

In addition to data required as model inputs, data are also needed to compare model outputs with observed values. Only by such comparisons can models be demonstrated as successful or otherwise. The field data need to be appropriate to the scale and objectives of the model. Unfortunately, these requirements are often ignored, usually because the data are not available. As an example, the objective of many models and model users is to predict how much sediment is removed from small catchments, ranging in size from a few hectares to one or two square kilometres. This is exactly the scale at which measured data are deficient. Lots of data exist for erosion plots, particularly those of $40 \, m^2$, and for larger river systems. Thus models like WEPP and EUROSEM are tested on their ability to predict soil loss from small plots, although they are designed to operate on small catchments. Some long-term data on erosion rates in the landscape are available for small catchments based on studies of the spatial distribution of caesium-137 (Ritchie & Ritchie, 2001) and other radioactive tracers. Process-based models can be run as continuous simulation models for hundreds of years, and the output averaged to compare with erosion rates obtained from tracer studies. However, even if the comparisons are good, they are unreliable because the models generally only consider sediment movement by rainsplash, unconcentrated overland flow and rills, whereas, in reality, assessments based on tracers incorporate the effects of all processes operating in the landscape including translocation of soils by tillage, gully erosion, wind erosion, mudflows and landslides.

Surprisingly few studies assess the ability of models to predict the timing and location of erosion, rather than the quantity. Yet field data on time and place are usually readily available. Aerial photography and field observations will identify where, in the landscape, rills and sediment deposition occur, and talks with the local population, especially farmers, will usually provide details of when the erosion took place. At present, perhaps too many model users are obsessed with predicting numbers when, for many practical purposes, it is enough to know where and how frequently erosion occurs. Users should be aware, however, that when using a model to predict the location of erosion, its sensitivity may differ from when it is used to predict the amount of erosion. LISEM, for example, is very sensitive to the size of the grid cells when used to predict the spatial patterns of erosion observed in the landscape (Jetten *et al.*, 2003).

Model users should also consider whether it is sufficient to be able to predict the direction and magnitude of changes in erosion: relative values rather than absolute numbers. If rainforest is cleared to make way for rubber or oil palm plantations or a tourist complex, environmental managers and planners will need to know by how much erosion and sediment delivery to adjacent watercourses will increase. Some models may well predict the percentage change in erosion closely enough for practical purposes, even though the absolute predictions of erosion rates are an order of magnitude adrift. Similarly, the ability to predict order-of-magnitude changes in erosion as a result of changes in rainfall amounts and intensity will be valuable in assessing the effects of climatic change.

2.8.3 Calibration

One approach to assigning values to data inputs which are hard to obtain or difficult to measure accurately is to determine what values are required to allow the model predictions of erosion to match observed values. This is the process of calibration. In the 1960s and 1970s calibration was common practice and widely used with hydrological models to determine parameters like Manning's *n*, rainfall interception and saturated hydraulic conductivity. Before adopting a calibrated value for a parameter, a check should be made to ensure that its value lies within the range of those observed in the field or obtained experimentally in the laboratory.

More recently, calibration has been viewed by many model users with some cynicism, largely because, as models have become more process-

oriented and more complex, it is a more difficult exercise. Many models now require input data for 20 or more parameters and it is virtually impossible to calibrate a model on more than about four parameters. As already noted, these models can produce identical predictions of erosion through different routes. Different parameters may need to be calibrated depending on whether erosion is limited by the rate of sediment detachment or the sediment transport capacity. Where the same parameter affects both sets of processes, it may take different calibrated values for each set. A further complication is that calibrated values usually pertain to a very specific set of conditions, so that those for soil properties when the soil is bare may not be the same as those when the soil has a crop or vegetation cover. Values for bare soil may also differ between winter and summer, or between the start and the end of a rainy season. In other words, as is the case with measured parameter values, calibrated values are likely to vary in time and space. There is also the issue that many of the process-based models give several outputs. EUROSEM, for example, provides for each storm a prediction of total runoff, total soil loss, peak runoff, peak sediment concentration, a storm hydrograph and a storm sediment graph. It is possible to calibrate the model to produce good predictions of total runoff and total soil loss without being able to reproduce acceptable hydrographs and sediment graphs (Quinton & Morgan, 1998). Overall, the logistics of obtaining appropriate sets of calibrated parameter values are such that many model users now take the view that calibration is an impossible exercise and has no meaning or validity for complex process-based models. Others, however, believe that calibration is vital and should be attempted by all model users (see Chapter 3 and sections 7.3.3 and 14.4).

2.8.4 Validation

Before erosion models can be used to develop policies or design management systems, the user must have confidence that the model outputs are realistic. This means that the chosen model must be validated. Validation is usually achieved by comparing the model predictions with field observations and applying some measure of goodness-of-fit. Where the comparisons are based on numbers, the goodness-of-fit is often assessed by statistical techniques, demonstrating that there is no significant difference between the predicted and observed values, or that there is a 1:1 relationship between them expressed by a best-fit regression equation with a slope close to 1.0 and which passes through zero. The performance measure most favoured at present by researchers is the Model Efficiency Coefficient (*MEC*) (Nash & Sutcliffe, 1970), defined by:

$$MEC = 1 - \frac{\sum \left(X_{obs} - X_{pred} \right)^2}{\sum \left(X_{obs} - \overline{X}_{obs} \right)^2} \qquad (2.9)$$

where X_{obs} is the observed value, X_{pred} is the value predicted by the model and $\overline{X}$ is the mean of a set of observed values. The *MEC* is a measure of the variance in the predictions from a 1:1 prediction line with the measured values. Thus, the closer *MEC* is to 1.0 in value, the better is the model performance. Values are rarely > 0.7 (Nearing, 1998) and a value > 0.5 is considered satisfactory (Quinton, 1997). Negative values indicate that the model predictions are poor and have a higher variation than the observed values. Where validation is based on the model predicting erosion in the right place in the landscape at the right time, comparisons are more subjective, although they can be expressed in terms of the number of observations being correctly predicted.

For model users, validation is a more complex issue than for model developers. The latter need to be able to demonstrate that their model is rational and gives reasonable predictions when compared with what is observed. Validation can therefore be based on large datasets obtained from research stations, although some caution is necessary where, as indicated above, these data are not at the same spatial or temporal scale as the model. When selecting a model, the user should

first check whether it has been validated by the developer at the appropriate scale. However, the user requires more than a generic validation. It is important to show that the model operates well in the conditions that apply in the study area. In some cases, it may be necessary to modify the model before it can be applied. Hessel (2002) had to make changes to LISEM before it could be used with the very high sediment concentrations found in the runoff on the loess soils of China. Fortunately, he was able to access suitable research data both to develop and validate the changes (see Chapter 12) but, in many parts of the world, local data suitable for validation are often non-existent or sparse. Although, ideally, the validation should be based on the principles and procedures described in Chapter 3, in reality it is often restricted to showing that the model outputs are close to the values obtained from another study area for which field data exist and which is geographically similar. The user invariably has to begin by searching for whatever data are available from whatever sources. These may be from erosion plots at research stations, but perhaps for periods of three years or less, from consultations with farmers and other land-users or from undergraduate studies at the local university. Sometimes the user may have to implement his or her own research either by undertaking a period of field observations during the season when erosion is most likely, or by interpreting aerial photography and mapping erosion features. In some cases, data availability may be so sparse that a user attempting to demonstrate that the model outputs for, say, a small catchment in Kalimantan, Indonesia, are sensible may be able to do no more than show that the model predictions for erosion plots under similar climate, soils, slopes and land covers at research stations in Java, Indonesia, or, even, in Malaysia, are close to measured values. Similarly, a user endeavouring to show that model predictions of erosion on cleared land along pipeline corridors in northern Argentina or west of Tbilisi in Georgia are valid, may only be able to show that model outputs are close to what is observed on slopes devoid of vegetation in similar climatic regions elsewhere in the world (see Section 14.7.4 for examples of using models hypothetically in the absence of data, and Section 15.3 for specific issues associated with studies of climatic change).

The availability of suitable data and the demands of potential users have meant that validation has been largely limited to model outputs such as the quantity of runoff and sediment leaving an area over a unit of time. This is satisfactory for relatively simple models where such data are all that are predicted, but for more complex process-based models it ignores the fact that the same model outputs can be obtained in a variety of ways. As already seen, with such models, predictions are made of the quantities of soil detached by raindrop impact, detached by runoff, transported by rainfall, transported by runoff, and deposited at numerous points in the landscape. If the soil loss from an area is controlled by the transport capacity of the runoff, the model needs only to get the transport capacity of the runoff from the lowest land unit in the landscape correct, and to ensure that the availability of detached particles on that unit either equals or exceeds transport capacity, to get a reasonable prediction. All the predictions of the other processes and what is happening elsewhere in the landscape could be wrong, but if they do not affect the output, this may never be known. Unfortunately, detailed observations of the rates of soil particle detachment, sediment transport and deposition on different land units within catchments are not generally available, and data on their particle-size distribution are even scarcer. It is therefore impossible to validate complex process-based models fully. The implications of this for making judgments on model outputs are explored in Chapter 7.

2.8.5 Uncertainty

The last decade has witnessed an increasing recognition that model predictions are subject to uncertainty. One source of uncertainty is the difficulty of determining input parameter values. Uncertainties arise from errors associated with the techniques of measurement, errors arising from high spatial and temporal variability which means that a parameter cannot adequately be expressed by a single value, errors involved in estimating the values of those parameters that cannot be easily measured, the use of guide values rather than measured values, and errors arising from predictions made by the various equations used to run the model. The relationship between these different sources of uncertainty is far from clear, so it is not obvious whether they are additive or multiplicative in their effect, or whether some cancel others out so that averaged overall, the errors have little influence. The procedures used to reduce the effects of error (see Chapter 4) require large numbers of datasets. For a generic evaluation of a particular model, these sets can be generated artificially by varying the values of input parameters randomly within the limits of what is likely to occur in field conditions. Multiple simulations can then be run from which the number of simulations needed to produce predictions within certain error limits can be determined (Quinton, 1997; Brazier *et al.*, 2007).

The problems with the above approach are, firstly, that the users need to know what level of error is acceptable for their purpose. Given that users are not always as familiar as model developers with the concepts of probability and uncertainty, they may have little idea of how to determine what is or is not acceptable. Some users known to the author would not consider a model worthwhile unless it was able to predict annual soil loss in tonnes per hectare to two decimal places, whereas in reality, some models will be performing well if they can predict to the correct order of magnitude. Secondly, a view of 'correct' prediction is based on the assumption that the measured or observed value is correct. In practice, this disregards errors involved in the measurement process. As already seen, most data used for model validation come from measurements made on erosion plots. As long ago as the late 1950s, Hudson (1957) drew attention to the sources of error associated with measurements from such plots and, whilst some improvements in the design of equipment have been made, the majority of these errors remain, and are still virtually unquantifiable (Morgan, 2005). Attempting to predict a value which is subject to an unknown amount of error is like aiming at a moving target. However, model users need to take account of the fact that it is unrealistic to expect a model to predict a value with greater certainty than the likely variability around the measured value (Nearing, 2000, 2006).

A third issue related to uncertainty arises directly from the different approaches to validation. Just because the levels of uncertainty associated with a model are known from a generic evaluation, it does not mean that the same levels will be achieved when the model is applied to a specific locality. Most erosion models are highly sensitive to variations in rainfall, a parameter that can be measured reasonably accurately. A generic analysis of uncertainty is likely to incorporate a wide range in rainfall, with annual totals between <100 and >3000 mm or storm totals between 1–3 mm and >400 mm. Within a given field or small catchment, the rainfall is likely to vary by much smaller amounts and uncertainties in prediction associated with rainfall will accordingly be small. In contrast, parameters that may contribute moderate to small amounts to uncertainties in generic predictions may be extremely important for predictions over small areas.

Arguably the main problem with uncertainty analysis at present is that an enormous number of computer simulations need to be carried out which, although they may reduce the level of variability around a prediction, may not make the prediction a closer fit to the observed value (Quinton, 1997). It seems unlikely that a user will undertake uncertainty analysis, but see Chapter 5 for an example of what can be achieved. Whether or not they carry out the analysis, users need to be aware of the level of uncertainty associated with predictions from a chosen model. It is therefore incumbent on model developers that they perform uncertainty analysis as part of the information they provide about their model in User Guides. Chapter 4 provides a more detailed discussion of the various types of uncertainty and the approaches that can be used to deal with them.

2.9   Some Practicalities

By concentrating on the problems associated with developing and validating models, the impression may be gained that models are either of limited value or simply too difficult to use. This, however, is far from the case. Erosion is one of the leading environmental problems worldwide, and controlling it requires a range of activities from the development of policies at national and international level to the design and implementation of conservation measures at the field, road bank or construction site level. Erosion control is relevant not just to the fate of sediment removed from the landscape, but also to reducing the associated risks of pollution from chemicals adsorbed to the sediment and increased flooding brought about by the reduction in the capacity of river channels and reservoirs as a result of sediment deposition. Policies, strategies and the design of soil protection works must all be based on sound scientific data. It takes far too long and is too expensive to obtain the necessary data from field measurement, so models are vital for filling that gap, provided that they too are based on good science and good data.

Enormous advances have been made in the development of models over the last 20 or more years, and there is now a large number from which to choose. Since these models range in complexity from simple black-box types to detailed process-based ones and operate at a whole range of temporal and spatial scales, it is important that the user defines carefully what is needed to address a particular problem. The model user needs to think along the same lines as the model developer. He or she must start by deciding what data are needed to deal with the problem, over what temporal and spatial scales and covering what processes. Thus it helps if the user develops a conceptual framework of what is required. Since this framework is essentially a conceptual model, this initial step should make it easier to research the available models and find ones with a similar conceptual base. It is important at this stage that the user is not misled into believing that detailed process-based models that yield numerical predictions of runoff and sediment yield for single storms with their associated hydrographs and sediment graphs must be the best. For many problems, numerical predictions at a storm level may not be needed; indeed, they may not even be helpful. Further, the data may not be available to run such models. As shown by the case studies in this volume, many problems can be dealt with by relatively simple models, which may even be more reliable in their predictions (Jetten *et al.*, 2003). Assessing the nature of the match between the conceptual framework of the problem and that of the model is vital for avoiding the problem of choosing an inappropriate model or, worse, misusing a model by attempting to use it for something for which it was not designed. Sometimes, where there is no obvious match between the problem and the model, it is necessary to choose the model that is the nearest approximation to what is required. The more distant the match, the greater is the likelihood of misuse and of the model not working. Users then need to be honest in their evaluation of the model results. If they are useable, that is a bonus; if they are not, it is not the fault of the model.

Having selected a model, the user must demonstrate that it is applicable to the conditions of the study area and to the temporal and spatial scale of the problem. Ideally, the model should be validated, at least for the outputs that are going to be used, with data from the area. Only where this is not possible should data from similar conditions elsewhere be used. The user needs to gain a good understanding and experience with the model, seeking advice from the model developer and from other users, in order to learn how best to set the model up to simulate the local conditions. This may be particularly important with distributed process-based models where knowledge of the flow paths taken by runoff and sediment over the landscape can inform the identification of the land units (see Chapter 13). Although the user may operate with single predicted values as model output, partly because developing, for example, soil conservation designs using multiple values is too complicated, the user should be aware of the likely levels of

uncertainty around, or the accuracy of, that value. The user should also be aware that there are variations in the accuracy of predictions depending upon the value being predicted. Studies with WEPP have shown that it has the tendency to overpredict at low values and underpredict at high values (Brazier *et al.*, 2000).

As more research takes place into the design of soil protection measures, it may be that users will appreciate that designing to deal with a mean or average condition is not the most appropriate. Users may well find they need models that give a mean and standard deviation, and that designs need to be made for the event which is either one or two standard deviations above the mean. Users should learn to treat the data produced from models in the same way as they would use data derived from observation and measurement. They need to ask the same questions about representativeness, reliability, frequency and probability. The more that users take advantage of what models can offer, the more they will appreciate that the thought processes and methodologies of users and model developers are very similar. This should aid collaboration. It might also encourage model developers to provide more information of value to users, for example, the objective of the model, the conditions for which it operates, the level of accuracy required in the input parameter values, and the errors associated with its predictions. Unfortunately, too few User Guides at present provide this basic information.

Acronyms

Many models are known by acronyms. Those mentioned in the text are:

AGNPS (Agricultural Nonpoint Source Pollution Model)

ANSWERS (Areal Nonpoint Source Watershed Environment Response Simulation)

CREAMS (Chemicals, Runoff and Erosion from Agricultural Management Systems)

EUROSEM (European Soil Erosion Model)

GAMES (Guelph model for evaluating the effects of Agricultural Management systems on Erosion and Sedimentation)

GUEST (Griffith University Erosion System Template)
LISEM (Limburg Soil Erosion Model)
MMF (Morgan-Morgan-Finney model)
PESERA (Pan-European Soil Erosion Risk Assessment)
RUSLE (Revised Universal Soil Loss Equation)
SHE (Système Hydrologique Européen)
USLE (Universal Soil Loss Equation)
WATEM (Water and Tillage Erosion Model)
WEPP (Water Erosion Prediction Project).

References

Auzet, A.V., Boiffin, J., Ludwig, B. & Guérif, J. (1995) Effects of agricultural land use on spatial and temporal distribution of soil erosion in small catchments: implications for modelling. In Boardman, J. & Favis-Mortlock, D. (eds), *Modelling Soil Erosion by Water*. Springer-Verlag, Berlin: 329–38.

Beasley, D.B., Huggins, L.F. & Monke, E.J. (1980) ANSWERS: a model for watershed planning. *Transactions of the American Society of Agricultural Engineers* 23: 938–44.

Beuselinck, L., Govers, G., Steegen, A., et al. (1999) Evaluation of the simple settling theory for predicting sediment deposition by overland flow. *Earth Surface Processes and Landforms* 24: 993–1007.

Beven, K.J. & Kirkby, M.J. (1979) A physically-based variable contributing area model of basin hydrology. *Hydrological Sciences Bulletin* 24: 43–69.

Biot, Y. 1990. THEPROM – an erosion productivity model. In Boardman, J., Foster, I.D.L. & Dearing, J.A. (eds), *Soil Erosion on Agricultural Land*. John Wiley & Sons, Chichester: 465–79.

Brazier, R.E., Beven, K., Freer, J. & Rowan, J.S. (2000) Equifinality and uncertainty in physically based soil erosion models: application of the GLUE methodology to WEPP. *Earth Surface Processes and Landforms* 25: 825–45.

Brazier, R.E., Beven, K.J., Anthony, S.G. & Rowan, J.S. (2007) Implications of model uncertainty for the mapping of hillslope-scale soil erosion predictions. *Earth Surface Processes and Landforms* 26: 1333–52.

Christian, C.S. & Stewart, G.A. (1968) *Methodology of integrated survey. Aerial surveys and integrated studies*. UNESCO, Paris: 233–80.

Crawford, N.H. & Linsley, R.K. (1966) *The Stanford Watershed Model Mark IV*. Department of Civil Engineering, Stanford University, Technical Report No. 39.

De Roo, A., Jetten, V., Wesseling, C. & Ritsema, C. (1998) LISEM: a physically-based hydrologic and soil erosion catchment model. In Boardman, J. & Favis-Mortlock, D. (eds), *Modelling Soil Erosion by Water*. Springer-Verlag, Berlin: 429–40.

Ellison, W.D. (1947) Soil erosion studies – Part I. *Agricultural Engineering* 28: 145–6.

Eyles, R.J. (1967) Laterite at Kerdau, Pahang, Malaysia. *Journal of Tropical Geography* 25: 18–23.

Favis-Mortlock, D., Boardman, J., Parsons, A.J. & Lascelles, B. (2000) Emergence and erosion: a model for rill initiation and development. *Hydrological Processes* 14: 2173–2205.

Fiener, P., Govers, G. & van Oost, K. (2008) Evaluation of a dynamic multi-class sediment transport model in a catchment under soil-conservation agriculture. *Earth Surface Processes and Landforms* 33: 1639–60.

Gobin, A., Govers, G. & Kirkby, M. (2006) Pan-European soil erosion assessment and maps. In Boardman, J. & Poesen, J. (eds), *Soil Erosion in Europe*. John Wiley & Sons, Chichester: 661–74.

Govers, G., Vandaele, K., Desmet, P., et al. (1994) The role of tillage in soil redistribution on hillslopes. *European Journal of Soil Science* 45: 469–78.

Hessel, R. (2002) Modelling soil erosion in a small catchment on the Chinese Loess Plateau. *Nederlandse Geografische Studies* 307.

Hoogmoed, W.B. & Stroosnijder, L. (1984) Crust formation on sandy soils in the Sahel. I. Rainfall and infiltration. *Soil and Tillage Research* 4: 5–24.

Hudson, N.W. (1957) The design of field experiments on soil erosion. *Journal of Agricultural Engineering Research* 2: 56–65.

Issa, O.M., Le Bissonnais, Y., Planchon, O., et al. (2006) Soil detachment and transport on field- and laboratory-scale interrill areas: erosion processes and the size-selectivity of eroded sediment. *Earth Surface Processes and Landforms* 31: 929–39.

Jetten, V. & de Roo, A.P.J. (2001) Spatial analysis of erosion conservation measures with LISEM. In Harmon, R.S. & Doe III, W.W. (eds), *Landscape Erosion and Evolution Modeling*. Kluwer, New York: 429–45.

Jetten, V., Govers, G. & Hessel, R. (2003) Erosion models: quality of spatial predictions. *Hydrological Processes* 17: 887–900.

Kirkby, M.J. (1980) Modelling water erosion processes. In Kirkby, M.J. & Morgan, R.P.C. (eds), *Soil Erosion*. John Wiley & Sons, Chichester: 183–216.

Kirkby, M.J. (1998) Modelling across scales: the MEDALUS family of models. In Boardman, J. & Favis-Mortlock, D. (eds), *Modelling Soil Erosion by Water*. Springer-Verlag, Berlin: 161–73.

Kirkby, M.J. & Neale, R.H. (1987) A soil erosion model incorporating seasonal factors. In Gardiner, V. (ed), *International Geomorphology 1986. Part II.* John Wiley & Sons, Chichester: 189–210.

Kirkby, M.J., Abrahart, R.J., Bathurst, J.C., *et al.* (2002) MEDRUSH: a basin-scale physically based model for forecasting runoff and sediment yield. In Geeson, N.A., Brandt, C.J. & Thornes, J.B. (eds), *Mediterranean Desertification: a Mosaic of Processes and Responses.* John Wiley & Sons, Chichester: 203–27.

Knisel, W.G. (1980) CREAMS: a field scale model for chemicals, runoff and erosion from agricultural management systems. *USDA Conservation Research Report* 26.

Lee, J.J. (1998) Cross-scale aspects of EPA erosion studies. In Boardman, J. & Favis-Mortlock, D. (eds), *Modelling Soil Erosion by Water.* Springer-Verlag, Berlin: 191–9.

Meyer, L.D. & Wischmeier, W.H. (1969) Mathematical simulation of the process of soil erosion by water. *Transactions of the American Society of Agricultural Engineers* 12: 754–8, 762.

Morgan, R.P.C. (2001) A simple approach to soil loss prediction: a revised Morgan-Morgan-Finney model. *Catena* 44: 305–22.

Morgan, R.P.C. (2005) *Soil Erosion and Conservation.* (3rd edn). Blackwell, Oxford.

Morgan, R.P.C., Morgan, D.D.V. & Finney, H.J. (1984) A predictive model for the assessment of soil erosion risk. *Journal of Agricultural Engineering Research* 30: 245–53.

Morgan, R.P.C., Rickson, R.J., McIntyre, K., *et al.* 1997. Soil erosion survey of the central part of the Swaziland Middleveld. *Soil Technology* 11: 263–89.

Morgan, R.P.C., Quinton, J.N., Smith, R.E., *et al.* (1998) The European Soil Erosion Model (EUROSEM): a dynamic approach for predicting sediment transport from fields and small catchments. *Earth Surface Processes and Landforms* 23: 527–44.

Morgan, R.P.C. & Duzant, J.H. (2008) Modified MMF (Morgan-Morgan-Finney) model for evaluating effects of crops and vegetation cover on soil erosion. *Earth Surface Processes and Landforms* 32: 90–106.

Moussa, R., Voltz, M. & Andrieux, P. (2003) Impacts of various scenarios of agricultural management on the hydrological behaviour of a farmed catchment during flood events. *International Association of Hydrological Sciences Publication* 278: 417–21.

Nash, J.E. & Sutcliffe, J.V. (1970) River flow forecasting through conceptual models. I. Discussion of principles. *Journal of Hydrology* 10: 282–90.

Nearing, M.A. (1998) Why soil erosion models overpredict small soil losses and underpredict large soil losses. *Catena* 32: 15–22.

Nearing, M.A. (2000) Evaluating soil erosion models using measured plot data: accounting for variability in the data. *Earth Surface Processes and Landforms* 25: 1035–43.

Nearing, M.A. (2006) Can soil erosion be predicted? In Owens, P.N. & Collins, A.J. (eds), *Soil Erosion and Sediment Redistribution in River Catchments.* CAB International, Wallingford: 145–52.

Nearing, M.A., Deer-Ascough, L. & Chaves, H.M.L. (1989a) WEPP model sensitivity analysis. In Lane, L.J. & Nearing, M.A. (eds), *USDA Water Erosion Prediction Project: hillslope model documentation.* USDA-ARS NSERL Report 2: 14.1–14.33.

Nearing, M.A., Foster, G.R., Lane, L.J. & Finckner, S.C. (1989b) A process-based soil erosion model for USDA-Water Erosion Prediction Project technology. *Transactions of the American Society of Agricultural Engineers* 32: 1587–93.

Negev, M. (1967) *A sediment model on a digital computer.* Stanford University, Report No. 76.

Quinton, J.N. (1997) Reducing predictive uncertainty in model simulations: a comparison of two methods using the European Soil Erosion Model (EUROSEM). *Catena* 30: 101–17.

Quinton, J.N. & Morgan, R.P.C. (1998) EUROSEM: an evaluation with single event data from the C5 Watershed, Oklahoma, USA. In Boardman, J. & Favis-Mortlock, D. (eds), *Modelling Soil Erosion by Water.* Springer-Verlag, Berlin: 65–74.

Renard, K.G., Foster, G.R., Weesies, G.A. & Porter, J.P. (1991) RUSLE: Revised Universal Soil Loss Equation. *Journal of Soil and Water Conservation* 46: 30–33.

Ritchie, J.C. & Ritchie, C.A. (2001) Bibliography of publications of [137]caesium studies related to erosion and sediment deposition. http://hydrolab.arsusda.gov/cesium137bib.htm

Rose, C.W., Williams, J.R., Sanders, G.C. & Barry, D.A. (1983) A mathematical model of soil erosion and deposition process. I. Theory for a plane element. *Soil Science Society of America Journal* 47: 991–5.

Rose, C.W., Yu, B., Hogarth, W.L., *et al.* (2003) Sediment deposition from flows at low gradients into a buffer strip – a critical test of re-entrainment theory. *Journal of Hydrology* 280: 33–51.

Rudra, R.P., Dickinson, W.T. & Wall, G.J. (1986) GAMES – a screening model of soil erosion and fluvial sedimentation on agricultural watersheds. *Canadian Water Research Journal* 11: 58–71.

Smith, R.E. (1979) *A kinematic model for surface mine sediment yield*. American Society of Agricultural Engineers Winter Meeting, Paper No. 79-2533.

Thornes, J.B. (1990) The interaction of erosional and vegetational dynamics in land degradation: spatial outcomes. In Thornes, J.B. (ed), *Vegetation and Erosion: Processes and Environments*. John Wiley & Sons, Chichester: 41–53.

Tollner, E.W., Barfield, B.J., Haan, C.T. & Kao, T.Y. (1976) Suspended sediment filtration capacity of simulated vegetation. *Transactions of the American Society of Agricultural Engineers* **19**: 678–82.

Torri, D., Regüés, D., Pelligrini, S. & Bazzoffi, P. (1999) Within-storm soil surface dynamics and erosive effects of rainstorms. *Catena* **38**: 131–50.

Van Muysen, W., Govers, G. & van Oost, K. (2002) Identification of important factors in the process of tillage erosion: the case of mouldboard tillage. *Soil and Tillage Research* **65**: 77–93.

Van Oost, K., Govers, G. & Desmet, P.J.J. (2000) Evaluating the effects of changes in landscape structure on soil erosion by water and tillage. *Landscape Ecology* **15**: 579–91.

Verstraeten, G. & Poesen, J. (2001) Modelling the long-term sediment trap efficiency of small ponds. *Hydrological Processes* **15**: 2797–819.

Verstraeten, G., van Oost, K., van Rompaey, A., *et al.* (2002) Evaluating an integrated approach to catchment management to reduce soil loss and sediment pollution through modelling. *Soil Use and Management* **18**: 386–94.

Webster, R. & Beckett, P.H.T. (1970) Terrain classification and evaluation using air photography: a review of recent work at Oxford. *Photogrammetria* **26**: 51–7.

Wicks, J.M. & Bathurst, J.C. (1996) SHESHED: a physically based, distributed erosion and sediment yield component for the SHE hydrological modelling system. *Journal of Hydrology* **175**: 213–38.

Wischmeier, W.H. (1978) Use and misuse of the Universal Soil Loss Equation. *Journal of Soil and Water Conservation* **31**: 5–9.

Wischmeier, W.H., Johnson, C.B. & Cross, B.V. (1971) A soil erodibility nomograph for farmland and construction sites. *Journal of Soil and Water Conservation* **26**: 189–93.

Wischmeier, W.H. & Smith, D.D. (1965) Prediction rainfall-erosion losses from cropland east of the Rocky Mountains. Guide for selection of practices for soil and water conservation. *USDA Agricultural Handbook* 282.

Wischmeier, W.H. & Smith, D.D. (1978) Predicting rainfall erosion losses. A guide to conservation planning. *USDA Agricultural Research Service Handbook* 537.

Wood, G.A., McHugh, M., Morgan, R.P.C. & Williamson, A. (2006) Estimating sediment generation from hillslopes in England and Wales: development of a management planning tool. In Owens, P.N. & Collins, A.J. (eds), *Soil Erosion and Sediment Redistribution in River Catchments*. CAB International, Wallingford: 217–27.

Young, R.A., Onstad, C.A., Bosch, D.D. & Anderson, W.P. (1989) AGNPS: a nonpoint-source pollution model for evaluating agricultural watersheds. *Journal of Soil and Water Conservation* **44**: 168–73.

3 Calibration of Erosion Models

V.G. JETTEN[1] AND M.P. MANETA[2]

[1]*Department of Earth Systems Analysis, International Institute of Geoinformation Science and Earth Observation, Enschede, The Netherlands*
[2]*Geosciences Department, University of Montana, Missoula, MT, USA*

3.1 Introduction

A dictionary may define the term calibration as "the act of checking or adjusting the accuracy of a measuring instrument, by comparison with a standard". For instance, a ruler can be calibrated against the standard length of 1 metre, which was defined in 1791 by the French Academy of Sciences as the 10^{-7} part of the distance along the Earth's surface from pole to Equator (later this definition changed). The mathematicians Delambre and Méchain supervised the measurements of the length of France along the meridian through Paris as a section of the circumference of the Earth. The standard metre, they deduced from their measurements, appeared to be only 0.2 mm short compared with a later satellite-based determination of the circumference of the Earth. This story has all the elements important for the calibration of an erosion model: the tuning of a model to predict observed (or perceived) 'true' values of erosion, using limited data to deduce the workings of the entire system, with errors that may occur for various reasons and with newer methods that may improve the result and improve our insight. Fortunately, calibration and improvement of erosion models has been a major focus of model users and model builders from the beginning of erosion modelling. This stems perhaps

from the fact that the first widely used soil loss model, the Universal Soil Loss Equation (USLE), was based on a regression analysis of data from several hundred experimental plots installed by the USDA Soil Conservation Service, and so a close link between theory and experimental data has always been a focal point. Later attempts at improving the modelling by including more process descriptions, and including spatial variability with the use of Geoinformation Science, have not changed that fact.

Jetten and Favis-Mortlock (2006) gave an overview of 16 water erosion models used in Europe at a range of scales from several square metres to the whole continent, and concluded that calibration and validation attempts are done for all models on all scales, which is encouraging. There exist several dozens of published water erosion models in the world, as well as models that have been made for a different purpose, such as river basin models that have a sediment component.

The purpose of the modelling closely dictates exactly what it is we are calibrating. The term 'erosion modelling' is used to indicate the total detachment of soil particles by rainfall and runoff in an area, but also to predict the soil loss from a specific area. In a wider context it is used to indicate the hazard or risk associated with the removal of topsoil, both onsite (fertility loss, gullying) and offsite (muddy floods and downstream pollution). Often erosion modelling is done to simulate the effects of a change in land use, application of soil conservation measures,

Handbook of Erosion Modelling, 1ˢᵗ edition. Edited by R.P.C. Morgan and M.A. Nearing. © 2011 Blackwell Publishing Ltd.

Table 3.1 Model acronyms and references.

Acronym	Full name	Web reference
AGNPS	AGricultural Non-Point Source pollution model (Young *et al.*, 1989)	http://www.wsi.nrcs.usda.gov/products/w2q/h&h/tools_models/agnps/index.html
ANSWERS	Areal Nonpoint Source Watershed Environment Response Simulation (Beasley *et al.*, 1980)	NA
CREAMS	Chemicals, Runoff, Erosion, and Agricultural Management Systems (Knisel, 1980)	NA
EPIC/APEX	Erosion-Productivity Impact Calculator Agricultural Policy/Environmental eXtender (Williams *et al.*, 1984)	http://www.brc.tamus.edu/simulation-models/epic-and-apex.aspx
EROSION3D	Erosion 2D and 3D models (Schmidt *et al.*, 1999)	http://www.bodenerosion.com/
EUROSEM	European Soil Erosion Model (Morgan *et al.*, 1998)	http://www.es.lancs.ac.uk/people/johnq/EUROSEM.html
FSM	Factorial Scoring Model (de Vente *et al.*, 2005)	NA
GLEAMS	Groundwater Loading Effects of Agricultural Management Systems (Leonard *et al.*, 1987)	http://www.tifton.uga.edu/sewrl/Gleams/gleams_y2k_update.htm
GUEST	Griffith University Erosion System Template (Misra & Rose, 1996)	NA
KINEROS2	KINEmatic Runoff and erOSion model (Woolhiser *et al.*, 1990)	http://www.tucson.ars.ag.gov/kineros/
LISEM	Limburg Soil Erosion Model (Jetten *et al.*, 1996)	http://www.itc.nl/lisem/
MMF	Morgan-Morgan-Finney model (Morgan, 2001)	NA
MUSLE	Modified Universal Soil Loss Equation (Williams, 1975)	NA
PESERA	Pan European Soil Erosion Risk Assessment (Kirkby *et al.*, 2004)	http://www.geog.leeds.ac.uk/research/groups/pesera
RUSLE	Revised Universal Soil Loss Equation (Renard *et al.*, 1991)	http://ww.iwr.msu.edu/rusle/
SCS-CN	Soil Conservation Service – Curve Number	NA
STREAM	Sealing Transfer Runoff Erosion Agricultural Modification (Cerdan *et al.*, 2002)	NA
TOPMODEL	Topographic model (Beven, 1997)	NA
USLE	Universal Soil Loss Equation (Wischmeier & Smith, 1978)	NA
WATEM/SEDEM	WAter and Tillage Erosion Model/Sediment Delivery Model (van Oost *et al.*, 2000)	http://geo.kuleuven.be/geography/modelling/erosion/watemsedem/index.htm
WEPP	Water Erosion Prediction Project (Flanagan & Nearing, 1995)	http://www.ars.usda.gov/Research/docs.htm?docid=10621

or a change in climate. Calibration aims to improve the predictive quality and fitness for use of erosion models for all these goals. However, we are usually not directly calibrating the processes we are attempting to predict. For instance, very often calibration is done on the discharge characteristics of a plot, catchment or basin, assuming that if we manage to predict the hydrograph correctly, this is sufficient to predict accurately the sediment dynamics in the catch-

ment. Whether this is feasible depends on a number of factors: availability and uncertainty of input data, the spatial complexity, the environmental setting, land use characteristics, and so on. From these objectives it seems logical to divide the models into three broad classes based on the spatial scale:

(I) A plot or field scale which is regarded as a single homogeneous spatial unit. Models at this scale vary from empirical/conceptual such as the

USLE and derivatives, to physically-based process models. In the former category, processes such as transport and deposition are often not included. Temporal scale varies from individual rainstorms to lumped annual values, with the latter making up the majority of published results.

(II) A catchment scale varying in size from less than 100 ha to several 100 km². Models operating at this scale are generally process-based models or hybrid models that are adaptations of the USLE with added process descriptions and methods for spatial water and/or sediment routing and accumulation. These vary from explicit modelling of detachment, transport and deposition, to the use of runoff coefficients and sediment delivery ratios. Temporal scales vary again from individual rainfall events to annual totals, although most calibration is done for individual events.

(III) A so-called 'large' scale with administrative boundaries, from provinces and parts of countries to continental. Models at this scale are partly physically based, but use variables derived from a DEM (digital elevation model) as proxies for slope angle, transport capacity and accumulation. Generally, sediment delivery ratios are used. Temporal scales vary from monthly to annual totals.

Note that the model acronyms and main references can be found in Table 3.1.

This chapter first gives an overview of calibration, using examples from all three categories. These are manual calibrations based on some goodness of fit parameters between predicted and observed runoff and soil loss. Secondly, several examples of more complex means of calibration are given based on Monte Carlo-type automatic parameter estimation approaches (such as GLUE and PEST). These are generally done to explore model sensitivity and to deal with the problem of equifinality (elaborated below). Finally, the role of spatial information is investigated as a possible way of improving model performance.

Three important remarks should be made at this point. Firstly, we only discuss calibration of models that predict water erosion. Wind erosion, tillage erosion and mass movement models are not considered, as they involve very different processes and require different methods of calibration. Secondly, it is impossible to give an exhaustive account of all calibration attempts. Instead, examples are given to highlight certain results. We recognize that this choice is subjective, but fortunately many different authors agree on the general state of the art in what we can and cannot do with erosion models. Thirdly, highlighted calibration problems of certain models are not a judgment on their quality. If anything, we consider it very positive if authors are honest about the performance of a model in the given circumstances and are willing to share less good results.

3.2 Calibration at Different Scales

In the last 15 years, several exercises have been held whereby erosion models were compared and tested in an orchestrated way using common datasets for calibration and validation. The IGBP-GCTE Soil Erosion Network and European COST 623 and COST 634 erosion networks have tested the fitness of erosion models to predict the consequences of climate change for erosion. The hillslope and catchment scale comparisons are briefly mentioned here. For evaluations, six field-based models and seven catchment-based models were examined using common datasets from various countries, split into a 'training set' for calibration and a 'testing set' for validation. Details about the datasets, the models and model specifics can be found in Boardman and Favis-Mortlock (1998) and in De Roo (1999). Nearing *et al.* (2005) compared the responses of seven models in two catchments (one in Belgium and one in the US) for climate change, expressed as a change in rainfall characteristics and vegetation cover. Apart from these, many individual model comparisons have been conducted. In particular, WEPP has been compared to other models (RUSLE, EPIC and others) for several environments (e.g. in Chile by Stolpe, 2005; Peru by Romero *et al.*, 2007; India by Pandey *et al.*, 2008a; Norway by Gronsten & Lundekvam, 2006; Italy by Pieri *et al.*, 2007; and Tunisia by Raclot & Albergel, 2006).

3.2.1   Calibration of field/plot scale models

Plot- and field-based research forms the basis of many erosion studies. The sediment loss from a plot or field (i.e. a small, relatively homogeneous spatial unit) is one of the most available experimental data. Experiments at this scale are relatively easy to set up, and a close link exists to farming units so that direct advice can be given to farmers for practices that will reduce soil loss. There are also clear disadvantages: the smaller the plot, the more rainsplash detachment is favoured over flow detachment, as larger rills and gullies will almost never occur on erosion plots. Also, deposition will not take place on smaller plots unless there is a break in slope angle or a change in vegetation. The objective of plot-based erosion modelling is usually to simulate soil loss from a single homogeneous unit, whereby a variety of land uses and soil conservation measures are often tested, such as mulching, minimum tillage, and green cover.

There are generally two types of models that are used: lumped models based on multiple regression type equations (such as the USLE and derivatives), and physically-based process models. The first calculate annual soil loss based on average rainfall kinetic energy and peak flow (in the case of RUSLE and EPIC) and use methods like the SCS Curve Number to derive a runoff coefficient. The second are models that calculate runoff from infiltration excess and simulate splash detachment, flow detachment and sediment transport (such as WEPP, EROSION 2D, PSEM2D and HEM). In addition, it should be noted that most catchment-based models can be applied as hillslope models.

In the GCTE hillslope comparison exercise (Favis-Mortlock, 1998), six models were tested using 73 years of data from seven sites in three countries. The sites from which the data were drawn range from plots of 0.01 ha to small catchments of just under 10 ha, with slopes varying from almost 1% to 18%, a range of soil types, and land use including both agriculture and natural vegetation. The results showed a large scatter between predicted and observed values, with differences of more than one order of magnitude for all discharge sizes. The scatter for the daily discharges was much larger than for the annual discharges. Moreover, in the continuous simulation a number of discharges were observed but not simulated, and, for certain rainfall events, runoff was simulated but not observed. This was especially the case for smaller events. Boardman and Favis-Mortlock (1998) concluded amongst other things that: calibration is almost always necessary and is most effective if the event(s) to be estimated lies inside the range of calibration events. For continuous simulation models, long-term average results are better simulated than results for individual time periods. In general, results are less good for shorter time periods, although there are exceptions. With these models it is more difficult to predict days when runoff or erosion does occur than days when it does not (when predicted independently).

The difficulty of predicting adequately both large and small events was remarked upon by Sadeghi et al. (2008) who used the process-based Hillslope Erosion Model (HEM) to simulate a series of erosion plots in Iran. They did not succeed in calibrating HEM using the erodibility factor alone. The non-linear response of the plots to more severe events could only be simulated by applying a non-linear regression between predicted and observed values. Risse et al. (1993) tested the USLE for over 1700 years of erosion on 208 plots and obtained a reasonable R^2 of 0.58, with strong overprediction of small events. Nearing (1998) also remarked that models seemed to have difficulties in predicting small-scale events. Nearing et al. (1999) and Nearing (2000) compared the soil loss of pairs of adjacent plots in the US, selecting plots that can be considered identical in terms of received rainfall, soils and surface conditions. One plot from a pair was considered as a predictor of soil loss of the adjacent plot. The coefficient of variance (CV) decreases from 150% for very small events (0.01 t ha^{-1}) to 1% for very large events (about 400 t ha^{-1}), which indicates the impossibility of correctly predicting soil loss of small events with a mathematical model that is always a simplified representation of reality. Although the low CV for large events is encouraging, small events are of much greater interest in recent studies of eroded agricultural lands as a source of polluted sediment: only small quantities of sediment are sufficient to pollute surface water if the clay particles are saturated with agrochemicals.

3.2.2 Calibration of catchment-scale models

The term 'catchment scale' is used rather loosely here to indicate a range of scales from catchments of roughly 1 km^2 to several hundreds of km^2. The difference between the 'plot' and 'catchment' scale is the detailed spatial characterization of a catchment, either using grid-based GIS systems or dividing the catchment into functional elements such as slope segments, channel segments and ponds.

Generally two types of models are used for catchment-based simulations: models that use breakpoint rainfall data to model individual rainfall events, or series of rainstorms, and models lumped in time that simulate annual or multi-annual erosion (Jetten and Favis-Mortlock, 2006). Most models at this scale are physically-based and attempt to simulate the individual hydrological and sediment transport processes.

There are multiple objectives at this scale: (i) to simulate soil loss from the catchment, e.g. to quantify downstream floods and sediment problems (Boardman *et al.*, 1994); (ii) to gain insight into the spatial distribution of sources and sinks in a catchment, e.g. to estimate the effect of conservation measures and agricultural policies (Jetten and De Roo, 2001; Lundekvam *et al.*, 2003; Hessel *et al.*, 2003a; Hessel and Tenge, 2008); (iii) to study connectivity between sources and sinks, e.g. to emphasize the role of tillage and wheeltracks (Takken *et al.*, 2001) and crusting (Le Bissonnais *et al.*, 2005); and (iv) to predict the impacts of future changes in, for example, climate and land use (Nearing *et al.*, 2005). The results can be very different for different climates and land uses. Semi-arid areas are known for their limited connectivity between hillslopes and stream beds, combined with mixed land use of

grazing (often overgrazing), small field sizes and torrential rainfall. On the other end of the spectrum, land reallocation in western Europe has caused the creation of uninterrupted large fields that produce runoff and sediment even at low slope angles. From the point of view of this multi-purpose modelling, proper calibration at this scale is perhaps the most difficult to achieve.

The GCTE catchment model comparison test was reported by Jetten *et al.* (1999). Six models were tested using five calibration and five validation events in a 40 ha catchment in The Netherlands. Calibration was mostly done using parameters that influence infiltration (such as initial moisture and saturated hydraulic conductivity). It appeared that the overall performance of the models was moderate for the calibration datasets, and less good for the validation datasets (see Fig. 3.1). Total discharge predictions were reasonable, and generally better than peak discharge, while both were better predicted than sediment discharge. It should be said that some models were not meant to be used at this scale, and therefore may not have performed well. Also, contributors were asked to find average calibration factors for the five training events, which were quite different in magnitude. Fig. 3.1 shows that some models have difficulty with predicting larger events when calibrated for small events. This may also have influenced the results. Whilst individual calibration of each rainstorm would probably have yielded better calibration results, it would still have been interesting to see how this would have affected predictions for the validation events. For example, Hessel *et al.* (2003b) analysed the performance of LISEM for a small catchment in northern China and concluded that it was impossible to find a single calibration set that would fit all measured rainfall events. From the discussions during the GCTE meetings it was clear that additional 'soft' information, in particular when modellers use knowledge of changes in soil structure as a result of agricultural activities and/or climate in their calibration strategy, can improve the quality of input data and model results.

Similar results were obtained in other model comparisons. Zhang *et al.* (1996) tested the WEPP

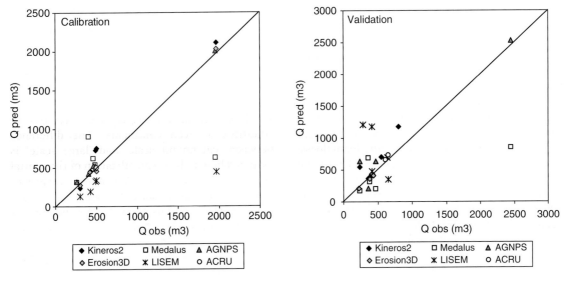

Fig. 3.1 Breakdown of the results of the GCTE catchment comparison per model (after Jetten *et al.*, 1999).

model and showed that even with optimized values for saturated hydraulic conductivity (K_{sat}), prediction was moderate especially for smaller events. Bathurst and Lukey (1998) compared observed and predicted soil loss from the SHE model for the Draix area (France) and showed results varying from accurate up to a factor 100 difference. Brochot (1998), on the other hand, obtained good results for the same area using a simple regression equation between soil loss, precipitation and infiltration, suggesting that model complexity and area complexity play a role in how well a model performs. De Roo (1993) compared the lumped USLE and MMF models with the spatially distributed ANSWERS and KINEROS models for three small Dutch catchments, and concluded that they performed equally well when the models were tested for the same type of output (annual soil loss).

From the above it is clear that more empirical models seem to perform equally well compared with more physically-based models. For example, RUSLE variants are used with some success when coupled to a Sediment Delivery Ratio (SDR) or runoff estimates using the SCS Curve Number method, to calculate the fraction of sed-

iment reaching a stream. Arhonditsis *et al.* (2002) simulated sediment production of a 194 km² catchment in Greece, and showed it was possible to use the USLE at this scale by adding an SDR term based on the distance (dk) between a sediment source area k and the channel: $SDR_k = dk^{-0.34}$ where 0.34 is the calibrated value. Furthermore, they changed the slope length exponent from 0.6 in the original USLE to 0.72 for their dataset. Tyagi *et al.* (2008) coupled the USLE, the SCS-CN method, Horton's infiltration equation and a reservoir routing method to simulate soil loss with 28 calibration and 26 validation events from three Indian and four US catchments. They showed good results for total sediment yield and moderate results for peak sediment yield, and concluded it was important that the calibration dataset included events that represented different antecedent moisture contents and crop growth stages. In other words, the wider the range of conditions in the calibration dataset, the better the validation results will be. This was also the outcome of the GCTE plot-scale comparison (Boardman and Favis-Mortlock, 1998). On the other hand, as noted earlier, Hessel *et al.* (2003b) showed that it was not possible to find a single calibration set that would enable the model LISEM to predict accurately all events in a small catchment in China (see also Chapter 12).

An alternative approach is to use decision support-based models that apply a rating to erosion factors. One example is STREAM, a physically-based distributed model that gives simulations lumped for single events. It uses observed crusting classes according to a system developed in northern France for silty loam (loess) soils (Cerdan *et al.*, 2002). These classes are related to infiltration rates and surface roughness values in a built-in dataset in the model. The strength of the model lies in the fact that an accurate image of runoff-contributing areas in a catchment is made by observing crusting classes, which allows good predictions in spite of the fact that a classified system is used.

A classified approach is also used by De Vente *et al.* (2008) who employ a Spatially Distributed Scoring Model (SPADS) to predict the annual sediment yield of 61 basins in Spain. This model is based on a rating with scores of 1–3 for vegetation cover, topography, lithology, annual rainfall, gully density and inverse distance from a river stream. This yields a catchment index which is then related to the area specific sediment yield (SSY) by a regression analysis. De Vente *et al.* compared this approach to predictions of SSY from the models WATEM/SEDEM and PESERA. After splitting the dataset into basins with high (>5%) and low (<5%) sediment delivery ratios, good results were obtained, where SPADS explained 67% of the variation and WATEM/SEDEM 48%, while PESERA did not perform well. The poor performance was attributed to the lack of gully erosion, channel erosion and channel sediment delivery processes in PESERA. However, the explained variance is not the most important result of this analysis; instead it is the ease with which specific data (such as gully information) can be included in a factorial analysis. It allows the easy testing of the relative importance of certain processes in a given area, which may shed some light on the performance of other erosion models. Also clear from this analysis is the fact that Sediment Delivery Ratios are very area-specific and cannot be used in different areas without a thorough analysis. This is significant for modelling at larger scales.

3.2.3 Calibration of large-scale models

There have been many attempts to calculate the soil loss from large areas, such as river basins, countries or even continents. The difference between 'catchment scale' and 'large scale' is assumed here to be a generalization of the spatial characteristics of an area. Where the catchment models explicitly characterize areas of detachment and accumulation, in large-scale models many processes happen inside a spatial element, such as grid cells of >1 km², first-order catchments or even administrative units. Two types of models seem to be used: a direct use of RUSLE with minor or major adaptations, and more sophisticated models that attempt to retain process descriptions for runoff, detachment, transport and deposition. There are several purposes for the model results at this scale, from showing average annual soil loss (often called erosion risk) to use by policy-makers (see Van der Knijff *et al.*, 2000) to predict the effects of soil conservation policies on the sediment loads of rivers and the siltation of dams and reservoirs. Consequently calibration at this scale varies from non-calibrated, through a non-quantitative verification of spatial erosion patterns, to a quantitative comparison of predicted and observed sediment levels in rivers or sedimentation behind dams.

It is clear that the USLE, originally meant to predict annual soil loss from a slope or field, cannot be used directly to predict soil loss on a large scale, unless adaptations are made. The original USLE is based on soil loss by rainfall energy modified with slope angle and slope length, which are used as a proxy for the flow detachment processes. The RUSLE and MUSLE improved this by adding a runoff factor to the driving force. There is usually no provision for deposition in these models. The popularity of the model on a variety of scales probably stems from the ease of use in a GIS, in particular the derivation of the slope length L and slope angle S from a DEM that are combined in the so-called 'LS factor'. They range from a direct slope angle and length calculation to methods using drainage network accumulation (as in the WATEM/SEDEM model). Also the availability of remote sensing images for vegetation properties has added to the popularity. When GIS gained popularity in the late 1990s as a main tool to use spatial data and convert a DEM to an LS map, the USLE proved a simple method to predict the soil loss in units of t ha^{-1} y^{-1} for each grid cell (see Mati *et al.*, 2000; Sanjay *et al.*, 2001; Bayramin *et al.*, 2007; Pandey *et al.*, 2008b). However, it is not clear what these large-scale soil loss maps depict. Are the values meant to be the soil loss from a pixel to the next downstream pixel, or the average soil loss of a field or a slope inside a pixel? If the former is meant, the model should have provision for accumulation of sediment over a network (see examples below); if the latter is meant, the properties of the pixel are seen as characterizing the fields/slopes inside the pixel, which may be questionable. This makes calibration difficult. At best, spatial patterns of erosion can and should be verified against reality. This was done for instance by Bou Kheir *et al.* (2006) who used USLE type parameters in a factorial approach to determine erosion susceptibility in five classes of an area in Lebanon, and checked these classes successfully against observed rill density.

An example of a more complex approach was given by Zhou and Wu (2008) who calculated the soil loss for several provinces in China using inverse distance interpolation for the rainfall data, an NDVI (Normalized Difference Vegetation Index) based assessment of the cover factor and a 30 m resolution DEM for the LS factor. Moreover, they introduced a sediment delivery ratio (SDR) to predict the amount of sediment entering the stream network. This permitted the authors to compare the measured and predicted annual sediment load for two years in six gauging stations, whereby the SDR was used to calibrate the results. They then interpreted changes in the SDR in relation to changes in annual rainfall. Takeuchi *et al.* (2009) adapted TOPMODEL successfully to simulate 23 peak runoff events and sediment loads from a 4423 km^2 catchment in

China. A different strategy is to model the transport capacity and compare that to the sediment load. This is done at this scale by the WATEM/SEDEM model and the PESERA model. Both models use a more complex relief factor as a proxy for both flow detachment and transport capacity, and therefore are capable of showing areas with net detachment and net deposition. PESERA, moreover, uses a SDR for the soil loss on the slopes nearest to the stream.

It is clear that such models are very difficult to calibrate. The necessary data at this scale do not exist; possibly sediment trapped in dams of large areas could serve as a calibration, but this has not been attempted yet as far as we know. Also, as noted above, it is not quite clear what pixel-based erosion values signify at this scale. Nevertheless, maps based on the output of these models are often used.

On this scale the poor model performance for small events is also encountered. For example, Lenzi and Di Luzio (1997) use AGNPS to predict total runoff and sediment loss from 20 events in a 77 km^2 catchment. They show that the runoff is predicted with an R^2 of 0.97 and the sediment loss with an R^2=0.72. However, these values are determined by two very large rainstorms for which the runoff is well predicted. If these are not considered, the explained variance for runoff and sediment loss drops to 0.31 and 0.28 respectively. Similar results are given for total runoff, peak runoff and sediment yield by Rode and Frede (1999), using AGNPS on two German catchments of 82 and 129 km^2 (23 and 35 events respectively). While from the above it seems that homogeneity is an issue for plots, this is perhaps even more so for catchments where rainfall variability plays an important role (Capolongo *et al.*, 2008).

The GCTE results, however, also show that many models have problems with the prediction of extreme events. This could be due to a number of reasons: firstly, the system may not behave the same for moderate and large events. During a heavy rainstorm connectivity may be dynamic and different than under normal circumstances, and no longer match the fixed connectivity of the model. Also many variables considered as static constants in the model are in fact dynamic, in particular those related to soil structure (Boiffin & Monnier, 1986; Torri *et al.*, 1999). Apart from these more obvious reasons, the predictive quality of the erosion models is strongly determined by spatial and temporal variability of the input parameters. Beven (2002) pointed out that surface and subsurface hydrology are largely dominated by local geometry and boundary conditions, rather than by the dynamics of the fluid itself. Consequently the influence of variability in the soil (surface) characteristics may be very large. Many studies have been done on the effects of spatial variability and error propagation, using sensitivity analyses based on Monte Carlo simulation and other methods (see e.g. De Roo *et al.*, 1992; Brazier *et al.*, 2000). The results show that most of the model parameters are stochastic in nature, and measurement errors and uncertainty as a result of spatial interpolation add considerable uncertainty to the model results (see Chapter 4 for further discussion, and Section 14.4 on the realism of calibrated parameter values).

3.3   Sensitivity of Process Models

Models used in catchment-scale predictions are often calibrated using saturated hydraulic conductivity (K_{sat}) and, for event-based models, initial soil moisture. This is logical as these parameters directly determine the infiltration excess and amount of runoff. The most common strategy is to calibrate the total runoff (without attempting to fit a hydrograph) (De Roo *et al.*, 1996; Quinton, 1997; Hessel *et al.*, 2003a; Takeuchi *et al.*, 2009). Sometimes a single parameter set can be used to calibrate all events in a season, but sometimes different values have to be used for each event.

This difficulty in finding the right combination of parameters to calibrate a series of events stems mostly from the fact that a different combination of input parameters can give the same result. Beven and Freer (2001) argued that "in mechanistic modelling of complex environmental systems, there are many different model structures and many different parameter sets within a chosen model structure that may be behavioural or acceptable in reproducing the observed behaviour of that system". They called this equifinality. This can simply be shown with a sensitivity analysis of two parameters that influence runoff: K_{sat} and a surface resistance parameter (such as Manning's n). A higher resistance slows down the runoff and permits more time for infiltration. It therefore influences the shape of the hydrograph. In order to show this, the model LISEM was run 100 times for a single event in the $1 \, km^2$ Ganspoel catchment in Belgium (Van Oost *et al.*, 2005), changing the K_{sat} and Manning's n from 0.2 to 2.0 with steps of 0.2 around best calibrated values. Fig. 3.2a shows the sensitivity of LISEM for combinations of n and K_{sat} for the total discharge (which varies non-linearly from $0 \, m^3$ in the lower right corner to $2300 \, m^3$ in the upper left corner). The black isoline connects all combinations of K_{sat} and n that give a runoff total close to the measured value of $253 \, m^3$. The central point (2) is the best fit. There is, however, a catch to this analysis, one that increases the equifinality unnecessarily. The same total runoff can be achieved with many different hydrographs. Fig. 3.2b shows three hydrographs for points 1 to 3 and the measured hydrograph. While none of the simulated graphs gives an exact fit, it is clear that the central point resembles the observed hydrograph most closely. This suggests that by considering only lumped values of total runoff or peak values without considering the shape of the hydrograph, uncertainty is unnecessarily added and the problem of equifinality is seemingly increased. This may have caused unnecessary scatter in the data when comparing model predictions with measured values.

3.4   Calibration Examples Based on Automatic Parameter Estimation

The difficulty in finding the right combination of input parameters to simulate a given output invites the use of automatic methods of finding the correct combination, based on statistical goodness-of-fit criteria between predicted model output and observed data. One of the advantages of such

(a)

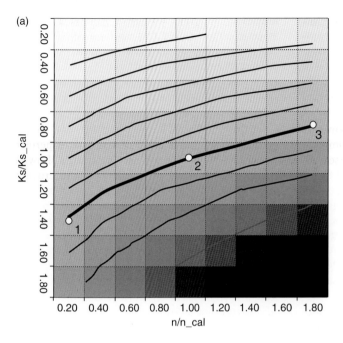

(b)

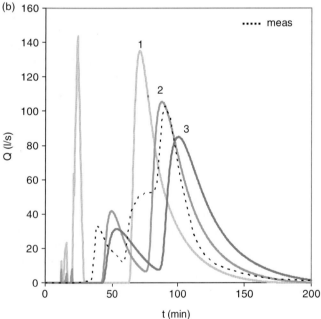

Fig. 3.2 (a) Sensitivity analysis of LISEM: relative changes in Manning's n (x-axis) and K_{sat} (y-axis) in steps of 0.2 around the calibrated values (1.0, 1.0) and resulting total discharge for a rainstorm event in the Ganspoel catchment in Belgium. Legend: white = 2300 m³, black = 0 m³ total discharge. The black isoline shows the combinations that give the measured discharge of 253 m³. (b) Simulated hydrographs for points 1, 2 and 3 in Fig. 3.2a, and the measured hydrograph (dotted line).

methods is that not only the observed values can be included in the analysis, but also their uncertainty. An example of this is the PEST (Parameter Estimation) system, which can be used as a shell coupled to virtually any model (Doherty, 2005). Maneta *et al.* (2007) used their own model to simulate the discharge over a period of two months of a small ephemeral stream in central Spain from a

1 km² catchment. The model is a continuous simulation model that calculates the complete water balance and sediment detachment and movement. The best set of parameters found by PEST are initial infiltration rate, horizontal and vertical K_{sat} and Manning's n of the slopes and the channel. In their experience, PEST needs to be run for 2 to 3 times Y^2, where Y is the number of parameters allowed to change. It is interesting that a complete water balance is simulated, enabling the researchers to use groundwater wells and TDR soil moisture sensors as extra information for calibration. They concluded that while the discharge was well predicted (PEST reported an $R^2 = 0.937$ among other statistics), the groundwater fluctuated more rapidly in reality than predicted by the model, while the measured soil moisture fluctuated more slowly. This led to additional research in the same area to investigate preferential flow, which appears to be one of the main soil hydrological processes (Van Schaik, 2009). Using a tool such as PEST has strengths and weaknesses: it is easy and powerful to use, and robust enough to find a best parameter set in most situations. It provides goodness of fit statistics and allows the simultaneous use of different measurement series (discharge, groundwater, etc.) to find a solution. However, the best fit may not be a realistic parameter set if more then one best fit exists in parameter space.

Vigiak *et al.* (2006) assessed the uncertainty of model prediction and predicted and observed spatial patterns of erosion in a 2 km² catchment in Tanzania. The erosion patterns were assessed with series of splash cups and Gerlach troughs, as well as with the Assessment of Current Erosion Damage method (ACED; Herweg, 1996). This resulted in a map with five classes of observed erosion from slight (only splash evidence) to very severe (large rills and gullies). They then combined the detachment and transport equations from the MMF erosion model with their own hydrological model (Vigiak *et al.*, 2005) to simulate erosion. They assumed that the sediment transport capacity determined the spatial distribution of erosion, and analysed the uncertainty of distributed model predictions using the Generalized Likelihood Uncertainty Estimation (GLUE) methodology (Beven & Freer, 2001;

Brazier *et al.*, 2000; see also Chapters 4 and 5). They applied the GLUE methodology using Monte Carlo simulations, with uniform parameter sampling. They compared each simulated and classified erosion pattern with the observed pattern using a weighted Kappa index (based on the fraction of correctly classified spatial units; Congalton, 1991). A Kappa index below 0 shows no agreement and a value of 1 indicates perfect agreement between patterns. The results showed that the parameters most affecting the erosion patterns were the re-infiltration length in their model affecting runoff discharge, followed by the power exponents α and γ used in the streampower-based transport capacity function ($TC = K\,Q^{\alpha} \sin(\beta)^{\gamma}$). Figure 3.3 shows that the best combinations were obtained with α = 1.5 and γ = 0.5, and a short re-infiltration length L. This resulted in a Kappa coefficient of above 0.50 (obtained in 277 of 6000 simulations) which can be considered as moderately good. Table 3.2 shows the contingency between predicted and observed classes.

Methods like PEST and GLUE usually do not result in one best fit, and therefore may be less suitable to show directly to decision-makers, without extensive interpretation. However, they offer a lot of insight into the behaviour of a model with a specific dataset, in particular by quantifying the uncertainty.

3.5 Spatial Calibration of Erosion Patterns

From the examples above, a picture emerges that the largest source of uncertainty is insufficient knowledge of spatial patterns of sources and sinks of water and sediment in the area under observation. Assuming that the calibrations discussed above used a best-possible parameterization of the model, it seems that specifying patterns of input parameters based on land use, landscape and/or geostatistics is not sufficient to obtain more than moderate results. This is confirmed by Walling *et al.* (2003) who emphasized that a more distributed approach is needed to provide spatially distributed predictions of soil erosion and sediment transport within a catchment. This raises the

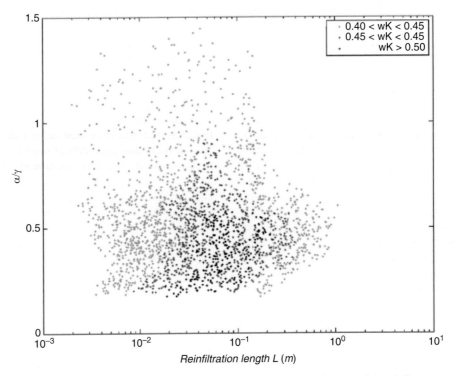

Fig. 3.3 Scatterplot of power α/γ ratio versus reinfiltration length L (in logarithmic scale) at different intervals of weighted Kappa (wK), from Vigiak *et al.* (2006).

Table 3.2 Contingency table of the ACED map (observed) and the average output of model behavioural simulations (predicted), from Vigiak *et al.* (2006). The diagonal represents the number of correctly classified units in each class (irrespective of their size).

Observed / Predicted	Very slight	Slight	Moderate	Severe	Very severe
Very slight	33	21	8	5	4
Slight	24	19	18	10	11
Moderate	16	21	30	16	8
Severe	4	9	16	14	26
Very severe	2	9	9	22	35

question of how to obtain detachment and deposition patterns to improve model calibration. There are several possibilities (not in any particular order of importance):

(i) *Soil surface crusting indices.* In areas where the soil is prone to crusting, good results have been obtained by coupling the effects of tillage and subsequent crusting by rainfall to analyse source areas of runoff. The model STREAM is a decision support based system linked to a classification of the degree of crusting in a catchment, based on extensive research in northern France (see Cerdan *et al.*, 2002; Le Bissonais *et al.*, 2005). It allows one to pinpoint the appearance of fields that are likely sources and sinks during a growing season. Jetten *et al.* (1996) also showed that while different patterns of runoff contributing areas can give the same discharge and sediment loss, these crusting classes help in choosing the likely source fields when there are limited point observations.

(ii) *Field erosion measurement*. Takken *et al.* (1999) attempted a spatial validation of LISEM using the measured erosion pattern in a $2.9\,km^2$ catchment in Belgium resulting from an extreme event on 8 June 1996. By measuring rills and other erosion and deposition features, they estimated sediment loss aggregated over 12 land use types. These were then compared with LISEM simulation results which were aggregated in the same way. The results show that the observed variation in erosion rates for different crop types was not well predicted, with an overprediction of erosion rates on fields with a high vegetation cover. Correlation between predicted and measured erosion rates per crop type at locations of the measured rill transects was weak. Predicted and observed patterns of deposition were generally similar, although important deposition against vegetation barriers and roads was not simulated, which resulted in overall discrepancies (Takken *et al.*, 1999). Hessel *et al.* (2003a) attempted a similar exercise but on a pixel-by-pixel basis for several events in a small catchment on the Chinese loess plateau (see Chapter 12). They showed the difficulties of comparing predicted and observed patterns, as there were both areas of good prediction but also large discrepancies. They attributed this to both simulation errors (e.g. LISEM has the tendency to overestimate deposition in certain grid cells) and inherent uncertainties in the dataset. For instance, if the DEM is not accurate enough to simulate a converging flow path in the right position compared with reality, a model can never produce correct erosion estimates on that location. Also the model simulates all detachment of soil particles, while in reality erosion is only visible once rills are being formed. Thus if a spatial comparison is attempted, an aggregation to fields or slope segments seems to be preferable over a pixel-based comparison.

(iii) *Farmers' knowledge*. Vigiak *et al.* (2007) compared the predicted erosion maps of five models to the field observations done with the ACED method (see above). One of the models was a decision tree based on farmers' knowledge in the area. This knowledge consists both of observed erosion features, and also of related factors such as poor crop development, soil colour changes and stoniness/rockiness. The decision tree appeared to be very capable of predicting the observed eroded area, indicating the strength of this kind of data for use in model comparisons and calibration.

(iv) *137-Caesium patterns and chemical proxies.* Several models have also been tested with ^{137}Cs data (De Roo & Walling, 1994; Walling *et al.*, 2003; Van Oost *et al.*, 2003; Owens & Walling, 1998; Porto *et al.*, 2003; see also Section 14.7.2). For example, Walling *et al.* (2003) compared the simulated erosion and deposition patterns of the models ANSWERS and AGNPS with the observed distribution derived from ^{137}Cs observations in two small catchments of 4.6 and $0.52\,km^2$ in the UK. Although both models produced catchment total runoff, peak runoff and sediment losses in agreement with the measured data of seven events, the predicted spatial patterns of soil redistribution and the sediment delivery ratios were very different for the two models. The AGNPS model predicted deposition zones at the bottom of the slopes, with highest erosion at mid-slope, in accordance with the ^{137}Cs-derived distribution, while ANSWERS showed a continuous increase of detachment from top to bottom of the slopes. Also Ritchie *et al.* (2005) showed that it is possible to derive quantitative erosion/deposition patterns from ^{137}Cs fallout in a semi-arid watershed in Arizona, which could then be used for model verification. Van Oost *et al.* (2003) included tillage displacement using the WATEM/SEDEM model to further explain ^{137}Cs patterns found in a Belgian catchment. Also other chemical elements may enable model verification. Van der Perk and Jetten (2006) showed that a pattern of Cu concentrations could be established in a French vineyard due to long-term use of copper as a fungicide.

A topography-based relative index of erosion and deposition explained between 32% and 56% of the variation in the soil Cu inventories. Finally, Zhang *et al.* (2008) used rare earth elements successfully as tracers to compare the detachment predicted by WEPP with the observed rill erosion on a series of plots.

(v) *Remote sensing images.* An obvious source of spatial erosion information is remote sensing imagery. In a review, Vrieling (2006) discussed the application of remote sensing to derive information on erosion and erosion-controlling factors. Direct information on erosion features (mostly rill systems and gullies) is only possible with high-resolution images (such as IKONOS, Quickbird and Worldview with resolutions of 0.2–1 m). It is possible to obtain information on controlling factors such as land cover, topography, soils and tillage practices, using various sensors. Using images for vegetation and crop-derived variables is fairly standard for deriving NDVI type indices. Multispectral or hyperspectral information is needed to derive soil properties (Shepherd & Walsh, 2002). LIDAR is increasingly used for altitude mapping and provides very detailed information on topography, which also enables gully mapping to be carried out (James *et al.*, 2007). Vrieling *et al.* (2008) employed a multiscale approach on a 100 km² grazing area in Brazil. They used low-resolution (250 m) high-frequency MODIS images to establish the annual period of likely erosion risk based on absence of land cover. A better estimate for that period was obtained using several ASTER images with 10 m resolution, and finally erosion features were detected using very high-resolution (0.6 m) Quickbird imagery.

3.6 Conclusions

This overview of calibration of models on different spatial and temporal scales shows that, although the objectives are different for each scale, the calibration procedures used are very similar. First of all, calibration is almost always done at all scales, which is encouraging. However, in spite of the fact that almost all models are spatially distributed, the success of a model is mostly expressed as the ability to predict discharge and soil loss lumped in space and time, with prediction of hydrographs as a second method. Occasionally simulated patterns are compared with observed patterns of, for instance, ^{137}Cs or soil loss per field in a catchment, but annual or seasonal totals are clear favourites. The real value of spatial modelling appears to be the use of the spatial output for purposes such as predicting the effect of land-use changes or conservation measures (Walling *et al.*, 2003, Jetten *et al.*, 2003), but the spatial output is rarely used for verification, except for the few studies mentioned above. Also interesting to note is the lack of validation results; while calibration fortunately is often done, at least in scientific literature, independent validation is far less common. Calibration is usually done when a model is used in a new area or for a new set of circumstances (e.g. effects of forest fire), for which adaptations to the model are needed. The authors then want to prove that their adaptations make the model perform better. This makes generalizations in model performance difficult because the same model is used in different variations, and also hybrids between models are created.

The more complex the area that is modelled, the more the uncertainty increases. One of the main influences on the predictive quality of erosion models seems to be the changing spatial patterns of sources and sinks, and their connectivity. Generally the larger the area, the more sources and sinks are included. A study of soil loss from small plots deals mainly with sources of sediment and does not have to consider connectivity issues, whereas a catchment study will have to include information on source areas such as crusted fields and sinks, valley floors and obstructions, as well as their connectivity. The soil loss from the catchment measured at the outlet is often only a small part of the amount of detached sediment. This may cause large uncertainties in the predictions, which is demonstrated by various studies using Monte Carlo type simulations. It may be, however, that by regarding only lumped values and disregarding more complex hydrograph information, we are increasing uncertainty unnecessarily.

When modelling larger areas with low spatial resolution, sources and sinks are not explicitly modelled because they occur within one spatial element (grid cell) and sediment delivery ratios are used. There is a lack of calibration, which is understandable in view of the lack of erosion/soil loss data on this scale. More effort should perhaps be made here because the model results at this scale are often used by governments and form the basis for national and international policies, perhaps more than the model results at more detailed scales. While in the scientific literature USLE variations are used in proper ways with adaptations to produce sediment delivery estimates (see examples above), the original USLE is still popular as a consultancy tool (see Chapter 8), although sometimes used completely outside its original purpose.

In terms of model quality and calibration, the earlier conclusions can be repeated here, and are since confirmed by many authors: calibration should always be done if possible, and the range of data on which the model is calibrated should be a large as possible. Different calibration datasets or strategies may be needed for large events and small events, and generally small events or totals are less well predicted. When comparing the performance for lumped results (e.g. annual soil loss), the more empirical models perform just as well as physically-based models that attempt to simulate the hydrological and erosion processes in detail. They do not perform any better, but are more attractive as they generally have lower data requirements. Under a given set of circumstances, some models perform better than others, but generalizations on model performance cannot be made. The same model is shown to perform well by one author and only moderately well by others.

Assuming that the calibrations discussed above used a best-possible parameterization of the model, it seems that specifying patterns of input parameters based on land use, landscape and/or geostatistics is not sufficient to obtain more than moderate results. Several authors argue that a more distributed approach is needed to provide spatially distributed predictions of soil erosion and sediment transport within a catchment. Different sources of spatial information are available: agricultural information, radioactive and chemical elements that can be used as tracers, and earth observation data of increasing temporal and spatial scales. Thus the necessary information seems to be available and accessible and should be used in the future (see Chapter 17).

References

Arhonditsis, G., Koulouri, M., Giourga, C. & Loumou, A. (2002) Quantitative assessment of agricultural runoff and soil erosion using mathematical modelling: applications in the Mediterranean region. *Environment Management* **30**: 434–53.

Bathurst, J.C. & Lukey, B. (1998) Modelling badlands erosion with SHETRAN at Draix, southeast France. In *Modelling Soil Erosion, Sediment Transport and Closely Related Hydrological Processes*, Proceedings of the International Association of Hydrological Sciences Vienna Symposium. IAHS Publication **249**: 129–36.

Bayramin, I., Erdogan, E.H. & Erpul, G. (2007) Use of USLE/GIS Methodology for predicting soil loss in a semiarid agricultural watershed. *Environmental Monitoring and Assessment* **131**: 153–61.

Beasley, D.B., Huggins, L.F. & Monke, E.J. (1980) ANSWERS: a model for watershed planning. *Transactions of the American Society of Agricultural Engineers* **23**: 939–44.

Beven, K.J. (1997) *Distributed Modelling in Hydrology: Applications of TOPMODEL*. John Wiley & Sons, Chichester.

Beven, K. (2002) Towards an alternative blueprint for a physically based digitally simulated hydrologic response modelling system. *Hydrological Processes* **16**: 189–206.

Beven, K. & Freer, J. (2001) Equifinality, data assimilation, and uncertainty estimation in mechanistic modelling of complex environmental systems using the GLUE methodology. *Journal of Hydrology* **249**: 11–29.

Boardman, J. & Favis-Mortlock, D. (1998) *Modelling soil erosion by water*. NATO ASI Series 1: Global Environmental Change, Volume 55. Springer-Verlag, Berlin: p. 531.

Boardman, J., Ligneau, L., De Roo, A.P.J. & Vandaele, K. (1994) Flooding of property by runoff from agricultural land in northwestern Europe. *Geomorphology* **10**: 183–96.

Boiffin, J. & Monnier, G. (1986) Infiltration rate as affected by soil surface crusting caused by rainfall. In *Assessment of Soil Surface Sealing and Crusting*, Callebaut, F., Gabriels, D. & De Boodt, M. (eds). *Proceedings of the Ghent Symposium 1985*: pp. 210–17.

Bou Kheir, R., Cerdan, O. & Abdallah, C. (2006) Regional soil erosion risk mapping in Lebanon. *Geomorphology* **82**: 347–59.

Brazier, R.E., Beven, K.J., Freer, J. & Rowan, J.S. (2000) Equifinality and uncertainty in physically-based soil erosion models: application of the GLUE methodology to WEPP – the Water Erosion Prediction Project for sites in the UK and US. *Earth Surface Processes and Landforms* **25**: 825–45.

Brochot, S. (1998) Approches globales pour l'estimation de l'érosion torrentielle - Apports des versants et production de sédiments. *Revue Ingénieries du Cemagref* **15**: 61–78.

Capolongo, D., Diodato, N., Mannaerts, C.M., *et al.* (2008) Analyzing temporal changes in climate erosivity using a simplified rainfall erosivity model in Basilicata (southern Italy). *Journal of Hydrology* **356**: 119–30.

Cerdan, O., Souchere, V., Lecomte, V., *et al.* (2002) Incorporating soil surface crusting processes in an expert-based runoff model: Sealing and transfer by runoff and erosion related to agricultural management. *Catena* **46**: 189–205.

Congalton, R.G. (1991) A review of assessing the accuracy of classifications of remotely sensed data. *Remote Sensing of Environment* **37**: 35–46.

De Roo, A.P.J. (1993) Modelling surface runoff and soil erosion in catchments using Geographical Information Systems. *Netherlands Geographical Studies* 157. Faculty of Geographical Sciences, Utrecht University, 294 pp.

De Roo, A.P.J. (1999) Modelling of soil erosion by water on a catchment scale. *Catena* **37**: 275–541.

De Roo, A.P.J., Hazelhoff, L. & Heuvelink, G.B.M. (1992) Estimating the effects of spatial variability of infiltration on the output of a distributed runoff and soil erosion model using Monte Carlo methods. *Hydrological Processes* **6**: 127–43.

De Roo, A.P.J. & Walling, D.E. (1994) Validating the ANSWERS soil erosion model using [137]Cs. In Rickson, R.J. (ed.), *Conserving Soil Resources: European Perspectives*. CAB International, Wallingford, UK: 246–63.

De Roo, A.P.J., Offermans, R.J.E. & Cremers, N.H.D.T. (1996) LISEM: A single event physically-based hydrologic and soil erosion model for drainage basins: II. Sensivity analysis, validation and application. *Hydrological Processes* **10**: 1119–26.

De Vente, J., Poesen, J. & Verstraeten, G. (2005) The application of semi-quantitative methods and reservoir sedimentation rates for the prediction of basin sediment yield in Spain. *Journal of Hydrology* **305**: 63–86

De Vente, J., Poesen, J., Verstraeten, G., *et al.* (2008) Spatially distributed modelling of soil erosion and sediment yield at regional scales in Spain. *Global and Planetary Change* **60**: 393–415.

Doherty, J. (2005) *PEST: Software for model-independent parameter estimation.* Watermark Numerical Computing, Australia. Available from: http://www.sspa.com/pest

Favis-Mortlock, D.T. (1998) Validation of field-scale soil erosion models using common datasets. In Boardman, J. & Favis-Mortlock, D.T. (eds), *Modelling Soil Erosion by Water*. NATO-ASI Series I-55. Springer-Verlag, Berlin: 89–128.

Flanagan, D.C. & Nearing, M.A. (eds) (1995) *USDA – Water Erosion Prediction Project Hillslope Profile and Watershed Model Documentation*. NSERL Report No. 10. USDA-ARS National Soil Erosion Research Laboratory, West Lafayette, IN 47907.

Gronsten, H.A. & Lundekvam, H. (2006) Prediction of surface runoff and soil loss in southeastern Norway using the WEPP Hillslope model. *Soil and Tillage Research* **85**: 186–99.

Herweg, K. (1996) *Assessment of Current Erosion Damage*. Centre for Development and Environment Institute of Geography, University of Berne, Berne, Switzerland.

Hessel, R. & Tenge, A. (2008) A pragmatic approach to modelling soil and water conservation measures with a catchment scale erosion model. *Catena* **74**: 119–26.

Hessel, R., Messing, I., Liding, C., *et al.* (2003a) Soil erosion simulations of land use scenarios for a small Loess Plateau catchment. *Catena* **54**: 289–302.

Hessel, R., Jetten, V., Liu, B., *et al.* (2003b) Calibration of the LISEM model for a small Loess Plateau catchment. *Catena* **54**: 235–54.

James, L.A., Watson D.G. & Hansen, W.F. (2007) Using LiDAR data to map gullies and headwater streams under forest canopy: South Carolina, USA. *Catena* **71**: 132–44.

Jetten, V. & De Roo, A.P.J. (2001) Spatial analysis of erosion conservation measures with LISEM. In Harmon, R. & Doe III, W. (eds), *Landscape Erosion and Evolution Modelling*. Kluwer Academic Plenum Publishing, New York: 429–45.

Jetten, V. & Favis-Mortlock, D. (2006) Modelling soil erosion in Europe. In Boardman, J. & Poesen, J. (eds), *Soil Erosion in Europe*. John Wiley & Sons, Chichester: 695–716.

Jetten, V., De Roo, A.P.J. & Favis-Mortlock, D.T. (1999) Evaluation of field-scale and catchment-scale soil erosion models. *Catena* **37**: 521–41.

Jetten, V.G., Boiffin, J. & De Roo, A.P.J. (1996) Defining monitoring strategies for runoff and erosion studies in agricultural catchments: a simulation approach. *European Journal of Soil Science* **47**: 579–92.

Jetten, V., Govers G. & Hessel, R. (2003) Erosion models. Quality of spatial predictions. *Hydrological Processes* **17**: 887–900.

Kirkby, M.J., Jones, R.J.A., Irvine, B., *et al.* (2004) Pan-European Soil Erosion Risk Assessment: The PESERA Map, Version 1, October 2003. Explanation of Special Publication Ispra 2004 No.73 (S.P.I.04.73). European Soil Bureau Research Report No.16, EUR 21176, 18pp. Office for Official Publications of the European Communities, Luxembourg.

Knisel, W.G. (ed.) (1980) *CREAMS – A Field Scale Model for Chemicals, Runoff, and Erosion from Agricultural Management Systems*. USDA Conservation Report No. 26.

Le Bissonnais, Y., Cerdan, O., Lecomte, V., *et al.* (2005) Variability of soil surface characteristics influencing runoff and interrill erosion. *Catena* **62**: 111–24.

Lenzi, M.A. & Di Luzio, M. (1997) Surface runoff, soil erosion and water quality modelling in the Alpone watershed using AGNPS integrated with a Geographic Information System. *European Journal of Agronomy* **6**: 1–14.

Leonard, R.A., Knisel, W.G. & Still, D.A. (1987) GLEAMS: Groundwater Loading Effects of Agricultural Management Systems. *Transactions of the American Society of Agricultural Engineers* **30**: 1403–18.

Lundekvam, H.E., Romstad, E. & Øygarden, L. (2003) Agricultural policies in Norway and effects on soil erosion. *Environmental Science and Policy* **6**: 57–67.

Maneta, M.P., Pasternack, G.B., Wallender, W.W., *et al.* (2007) Temporal instability of parameters in an event-based distributed hydrologic model applied to a small semiarid catchment. *Journal of Hydrology* **341**: 207–21.

Mati, B.M., Morgan, R.P.C., Gichuki, F.N., *et al.* (2000) Assessment of erosion hazard with the USLE and GIS: A case study of the Upper Ewaso Ng'iro

North basin of Kenya. *International Journal of Applied Earth Observation and Geoinformation* **2**: 78–86.

Misra, R.K. & Rose, C.W. (1996) Application and sensitivity analysis of process-based erosion model GUEST. *European Journal of Soil Science* **47**: 593–604.

Morgan, R.P.C. (2001) A simple approach to soil loss prediction: a revised Morgan–Morgan–Finney model. *Catena* **44**: 305–322.

Morgan, R.P.C., Quinton, J.N., Smith, R.E., *et al.* (1998) The European Soil Erosion Model (EUROSEM): a dynamic approach for predicting sediment transport from fields and small catchments. *Earth Surface Processes and Landforms* **23**: 527–44.

Nearing, M.A. (1998) Why soil erosion models over-predict small soil losses and under-predict large soil losses. *Catena* **32**: 15–22.

Nearing, M.A. (2000) Evaluating soil erosion models using measured plot data: accounting for variability in the data. *Earth Surface Processes and Landforms* **25**: 1035–43.

Nearing, M.A., Govers, G. & Norton, L.D. (1999) Variability in soil erosion data from replicated plots. *Soil Science Society of America Journal* **63**: 1829–35.

Nearing, M.A., Jetten, V., Baffaut, C., *et al.* (2005) Modelling response of soil erosion and runoff to changes in precipitation and cover. *Catena* **61**: 131–54.

Owens, P.N. & Walling, D.E. (1998) The use of a numerical mass-balance model to estimate rates of soil redistribution on uncultivated land from ^{137}Cs measurements. *Journal of Environmental Radioactivity* **40**: 185–203.

Pandey, A., Chowdary, V.M., Mal, B.C. & Billib, M. (2008a) Runoff and sediment yield modelling from a small agricultural watershed in India using the WEPP model. *Journal of Hydrology* **348**: 305–19.

Pandey, A., Dabral, P.P. & Baithuri, N. (2008b) Soil erosion assessment in a hilly catchment of north eastern India using USLE, GIS and remote sensing. *Water Resources Management* **22**: 1783–98.

Pieri, L., Bittelli, M., Wu, J.Q., *et al.* (2007) Using the Water Erosion Prediction Project (WEPP) model to simulate field-observed runoff and erosion in the Apennines mountain range, Italy. *Journal of Hydrology* **336**: 84–97.

Porto, P., Walling, D.E., Tamburino, V. & Callegari, G. (2003) Relating caesium-137 and soil loss from cultivated land. *Catena* **53**: 303–26.

Quinton, J.N. (1997) Reducing predictive understanding in model simulations: a comparison of two methods using the European Soil Erosion Model (EUROSEM). *Catena* **30**: 101–7.

Raclot, D. & Albergel, J. (2006) Runoff and water erosion modelling using WEPP on a Mediterranean cultivated catchment. *Physics and Chemistry of the Earth. B. Hydrology, Oceans and Atmosphere* **31**: 1038–47.

Renard, K.G., Foster, G.R., Weesies, G.A. & Porter, J.P. (1991) RUSLE – Revised Universal Soil Loss Equation. *Journal of Soil and Water Conservation* **46**: 30–33.

Risse, L.M., Nearing, M.A., Nicks, A.D. & Laflen, J.M. (1993) Assessment of error in the Universal Soil Loss Equation. *Soil Science Society of America Journal* **57**: 825–33.

Ritchie, J.C., Nearing, M.A., Nichols, M.H. & Ritchie, C.A. (2005) Patterns of soil erosion and redeposition on Lucky Hills Watershed, Walnut Gulch Experimental Watershed, Arizona. *Catena* **61**: 122–30.

Rode, M. & Frede, H.-G. (1999) Testing AGNPS for Soil Erosion and Water Quality Modelling in Agricultural Catchments in Hesse (Germany). *Physics and Chemistry of the Earth. B. Hydrology, Oceans and Atmosphere* **24**: 297–301.

Romero, C.C., Stroosnijder, L. & Baigorria, G.A. (2007) Interrill and rill erodibility in the northern Andean Highlands. *Catena* **70**: 105–13.

Sadeghi, S.H.R., Azari, M. & Ghaderi Vangah, B. (2008) Field evaluation of the Hillslope Erosion Model (HEM) in Iran. *Biosystems Engineering* **99**: 304–11.

Sanjay, K.J., Kumar, S. & Varghese, J. (2001) Estimation of soil erosion for a Himalayan watershed using GIS technique. *Water Resources Management* **15**: 41–54.

Schmidt, J., van Werner, M. & Michael, A. (1999) Application of the EROSION 3D model to the CATSOP watershed, The Netherlands. *Catena* **37**: 521–41.

Shepherd, K.D. & Walsh, M.G. (2002) Development of reflectance spectral libraries for characterization of soil properties. *Soil Science Society of America Journal* **66**: 988–99.

Stolpe, N.B. (2005) A comparison of the RUSLE, EPIC and WEPP erosion models as calibrated to climate and soil of south-central Chile. *Acta Agriculturae Scandinavica: Section B, Soil and Plant Science* **55**: 2–9.

Takeuchi, K., Wang, G., Hapuarachchi, P., *et al.* 2009. Estimation of soil erosion and sediment yield during individual rainstorms at catchment scale. *Water Resources Management* **23**: 1447–65.

Takken, I., Beuselinck, L., Nachtergaele, J., *et al.* (1999) Spatial evaluation of a physically-based distributed erosion model (LISEM). *Catena* **37**: 431–47.

Takken, I., Govers, G., Jetten, V., *et al.* (2001) Effects of tillage on runoff and erosion patterns. *Soil and Tillage Research* **61**: 55–60.

Torri, D., Regués, D., Pellegrini, S. & Bazzoffi, P. (1999) Within-storm soil surface dynamics and erosive effects of rainstorms. *Catena* **38**: 131–50.

Tyagi, J.V., Mishra, S.K., Singh, R. & Singh, V.P. (2008) SCS-CN based time-distributed sediment yield model. *Journal of Hydrology* **352**: 388–403.

Van der Knijff, J.M., Jones, R.J.A. & Montanarella, L. (2000) *Soil Erosion Risk Assessment in Europe.* Joint Research Centre ISPRA, Soil Bureau, 34 pp.

Van der Perk, M. & Jetten, V.G. 2006. The use of a simple sediment budget model to estimate long-term contaminant export from small catchments. *Geomorphology* **79**: 3–12.

Van Oost, K., Govers, G. & Desmet, P.J.J. (2000) Evaluating the effects of changes in landscape structure on soil erosion by water and tillage. *Landscape Ecology* **15**: 579–91.

Van Oost, K., Govers, G. & Van Muysen, W. (2003) A process-based conversion model for Caesium-137 derived erosion rates on agricultural land: an integrated spatial approach. *Earth Surface Processes and Landforms* **28**: 187–207.

Van Oost, K., Govers, G., Cerdan, O., *et al.* (2005) Spatially distributed data for erosion model calibration and validation: the Ganspoel and Kinderveld datasets. *Catena* **61**: 105–21.

Van Schaik, N.L.M.B. (2009) Spatial variability of infiltration patterns related to site characteristics in a semi-arid watershed. *Catena* **78**: 36–47.

Vigiak, O., Okoba, B.O., Sterk, G. & Groenenberg, S. (2005) Modelling catchment-scale erosion patterns in the East African Highlands. *Earth Surface Processes and Landforms* **30**: 183–96.

Vigiak, O., van Loon, E. & Sterk, G. (2006) Modelling spatial scales of water erosion in the West Usambara Mountains of Tanzania. *Geomorphology* **76**: 26–42.

Vigiak, O., Sterk, G., Romanowicz, R.J. & Beven, K.J. (2007) A semi-empirical model to assess uncertainty of spatial patterns of erosion. *Catena* **66**: 198–210.

Vrieling, A. (2006) Satellite remote sensing for water erosion assessment: a review. *Catena* **65**: 2–18.

Vrieling, A., de Jong, S.M., Sterk, G. & Rodrigues, S.C. (2008) Timing of erosion and satellite data: A multi-resolution approach to soil erosion risk mapping. *International Journal of Applied Earth Observation and Geoinformation* **10**: 267–81.

Walling, D.E., He, Q. & Whelan, P.A. (2003) Using ^{137}Cs measurements to validate the application of the

AGNPS and ANSWERS erosion and sediment yield models in two small Devon catchments. *Soil and Tillage Research* **69**: 27–43.

Williams, J.R. (1975) Sediment yield prediction with universal equation using runoff energy factor. In *Present and Prospective Technology for Predicting Sediment Yields and Sources*. US Department of Agriculture, Agricultural Research Service Publication ARS-S-40: pp. 118–24.

Williams, J.R., Jones, C.A. & Dyke, P.T. (1984) A modeling approach to determining the relationship between erosion and soil productivity. *Transactions of the American Society of Agricultural Engineers* **27**: 129–44.

Wischmeier, W.H. & Smith, D.D. (1978) *Predicting Rainfall Erosion Losses: a Guide to Conservation Planning*. Agriculture Handbook No. 537, USDA Science and Education Administration, Washington, DC.

Woolhiser, D.A., Smith, R.E., Goodrich, D.C. (1990) *KINEROS: A Kinematic Runoff and Erosion Model. Documentation and User Manual*. USDA-Agricultural Research Service ARS-77: pp. 130.

Young, R.A., Onstad, C.A., Bosch, D.D., Anderson, W.P. (1989) AGNPS: a nonpoint-source pollution model for evaluating agricultural watersheds. *Journal of Soil and Water Conservation* **44**: 168–73.

Zhang, Q., Lei, T. & Zhao, J. (2008) Estimation of the detachment rate in eroding rills in flume experiments using an REE tracing method. *Geoderma* **147**: 8–15.

Zhang, X.C., Nearing, M.A., Risse, L.M. & McGregor, K.C. (1996) Evaluation of runoff and soil loss predictions using natural runoff plot data. *Transactions of the American Society of Agricultural Engineers* **39**: 855–63.

Zhou, W. & Wu, B. (2008) Assessment of soil erosion and sediment delivery ratio using remote sensing and GIS: a case study of upstream Chaobaihe River catchment, north China. *International Journal of Sediment Research* **23**: 167–73.

4 Dealing with Uncertainty in Erosion Model Predictions

K.J. BEVEN[1] AND R.E. BRAZIER[2]

[1]*Lancaster Environment Centre, University of Lancaster, Lancaster, UK; GeoCentrum, Uppsala University, Uppsala, Sweden; ECHO/ISTE, EPFL, Lausanne, Switzerland*
[2]*School of Geography, University of Exeter, Exeter, UK*

4.1 Why Worry About Uncertainty in Erosion Models?

It is probably still the case that the most widely used erosion model in practice today is the Universal Soil Loss Equation (USLE; Wischmeier & Smith, 1960, 1978) and its revisions and variants, either as a stand-alone predictor or as a component of other models (e.g. the SWAT model; Arnold *et al.*, 1998; Gassman *et al.*, 2007). The USLE is an empirical relationship, originally derived from a limited dataset from erosion plot studies at 49 sites in the US, but later applied all over the world (Nearing *et al.*, 2005). It has this in common with the Soil Conservation Service (SCS) runoff generation model of Mockus (1949) with which it has often been coupled (see Beven, 2001a, for further discussion of this model). In scientific terms, these equations make no real attempt to represent the *processes* involved in runoff generation or sediment mobilization, transport and deposition, and consequently are intellectually somewhat dissatisfying even if we could be secure in their predictions – but they also can be dissatisfying in that respect as well, as will be seen below.

There has therefore been a tendency to develop process-based erosion models, a tendency that has been reinforced by the much greater computer

Handbook of Erosion Modelling, 1ˢᵗ edition. Edited by R.P.C. Morgan and M.A. Nearing. © 2011 Blackwell Publishing Ltd.

power now available, and the way in which it has made spatial datasets and geographical information systems more accessible. This tendency had earlier been seen in the development of distributed hydrological models, for example the many descendants of the Blueprint described in the seminal paper of Freeze and Harlan (1969); indeed, one of the reasons used in justifying the development of distributed models was that they would provide spatial predictions of runoff amounts and velocities that could be used in the prediction of sediment transport and other water quality variables (e.g. Freeze, 1978; Beven & O'Connell, 1982; Beven, 1985; Refsgaard & Abbott, 1996). Examples of such process-based models include WEPP (Nearing *et al.*, 2005) and EUROSEM (Morgan *et al.*, 1993; Quinton, 1994). Soil erosion components also appear in other distributed models such as SHETRAN (Ewen *et al.*, 2000; Lukey *et al.*, 2000).

Any distributed process-based model of soil erosion and sediment transport is necessarily dependent upon some underlying distributed hydrological model. But the limitations of distributed hydrological models became recognised soon after they started to become more widely available (e.g. Beven, 1989, 1996b; Grayson *et al.*, 1992). There are, indeed, important philosophical issues about the possibility of developing general process-based models of environmental systems, as discussed in Beven (2001b, 2002a, 2009). The fact is that it is very difficult to be secure about the equations that are used to describe complex interacting processes in a spatially heterogeneous and temporally non-stationary environment. In addition, it is very difficult to be secure about the effective parameter values that appear in those equations and that must be specified in any application of such models, since they cannot generally be measured directly. It is also often the case that we cannot be secure about the input data that are used to drive the model, particularly when spatial patterns of the inputs might be important. Thus, we should expect that the definition and application of such process-based models will necessarily be uncertain (Beven, 2002a,b, 2006a). Since such models are often used in practice to support decisions, it follows that an appropriate assessment of uncertainty might be important to the decision-making process (Pappenberger & Beven, 2006; Beven, 2009).

The very few studies of the uncertainty of soil erosion models are reviewed below. There have also been relatively few studies of uncertainty in the predictions of fully distributed process-based hydrological models. This is, at least in part, because it seems to be too great a problem to address when such models can still require significant run times and involve such a large number of sources of uncertainty, including alternative formulations of the process equations. Thus, there are uncertainties in the descriptive equations of the processes; there are uncertainties in the parameter values required for every spatial element; there are uncertainties in the measurements of the required inputs and the need to interpolate the measurements in space and time; and there is uncertainty in the observations used to evaluate model predictions. Since each of these sources of uncertainty can be difficult to characterize *a priori*, it follows that there is significant scope for uncertainty about uncertainty estimation.

4.2 Uncertainty About Uncertainty Estimation

In trying to assess the uncertainty in model predictions we should distinguish between two types of simulation applications (we will not deal here with techniques for data assimilation to reduce uncertainty in real-time forecasting (see Beven, 2009), since this is not common in applications of erosion models). In the first type of simulation application, there are no data with which to evaluate the model predictions so that only a forward uncertainty analysis is possible. The results will then depend totally on the assumptions made about the different sources of uncertainty in the modelling process. The second type of application is where there are some data with which to evaluate the model predictions. These data can therefore be used to constrain the range of feasible models and therefore the resulting prediction uncertainty. The results will then depend strongly on how the data are incorporated into the evaluation process. Different assumptions can be made. Here we will differentiate between forward uncertainty estimation, a formal Bayesian statistical approach and the Generalised Likelihood Uncertainty Estimation (GLUE) methodology proposed by Beven and Binley (1992).

4.2.1 *Forward uncertainty estimation*

In forward uncertainty estimation, each source of uncertainty must be defined in some way. The prediction uncertainties then follow deductively from the assumed definitions. The most common form of forward uncertainty estimation is to assume that a model structure is known, that the input data are known, and to define statistical distributions for the model parameters on the basis of prior knowledge. In simple cases, where a model can be transformed to be linear in its parameters (such as the USLE), the uncertainty can be propagated analytically; but in the case of nonlinear hydrological or erosion models it is necessary to use approximate numerical methods in a forward uncertainty analysis, such as Monte Carlo simulation. This is often combined with Latin Hypercube sampling to ensure that the parameter space is sampled efficiently (see Beven, 2009).

Clearly, the resulting prediction uncertainties depend only on the assumptions made about sources of uncertainty. Thus it is important that the assumptions should be realistic in a particular application. This is the main problem with

this approach. It is simply not easy to make dis-
tributional assumptions about model parameters,
even for models that can be interpreted physically. One reason is that what is required by the
model are effective values of parameters, which
might be different from, or might not be commensurate with, values that could be measured
in the field due to scale effects (see Chapter 5),
non-linearities and non-stationarities (Beven,
2006a). It is then even more difficult to make
assumptions about the co-variation to be expected
between different effective parameters. We often
suspect that parameters might interact within
a model structure, but defining that interaction
a priori will be difficult. Neglecting such
co-variation will overestimate the output uncertainty. If the interaction can be defined, however,
there are simple ways of generating co-varying
parameter values, such as the use of copula sampling (again, see Beven, 2009).

The results of any forward uncertainty analysis will also be conditional on the decisions made
about model structure and input data. Where
there are no evaluation data available this will
have to be accepted, but it is far better if such
data can be used at least to check if the resulting
predictions are within the range of the observations. Experience suggests that this will not
always be the case.

4.2.2 Bayesian uncertainty estimation

Where some evaluation data are available then
it might be possible to use a formal statistical
framework to evaluate predictive uncertainty.
This requires assumptions about the nature of
the residual model error, ε_r in

$$O_{x,t} = M(\underline{\theta})_{x,t} + \varepsilon_r \qquad (4.1)$$

where O is some observation, $M(\underline{\theta})$ is the equivalent output from a model with parameter vector
$\underline{\theta}$, $\underline{x}$ are space coordinates and t is time. In particular, it is necessary to assume that a model of the
residual errors can be found such that, having taken
out any structure in the residuals, only random

variation needs to be considered (e.g. Kennedy &
O'Hagan, 2001).

In the simplest possible case, for a model that
performs very well and for observations that are
not biased in any way, then checking for structure in the residuals will show that they are independent and identically distributed and can be
represented by a Gaussian distribution with zero
mean and constant variance. This then implies
that a simple Gaussian likelihood function can be
used to evaluate the predictions of any parameter
set in the model. Within a formal Bayesian framework, Bayes' equation is then used to update the
parameter estimates as more data are added in
the form:

$$P(\underline{\theta}|\underline{O}) \propto P_o(\theta)\, L(\underline{O}|\theta) \qquad (4.2)$$

or in words: the posterior probability of a parameter set conditional on a set of observations $\underline{O}$,
$P(\underline{\theta}|\underline{O})$, is proportional to the product of the
assumed prior distribution of the parameters,
$P_o(\theta)$, and the likelihood of predicting the observations with the model conditional on the parame-
ters, $L(\underline{O}|\theta)$. The Bayes approach allows for the
subjective choice of the prior distribution of
parameters (and their co-variation) that is required
for a forward uncertainty analysis, but application of Bayes' equation will normally reduce the
uncertainty in the predictions and make the posterior parameter distributions better defined as
more data are added. In fact, it can be used iteratively
as each new set of observations is added, the posterior distribution up to now becoming the prior
distribution for the additional conditioning provided by the new observations. Where we are not
too sure about parameter distributions in starting
off the process, we can choose non-informative
prior distributions.

There are two important issues in applying
Bayesian uncertainty estimation. The first is
choosing an appropriate likelihood function in
Equation 4.2; the second is integrating that likelihood over the parameter space which generally
has to be done numerically. The difficulty in
choosing a likelihood function arises because

model residuals often show complex structures that can involve correlation in time and/or space; heteroscedasticity (changing variance with magnitude of prediction); non-Gaussian distributions; and non-stationary bias. Some of these characteristics can be allowed for by modelling or transforming the residuals to allow a simple likelihood function to be used. There is a danger in doing so, however. If the model of the residuals is too simple, then it will result in overconditioning of the parameter distribution (see, for example, Beven, 2006a; Beven *et al.*, 2008).

The problem of integrating the likelihood function over the parameter space is directly related to the choice of likelihood function. This is because the shape of the response surface in the parameter space will reflect that choice: in particular the 'peakiness' of the surface. When the posterior parameter distribution is strongly conditioned by the observations and the assumed likelihood function, then the region of high likelihood might be very local in the parameter space. If only a small number of parameters is being considered then this is not a problem, but the higher the dimensions of the parameter space, the more difficult it becomes to find a local high likelihood region, especially if non-informative prior distributions, or incorrect prior distributions, have been assumed.

The most common technique used to integrate the likelihood function is the range of Monte Carlo Markov Chain (MCMC) methods (e.g. Gamerman, 1997; Beven, 2009). In MCMC methods, an initial sample is used to guide the next set of samples in a way that should lead to samples with a density in the parameter space proportional to the likelihood. The greater the number of samples used, the better the integration of the likelihood function. Tests for convergence of the posterior distribution can be used as more samples are added. Sampling (and consequently rates of convergence) can be controlled by parameters of the MCMC algorithm. For complex cases with high parameter dimensions, convergence may take a very large number of samples with sudden jumps from one part of the parameter space to another. This is indicative that, for these com-

plex cases, the likelihood surface may not be simple, but may involve a number of different peaks in the surface. Where the nature of the correct definition of likelihood function is in doubt, there is also no guarantee that the optimum for one likelihood function will be in the same region as an alternative likelihood function. This is significant because the validity of a likelihood function is generally checked with respect to the residuals produced by the model with parameter values at the peak of a likelihood function. It is therefore possible that the structure of the residuals might support the choice of likelihood function, or not. If not, then the series of residuals might suggest an alternative form of likelihood function, but since the maximum likelihood model is then likely to have quite different parameters, it might also have different residual characteristics again. It is therefore easier to reject a likelihood function than to find the 'best' likelihood function. There are examples in the hydrological literature where the residuals clearly do not support the chosen likelihood function (but see Engeland *et al.*, 2005, for an example of good practice in checking residual structures).

This issue of an appropriate likelihood function is important because of the way in which the assumption of randomness of residuals in statistical inference implies that every residual is informative. The result is then to provide very strong conditioning to the likelihood (make the likelihood surface very peaked) as the number of residuals included in the evaluation increases (see Mantovan & Todini, 2006; Beven *et al.*, 2008). When time series of data are available (as is often the case when models are evaluated with discharge data and continuous turbidity measurements, for example) then the conditioning can be very strong indeed and the resulting marginal posterior parameter distributions are well defined with small variance.

Now this could be interpreted as indicating that the model is very well defined (even if the residual variance might still be quite high), but this is dangerous because in most environmental modelling it is not true that every residual can be considered informative because of the complex sources of error. For example, input data (such as rainfall) can be very well estimated in some storms, but quite poorly estimated in others because of poor siting of rain-gauges or anomalies in radar rainfall data. These are generally non-stationary errors rather than random errors, and when fed through the model they are processed non-linearly to give error series with complex bias and correlation characteristics.

Thus, for this type of modelling there are good reasons to believe that in real applications, with poorly characterized input and model structural errors, the statistical assumptions lead to over-conditioning of the likelihood surface because of an overestimation of the information content of a series of residuals. Relaxing the assumptions to make a more realistic assessment of residual information, however, means either making further statistical (and difficult to justify) assumptions about the nature of all sources of uncertainty in the modelling process, or moving away from the formal statistical definition of likelihood. This was the original issue that led to the development of the GLUE methodology first outlined in Beven and Binley (1992) following the observation that use of the performance measures used in the optimization of hydrological models often suggested that there were many different models that gave similar levels of performance. This led to a rejection of the concept of the optimum model in favour of the equifinality thesis in model calibration (Beven, 1993, 1996a, 2001c; Beven & Freer, 2001) that is the basis for the GLUE methodology. The issue can be illustrated by the application of soil erosion models to simulate real datasets (Section 4.6).

4.2.3 Generalised Likelihood Uncertainty Estimation (GLUE)

The GLUE methodology is based on the use of informal likelihood measures in model evaluation. Many such measures can still be given a probabilistic interpretation (see Smith et al., 2008) and there is no reason why formal statistical likelihoods and Bayes combination of likelihoods based on random residual assumptions cannot be

used in GLUE (see Romanowicz et al., 1994, 1996). It is general in that respect. However, it is the potential to use informal likelihood measures (including fuzzy and binary measures) and different ways of combining likelihoods that makes the GLUE methodology of interest. It is much more flexible, but at the expense of adding choice to the user and without the claims to objectivity of formal Bayesian methods. However, as noted above, that objectivity only holds if the assumptions of the formal analysis are correct.

In its origins, GLUE is a development of the Hornberger-Spear-Young General or Regionalised Sensitivity Analysis (see Hornberger & Spear, 1981; Hornberger et al., 1985; Spear, 1997). This form of sensitivity analysis was based on making a large number of Monte Carlo realizations of a model, then dividing the results into those that gave acceptable simulations of the data available (the behavioural models) and those that did not (the non-behavioural models). A comparison of the parameter distributions for the behavioural and non-behavioural sets of models then could be interpreted in terms of the sensitivity of the results to individual parameters that had been varied in the model realizations.

The GLUE methodology adds a likelihood weighting and prediction step to this in which each of the non-behavioural models is given a likelihood of zero, while each of the models in the behavioural set is given a likelihood based on how well it has performed in the evaluation process. Scaling the likelihoods such that the cumulative sum is equal to 1 allows any likelihood-weighted predictions over all models in the behavioural set to be interpreted probabilistically. If an informal likelihood is used, this has a quite different interpretation to the formal statistical probability, being the probability contributed to the range of predictions by that model, conditional on the assumed likelihood measure (see discussion in Beven, 2009).

The choice of a likelihood in GLUE can be quite subjective, and many people see this as reason enough to reject it as an uncertainty estimation methodology. A number of studies have shown that for well-defined *hypothetical* examples where the input data and model structure are known to be correct and random residuals are created by construction, then an arbitrary choice of informal likelihood will not reproduce the results of a formal, and in hypothetical cases objective, statistical analysis (Mantovan & Todini, 2006; Stedinger *et al.*, 2008). There is then a suggestion that if the use of informal likelihood measures in GLUE does not work for such hypothetical examples, then why should we expect them to work in real applications (see also the series of discussions in Beven, 2006b; Todini & Mantovan, 2007; Hall *et al.*, 2007, Montanari, 2007; Andréassian *et al.*, 2007).

A response to these arguments is provided by Beven (2006a) who discusses the differences between ideal and non-ideal (real application) cases, and Beven *et al.* (2008) who show that even mild departures from the correct assumptions of a hypothetical case can lead to bias in parameter estimates. The case against the use of informal likelihoods is not so clear-cut when there is a danger of overconditioning of parameters if the strong assumptions of a formal likelihood are not met. In real applications, therefore, when input errors and model structural errors might be important, we need to be more circumspect about the objectivity of formal statistical inference. The GLUE methodology then has some features that can be advantageous (and, as noted above, can incorporate formal methods as special cases when it is felt that the strong assumptions can be justified).

One such feature is the implicit handling of residual errors, in all their complexity. When a particular realization of a model is compared with observations, the residual errors are known exactly. In formal statistical methods, a model of those errors is then proposed and the parameters identified. This model is then assumed to hold when the model is used for prediction. If, as in most applications of GLUE, the treatment of the residual error is left implicit, a similar assumption holds; that the characteristics of the errors for a model belonging to the behavioural set and used in prediction are similar (in all their complexity) in prediction. If a behavioural model is overpredicting in evaluation under certain condi-

tions, then we expect it to overpredict in other similar conditions. Likewise, we expect it to underpredict where it has underpredicted in the past. Thus, in weighting the predictions of that model, there is also effectively a weighting of the residual errors implied by that model. If a model is a good representation of the system response then we might expect some model realizations to underpredict and others to overpredict in different parts of the evaluation period in some consistent way. It is then likely that in prediction the set of models will bracket the observations (as demonstrated for hypothetical examples by Beven *et al.*, 2008; Smith *et al.*, 2008).

The more interesting case occurs when the model cannot bracket the observations in calibration (and we therefore should not expect it to do so in evaluation or prediction). This suggests that there is some lack of knowledge about the data or processes that is leading to (possibly non-stationary) bias in the model predictions (see Section 4.4 below). There might be many reasons for this, most importantly model structure error or input error, but experience suggests it is a generic problem in the application of environmental models. These are just the conditions when an evaluation of residuals might reveal difficulties in formulating a simple statistical error model in the formal Bayes approach. It led Beven (2006a) to suggest a different way of approaching model evaluation by defining limits of acceptability prior to making any model runs. The limits of acceptability might be based purely on the user requirements (how good do the predictions need to be?), or on a consideration of the errors in both input data and the observations with which the model is being compared. The idea is then that models are treated as members of the behavioural set if their predictions lie within the limits of acceptability, and are considered non-behavioural if they lie without. Behavioural models can be given a likelihood based on their performance within the limits of acceptability. There is only limited experience with this approach (although it was already being used in, for example, Freer *et al.*, 2004; Page *et al.*, 2007; Iorgulescu *et al.*, 2005, 2007), but it seems that it can be rather difficult to find models that consistently lie within limits of acceptability, perhaps because there are 'outliers' in either input data or evaluation observations (which can be difficult to detect) or because of fundamental error in model structures (see Blazkova & Beven, 2009; Dean *et al.*, 2009; Krueger *et al.*, 2009; Liu *et al.*, 2009). What this approach does allow is the rejection of all the models tried if necessary, whereas in statistical approaches, error can always be accommodated by expanding the variance of the residual error model.

4.3   Model Evaluation as Hypothesis Testing in the Face of Uncertainty

In the previous section we have described two fundamentally different approaches to uncertainty estimation. They overlap, in that GLUE can be used with formal statistical likelihood measures if the strong assumptions can be justified, but they are essentially based on rather different philosophies. In formal statistical methods, the primary aim (relaxed somewhat in Bayesian MCMC methodology) is to find the maximum likelihood model and allow for random error by a statistical distribution. In GLUE the aim is to make predictions with all models that can be considered behavioural and allow for the errors associated with each model, with all their complexity, implicitly. The latter is therefore more rejectionist in its approach, in that it is possible that all the models tried might be rejected (e.g. Choi & Beven, 2007; Dean *et al.*, 2009), while if the set of behavioural models is large then they can be considered as multiple working hypotheses about how the system responds (Beven, 1996a, 2002a, 2008, 2009). It is readily seen that this can easily be extended to consideration of multiple model structures which might all have parameter sets that can be accepted as behavioural in evaluation (or not!), but which might imply quite different process controls and predictions. It follows that, given the limitations of the type of input and observational data that are available, it will not necessarily be clear that one model is distinctly better than another. This can be handled

naturally within the GLUE framework: a particular model is either behavioural or not according to the criteria set by the user. Within the statistical framework this has led to the development of Bayesian model averaging or Bayesian melding techniques (although these generally require that the same form of error model be used for all the model structures considered (and for all parameter sets within those model structures) to allow a common basis for assessing likelihood).

4.4   The Information Content of Observations in Constraining Uncertainty

The discussion so far has revealed that there might be different ways of evaluating likelihood in assessing the performance of a model. Formal statistical likelihood follows from assumptions about the error model; informal likelihood measures are a more subjective choice. Mantovan and Todini (2006) stressed how the formal framework allows the objective evaluation of the information content of observations in conditioning parameter distributions in ideal cases, but in the response of Beven *et al.* (2008) it is shown how this can lead to an overestimation of information content in even mildly non-ideal cases. There is then the possibility of making a Type II error, of giving a good model a low likelihood because of a particular realisation of the errors, particularly the input errors. The global informal likelihood measures in that study seemed to underestimate the information content of observations. This safeguards against Type II errors but makes a Type I error (of accepting a poor model because of uncertainty) more likely. In real applications, with an expectation of input and model structural error, there are good reasons to expect that the real information content of observations will be overestimated by formal likelihood measures.

This is essentially a problem of differentiating between aleatory (statistical) and epistemic (knowledge) uncertainties. The first we can happily treat in terms of probabilities and formal likelihoods. They should be 'well-behaved' in error structure (effectively by definition!). The same is not necessarily the case for epistemic uncertainties, which represent all the different factors affecting the accuracy and uncertainty with which we can make predictions, but about which we cannot make very strong statements. Epistemic uncertainties (in respect of model structures, effective parameter variables, input data and boundary conditions) will be important in real applications.

Unfortunately, there is no equivalent theory for the information content of data in the face of epistemic uncertainties. If we choose to represent epistemic uncertainties in terms of probabilities and formal likelihoods, we will necessarily be making an approximation. In some cases this approximation might be useful in compensating for unknown sources of uncertainty (for example in the data assimilation strategies for real-time forecasting), while in others it might lead to misleading estimates of parameters and predictions (and the choice of an over-strong formal likelihood might then be *incoherent* (Beven *et al.*, 2008)). We should also recognise that we might need to expect the unexpected: the result of unknown unknowns that we do not yet recognise, and will not recognise until something unexpected is observed (and model predictions fail). There is no real way to allow for the unexpected, of course, but we might learn more about our science from studying model failures than simply compensating for them with a statistical error model.

The limits of acceptability approach discussed above offer some interesting possibilities in this respect. If we think about the value of observations in conditioning models as multiple hypotheses, then those observations that allow the maximum differentiation between hypotheses (the greatest rejection of models) will be the most informative (as long as there is no evidence that those observations might themselves be misleading). Thus, periods of observations that are similar to types of response seen previously will have little marginal information content (note that this is not the case under statistical assumptions; additional measurements are always assumed to

add information multiplicatively within the Bayes framework). Periods of observations that are quite different to the types of response seen previously, on the other hand, would have much greater information content in differentiating between different models as hypotheses. Much more research is required on appropriate methods for evaluating models as hypotheses, given observations in future.

4.5 Review of Uncertainty Analysis of Soil Erosion Models

Although uncertainty analysis and even robust examples of more traditional evaluations of soil erosion models are rare, there are some examples from the literature where attempts have been made to understand the quality of soil erosion predictions or the components within erosion models to which predictions are most sensitive. This assessment of uncertainty in soil erosion models is in its infancy, unlike in the parallel field of hydrological modelling, but nonetheless a review is warranted, as already the approaches employed exhibit quite different interpretations of how to assess model performance and therefore demonstrate how valid model predictions are. Broadly speaking, the approaches taken can be described as 'sensitivity analyses', 'forward error analyses' or 'uncertainty analyses'. The following section provides a discussion of the current state-of-the-art in assessing the quality of erosion predictions, and begins with the earliest work first in an attempt to trace the chronological progression of uncertainty analysis of erosion models.

4.5.1 *Sensitivity analysis of erosion models*

By far the most common form of model exploration employed in the soil erosion modelling literature is that of sensitivity analysis. The following section summarises the development of this technique and how it has been used to improve understanding of soil erosion models and how (in a direct sense) input parameters affect model output, but also (indirectly) how model structure imposes constraints upon how well soil erosion models can perform. The review is restricted to first-order variance or univariate techniques as, despite the shortcomings of these approaches (Summers et al., 1993; Saltelli et al., 2004; Pappenberger et al., 2006, 2008; Beven, 2009), they are the most common methods used to assess the performance of soil erosion models.

Despite the many hundreds of scientific papers written about the Universal Soil Loss Equation (USLE), very little research has been performed which provides a robust evaluation of model performance against observed data, whilst considering model uncertainty. Indeed, Risse et al. (1993) stated: "Although nearly three decades of widespread use have confirmed the reliability of the Universal Soil Loss Equation, very little work has been done to assess the error associated with it". Such a statement might be considered to be somewhat paradoxical if we propose that a reliable prediction should be one where error is both explicit and ideally minimal. Nonetheless, Risse et al. (1993) demonstrated that the USLE performs reasonably well when predicting soil loss from the plots that it was originally formulated on – Nash-Sutcliffe efficiencies of 0.58 when comparing annual soil loss predictions with observations, improving to 0.75 when average annual data are considered. The model predictions were most sensitive to the topographic factor (LS) and the cover and management factor (C), largely due to the multiplicative nature of the model structure which ensures that any error associated with parameterisation is multiplied through to the model output. Sonneveld and Nearing (2003) developed this point further, by illustrating that the "… modest correlation between observed soil losses and model calculations, even with the same data that was used for its calibration … raises questions about its mathematical model structure and the robustness of the assumed parameter values that are implicitly assigned to the model". To address these points, Sonneveld and Nearing (2003) conducted a validation of the model which tested the sensitivity of the parameters and demonstrated that the "USLE model is not very robust", which undermines the widespread use of such a tool within erosion modelling, despite (and perhaps because of) its desirably simple model structure. Such a conclusion is further underlined by the work of Boomer et al. (2008) who found that the USLE and its successor, the R (Revised) USLE2, failed to predict sediment yields adequately in 101 catchments in the Chesapeake Bay area, indicating the need to evaluate such models with appropriate observed data before they are useful as land management tools.

In response to the conceptually simple structure of the USLE and the vast improvements in computational speed and processing capability that have evolved since the late 1980s, physically-based erosion models such as the Water Erosion Prediction Project (WEPP), hailed as the "new generation of erosion prediction technology" (Laflen et al., 1991), have been developed. With such models came an increase in both the number of parameters that were being used to model soil erosion and the consequent uncertainty that was associated with their output. Early efforts to evaluate model performance generally took the form of univariate sensitivity analyses, where parameters were varied one at a time, around their 'estimated' values (see Nearing et al., 1990; Tiscareno-Lopez et al., 1995), in an effort to identify which parameters the model output was most sensitive to. The goal of univariate sensitivity analyses in this context is to understand which parameters need to be estimated most carefully and which parameters are more redundant within the model structure, and therefore require less attention. Invariably, parameters controlling the hydraulic conductivity and the erodibility of soils tend to exert the strongest influence on model output, although parameters describing rainfall characteristics, particularly rainfall intensity, are also often found to exhibit sensitivity. Similar findings, describing the influence of the saturated hydraulic conductivity parameter, were also presented after sensitivity analysis of the Limburg Soil Erosion Model, LISEM (De Roo et al., 1996), highlighting the importance of predicting hydrology correctly in erosion models.

The Chemicals, Runoff, and Erosion from Agricultural Management Systems (CREAMS) model (Knisel, 1980), parts of which underlie the WEPP model (see later discussion), and its companion the Groundwater Loading Effects of Agricultural Management Systems (GLEAMS) (Leonard *et al.*, 1987), were also products of the 1980s revolution in 'physically-based' model development, and were subjected to sensitivity analyses (Silburn & Loch, 1989) which focused on the erosion–sedimentation part of the model. Despite early efforts to ensure that CREAMS/ GLEAMS were physically-based and therefore did not need to be calibrated, the sensitivity analyses show similar limitations to the empirically-based USLE, perhaps not surprisingly as parts of the USLE were incorporated into the erosion component of the model. The model was shown to be sensitive to a variety of parameters including the USLE erodibility and vegetation cover parameters, peak runoff rate, storm erosivity, slope length and sediment size distribution (Silburn & Loch, 1989), demonstrating the legacy of former model development in the new generation of erosion prediction tools. In addition, Silburn and Loch (1989) found parameter interaction to play a role in controlling model output, although this was not adequately assessed via the univariate sensitivity analysis.

The Système Hydrologique Européen (SHE) soil erosion model was also subject to a sensitivity analysis (Wicks *et al.*, 1992), although the goals of this work were more explicitly to understand whether the model could be applied without calibration. Results demonstrated that the model could predict overland flow response to simulated rainfall on saturated soil, but was less able to predict flow generated on dry soil, and indeed required detailed calibration to predict erosion or sediment yield as the erosion submodel was very sensitive to correct representation of the hydrology (Wicks *et al.*, 1992).

Veihe and Quinton (2000) and Veihe *et al.* (2000) analysed the EUROSEM model using two techniques – a Monte Carlo-based approach which is discussed in the following sections, and a simple sensitivity analysis which varied parameters by ±10%. Results of the latter approach indicated that the presence of rock fragments in the soil has a significant impact on runoff generation and subsequent soil erosion predicted by the model. However, Veihe and Quinton (2000) recognised that such simple sensitivity analyses do not inform the user about the distribution of output from the model, and as such they advised that more complex, multiple parameter variation is carried out using Monte Carlo techniques.

In part to address the need to understand the influence of multiple model parameters, recently a more complex form of sensitivity analysis has been developed by Wei *et al.* (2007) using the Rangeland Hydrology and Erosion Model (RHEM) based on the WEPP model (Nearing *et al.*, 1989). In this exercise, the authors selected 14 of the model parameters relating to the hydrology and erosion subcomponents of the model and varied these parameters by 5%, one at a time, to calculate a local sensitivity index at each of 10,000 randomly selected points in the 14-dimensional parameter space (as defined by pre-determined parameter ranges (Wei *et al.*, 2007). The 'localized' sensitivity indices were then summarized to compare the relative importance of each parameter to model output and, via classification of the local sensitivities of each parameter, to illustrate the distributions of parameter sensitivity within the model. Wei *et al.* (2007) reported that results were in agreement with those of Tiscareno-Lopez *et al.* (1995), who performed a similar (although less complex) univariate sensitivity analysis on the WEPP model.

Two lessons can be learnt from this work:
(1) As with all univariate sensitivity analyses, the neglect of the interaction of parameters in combination with each other severely undermines the conclusions of the research. By only varying one parameter at a time, it is assumed *implicitly* that there is no interaction between parameters, when this is clearly not the case. Wei *et al.* (2007) attempted to deal with this problem via regression analysis of parameters to explore what the relationships between parameters may be. However, this is only performed between pairs of parameters, ignoring the other 12 at any one time, so does not constitute a full analysis of parameter interaction.

(2) Although some effort was made by Wei *et al.* (2007) to *relate* the model sensitivity to observed soil loss, this was not done in a structured sense, so that there is no indication of how well the model performs against the real world data (evaluation), just a suggestion that there is more model sensitivity in the predictions made of low soil loss. Further progress is made with analysis of the RHEM in Wei *et al.* (2008) (see below).

Finally, an example of a recent sensitivity analysis was conducted by Morgan and Duzant (2008) on the Morgan-Morgan-Finney (MMF) model. Improvements to the original MMF model were made in order to account for the effects of vegetation cover on soil erosion predictions. The approach was similar to that of Nearing *et al.* (1990) – as described above – which holds all parameters at a set value whilst varying one at a time to study the effects upon model output. Results were ranked in terms of relative influence on model output and were deemed to be acceptable as they conformed to previous analysis of the MMF model (Morgan *et al.*, 1984). The main criticism here (in addition to those described in the previous paragraph) is that unlike the recent modifications to the model that Morgan and Duzant (2008) provide, the *approach* used to assess the model performance is more than 20 years old and has been superseded by a range of methods described below that provide a far more informative perspective on model uncertainty.

As has been demonstrated, sensitivity analyses may serve as a useful initial exploration of the importance of each parameter within a model structure, as many of the above-mentioned models contain dozens of parameters, not all of which can be well estimated. Model complexity, however, is not easily explored simply by varying single parameters, and indeed it has been argued that a meaningful exploration of the multidimensional parameter space that these models occupy demands the variation of multiple parameters in parallel with each other (Saltelli *et al.*, 2004; Pappenberger *et al.*, 2006, 2008; Beven, 2009). Univariate approaches provide no assessment of the interaction of parameters or the levels of autocorrelation between parameters, which may

exert more (or a different) control on model output than the variation of a single parameter in isolation. In addition, many sensitivity analyses are carried out without recourse to observed data as forward error or uncertainty analyses. Such exercises can be conducted as useful numerical experiments to underpin more robust evaluation or even calibration of less certain parts of the erosion model. For example, Wainwright *et al.* (2008a,b,c) calibrated sensitive hydrological parameters to test performance of different sediment models for the correct hydrological reasons. However, ideally, sensitivity of a model to variations in parameter values should be assessed via comparison of model output with observed data, in order to relate model predictions to the real world (Brazier *et al.*, 2000). Therefore, the types of sensitivity analyses described above are perhaps best used as the starting point of model evaluation, as just one of the tools that an erosion modeller may employ to explore model capability and test model performance.

4.5.2   *Forward error analysis of erosion models*

Numerous papers describe the 'error' associated with model predictions with respect to some objective function of model performance when compared with observed data. Examples of such an approach to understanding 'error' in the USLE and derivatives are illustrated in Risse *et al.* (1993), Spaeth *et al.* (2003) and Tiwari *et al.* (2000). Very few authors, however, try to analyse the sources of this error within the model parameters, structure or observed data. It is this *analysis* of error that the following section is concerned with – as an attempt to *understand* where a model might be in error and why that model makes erroneous predictions, in order to improve model predictions through reduction of that error.

Examples of the error analysis of erosion models in the literature are rare despite the tools available to model error, particularly in GIS-based models (Heuvelink, 1998; Forier & Canters, 1996). Wang *et al.* (2000a,b, 2002) attempted to quantify the spatial error of the topographic factor

(*LS*) and the rainfall-runoff erosivity (*R*) factor in RUSLE. Each of these studies described the spatial variability in one of the six model parameters by using experimental semi-variograms to define the variability, which is then input to the model. Thus, an effort is made to quantify one type of error, for one parameter at a time, albeit over the entire space being modelled. Results demonstrate that it is important to understand the spatial error associated with estimates of input parameters to models such as the RUSLE, and that if such 'uncertainty' is not considered then it is impossible to 'construct error budgets for the predictions of soil loss' (Wang *et al.*, 2002). Such work definitely represents progress in the modelling of error associated with erosion model output. However, both of these numerical experiments were carried out in isolation of each other – the error surrounding the *R* factor and its influence on model output was not assessed alongside that of the *LS* factor to produce a combined error associated with the interaction of both parameters. Moreover, the error associated with the remaining four parameters in the model was not assessed at all. The next step might be to extend the approach to include a full error analysis, conducted to assess the total error associated with parameterisation of the model, in order that the user can understand the dominant source of error from input to the model.

The work of Mokrech (2001) described an in-depth approach to forward modelling uncertainty of the Thornes erosion model (Thornes, 1990) by understanding the propagation of parameter error within the model structure. The Thornes model is parsimonious, containing only four real parameters, and as such is a good choice of model to apply at large spatial scales, but also to understand how the inevitable error that surrounds often poorly constrained parameters may influence model output using a forward uncertainty analysis.

Two approaches are suggested: (1) classical error propagation, whereby error associated with each parameter is moved through the model in an analytical sense; and (2) Monte Carlo simulation, whereby a stochastic description of the form of model error is described *a priori* for each parameter, in this example assuming a Gaussian distribution, to provide a range of parameter values for each parameter within the model. The former approach does not allow for a robust assessment of the error associated with non-linear models when compared to the latter, which is employed using many realisations of the model to provide a distribution of model outputs consistent with the assumptions made about sources of uncertainty in the forward analysis. This error propagation approach is employed in the Mokrech (2001) study without consideration of spatial autocorrelation or covariation in the varied parameters. As Heuvelink (1998) recognized, this consequently provides a rather narrow description of model error. To overcome this, Mokrech tried to represent the distribution function of the model error in a spatially explicit way; however, such an approach still does not take into account autocorrelation or interaction of parameters, and so may not map the true uncertainty associated with model predictions very well.

The two approaches of analytical error propagation and Monte Carlo simulation are compared, and it is concluded that the analytical approach suggests the greater 'uncertainty' in the model output, although this term is used somewhat loosely here as this uncertainty is not assessed against any observed data (such as hydrographs or sediment yield data). Finally, it is concluded that the spatial scale of the input data (a 30-m grid is used) may also affect the output uncertainty surrounding the model predictions. Such a conclusion is not surprising as no attempt is made to understand the quality of the spatial data at representing the dominant processes that operate at this spatial scale, and whether or not the processes that can be represented by 30-m data are actually the processes responsible for the erosion being modelled (see Brazier *et al.*, 2005, and Chapter 5 for a discussion of this).

When compared with the first-order sensitivity analyses described in Section 4.5.1, the forward error analyses detailed above represent a significant advance in understanding soil erosion models. Not only is consideration given to error associated with parameters, but it is also done in a spatial framework allowing for some representation of the effects of both spatial variability and the heterogeneity of error to be incorporated into predictions. The approach of Mokrech (2001) shows the most promise, as numerous parameters and therefore parameter interactions are considered in parallel. However, the work is still best described as forward error or forward uncertainty analysis as the model predictions are not evaluated against observed data, such that the 'reality' of the model uncertainty that is presented is not tested.

4.5.3 Uncertainty analysis of soil erosion models

The following section describes a range of approaches to model evaluation that are described in the literature as 'uncertainty analyses'. As will be seen, however, very few of these numerical experiments are actually uncertainty analyses in the true sense of the term, as they do not evaluate model performance with respect to observed data. The results of each approach are therefore presented alongside a summary of the methods employed, to demonstrate how robust and meaningful each model evaluation is.

Early approaches to evaluate model uncertainty were conducted on the WEPP model by Chaves and Nearing (1991) and Tiscareno-Lopez et al. (1995). Chaves and Nearing (1991) explored the effect of 60 scenarios (random parameterizations across the 28-dimensional parameter space) on the output of the model – quantified in terms of the coefficient of variation of peak runoff rate, soil loss, sediment yield and sediment enrichment ratio. The authors found that maximum coefficients of variation for the four outputs were 196, 267, 323, and 47%, respectively, demonstrating a wide range of uncertainty associated with model output, but not evaluating this uncertainty against any observed data. In contrast, Tiscareno-Lopez et al. (1995) did employ observed data from the semi-arid Walnut Gulch Experimental Watershed to assess model uncertainty. Parameters were varied to reflect the typical variance associated with

parameterization, the effects of which were assessed against the mean square error of model output. In this example, results showed that the model was able to simulate the watershed dynamics well. Thus, these two model evaluations illustrate well how subtly different approaches to uncertainty evaluation can yield very different results, and that the choice of the uncertainty analysis technique may be crucial in determining the degree of confidence that the user may have in the model. Section 4.6 revisits this issue by summarising a further uncertainty analysis of the WEPP model (Brazier et al., 2000), and demonstrating a potential way forward for the uncertainty analysis of erosion models.

Recognising the limitations of previous first-order sensitivity analyses of the USLE (see above), Hession et al. (1996) conducted a two-stage uncertainty analysis of the model. The authors attempted to address epistemic uncertainty "due to incomplete understanding or inadequate measurement of system properties" and uncertainty associated with model stochastic variability which is "due to random variability of the natural environment". It is argued that the former may be reduced, through better measurement or constraint of parameter values, but the latter may always be present in model predictions and cannot be reduced. Both types of uncertainty were incorporated into the parameter distributions that were sampled in a Monte Carlo-based analysis of model performance against observations made over a 27-year period on an experimental plot in Guthrie, Oklahoma, US. Figure 4.1 (after Hession et al., 1996) illustrates that confidence limits based on 10th and 90th percentiles describing the uncertainty associated with model predictions capture some of the observed soil erosion behaviour over the 27-year period. However, the model does not do well at predicting the 30% of events that are small ($<$ca. $3\,kg\,m^{-2}$) or the 15% that are large ($>17\,kg\,m^{-2}$). The authors did not present data describing the performance of the model in terms of any objective function against observations, but the results demonstrate well that model uncertainty must be assessed in applications of models such as the USLE, if we are to

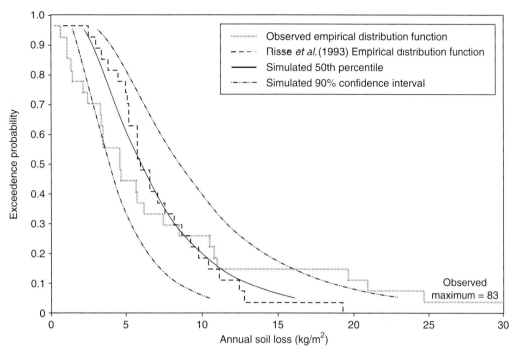

Fig. 4.1 Summary of the distribution of complementary cumulative distribution functions from multiple model runs of the USLE, describing uncertainty due to lack of knowledge (epistemic uncertainty) (after Hession *et al.*, 1996).

be honest about model predictions/capabilities (Hession *et al.*, 1996).

Parallel developments of the European Soil Erosion Model (EUROSEM; Morgan *et al.*, 1993) also involved a first attempt at evaluating model uncertainty through the work of Quinton (1994, 1997) which explored two parameterization methods to reduce model uncertainty. The first method – Improved Parameter Set Selection (IPSS) – selected four sensitive parameters (whilst all others were held constant) and varied each parameter five times, to give a total of 625 model runs to explore model uncertainty and how it might be affected by differing parameterization strategies. Evaluation of the model runs against a 'training' dataset was then used to select optimal parameter values, which could be used for subsequent model runs in an effort to reduce model uncertainty. The second method used observed hydrographs

to generate parameter values for the constant saturated hydraulic conductivity parameter, which resulted in better prediction of the event hydrograph, although not the sedigraph or erosion predictions (see Fig. 4.2).

Quinton (1997) concluded that, using these techniques, it is possible to reduce the width of the uncertainty bounds describing model performance, but that in so doing, the observed data that are being predicted more often fall outwith the revised uncertainty bounds. The author also calls for erosion model applications to be accompanied by uncertainty analysis, and stresses that "... not doing so misleads the user into believing that the model output is more certain than is actually the case" (Quinton, 1997). This approach is partly demonstrated with the EUROSEM model in Folly *et al.* (1999), Veihe and Quinton (2000) and Veihe *et al.* (2000),

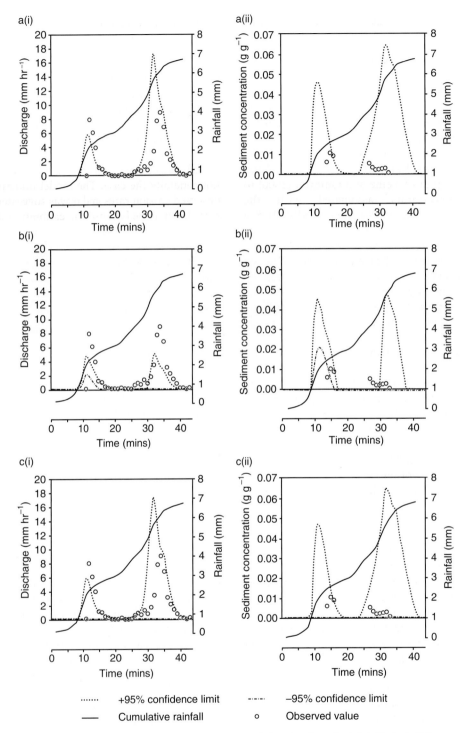

Fig. 4.2 Comparisons of simulations of hydrographs (a(i), b(i), c(i)) and sedigraphs (a(ii), b(ii), c(ii)) for an event on the 20 July 1992 at Woburn Experimental Farm, Bedford, England, for the (a) blind (b) IPSS and (c) constant saturated hydraulic conductivity cases (after Quinton, 1997).

although as is discussed above, these papers describe what is more appropriately termed sensitivity analysis, not uncertainty analysis. Such a call is better answered by a holistic framework as demonstrated in Chapter 5, where a modified version of the GLUE approach is used to evaluation predictions from EUROSEM.

Hantush and Kalin (2005) took a similar approach to that of Veihe and Quinton (2000) to analyse uncertainty associated with the KINEROS2 model (a development of KINEROS, a kinematic runoff and erosion model; Woolhiser *et al.*, 1990). The model is evaluated against observed data from a 33.6 ha catchment, known as W-2, which is located near Treynor, Iowa, US. Monte Carlo simulations were used to generate exceedance probability curves (for comparison, these are akin to the cumulative distribution function (cdf) curves of Veihe & Quinton, 2000, or the cumulative distribution curves of Brazier *et al.*, 2000) and uncertainty bounds based on 25th, 50th and 75th percentiles (see Fig. 4.3). The authors demonstrate that the KINEROS2 model provides more reliable predictions for larger events and less reliable predictions of the dynamics of smaller erosion events.

These model findings relate well to observations made by other authors. For example, Nearing (2000) and Nearing *et al.* (1999) reported greater variability in the response of replicate plots to low-magnitude, high-frequency events, than their larger, rarer counterparts. Thus, in a sense, it is unrealistic to expect models such as KINEROS2 to make more certain predictions than the empirical data they are being evaluated against. However, when the model is applied to catchments for which calibration/evaluation data do not exist, it is suggested that runoff and sediment yield predictions will only be "… within order of magnitude of accuracy" (Hantush & Kalin, 2005).

A further analysis of the uncertainty associated with the erosion predictions of KINEROS2 is illustrated by Martinez-Carreras *et al.* (2007), in the context of erosion occurring in a semi-arid badland environment in the Ca l'Isard catchment (1.32 km²) in the Eastern Pyrenees. First, calibration data were collected in a small subcatchment to aid parameterization of the model prior to application at the catchment scale. Subsequently, the model was run against four years of observed data using the GLUE approach. Extending the argument of Hantush and Kalin (2005), it might be thought that as calibration data were available, we would expect model performance to improve; however, this was not generally the case. The model underpredicted observed erosion rates and it was suggested that in part this may be because the calibration data that were used were not suitable. It is very important in exercises of this type that the observed and predicted variables be *commensurate* (i.e. have the same meaning). This is not always the case, and the assumption of Hantush and Kalin (2005) and Martinez-Carreras *et al.* (2007), as well as many other authors, that model calibration improves model prediction, may be true for specific sites or model applications, but where the calibration is affected by observational limitations there is no guarantee that the effective parameter values determined by calibration will be transferable to sites where no calibration data are available (see also sections 2.8.2 and 2.8.3).

As has been argued above, erosion models may have many sources of error, all of which are likely to contribute to the uncertainty that surrounds model predictions. Jetten *et al.* (2003) recognised that the current generation of erosion models are very rarely (if ever) tested against datasets describing not only the catchment outlet hydrograph and sediment yield, but also the *patterns* of erosion within the catchment. In their overview of the Global Change and Terrestrial Ecosystems Focus 3 programme, Jetten *et al.* (2003) acknowledged that erosion models are only "moderately good" at predicting outlet hydrographs and are "not very good for net soil loss". The authors called for better description of the spatial distribution of soil erosion and better use of these data to evaluate soil erosion models (see also Section 3.5).

Incorporating spatial erosion data for model evaluation was central to the work of Van Oost *et al.* (2005), who presented a dataset describing the spatial variability in soil erosion from the

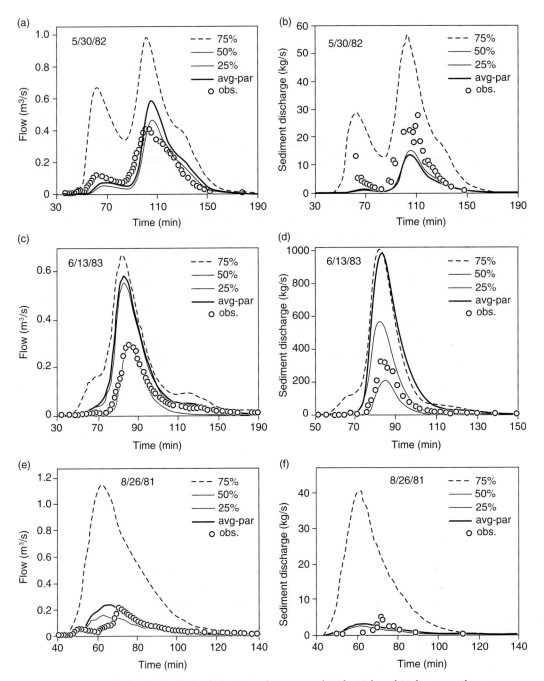

Fig. 4.3 Comparison with observed values of Monte Carlo-generated 75th, 50th and 25th percentiles, and simulated model flow rates and sediment discharges based on average model parameters (avg-par). Monte Carlo simulations of the KINEROS2 model (after Hantush & Kalin, 2005).

Ganspoel and Kinderveld catchments over a three-year period. The data were used to evaluate the WaTEM/SEDEM model (Van Oost *et al.*, 2000; Van Rompaey *et al.*, 2001) in a Monte Carlo-based analysis where two of the five model parameters were varied simultaneously to generate 5000 random parameter sets. This analysis therefore improved upon the univariate sensitivity analyses described above, but might be best described as a bivariate uncertainty analysis, as all other parameters were held at 'reasonable' values. Exported sediment predictions were compared against observations and evaluated using a simple likelihood function for single events. The approach demonstrated that uncertainty surrounding predictions was wide when only one likelihood measure was used to evaluate the model performance. A further measure was then introduced, evaluating the model predictions against deposited sediment, and then the product of the two likelihood measures was combined. Not surprisingly, more models were rejected using the combined likelihood approach, which was reasonable as the model was being more robustly tested. However, the authors argued that this approach "... allowed us to narrow substantially the uncertainty associated with the event rainfall erosivity parameter", as can be seen when comparing Fig. 4.4A with Fig. 4.4B. Although this may appear to be true, it should be noted that many combinations of parameters still provided a poor goodness-of-fit with the same value of rainfall erosivity (approximately 0.004) as giving the best model predictions. In addition, the authors failed to make the next step, which is to describe the uncertainty around the single event observations in terms of uncertainty bounds or confidence limits. In addition, we are only shown an example of one event and the way in which uncertainty associated with one parameter may be constrained, which begs the question 'how does uncertainty vary between different events?' Clearly an understanding of how uncertainty in erosion predictions varies between different storms would be useful, as would the explicit representation of uncertainty bounds around each model parameter.

A further useful point was made by Van Oost *et al.* (2005) concerning the use of the spatial pattern of erosion in a catchment as evaluation data. The authors suggested that these 'soft' or 'qualitative' data might be useful in model evaluation, although they did not make any use of this approach themselves. It is suggested here that evaluation of erosion models on more than just plot or catchment outlet data is desirable. In the same way as in Brazier *et al.* (2000), we tried to evaluate erosion predictions for 'the correct hydrological reasons', as it would be useful to see sediment yields predicted in a way that reflects the correct spatial patterns of erosion.

Unfortunately, despite the recognition of the uncertainty associated with model predictions, many models are still applied 'off-the-shelf' without consideration of the implications of model uncertainty. There are many examples of such an approach to erosion modelling. All of the model analyses reviewed here demonstrate that model users *should* evaluate model uncertainty, yet it is clear from the literature that *most* applications of erosion models do not do this. Just one example of this problem is illustrated by Mati *et al.* (2006), who applied the EUROSEM model to the Embori and Mukogodo catchments, Kenya. EUROSEM has been shown to make reasonable (although uncertain) predictions (see above), but here it is applied to predict erosion without any consideration of this uncertainty. While we must allow that models are used for different purposes including prediction, experimentation to learn about process representation, to guide mitigation or soil conservation, or perhaps to assess the impact of land use change, it is clear that model developers (as well as model users) must learn from previous lessons and start to incorporate estimation of uncertainty into their model applications. This incorporation of previous research into our application and development of contemporary erosion models is especially important if we expect to make any progress in developing 'better' erosion models (see Chapter 5).

Vigiak *et al.* (2006) applied the MMF model (Morgan *et al.*, 1984) to a small catchment in the West Usambara Mountains (Kwalei catchment,

(a)

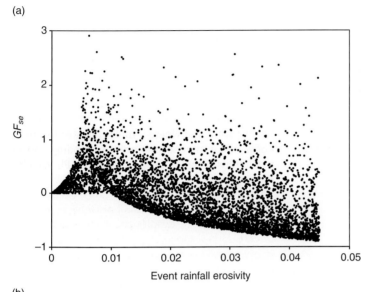

(b)

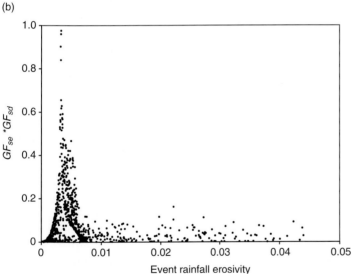

Fig. 4.4 Results of fitting the WaTEM/ SEDEM model to May 1977 data for the Ganspoel, varying just two model parameters. Each point represents a model run. The GF values indicate the goodness of fit to the observed data. (a) Model evaluation based on comparison of the simulated and observed sediment export at the outlet of the catchment. (b) The model is evaluated by using a combined GF value, using both observed sediment export and within-catchment deposition (after Van Oost et al., 2005).

Tanzania) in a framework that evaluated the spatial distribution of erosion, rather than the sediment yield from the catchment outlet. This work presents an interesting alternative to the sensitivity analysis performed by Morgan and Duzant (2008), discussed above, and the more standard approaches of fitting to catchment outlet hydrographs and sedigraphs that are common in the literature. In a sense, the work answers the call of

Van Oost et al. (2005), by basing the Monte Carlo analysis of the model on data collected throughout the catchment, describing the *patterns* of erosion to test the uncertainty associated with the sediment transport parameters. In this example, more than 75% of the erosion patterns can be explained by the model, and 64% of the severely eroded fields were predicted by the model. However, as with all soil erosion models, the

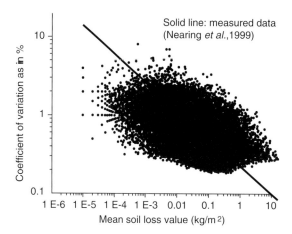

Fig. 4.5 Coefficient of variation (*CV*) for measured soil loss from the output distribution at a point against the expected predicted soil loss value. The corresponding relationship developed from measured data (Nearing *et al.*, 1999) is also shown as a single log–log line (after Wei *et al.*, 2008).

uncertainty associated with the soil erosion predictions is high, particularly where observed erosion rates were low (or observed as deposition), and small changes in sediment transport parameters led to large differences in erosion/deposition rate predictions. It is suggested that this finding is a general problem for spatially distributed erosion models, as appropriate definition of sediment transport parameters (which may be highly variable in time and space) is exceedingly difficult.

The final example of uncertainty analysis that will be reviewed here is that of Wei *et al.* (2008) who employed a "dual-Monte Carlo approach to estimate model uncertainty" associated with predictions from the Rangeland Hydrology and Erosion Model (RHEM). Model predictions are evaluated against event-based observations of three storms using a two-stage Monte Carlo procedure to explore the implications of parameter uncertainty on model predictive uncertainty. Results demonstrate that predictive uncertainty is significant and realistic when compared to the coefficient of variation around mean soil loss observations described by Nearing *et al.* (1999), as is shown in Fig. 4.5.

Interestingly, predictive uncertainty increased with the magnitude of erosion predictions (as seen in Fig. 4.6), whereas variability of observed erosion was greatest for smaller storms described in the Nearing *et al.* (1999) dataset.

Using these results, Wei *et al.* (2008) argue that it is useful to transfer the uncertainty associated with predictions to sites where observed data are not available, and where it might be most useful to know about the uncertainty surrounding model predictions to guide decision-making. Philosophically, this approach is similar to that of Brazier *et al.* (2001) who mapped the uncertainty associated with predictions from the WEPP model on to the wider landscape to illustrate the implications of model uncertainty for the spatial prediction of erosion. In the Wei *et al.* (2008) example, regression equations are used to describe the confidence limits (analogous to the uncertainty bounds of Brazier *et al.*, 2000), in relation to two input parameters – rainfall amount and saturated hydraulic conductivity – and two outputs – soil loss and runoff depth. It is argued that these regression equations can then be used to transfer the description of model uncertainty to wider predictions, without recourse to further Monte Carlo model runs.

The approach has some merit, and indeed demonstrates a potential way forward that model users may take in order to illustrate how good model predictions may be. However, it is likely that confidence limits would be different if model uncertainty was assessed specifically at each site (through Monte Carlo simulations), so it is not yet clear how meaningful such transferral of uncertainty bounds actually is, without supporting data (and model evaluation) to test the assumption that predictive uncertainty is transferable. It might be argued that model predictions are not transferable (without validation), so perhaps this approach needs further evaluation before it is widely used. A further point to consider is that Wei *et al.* (2008) do not try to assess the full range of uncertainty associated with model predictions; some consideration of model structural uncertainty and uncertainty associated with observations *alongside* the assessment of parameter uncertainty presented here would be very useful.

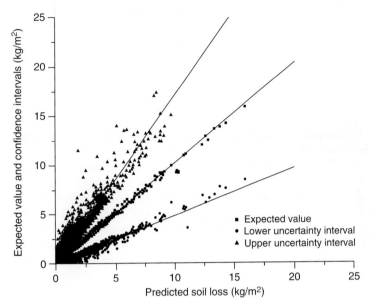

Fig. 4.6 Expected prediction and uncertainty intervals vs. the predicted value from the RHEM model. Expected values were computed as the mean of 1000 model runs, providing randomly generated values for each model parameter to compare with predicted soil loss from three sample storms at three separate locations (after Wei *et al.*, 2008).

Section 4.5 has reviewed a wide range of ways in which model developers have attempted to describe model sensitivity, and the error or uncertainty associated with erosion model predictions. All of these approaches demonstrate some merit, although none can be described as a holistic treatment of model uncertainty by quantifying both input uncertainty (through parameterization and parameter interactions) and model structural uncertainty: the dominant sources of epistemic uncertainty discussed in Section 4.4. In addition, few models have been adequately evaluated against observed data, to demonstrate how well predictions relate to real-world observations or to show how well observations are captured within uncertainty bounds. The following section summarizes an attempt to deal with these problems by using the GLUE approach to evaluate the WEPP model against data from both the UK and the US (Brazier *et al.*, 2000). Although it is by no means a complete assessment of model uncertainty, it serves as a good example of what can be done to assess the quality of model predictions, and to understand the sources of model uncertainty which may help

model developers to constrain uncertainty in future model development.

4.6 Case Study: Using WEPP to Predict UK and US Erosion Data

Many papers have been written describing the WEPP model and its applications (more than 100 to date), yet very few (as reviewed above) have attempted to evaluate the uncertainty associated with model predictions. If process-based soil erosion models proliferate, in the same manner as their forebears, the USLE family of models, and this proliferation is not accompanied by assessment of predictive uncertainty, then the field of erosion modelling will not move forward in terms of reducing this predictive uncertainty – a scenario that is undesirable for both model developers and model users alike.

The WEPP model was analysed by Brazier *et al.* (2000) against data that were collected in large plot-scale experiments in both the UK, at the Woburn Erosion Reference Experiment (Catt *et al.*, 1994), and the US, at the Holly Springs site, Mississippi, as part of the USDA/ARS USLE database (Zhang, personal communication, 1999). The GLUE approach was used to explore parameter and model structural uncertainty against 12 years of runoff and erosion data at the two sites by running the model in continuous simulation mode and varying the 16 most sensitive model parameters simultaneously. No *a priori* assumptions were made about parameter distributions, so parameter values were randomly sampled from uniform distributions with initially wide minima and maxima. Approximately 5.7 million model runs were carried out in order to explore, as fully as possible, the 16-dimensional model parameter space. All other parameters within the model were held constant. Each model prediction was evaluated against observed data using a combined-likelihood function that assessed the model performance in terms of its ability to predict both runoff and erosion, or to make good erosion predictions for the correct hydrological reasons. Behavioural parameter sets or model structures that satisfied predetermined criteria for goodness of fit were then retained to describe the uncertainty surrounding model predictions. It should be noted here that the predetermination of what is an acceptable model prediction is clearly subjective, although every effort was made in this work to ensure that the criteria were neither too strict, resulting in the rejection of all models as non-behavioural, nor too relaxed, resulting in the acceptance of all models as behavioural, which is clearly not the case.

Results show, for both the UK and US sites, that the model consistently overpredicted small events and underpredicted large events, leading to the overestimation of observed data during years when little erosion occurred and the underestimation of observed data during years when erosion was significant. Figure 4.7 illustrates the modelled erosion predictions for both sites (shown here as minimum, maximum and median predictions to describe uncertainty bounds). Predictions were highly uncertain. Although most years of observations were captured within the model uncertainty bounds, certain years,

notably the dry year of 1963 on the Holly Springs plots, were overpredicted by a factor of seven, by even the minimum uncertainty bounds.

Brazier *et al.* (2000) also showed that model predictions that are produced through the random generation of parameter values are better than those produced by either estimation or optimization techniques presented by previous authors (Zhang *et al.*, 1996), and a common approach when parameterizing erosion models. This is an important finding, as it underlines how poorly the model structure is able to describe the system that is being modelled. It also questions the approach of 'estimating' parameter values in general, without considering the uncertainty that these estimations implicitly incorporate into model predictions. It is clear that a better approach would be to quantify parameter uncertainty, so as to be explicit about how well the effective parameter values of erosion models can be estimated and thus what levels of uncertainty we might expect, given our current understanding and representation of the processes of erosion in models.

Finally, in common with all of the other approaches discussed herein, this work shows that model predictions are far from precise, and that even with the 'new generation' of erosion prediction technology that WEPP was heralded to be, much work is still to be done to provide us with reliable or certain predictions of erosion. Qualification of model predictions, through uncertainty analysis, must not be seen as a negative process. Far from it, all of the above approaches allow us varying degrees of insight into the performance of erosion models, which can direct model developers towards areas of process-understanding that need to be improved and better represented within soil erosion models. Analysis of model uncertainty is also a vital tool in the toolkit of model users. Decision-makers or land managers who apply erosion models to foresee the effects of management decisions must do so with some consideration of uncertainty. Ideally this would be done within a framework such as GLUE (see Chapter 5 for a further example), or at the very least with some

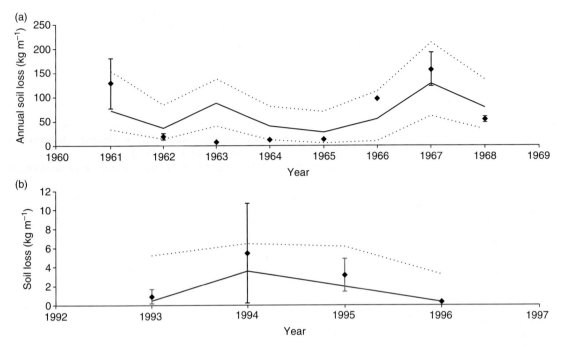

Fig. 4.7 Observed and predicted annual soil loss from the (a) Holly Springs plots, US, and (b) the Woburn Erosion Reference Experimental plots, UK (kg m⁻¹). Observed data are shown as diamonds, with error bars describing range of replicate values, dotted lines describe minimum and maximum predictive uncertainty, and the solid line describes the median of model predictions (after Brazier *et al.*, 2000).

consideration of uncertainty, otherwise the credibility of erosion modelling will suffer.

4.7   The Future

Whenever erosion models are compared with observations, it is generally found that even when allowing for uncertainties in model parameters, there are often observations that cannot be matched by the model predictions. This suggests that modelling results are subject to important epistemic uncertainties that have not been taken into account. The most important uncertainties in this respect might be: commensurability and bias uncertainties in model inputs; commensurability in the observable variables relative to what is being predicted; commensurability in effective

parameter values relative to estimates derived directly from field measurements; and, of course, the representation of erosion, transport and deposition processes in the model.

If erosion models fail in this way, it is worth emphasizing that this is an opportunity to reconsider modelling strategies to make them more realistic as hypotheses about how the sediment system functions (e.g. Beven, 2002a,b, 2008, 2009). What then needs to be reconsidered? Are the input data used to drive the model commensurate or accurate enough? Are the evaluation observations commensurate or accurate enough? Do the nature of the failures give an indication of deficiencies in the model process representations? Are model structures developed under a limited range of applications being applied too widely elsewhere? These types of questions should be at the heart of a modelling process treated as learning about the responses of the places at which models are applied, a process that places emphasis on making data available to allow the uncertainty in model predictions to be constrained (Beven, 2007).

One issue to be addressed in the future is developing guidelines on good modelling practice and the value of different types of data in constraining uncertainty (Pappenberger & Beven, 2006; Beven *et al.*, 2008). It does seem, however, that there will be uncertainty about uncertainty estimation methods for some time to come. Clear differences of opinion exist between those who insist that formal statistical methods of uncertainty estimation are the only way to evaluate model predictions, and those who accept that epistemic uncertainties undermine the theoretical rigour of formal statistics, and that consequently it will be worthwhile pursuing other, more exploratory methodologies.

It should, however, be stressed that uncertainty estimation should not be considered as the end point of an analysis. Since, when used with erosion models, it will often reveal the failure of models to match observations, it demands improvements in both models and data to check whether such errors represent Type I, Type II or Type III modelling errors. Types I and II have been mentioned earlier, and can be constrained by the collection of more data to refine the characterization of the system and constrain the uncertainties further. Type III errors are the unknown unknowns, the missing processes, the epistemic errors that we had not previously considered important. Addressing Type III errors implies future creativity from modellers, based on a careful analysis of model failures.

References

Andréassian, V., Lerat, J., Loumagne, C., *et al.* (2007) What is really undermining hydrologic science today? *Hydrological Processes* 21: 2819–22.

Arnold, J.W., Srinivasan, R., Muttiah, R.S. & Williams, J.R. (1998) Large area hydrologic modeling and assessment, Part I: model development. *Journal of American Water Resources Association* 34: 73–89.

Beven, K.J. (1985) Distributed modelling. In Anderson, M.G. & Burt, T.P. (eds), *Hydological Forecasting*. John Wiley & Sons, Chichester: 405–35.

Beven, K.J. (1989) Changing ideas in hydrology: the case of physically based models. *Journal of Hydrology* 105: 157–72.

Beven, K.J. (1993) Prophecy, reality and uncertainty in distributed hydrological modelling. *Advances in Water Resources* 16: 41–51.

Beven, K.J. (1996a) Equifinality and uncertainty in geomorphological modelling. In Rhoads, B.L. & Thorn, C.E. (eds), *The Scientific Nature of Geomorphology*. John Wiley & Sons, Chichester: 289–313.

Beven, K.J. (1996b) A discussion of distributed modelling. In Refsgaard, J.-C. & Abbott, M.B. (eds), *Distributed Hydrological Modelling*. Kluwer, Dordrecht: 255–78.

Beven, K.J. (2001a) *Rainfall-Runoff Modelling: The Primer*. John Wiley & Sons, Chichester.

Beven, K.J. (2001b) Dalton Medal Lecture: How far can we go in distributed hydrological modelling? *Hydrology and Earth System Sciences* 5: 1–12.

Beven, K.J. (2001c) Calibration, validation and equifinality in hydrological modelling. In Anderson, M.G. & Bates, P.D. (eds), *Model Validation: Perspectives in Hydrological Science*. John Wiley & Sons, Chichester: 43–55.

Beven, K.J. (2002a) Towards a coherent philosophy for environmental modelling. *Proceedings of the Royal Society London A* 458: 2465–84.

Beven, K.J. (2002b) Towards an alternative blueprint for a physically-based digitally simulated hydrologic response modelling system. *Hydrological Processes* 16: 189–206.

Beven, K.J. (2006a) A manifesto for the equifinality thesis. *Journal of Hydrology* 320: 18–36.

Beven, K.J. (2006b) On undermining the science? *Hydrological Processes* 20: 3141–6.

Beven, K.J. (2007) Working towards integrated environmental models of everywhere: uncertainty, data, and modelling as a learning process. *Hydrology and Earth System Science* 11: 460–67.

Beven, K.J. (2008) On doing better hydrological science. *Hydrological Processes* 22: 3549–53.

Beven, K.J. (2009) *Environmental Modelling: An Uncertain Future?* Routledge, London.

Beven, K.J. & O'Connell, P.E. (1982) On the role of physically-based distributed models in hydrology. *Institute of Hydrology Report*, No. 81, Wallingford, UK.

Beven, K.J. & Binley, A.M. (1992) The future of distributed models: model calibration and uncertainty prediction. *Hydrological Processes* **6**: 279–98.

Beven, K.J. & Freer, J. (2001) Equifinality, data assimilation, and uncertainty estimation in mechanistic modelling of complex environmental systems. *Journal of Hydrology* **249**: 11–29.

Beven, K.J., Smith, P.J. & Freer, J. (2008) So just why would a modeller choose to be incoherent? *Journal of Hydrology* **354**: 15–32.

Blazkova, S. & Beven, K.J. (2009) Uncertainty in flood estimation. *Structure and Infrastructure Engineering* **5**: 325–32. DOI: 10.1080/15732470701189514

Boomer, K.B., Weller, D.E. & Jordan, T.E. (2008) Empirical models based on the Universal Soil Loss Equation fail to predict sediment discharges from Chesapeake Bay catchments. *Journal of Environmental Quality* **37**: 79–89.

Brazier, R.E., Beven, K.J., Freer, J. & Rowan, J.S. (2000) Equifinality and uncertainty in physically-based soil erosion models: application of the GLUE methodology to WEPP – The Water Erosion Prediction Project for sites in the UK and US. *Earth Surface Processes and Landforms* **25**: 825–45.

Brazier, R.E., Beven, K.J., Anthony, S., *et al.* (2001) Implications of complex model uncertainty for the mapping of hillslope scale soil erosion predictions. *Earth Surface Processes and Landforms* **26**: 1333–52.

Brazier, R.E., Heathwaite, A.L. & Liu, S. (2005) Scaling issues relating to P transfer from land to water. *Journal of Hydrology* **304**: 330–42.

Catt, J.A., Quinton, J.N., Rickson, R.J. & Styles, P.D.R. (1994) Nutrient losses and crop yields in the Woburn Erosion Reference Experiment. In Rickson, R.J. (ed.), *Conserving Soil Resources: European Perspectives*. CAB International, Wallingford: 94–104.

Chaves, H.M.L. & Nearing, M.A. (1991) Uncertainty analysis of the WEPP soil-erosion model. *Transactions of the American Society of Agricultural Engineers* **34**: 2437–45.

Choi, H.T. & Beven, K.J. (2007) Multi-period and multi-criteria model conditioning to reduce prediction uncertainty in distributed rainfall-runoff modelling within GLUE framework. *Journal of Hydrology* **332**: 316–36.

Dean, S., Freer, J.E., Beven, K.J., *et al.* (2009) Uncertainty assessment of a process-based integrated catchment model of phosphorus (INCA-P). *Stochastic Environmental Research and Risk Assessment* **23**: 991–1010. DOI: 10.1007/s00477-008-0273-z

De Roo, A.P.J., Offermans, R.J.E. & Cremers, N.H.D.T. (1996) LISEM: A single-event, physically based hydrological and soil erosion model for drainage basins 2. Sensitivity analysis, validation and application. *Hydrological Processes* **10**: 1119–26.

Engeland, K., Xu, C-Y. & Gottschalk, L. (2005) Assessing uncertainties in a conceptual water balance model using Bayesian methodology. *Hydrological Sciences Journal* **50**: 45–63.

Ewen, J., Parkin, G. & O'Connell, P.E. (2000) SHETRAN: Distributed river basin flow and transport modeling system. *Journal of Hydrologic Engineering ASCE* **5**: 250–58. DOI: 10.1061/(ASCE)1084-0699(2000)5:3(250)

Folly, A., Quinton, J.N. & Smith R.E. (1999) Evaluation of the EUROSEM model using data from the Catsop watershed, The Netherlands. *Catena* **37**: 507–19.

Forier, F. & Canters, F. (1996) A user-friendly tool for error modelling and error propagation in a GIS environment, In Mowrer, H.T., Czaplewski, R.L. & Hamre, R.H. (eds), *Spatial Accuracy Assessment in Natural Resources and Environmental Sciences*. USDA Forest Service Technical Report RM-GTR-277, pp. 225–34. URL: http://orca.vub.ac.be/~frforier/artikel1.html

Freer, J.E., McMillan, H., McDonnell, J.J. & Beven, K.J. (2004) Constraining dynamic TOPMODEL responses for imprecise water table information using fuzzy rule based performance measures. *Journal of Hydrology* **291**: 254–77.

Freeze, R.A. (1978) Mathematical models of hillslope hydrology. In Kirkby, M.J. (ed.), *Hillslope Hydrology*. John Wiley & Sons, Chichester: 177–225.

Freeze, R.A. & Harlan, R.L. (1969) Blueprint for a physically-based, digitally-simulated hydrologic response model. *Journal of Hydrology* **9**: 237–58.

Gamerman, D. (1997) *Markov Chain Monte Carlo: stochastic simulation for Bayesian inference*. Chapman & Hall/CRC, Boca Raton, LA.

Gassman, P.W., Reyes, M.R., Green, C.H. & Arnold, J.G. (2007) The soil and water assessment tool: historical development, applications, and future research directions. *Transactions of the American Society of Agricultural and Biological Engineers* **50**: 1211–50.

Grayson, R.B., Moore, I.D. & McMahon, T.A. (1992) Physically based hydrologic modeling, 2. Is the concept realistic? *Water Resources Research* **28**: 2659–66.

Hall, J., O'Connell, E. & Ewen, J. (2007) On not undermining the science: Discussion of invited commentary by Keith Beven, Hydrological Processes, 20, 3141–3146 (2006). *Hydrological Processes* **21**: 985–8.

Hantush, M.H. & Kalin, L. (2005) Uncertainty and sensitivity analysis of runoff and sediment yield in a small agricultural watershed with KINEROS2. *Hydrological Sciences Journal* **50**: 1151–71.

Hession, W.C., Storm, D.C. & Haan, C.T. (1996) Two-phase uncertainty analysis: an example using the Universal Soil Loss Equation. *Transactions of the American Society of Agricultural Engineeers* **39**: 1309–19.

Heuvelink, G.B.M. (1998) *Error Propagation in Environmental Modelling with GIS*. Taylor & Francis Ltd., London.

Hornberger, G.M., Beven, K.J., Cosby, B.J. & Sappington, D.E. (1985) Shenandoah Watershed Study: Calibration of a topography-based, variable contributing area hydrological model to a small forested catchment. *Water Resources Research* **21**: 1841–50.

Hornberger, G.M. & Spear, R.C. (1981) An approach to the preliminary analysis of environmental systems. *Journal of Environmental Management* **12**: 7–18.

Iorgulescu, I., Beven, K.J. & Musy, A. (2005) Data-based modelling of runoff and chemical tracer concentrations in the Haute-Mentue (Switzerland) Research Catchment. *Hydrological Processes* **19**: 2257–574.

Iorgulescu, I., Beven, K.J. & Musy, A. (2007) Flow, mixing, and displacement in using a data-based hydrochemical model to predict conservative tracer data. *Water Resources Research* **43**: W03401. DOI: 10.1029/2005WR004019

Jetten, V., Govers, G. & Hessel, R. (2003) Erosion models: quality of spatial predictions. *Hydrological Processes* **17**: 887–900. DOI: 10.1002/hyp.1168

Kennedy, M.C. & O'Hagan, A. (2001) Bayesian calibration of mathematical models. *Journal of the Royal Statistical Society* **D63**: 425–50.

Knisel, W.G. (1980) *A field-scale model for chemicals, runoff and erosion from agricultural management systems*. Conservation Report No. 26. USDA-SEA, Washington, DC.

Krueger, T., Quinton, J.N., Freer, J., *et al.* (2009) Uncertainties in data and models to describe event dynamics of agricultural sediment and phosphorus transfer. *Journal of Environmental Quality* **38**: 1137–48. DOI: 10.2134/jeq2008.0179

Laflen, J.M., Lane, L.J. & Foster, G.R. (1991) WEPP – A new generation of erosion prediction technology. *Journal of Soil and Water Conservation* **46**: 34–8.

Leonard, R.A., Knisel, W.G. & Still, D.A. (1987) GLEAMS: Groundwater Loading Effects on Agricultural Management Systems. *Transactions of the American Society of Agricultural Engineers* **30**: 1403–28.

Liu, Y., Freer, J.E., Beven, K.J. & Matgen, P. (2009) Towards a limits of acceptability approach to the calibration of hydrological models: extending observa-

tion error. *Journal of Hydrology* **367**: 93–103. DOI: 10.1016/j.jhydrol.2009.01.016

Lukey, B.T., Sheffield, J., Bathurst, J.C., *et al.* (2000) Test of the SHETRAN technology for modelling the impact of reforestation on badlands runoff and sediment yield at Draix, France. *Journal of Hydrology* **235**: 44–62.

Mantovan, P. & Todini, E. (2006) Hydrological forecasting uncertainty assessment: incoherence of the GLUE methodology. *Journal of Hydrology* **330**: 368–81.

Martinez-Carreras, N., Soler, M., Hernandez, E. & Gallart, F. (2007) Simulating badland erosion with KINEROS2 in a small Mediterranean mountain basin (Vallcebre, Eastern Pyrenees). *Catena* **71**: 145–54.

Mati, B.M., Morgan R.P.C. & Quinton, J.N. (2006) Soil erosion modelling with EUROSEM at Embori and Mukogodo catchments, Kenya. *Earth Surface Processes Landforms* **31**: 579–88.

Mockus, V. 1949. Estimation of total (and peak rates of) surface runoff for individual storms. *Exhibit A, Appendix B, Interim Survey Report, Grand (Neosho) River Watershed*. US Department of Agriculture: Washington, DC.

Mokrech, M. (2001) *The management of uncertainty in geographic information systems: a case study using a soil-erosion model*. Unpublished PhD thesis, Geography Department, King's College, University of London.

Montanari, A. (2007) What do we mean by 'uncertainty'? The need for a consistent wording about uncertainty assessment in hydrology. *Hydrological Processes* **21**: 841–5.

Morgan, R.P.C. & Duzant, J.H. (2008) Modified MMF (Morgan-Morgan-Finney) model for evaluating effects of crops and vegetation cover on soil erosion. *Earth Surface Processes and Landforms* **32**: 90–106.

Morgan, R.P.C., Morgan, D.D.V. & Finney, H.J. (1984) A predictive model for the assessment of erosion risk. *Journal of Agricultural Engineering Research* **30**: 245–53.

Morgan, R.P.C., Quinton, J.N. & Rickson, R.J. (1993) *EUROSEM: A User Guide*. Silsoe College, Cranfield University.

Nearing, M.A. (2000) Evaluating soil erosion models using measured plot data: accounting for variability in the data. *Earth Surface Processes and Landforms* **25**: 1035–43.

Nearing, M. A., Foster, G.R., Lane, L.J. & Finkner, S.C. (1989) A process-based soil erosion model for USDA Water Erosion Prediction Project technology. *Transactions of the American Society of Agricultural Engineers* **32**: 1587–93.

Nearing, M.A., Deer-Ascough, L. & Laflen, J.M. (1990) Sensitivity analysis of the WEPP hillslope profile erosion model. *Transactions of the American Society of Agricultural Engineers* **33**: 839–49.

Nearing, M., Govers, G. & Norton, L.D. (1999) Variability in soil erosion data from replicated plots. *Soil Science Society of America Journal* **63**: 1829–35.

Nearing, M., Renard, K. & Nichols, M. (2005) Erosion prediction and modelling. In Anderson, M.G. (ed.), *Encyclopaedia of Hydrological Sciences.* John Wiley & Sons, Chichester: 1221–7.

Page, T., Beven, K.J. & Freer, J. (2007) Modelling the chloride signal at the Plynlimon Catchments, Wales using a modified dynamic TOPMODEL. *Hydrological Processes* **21**: 292–307.

Pappenberger, F. & Beven, K.J. (2006) Ignorance is bliss: 7 reasons not to use uncertainty analysis. *Water Resources Research* **42**: W05302. DOI: 10.1029/2005WR004820

Pappenberger, F., Beven, K.J., Ratto, M. & Matgen, P. (2008) Multi-method global sensitivity analysis of flood inundation models. *Advances in Water Resources* **31**: 1–14.

Pappenberger, F., Iorgulescu, I. and Beven, K.J. (2006) Sensitivity analysis based on regional splits (SARS) and regression trees. *Environmental Modelling and Software* **21**: 976–90.

Quinton, J.N. (1994) *The validation of physically-based erosion models with particular reference to EUROSEM.* PhD thesis, Cranfield University, Silsoe, Bedford, UK.

Quinton J.N. (1997) Reducing predictive uncertainty in model simulations: a comparison of two methods using the European soil erosion model (EUROSEM). *Catena* **30**: 101–17.

Refsgaard, J-C. & Abbott, M.B. (1996) The role of distributed hydrological modelling in water resources management. In Refsgaard, J-C. & Abbott, M.B. (eds), *Distributed Hydrological Modelling.* Kluwer, Dordrecht: 1–16.

Risse, L.M., Nearing, M.A., Nicks, A.D. & Laflen, J.M. (1993) Error assessment in the Universal Soil Loss Equation. *Soil Science Society of America Journal* **57**: 825–33.

Romanowicz, R., Beven, K.J. & Tawn, J. (1994) Evaluation of predictive uncertainty in non-linear hydrological models using a Bayesian approach. In Barnett, V. & Turkman, K.F. (eds), *Statistics for the Environment II. Water Related Issues.* John Wiley & Sons, Chichester: 297–317.

Romanowicz, R., Beven, K.J. and Tawn, J. (1996) Bayesian calibration of flood inundation models. In Anderson, M.G., Walling, D.E. & Bates, P.D. (eds), *Floodplain Processes.* John Wiley & Sons, Chichester: 333–60.

Saltelli, A., Tarantola, S., Campolongo, F. & Ratto, M. (2004) *Sensitivity Analysis in Practice: A Guide to Assessing Scientific Models.* John Wiley & Sons, Chichester.

Silburn, D.M. & Loch, D.J. (1989) Evaluation of the CREAMS model. 1. Sensitivity analysis of the soil–erosion sedimentation component for aggregated clay soils. *Australian Journal of Soil Research* **27**: 545–61.

Smith, P.J., Tawn, J. & Beven, K.J. (2008) Informal likelihood measures in model assessment: theoretic development and investigation. *Advances in Water Resources* **31**: 1087–1100.

Sonneveld, B.G.J.S. & Nearing, M.A. (2003) A nonparametric/parametric analysis of the Universal Soil Loss Equation. *Catena* **52**: 9–21.

Spaeth, K.E., Pierson, F.B., Weltz, M.A. & Blackburn, W.H. (2003) Evaluation of USLE and RUSLE estimated soil loss on rangeland. *Journal of Range Management* **56**: 234–46.

Spear, R.C. (1997) Large simulation models: calibration, uniqueness and goodness of fit. *Environmental Modelling Software* **12**: 219–28.

Stedinger, J.R., Vogel, R.M., Lee, S.U. & Batchelder, R. (2008) Appraisal of the generalized likelihood uncertainty estimation (GLUE) method. *Water Resources Research* **44**: W00B06. DOI: 10.1029/2008WR006822

Summers, J.K., Wilson, H.T. & Kou, J. (1993) A method for quantifying the prediction uncertainties associated with water quality models. *Ecological Modelling* **65**: 161–76.

Thornes, J.B. (1990) The interaction of erosional and vegetation dynamics in land degradation: spatial outcomes. In Thornes, J.B. (ed.), *Vegetation and Erosion.* John Wiley & Sons, Chichester: 41–53.

Tiscareno-Lopez, M., Weltz, M.A. & Lopes, V.L. (1995) Assessing uncertainties in WEPP's soil-erosion predictions on rangelands. *Journal of Soil and Water Conservation* **50**: 512–16.

Tiwari, A.K., Risse, L.M. & Nearing, M.A. (2000) Evaluation of WEPP and its comparison with USLE and RUSLE. *Transactions of the American Society of Agricultural Engineers* **43**: 1129–35.

Todini, E. & Mantovan, P. (2007) Comment on: 'On undermining the science?' by Keith Beven. *Hydrological Processes* **21**: 1633–8.

Van Oost, K., Govers, G. & Desmet, P. (2000) Evaluating the effects of changes in landscape structure on soil erosion by water and tillage. *Landscape Ecology* **15**: 577–89.

Van Oost, K., Govers, G., Cerdan, O., *et al.* (2005) Spatially distributed data for erosion model calibration and validation: the Ganspoel and Kinderveld datasets. *Catena* **61**: 105–21.

Van Rompaey, A.J.J., Verstraeten, G., Van Oost, K., *et al.* 2001. Modelling mean annual sediment yield using a distributed approach. *Earth Surface Processes and Landforms* **26**: 1221–36.

Veihe, A. & Quinton, J.N. (2000) Sensitivity analysis of EUROSEM using Monte Carlo simulation I: hydrological, soil and vegetation parameters. *Hydrological Processes* **14**: 915–26.

Veihe, A., Quinton J.N. & Poesen, J. (2000) Sensitivity analysis of EUROSEM using Monte Carlo simulation II: the effect of rills and rock fragments. *Hydrological Processes* **14**: 927–39.

Vigiak, O., Sterk, G., Romanowicz, R.J. & Beven, K.J. (2006) A semi-empirical model to assess uncertainty of spatial patterns of erosion. *Catena* **66**: 198–210.

Wainwright, J., Parsons, A.J., Müller, E.N., *et al.* (2008a) A transport distance approach to scaling erosion rates: 1. Background and model development. *Earth Surface Processes and Landforms* **33**: 813–26. DOI: 10.1002/esp.1624

Wainwright, J., Parsons, A.J., Müller, E.N., *et al.* (2008b) A transport distance approach to scaling erosion rates: 2. Sensitivity and evaluation of MAHLERAN. *Earth Surface Processes and Landforms* **33**: 962–84. DOI: 10.1002/esp.1623

Wainwright, J., Parsons, A.J., Müller, E.N., *et al.* (2008c) A transport distance approach to scaling erosion rates: 3. Evaluating scaling characteristics of MAHLERAN. *Earth Surface Processes and Landforms* **33**: 1113–28. DOI: 10.1002/esp.1622

Wang, G., Fang, S., Gertner, G.Z. & Anderson, A.B. (2000a) Uncertainty propagation and partitioning in spatial prediction of topographical factors of RUSLE. In Heuvelink, G.B.M. & Lemmens, M.J.P.M. (eds), *Proceedings of the Fourth International Symposium on Spatial Accuracy Assessment in Natural Resources and Environmental Sciences*. Delft University Press, Amsterdam.

Wang, G., Gertner, G.Z., Parysow, P. & Anderson, A.B. (2000b) Spatial prediction and uncertainty analysis of topographical factors for the Revised Universal Soil Loss Equation (RUSLE). *Journal of Soil and Water Conservation* **55**: 373–82.

Wang G., Gertner G., Singh, V., *et al.* (2002) Spatial and temporal prediction and uncertainty of soil loss using the Revised Universal Soil Loss Equation: a case study of the rainfall–runoff erosivity R factor. *Ecological Modelling* **153**: 143–55.

Wei, H., Nearing, M.A. & Stone J.J. (2007) A comprehensive sensitivity analysis framework for model evaluation and improvement using a case study of rangeland hydrology and erosion model. *Transactions of the American Society of Agricultural and Biological Engineers* **50**: 945–53.

Wei, H., Nearing, M.A., Stone J.J. & Breshears D.D. (2008) A dual Monte Carlo approach to estimate model uncertainty and its application to the rangeland hydrology and erosion model. *Transactions of the American Society of Agricultural and Biological Engineers* **51**: 515–20.

Wicks, J.M., Bathurst, J.C. & Johnson, C.W. (1992) Calibrating the SHE soil-erosion model for different land covers. *Journal of Irrigation and Drainage Engineering ASCE* **118**: 708–23.

Wischmeier, W.H. & Smith, D.D. (1960) A universal soil-loss equation to guide conservation farm planning. *Transactions of the 7th International Congress of Soil Science*. International Society of Soil Science, Amsterdam: 418–25.

Wischmeier, W.H. & Smith, D.D. (1978) Predicting rainfall erosion losses. A guide to conservation planning. *Agriculture Handbook No. 537*. USDA-SEA, US Government Printing Office, Washington, DC.

Woolhiser, D.A., Smith, R.E. & Goodrich, D.C. 1990. KINEROS – a kinematic runoff and erosion model: documentation and user manual. USDA Agricultural Research Service, ARS-77.

Zhang, X.C., Nearing, M.A., Risse, L.M. & McGregor, K.C. (1996) Evaluation of WEPP runoff and soil loss predictions using natural runoff plot data. *Transactions of the American Society of Agricultural Engineers* **39**: 855–63.

5 A Case Study of Uncertainty: Applying GLUE to EUROSEM

J.N. QUINTON[1], T. KRUEGER[2], J. FREER[3],
R.E. BRAZIER[4] AND G.S. BILOTTA[5]

[1]*Lancaster Environment Centre, University of Lancaster, Lancaster, UK*
[2]*School of Environmental Sciences, University of East Anglia, Norwich, UK*
[3]*School of Geographical Sciences, University of Bristol, Bristol, UK*
[4]*School of Geography, University of Exeter, Exeter, UK*
[5]*School of Environment and Technology, University of Brighton, Brighton, UK*

5.1 Introduction

At the time of writing, it is 10 years since Morgan *et al.* (1998a) published their paper describing the European Soil Erosion Model (EUROSEM). Since then the model has been downloaded by over 700 scientists from 64 countries worldwide. It remains one of the few truly dynamic and distributed soil erosion models.

The concepts behind EUROSEM date back over 20 years. It was first proposed at the Workshop on Erosion Assessment and Modelling held in Brussels in December 1986. At that workshop, Chisci and Morgan (1986) were critical of the models being developed in the US and believed that there was a potential to utilise process-based research from within the European Community to develop something better. They set out six design requirements: (1) to enable the risk of erosion to be assessed; (2) to be applicable at field and catchment scales; (3) to allow the contribution of sediment and solutes from the land surface to water bodies to be determined; (4) to provide reliable estimates of erosion and solute concentrations for comparison with acceptable standards; (5) to oper-

ate on an event basis; and (6) to be useful as a design tool for selecting soil protection measures. This was the basis for a collaborative research programme to turn these requirements into reality. In all the EUROSEM project ran from 1988 through to 1994, supported by two tranches of funding from the EU, and was also a core component of two further EU-funded projects: Modelling Within Storm Sediment Dynamics (MWISED) and Soil Productivity Indices and Erosion Sensitivity (SPIES). The first operational versions of the model were released in 1992, with further releases in 1994 and 1998. The model has also influenced the development of other erosion models, notably the Limburg Soil Erosion Model (LISEM: De Roo *et al.*, 1996). An Italian team lead by Lorenzo Borselli is currently developing a new version of the model, incorporating new process descriptions originating from more recent process research and the outputs of the MWISED project (Borselli *et al.*, 2008).

There have been a number of attempts to evaluate EUROSEM against measured data. Quinton (1994, 1997) evaluated the model against data from the Woburn Erosion Reference Experiment (Quinton & Catt, 2004). Quinton selected parameters to which the model was most sensitive and sampled four values of each from measured distributions and applied the model to single storms. The results demonstrated, for the first time, that

Handbook of Erosion Modelling, 1st edition. Edited by R.P.C. Morgan and M.A. Nearing. © 2011 Blackwell Publishing Ltd.

for an event-based erosion model the uncertainty bands (in this case 95% confidence intervals) around the predictions could be expected to be large, although the model in most cases was able to encompass the observed data. To try to reduce the width of the 95% confidence intervals, Quinton went on to select only those parameter sets which passed a performance criteria for a series of training storms, before applying the better performing parameter sets to the new observations. This had the desired effect of reducing the 95% confidence interval width, but resulted in the model failing to encompass much of the observed data. Other published studies in which EUROSEM has been evaluated are summarized in Table 5.1. These have compared the model output with data collected under natural rainfall (Quinton, 1997; Quinton & Morgan, 1998; Folly *et al.*, 1999; Cai *et al.*, 2005; Mati *et al.*, 2006) or against data collected with rainfall simulators (Veihe *et al.*, 2001; Cai *et al.*, 2005). The majority of these evaluations have adopted the common practice of splitting the dataset and using half to train or calibrate the model, and the second half of the dataset for evaluation. In this procedure the model is parameterized on the basis of field measurements, estimates and look-up tables in the model's user guide (Morgan *et al.*, 1998b). Parameters to which the model is sensitive are then changed so that the model output matches the observed data for the training datasets. Once an acceptable fit has been achieved (or the best fit possible), these parameter values are then used as the basis for applying the model to a second series of observed data either from the same site or from an entirely different location, to test the transferability of the model.

So why re-examine the model's performance at this relatively mature stage of development? A number of reasons for revisiting model evaluation are discussed in Chapter 4, not least that iterative improvements of both models and techniques to evaluate models are vital if we are to develop better predictions of soil erosion. There is also a growing understanding amongst many environmental scientists that developing more robust analyses of model predictions requires an evaluation of the uncertainties involved in the modelling process (Beven & Freer, 2001; Krueger *et al.*, 2007).

Approaches exist that address the difficulty of both defining physically-based models of erosion dynamics and evaluating their uncertainties. For example, Tayfur *et al.* (2003) have explored the simulation of erosion dynamics using fuzzy modelling approaches. Others have chosen to concentrate on the uncertainties in the data directly, for example using bootstrapping methods to assess rating curve and sediment load dynamics, showing considerable uncertainties in this information (Rustomji & Prosser, 2001). In a series of papers, Krueger *et al.* (2007, 2009, 2010) demonstrate the importance of taking into account uncertainties in hydrological measurements as well as those associated with models. Hydrographs are not error-free, and Krueger *et al.* (2009) demonstrate that errors may be significant (see Section 5.4). In such cases, quantifying uncertainties in data may be important to define correctly the appropriate metric (objective function or performance measure) to assess the quality of model simulations. However, characterizing measurement uncertainty in the form of probability distributions, fuzzy numbers, error intervals, or the like, requires extended measurement efforts if the uncertainty estimates are to bear any relation to actual properties in the field.

In this chapter we adopt the extended Generalised Likelihood Uncertainty Estimation (GLUE) methodology (Beven, 2006; see also sections 4.5 and 4.6) where the evaluation of the model predictions takes into account both the uncertainties in the model and the observed data. Although the examination of model uncertainty for highly parameterized hydrological models is now becoming common in the hydrological literature, to our knowledge this marks one of only a few attempts to vary a large number of parameters in a highly parameterized erosion model (see Brazier *et al.* (2000), Hantush & Kalin (2005) and Wei *et al.* (2008) for other examples). In this study, varying all of EUROSEM's parameters was still beyond our computational power, but such a comprehensive treatment of parameter uncertainties is not necessary given our understanding of previously published parameter sensitivities. Here, we take the next step towards our longer-term objective of evaluating erosion models with

Table 5.1 Published evaluations of the European Soil Erosion model.

Author	Country	Scale and data source	Evaluation methodology	Results
Quinton (1997)	UK	Plot scale (0.1 ha), event-based data (six hydrographs and sedigraphs) from arable agriculture.	Four parameters sampled from measured distributions and model run for all combinations of parameters. Evaluated based on observed data coincidence with 95% confidence intervals.	Observed data within 95% confidence limits, but confidence limits wide.
Quinton & Morgan (1998)	US	4.78 ha watershed in Ohio, event totals.	7 training storms and 4 test storms. Comparison of event totals and a visual comparison of hydrographs.	Performed best on higher magnitude events, and poorly on dynamic data. Hampered by low slopes (<1%) and poorly described geometry. It later became apparent that there were obstructions in the channel of this watershed that had not been documented prior to the simulation study.
Folly et al. (1999)	Netherlands	45 ha catchment, loess soil, mixed agriculture and natural rainfall.	Calibrated against five storms of varying magnitude and evaluated against five more storms.	Performed better with short duration events. Lack of crusting routines identified as problematic.
Veihe et al. (2001)	Costa Rica	2.5 × 1.5 m rainfall simulator plots on Dystrustept with maize, grass and no cover, and Usthotent with no cover.	Calibrated against one plot for each land use and model applied to four remaining plots in each land use. Uncertainty considered by modifying key parameters within measured ranges.	Better performance on bare plots. Poor evaluation results on vegetated surfaces. Attributed to spatial variability in hydraulic characteristics and differences between calibration and evaluation plots.
Veihe et al. (2001)	Nicaragua	22.8 × 1.5 m natural rainfall plots on bare Cambisol.	Model calibrated against total runoff and total soil loss data for events in 1993 and applied to events in 1994 and 1995.	Good performance on annual totals.
Veihe et al. (2001)	Mexico	1 × 1 m rainfall simulator plots on Bare Vitrand.	Calibrated on three plots and evaluated against results from remaining plots.	Total discharge and soil loss simulated reasonably well. Time to start of erosion poorly simulated and attributed to lack of soil crusting algorithm.
Cai et al. (2005)	China	10 × 2 m rainfall simulator plots on sandy clay loam soil with four treatments: uncultivated; contour only; contour + hedgerows; and contour + hedgerows + fertilizer.	Calibrated against one experimental run for each treatment and evaluated against two other runs.	Problems with timings and rates of erosion, but excellent fit to total runoff and erosion.
Mati et al. (2006)	Kenya	10 × 2 m erosion plots within the Embori (barley) and Mukogodo (bare and grass) catchments.	Model evaluation carried out with no calibration. Ten events at Embori and 24 plot events at Mukodoogo.	Good correlations between observed and predicted values at Embori. Model could not reproduce runoff and erosion at Mukogodo under different vegetation covers.

all uncertainties evaluated. This objective seems desirable if model capabilities and weaknesses are to be properly benchmarked, therefore guiding us to improving models and taking improved field measurements which ultimately allow uncertainties in model predictions to be reduced.

5.2 Description of the European Soil Erosion Model (EUROSEM) Structure and Operation

EUROSEM is a process-based model. It uses mathematical expressions to represent the processes of erosion that take place over a single event. The model is applicable to plots, fields and small catchments, but because of the number of parameters required, it is not suitable for application to larger catchments. Each catchment is split into elements for which uniform properties are assumed; these are then linked together to form a network of planes and channels (Fig. 5.1). Each element requires 37 parameters that describe its soil, vegetation, micro-topography, size and slope. Rainfall is entered as break-point data (time-dependent) and different rain gauges can be assigned to different elements within the catchment, allowing the simulation of a storm passing across the study area. EUROSEM is event-based and uses short time steps (e.g. one minute) to model the processes of soil erosion. Simulations rely on the generation of sediment and runoff

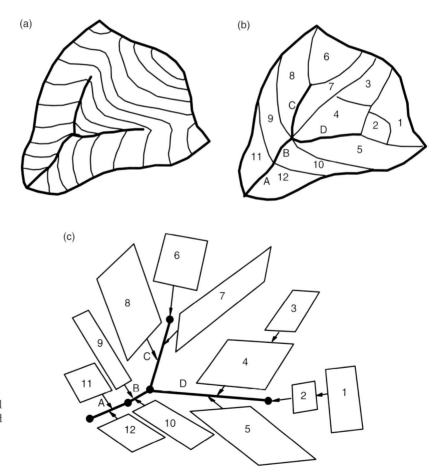

Fig. 5.1 Illustration of the decomposition of a natural catchment into a simulated network of elements and channels (from Morgan *et al.*, 1998b).

Table 5.2 Operating equations of EUROSEM (from Quinton, 1997, where details of the sources cited in column 4 can be found).

No.	Model subroutine and key equation	Definition of terms	Source
1	Interception $I_c = RP_c$	I_c is the depth of rainfall intercepted (mm), R the rainfall (mm) and P_c the percentage canopy cover expressed as a ratio.	Merriam (1973)
2	Interception storage $I_s = I_x \left(1 - e^{\frac{R_c}{I_x}}\right)$	I_s is the interception storage (mm), I_x the maximum interception storage (mm) and R_c the cumulative rainfall (mm).	van Elewijck (1989a,b)
3	Stemflow for grasses $S_f = 0.5 T_{if} \left(\cos. P_a . \sin^2 P_a\right)$	P_a is the average acute angle (degrees) of the plant stems to the ground surface and T_{if} (mm) is the temporarily intercepted throughfall.	
4	and for other plant species $S_f = 0.5 T_{if} \left(\cos. P_a\right)$		
5	Infiltration $f_c = K_s \left(1 - e^{-F/B}\right)^{-1}$	f_c is the infiltration capacity (mm h^{-1}), F the amount of rain already absorbed into the soil (mm), and K_s is the saturated hydraulic conductivity (mm h^{-1}); B and G are defined below.	Smith & Parlange (1978)
6	where $B = G(\theta_s - \theta_i)$	$(\theta_s - \theta_i)$ is the saturation deficit and G is the capillary drive (mm).	
7	and $G = \frac{1}{K_{s,\infty}} \int_{0}^{\infty} K(\psi) \, d(\psi)$	$K(\psi)$ is a hydraulic conductivity function and ψ is the soil matric potential.	Holtan (1961)
8	Modification of infiltration by vegetation $K_{sv} = K_s \left(1 - P_b\right)$	P_b is the plant basal area ratio, K_{sv} is the modified value of saturated hydraulic conductivity.	Woolhiser et al (1989)
9	Surface runoff rating equation $Q = a \, h^m$	Q is discharge (m^2 s^{-1}) and h is the depth of flow (m) a and m are defined below.	
10	where $a = \frac{s^{0.5}}{n}$	s is the slope (m m^{-1}) and n the value of Manning's n	
11	and $m = \frac{5}{3}$		
12	Surface runoff continuity equation $\dfrac{\partial h}{\partial t} + \dfrac{\partial Q}{\partial x} = q_{(x,t)}$	q is the lateral inflow rate (m^3 s^{-1}).	Woolhiser et al. (1989)
13	Soil detachment by raindrop impact $D_s = kK_e e^{-bh}$	K_e is the kinetic energy of the rain reaching the ground surface (J m^{-2}), k the detachability of the soil (g J^{-1}), h is the depth of surface water (m) and b is an exponent (1 to 3).	
14	Kinetic energy: of direct throughfall $K_e(DT) = 8.95 + (8.44 \log I)$	$K_e(DT)$ represents the kinetic energy (J m^{-2}) per mm of direct throughfall and I is the rainfall intensity (mm h^{-1}).	Brandt (1989)
15	of leaf drainage $K_e(LD) = (15.8.P_h^{0.5}) - 5.87$	$K_e(LD)$ is the kinetic energy (J m^{-2}) per mm of leaf drainage and P_h the height of the plant canopy (m).	Brandt (1989)
16	Soil detachment by flow $D_F = \beta w v_s (C_m - C)$	C_m is the equilibrium sediment concentration in the flow, C the actual sediment concentration, β the resistance of the soil to detachment, and v_s the settling velocity of the particles (m s^{-1}).	Smith et al. (1995)
17	where $C_m = \alpha(Su - 0.4)^\gamma$	S is the slope of the land (m m^{-1}), u is the flow velocity (m s^{-1}) and α and γ are coefficients which vary with the median particle size of the soil	Govers (1990)
18	Sediment continuity equation $\dfrac{\partial(AC)}{\partial t} + \dfrac{\partial(QC)}{\partial x} - e(x,t) = q_s(x,t)$	A is the cross-sectional area of the flow (m^2), q_s is the lateral input or extraction of sediment per unit length of flow (m^3 s^{-1}), e is the net pick-up rate of sediment from the bed per unit length of flow (m^3 s^{-1}) and x is the horizontal distance (m).	Bennett (1974), Kirkby (1980), Woolhiser et al. (1989)

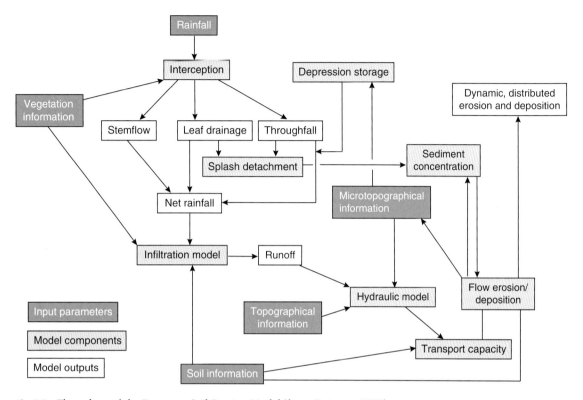

Fig. 5.2 Flow chart of the European Soil Erosion Model (from Quinton, 1997).

and its routing through the virtual landscape. This routing relies upon solving the dynamic mass balance equation (equation 18, Table 5.2) numerically, and allows the volume of water and concentration of sediment passing any given point in the landscape at any given time through the storm to be derived. Individual processes are represented by a series of algorithms (Table 5.2). These processes are linked together as represented in the model flow chart (Fig. 5.2). Rain falling on the plant canopy is split into throughfall, storage and leaf drainage. The drainage term is split between stemflow and leaf-fall. The kinetic energy of the throughfall and the leaf fall are calculated separately. The kinetic energy is then used to calculate the splash detachment. Water falling on the soil surface infiltrates into the soil at a rate calculated using the Smith-Parlange equation (equation 5, Table 5.2). The water that does not infiltrate is stored in

depressions on the soil surface; once these are filled to capacity they can over-top and runoff is routed over the surface using the kinematic wave equation. The moving water may erode the soil surface, carry sediment and deposit it. These processes are modelled as a continuous exchange of material between the flow and the surface through both erosion and deposition (equation 16, Table 5.2), with the transport capacity of the flow described as the point of balance between two continuously interacting processes.

EUROSEM also has the capability to simulate flow and erosion processes in concentrated flow paths, such as rills. However, such flow paths need to be predefined for each planar element: the model cannot generate rills. The processes which are active in the flow paths and in the areas between them are controlled by the properties of those areas: there is no artificial distinction between inter-rill and rill areas, and these terms are only used to distinguish between the hydraulic properties of the different areas. Shallow flows and low slopes on the areas feeding the rills with water and sediment tend to produce less sediment than the concentrated flow paths where water is deeper and moving more quickly.

5.3   Methodology

5.3.1   Field site

The field site was a fodder field (Fig. 5.3) with an estimated area of 5.44–6.81 ha, depending on whether a potential contributing area upslope is included or not, within the Den Brook catchment, Devon, UK (UK grid reference: SX 67712 99685). Note that, as is often the case in agricultural catchments, delineating the catchment boundaries and area is itself an uncertain process – this is explored more fully in Section 5.5. The catchment is described in more detail by Page *et al.* (2005). Den Brook has an estimated area of 46.43 ha, soils are poorly draining, often-waterlogged clays (clayey non-calcareous pelostagnogley (Avery, 1980), a Typic Haplaquept (USDA, 1975) of the Hallsworth Series. Land use at Den Brook is intensively managed grassland, with the grass used for grazing animals, silage and fodder production. Increasingly, farmers in the UK are incorporating fodder fields into their grassland farms to provide food for the stock over the winter months. Such fields are vulnerable to soil erosion because of poor canopy cover during the winter and spring, combined with row cultivation. At Den Brook, kale (*Brassica oleracea*) was grown with grass undersown. Sheep and lambs were allowed to graze the field during the winter months. Den Brook receives substantial amounts of rainfall, with a 40-year mean annual total of 1050 mm (Page *et al.*, 2005).

Rainfall was measured using a tipping-bucket rain gauge (Rainwise, Bar Harbor, ME) which recorded the total number of tips per minute (each tip equivalent to 0.254 mm rainfall). Discharge from the fodder field was measured as it flowed over a V-notch weir at the lowest corner of the field. Depth of water flowing through

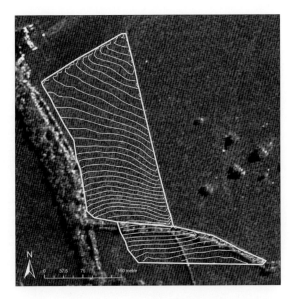

Fig. 5.3 Outline of the simulated field with 1 m contours. The measurement flume is on the northeast corner of the field. The contributing area considering the field boundaries is 5.44 ha, but it is likely that the area further upslope overflows into the field during large storm events, making the contributing area 6.81 ha. The NEXTMap Britain orthorectified radar image Intermap Technologies (2007) was provided courtesy of NERC via the NERC Earth Observation Data Centre (NEODC).

the weir was recorded using a bubble meter logged by an ISCO 24 bottle pumped sampler, which also took water samples from which suspended sediment concentrations were determined gravimetrically. The weir was calibrated by the bucket and stopwatch method and the uncertainties surrounding measurements at the same stage quantified using the fuzzy rating curve approach of Krueger *et al.* (2010) (an extension of a concept presented by Pappenberger *et al.*, 2006). Suspended sediment concentration uncertainty was not quantified in this study due to the absence of field measurements to do so, but the potential for this type of measurement uncertainty was acknowledged in the course of model evaluation.

Table 5.3 Parameter values used for the EUROSEM simulations. The 'best estimate' reflects parameter values derived from measurements and those estimated from the EUROSEM manual (Morgan *et al.*, 1998b). A 'Y' indicates that the parameter was sampled for the GLUE analysis, an 'N' indicates that the parameter was held constant at its best estimate. Minimum and maximum values were derived from field measurements and from the EUROSEM manual and were used to construct the uniform distributions from which parameter values were sampled for the GLUE analysis.

Parameter	Best estimate	Sampled	Minimum	Maximum
Length of element (m)	360	Y	300	900
Width of element (m)	125	N		
Slope (%)	0.08	N		
Manning's *n* inter-rill	0.02	Y	0.01	0.035
Saturated hydraulic conductivity (mm hr^{-1})	0.61	Y	0.1	1
Capillary drive (mm)	806	Y	0.1	1000
Porosity (v v^{-1})	0.44	Y	0.3	0.6
Initial moisture content (% saturated moisture content)	0.32	Y	0.1	1
Saturated moisture content (% porosity)	0.34	Y	0.5	1
Maximum interception depth (mm)	0.01	N		
Random roughness (mm)	20.44	Y	1	30
Inter-rill slope (%)	0.08	N		
Vegetation cover (%)	15	N		
Plant angle (degrees)	55	N		
Plant basal area (%)	5	N		
Plant height (m)	0.03	N		
D50 (μm)	30	Y	20	250
Erodibility (g J^{-1})	1.9	Y	0.1	5
Cohesion (kPa)	13.7	Y	10	50

5.3.2 Model parameterization

EUROSEM was evaluated separately for two events: 17 November 2006 (event 1) and 2 February 2007 (event 2). The two events were chosen to represent contrasting storm types, a single-peaked event (event 1) and a double-peaked event (event 2). For each of the events the soils were near saturation due to preceding rainfall, and vegetation cover was poor. Model parameterization for GLUE simulations requires the definition of ranges, which are then sampled (see Section 5.3.3). For the purposes of this study these definitions were based on a combination of field measured values and those suggested for similar sites in the EUROSEM user guide (Morgan *et al.*, 1998b) (Table 5.3). Parameters which previous EUROSEM model studies had shown to produce a sensitive response in simulated model outputs (Veihe & Quinton, 2000; Veihe *et al.*, 2000) were

sampled as uniform distributions, with minima and maxima reflecting our *a priori* uncertainty about feasible values. All other parameters were set at default 'best estimate' values. Rainfall was described using 1-minute data derived from the rain-gauge record.

5.3.3 The Generalised Likelihood Uncertainty Estimation (GLUE) methodology

GLUE was set up in this paper as summarized in Table 5.4 (following Krueger *et al.*, 2009). In brief, initial (prior) distributions of parameters (noted above) are determined, either from measured data, literature data or, as in this study, our own expert opinion. These distributions are then randomly sampled and a model simulation is run, varying all chosen (sensitive) parameters at the same time. The model results are compared with the measured data (discharge and sediment)

Table 5.4 Generalised Likelihood Uncertainty Estimation (GLUE) setup. All subjective assumptions are made explicit: the choice of the prior parameter distribution (see Table 5.3 for its bounds); the choice of performance measures; the number of parameter sets sampled; the initial performance thresholds; and whether or not input and output uncertainties are acknowledged.

Parameter sets	Prior parameter distribution	Input uncertainty
10^9	uniform	no

Output uncertainty		Initial performance threshold	
Discharge	Suspended solids	Discharge	Suspended solids
yes	no	0.4	150

Performance measure for discharge

$$\overline{D}_1 = \frac{\sum_{i=1}^{N} \left| \dfrac{Q_{sim,i} - Q_{obs,i}}{\sup(Q_{obs,i}) - \inf(Q_{obs,i})} \right|}{N}$$

where $Q_{sim,i}$ and $Q_{obs,i}$ are simulated and observed discharges at time-step $i = 1, \ldots, N$, $\sup(Q_{obs,i})$ and $\inf(Q_{obs,i})$ are upper and lower interval bounds and

$$Q_{sim,i} - Q_{obs,i} = \begin{cases} Q_{sim,i} - \sup(Q_{obs,i}) & \text{if} \quad Q_{sim,i} > \sup(Q_{obs,i}) \\ 0 & \text{if} \quad \inf(Q_{obs,i}) \leq Q_{sim,i} \leq \sup(Q_{obs,i}) \\ Q_{sim,i} - \inf(Q_{obs,i}) & \text{if} \quad Q_{sim,i} < \inf(Q_{obs,i}) \end{cases}$$

Performance measure for suspended solids

$$MAE = \frac{\sum_{i=1}^{N} |C_{sim,i} - C_{obs,i}|}{N}$$

where $C_{sim,i}$ and $C_{obs,i}$ are simulated and observed concentrations at time-step $i = 1, \ldots, N$.

independently and performance statistics calculated. The performance statistics are compared with a predefined performance threshold (see below). If the simulation exceeds the required performance threshold then the parameter set is retained and declared behavioural (or good) at simulating the observed behaviour; if it is lower than the performance threshold the parameter set is rejected. The procedure is then repeated a large number of times (one billion randomly selected parameter sets per event, in this study). This process is called Monte Carlo simulation. The retained parameter sets, weighted by their performance, are then used to construct new (posterior) parameter distributions. The model output for the multiple runs can be used to demonstrate uncertainty in model output and model parameters. The general framework for refining simulations with new data can be summarized as follows:

• The initial uniformly weighted prior parameter sets (from the Monte Carlo sampling) are weighted (updated) using the performance statistic calculated from the 1st storm event information (in this case discharge and sediment). These provide the first iteration of posterior parameter distributions conditioned on this event information, in that only the good simulations (as defined by the performance measure) are retained for further simulations.

• These posterior parameter distributions can then be subsequently updated using data from second and/or additional events in the same way that the initial uniform prior distributions were updated previously (Beven & Freer, 2001).

This methodology means that as new data become available, the posterior parameter distributions and prediction uncertainties can be continuously refined. If any of the sampled parameter sets continuously provide good simulations for available data (events) they will be retained as simulators of the observed system, but importantly their 'weight' will evolve dependent on the total number of simulations left and their individual ranking regarding their overall predictive capability.

Selecting a model performance measure and a threshold of model rejection is a difficult, sometimes arbitrary procedure. Mismatches between simulations and observations need to be tolerated when accommodating the effects of input data uncertainty in conjunction with other data uncertainties not accounted for (see discussion in Beven, 2006). Simulation performance is often in the eye of the beholder: what appears to be a good simulation to one model user is a poor simulation to another, depending on the level of confidence they are willing to accept in model output. Also, since models often fail to simulate all aspects the system dynamics equally well, the choice of performance measure reflects a judgment on what are the most important dynamics to get right. These considerations should reflect the priorities of the model application and be explicitly defined. We suggest that objective model benchmarks are often missing when setting up performance metrics, and we feel that a more thoughtful approach should be applied. Ultimately the user of the model predictions needs to define the performance limits for the model simulations. Therefore performance measures should, where objective assessments can be made, consider the quality of the observed data and so the potential uncertainties in the data series. In this study we are explicit about the model performance and its relation to the observed data for which we have quantified the uncertainties from field measurements where possible. To assess model performance we used a time-step based absolute deviation between simulations and observations of both discharge (Q) and suspended sediment concentration (SS). In the case of Q, this deviation was normalized by the measurement uncertainty

interval for each time-step which served as a model-independent benchmark. No such benchmark was available, or could reasonably be assumed, for SS. Being pragmatic, we therefore apply more typical forms of model performance when comparing with the SS information, but ensuring we are more relaxed about the quality of the model fit because we may mis-specify the error characteristics in the data. The distribution of deviations across the time-steps was summarized by mean deviations and percentiles. For Q, the mean absolute deviation $\bar{D}_{\|}$ (Table 5.4) was ultimately used as the performance measure sought to be minimized. This implies the objective of seeking model simulations that are, on average, close to or within the observed intervals across all time-steps, although occasional large deviations may be tolerated. This is a reasonable starting objective for the purpose of this study, although additional performance measures should ideally be used to aid model diagnostics (Freer *et al.*, 1996; Clark *et al.*, 2008). The corresponding performance measure for SS without measurement of uncertainty intervals is the mean absolute error *MAE* (Table 5.4). Performance thresholds for both measures were set initially for the purpose of visual model diagnostics, which evaluated whether the retained model realizations could be considered 'behavioural' in the sense of Beven (2006). For further discussion of this type of exploratory analysis, the reader is referred to Krueger *et al.* (2009). The model evaluation against the Q and SS data was carried out sequentially, so that the sampled parameter sets were first rejected or retained based on the $\bar{D}_{\|}$ performance threshold of 0.4 for Q, and the remaining sets were then updated in the same way based on the *MAE* performance threshold of 150 for SS. This reflects our understanding of hydrology being the dominant driver of erosion, and hence hydrology should be modelled sufficiently well before further erosion processes are evaluated to ensure that the model simulates erosion 'for the right reasons' (Quinton, 1994; Brazier *et al.*, 2000). The parameter sets retained for the first event were further updated by conditioning on the second event.

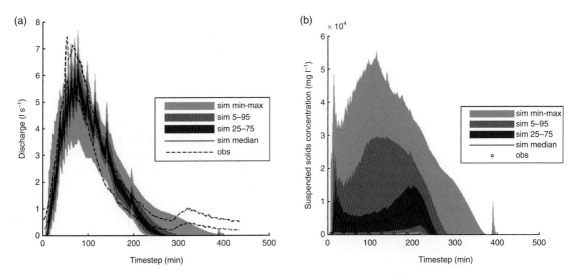

Fig. 5.4 Model performance for event 1 (17 November 2006) for (a) discharge and (b) sediment concentration. The simulations comprise those sampled parameter sets that yielded results better than the discharge performance threshold of $\bar{D}_{\parallel} = 0.4$, with each parameter set weighted by $\bar{D}_{\parallel}$ to form the GLUE likelihood distributions of simulations. Note the sedigraph is shown before conditioning on the sediment data, as the latter led us to reject all parameter sets sampled based on the sediment performance threshold of $MAE = 150$.

5.4 Results

Figure 5.4 shows the simulated and observed hydrograph and sedigraph for event 1, where the simulations comprise the sampled parameter sets yielding results better than the discharge performance threshold of $\bar{D}_{\parallel} = 0.4$ (note, as mentioned before, that the choice of this performance threshold is arbitrary at this stage but, as will be seen later, no better performance than $\bar{D}_{\parallel} = 0.3$ could be achieved for event 1 and this puts our initial threshold into perspective). For the discharge the model response follows the observed interval closely, apart from the initial Q and the final rise at approximately 350 minutes. The model response is also notably more 'spiky' than the observed one, reacting more quickly to rainfall. The sediment simulations, on the other hand, are, on average, one order of magnitude higher than the observed concentrations. Therefore, none of the parameter sets were retained further when evaluated against the SS performance threshold of $MAE = 150$, leading us to reject all the models. Note this MAE

performance threshold might in fact appear too relaxed by conventional expectations of model performance, but errors in event sediment data are likely to be high, and so this threshold tries to reflect the uncertainty in the data with no direct knowledge regarding the error characteristics. Importantly within our extended GLUE uncertainty framework, based on the evidence thus far presented, all models have been rejected and so further constraining of model simulations, as described previously, is not possible. However, for completeness, discussion and assessment of the variation in model dynamics for the different events, we present, independently, simulations from the same set of sampled parameter sets for the second storm below.

For the second, twin-peaked, event (Fig. 5.5), the EUROSEM discharge output again does relatively well at following the observed data, although it drops below the observed interval in between the consecutive peaks and underestimates the final rise of the second peak, resulting in a mismatch of shape between the observations and

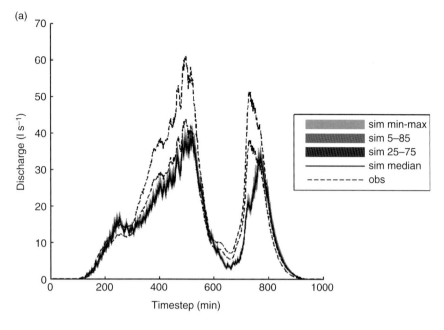

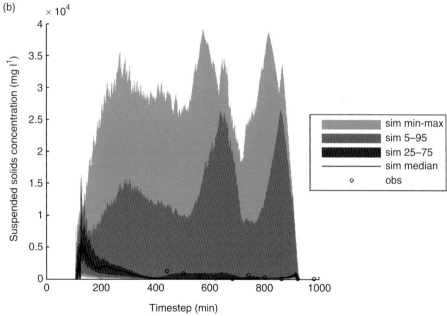

Fig. 5.5 Model performance for event 2 (2 February 2007) for (a) discharge and (b) sediment concentration. The simulations comprise those sampled parameter sets that yielded results better than the discharge performance threshold of $\bar{D}_{\parallel} = 0.4$, with each parameter set weighted by $\bar{D}_{\parallel}$ to form the GLUE likelihood distributions of simulations. Note the sedigraph is shown before conditioning on the sediment data, as the latter led us to reject all parameter sets sampled, based on the sediment performance threshold of $MAE = 150$.

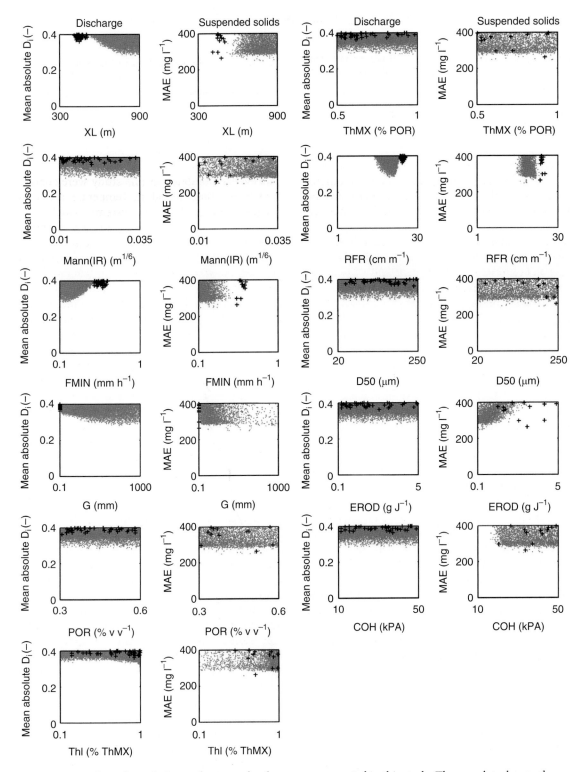

Fig. 5.6 Dotty plots of simulation performance for the parameters varied in this study. The grey dots denote the 17 November 06 event and the black crosses the 2 February 07 event. The parameter sets are those retained based on the 0.4 Q performance threshold applied to both events separately.

simulations of the second peak. However, the timing of the main peak is well simulated and so is the initial rise of the hydrograph. The 'spikiness' of the modelled discharge response is again apparent. In terms of the *SS* simulations, results are not different to event 1 and, again, no parameter set survived the test against the sediment data.

Figure 5.6 shows the relationship between model performance for *Q* and *SS* and parameter values for both events, with only those simulations which pass the $\bar{D}_{\parallel} = 0.4$ *Q* performance criterion displayed. It is clear from Fig. 5.6 that many more discharge simulations performed better for event 1 than for event 2. However, it is also clear that no sediment simulations performed better than the *MAE* threshold of 150 mg l⁻¹. Although rejected as non-behavioural, by showing the model realizations with respect to suspended sediment error we are able to see which parameters the model is sensitive to. Few of the parameters have much influence over the model's performance for the simulations we ran. Key parameters for discharge are: saturated hydraulic conductivity (*FMIN*), field length (*XL*; a surrogate for catchment area), capillary drive (*G*) and surface roughness (*RFR*). For sediment we can add cohesion (*COH*) and erodibility (*EROD*) to the list. The dotty plots show different responses for the two events. The better-performing values of saturated hydraulic conductivity, surface roughness and erodibility are higher for event 2 than for event 1, while those of slope length and capillary drive are lower for event 2 than for event 1. Given that the true area of the field lies somewhere within the sampled range of slope length times slope width, which might not coincide with the area of highest performance, it is clear that the slope length parameter can compensate for model deficiencies here. Indeed, for a true area corresponding to a slope length towards the low end of the sampled range, model performance would be worse than the chosen $\bar{D}_{\parallel} = 0.4$ threshold. It is also clear that updating the posterior parameter distribution of event 1 with that of event 2 results in rejection of all sampled parameter sets due to the non-overlapping distributions in Fig. 5.6, i.e. the two events cannot be simulated better than

$\bar{D}_{\parallel} = 0.4$ with the same parameter sets. Lowering the performance threshold would result in overlapping parameter distributions, but at the expense of simulation accuracy of individual events.

5.5   Discussion

The events simulated in this study were characteristic of those found in temperate areas in response to frontal rainfall systems, which typically generate long-duration but low-intensity rainstorms. Simulating such events required parameter sets that minimized both infiltration and erosion, since runoff had to be generated, but erosion rates were low, probably due to the cohesive nature of the soil. Most of the parameter values identified are sensible for the conditions described above. However, as the better-performing parameter regions for the two simulated events did not always overlap, we cannot identify consistent parameter sets for the conditions at Den Brook even if variability in initial conditions is taken into account. Therefore we would not recommend 'optimizing' model parameters as a general approach, as we also demonstrated that many parameter values gave similarly good simulations. Some of the most sensitive parameters had non-overlapping behavioural distributions for the two events (when considering only discharge), particularly *RFR*, *FMIN*, *G* and *XL*. This indicates that both events could not be described with the same sets of parameters using our behavioural threshold, unless the level of model performance for individual events was compromised. This may be due to errors in the model structure, the input data (which in this study is not accounted for) or real variations in hydraulic and soil surface microtopographical characteristics between events. As EUROSEM is an event-based model, it relies on the user to recognize changes in parameters in order to take count of surface changes. However, the changes in the optimal values of *FMIN* and *RFR* are counter-intuitive: both increase between event 1 and 2 when it would be expected that the soil surface would become smoother and less permeable due to greater exposure to raindrop impact. It is possible that the presence of sheep on the field during the simulation period may have significantly changed the soil structure between the two events. However, it is worth noting that none of the current generation hydrological models is fit for simulating the transient nature of some of the hydraulic processes at the field scale and beyond.

Another area which warrants further discussion is the difference in identified slope length for both events. Slope length controls the catchment area and therefore the volume of water falling on the catchment, as well as influencing the sediment concentration (Smith *et al.*, 1995). As EUROSEM does not simulate the flows of water in the subsurface, soil hydraulic properties across the field are uniform until water begins to flow, so runoff is generated at all points across the element at the same point in time. It is likely in the simulated storms that we have variable source areas which cannot be simulated internally by EUROSEM, and that by using different slope lengths we are able to provide a representation of this. For event 2, the identified XL parameter region around 400 m (Fig. 5.6), together with the assumed field width of 125 m, matches the actual area of the simulated field of 5.44 ha (Fig. 5.3). In contrast, for event 1, the identified XL in excess of 800 m (Fig. 5.6) yields a contributing area beyond the 6.81 ha which could be considered an extended contributing area when large storms may result in an overflow of an otherwise disconnected area upslope (Fig. 5.3). However, by compromising model performance, it is possible to match this extended area with a slope length of around 550 m, which is still below the Q performance threshold of $\bar{D}_{\parallel} = 0.4$ (Fig 5.6). It was valuable for this study to allow the slope length to vary in order to analyse the site characteristic of a potentially highly variable contributing area. However, we do not recommend using slope length as a means for compensating for model structural error without carefully considering the field situation (see Chapter 13 for a further example of the importance of using field evidence for determining the effective contributing area when parameterizing erosion models).

In contrast to the reasonable hydrological simulations (at least for individual events and with a high level of model flexibility in adjusting slope length), EUROSEM failed to simulate the measured sediment concentration data within an order of magnitude of the observed data. One possible explanation for the model's difficulties may lie with the processes operating at Den Brook. Although runoff volumes in the catchment are comparable with those measured at other UK sites, sediment concentrations observed for the events simulated in this study are at the lower end of sediment concentrations measured from similar field areas in the UK. In work on unbounded hillsides at three sites in the UK, Deasy *et al.* (2008) reported mean annual sediment concentrations of between 132 and 7226 mg l^{-1}. The lower concentrations of this range were associated with clay soils under arable agriculture, but on steeper slopes than those at Den Brook. It is possible that EUROSEM does not simulate low-energy erosion processes well, or that processes are occurring, such as the mobilization of soil particles by physio-chemical dispersion, that EUROSEM does not simulate. If this were the case it would be right to question whether applying EUROSEM to Den Brook represents a fair test of the model, and it might be concluded that it was not. However, it does illustrate how physically-based models force us to think critically about the system that we are trying to simulate and suggests that if we wish to simulate erosion in areas such as Den Brook, we will need to develop new descriptions of erosion processes that are better suited to such environments.

Whilst this chapter was not principally about model uncertainty, it does illustrate that a large number of different parameter sets, from often distinctly different regions in the sampled parameter space, are able to simulate the observed hydrological response, but not the observed sediment dynamics. The authors therefore echo the call of previous authors (Beven & Binley, 1992; Quinton, 1997; Brazier *et al.*, 2000) for erosion modellers to adopt uncertainty representation techniques, which are now prevalent in hydrological modelling. Without exploring the model simulations in detail, by sampling the possible parameter space, we would not be able to fully state the reasons why the model did not provide behavioural simulations over these two events. Such studies need to be encouraged to benchmark the predictive capability of our models and improve and refine their conceptualization of soil erosion processes. The simulations reported in this chapter took approximately 5.5 days to run on a high performance cluster. This process was repeated a number of times as we learnt from our initial attempts. Overall the simulations took in the order of five months to complete on a part-time basis. Such an overhead is unthinkable outside of a research arena. However, computers are getting faster – a study such as this would have been unfeasible when Quinton carried out his initial uncertainty evaluation of EUROSEM in 1997 – and we can expect the time needed for such simulations to fall. Thus model *developers* should be encouraged to quantify erosion model performance and be explicit about the quality of model predictions against observed data. However, for the land manager or model *user*, deploying the methods we describe here is not viable. However, this is not to say that model users should not try to incorporate uncertainty into their model application. We suggest that, at a minimum, model users should:

• assess the uncertainty in their measurements and express them in comparison with model output;
• be guided by reports of model sensitivity to vary as large a number of parameters as can be afforded in order to generate maximum, minimum and most likely model predictions;
• decide in advance what makes an acceptable prediction, and reject and refine the model if it does not meet these conditions;
• avoid using models to underpin decision-making without consideration of uncertainty where at all possible.

5.6   Conclusion

This chapter described an application of EUROSEM to a small catchment in Devon, UK. The chapter demonstrates that EUROSEM is sensitive to some

of its parameters and insensitive to many others. The use of Generalised Likelihood Uncertainty Estimation (GLUE) illustrated the uncertainty in model output. It was important to note that measurements of observed data are also not error-free, and that while we were able to incorporate such error into the assessment of the hydrological performance, such information was not available for the sediment dynamics. In general EUROSEM was able to simulate the hydrology, but the two events studied required different parameter sets. Sediment concentrations were simulated an order of magnitude higher than observed. We suggest that this may be because EUROSEM does not simulate well the erosion processes occurring in low-energy rainfall and overland flow, and that process descriptions within the model may need to be revisited. Given the amount of computer time required for this work, we suggest that this method may not always be appropriate for land managers or model users, but we do encourage them to include model and data uncertainty estimation when evaluating models and to consider model uncertainty when making decisions about land management. Although two billion simulations were carried out for this study, considerably fewer simulations would still provide an effective framework for evaluating uncertainty when modelling many land management scenarios.

Acknowledgements

This research was part-funded by the UK Department for Environment, Food and Rural Affairs (Defra) with the project number PE0120, the UK Natural Environment Research Council (NERC), Flood Risk from Extreme Events (FREE) programme with the grant number NE/E002242/1, and the UK Research Councils, Rural Economy and Land Use (RELU) programme with the grant number RES-229-25-0009-A.

References

Avery, B.W. (1980) A *Soil Classification for England and Wales*. Rothamsted Experimental Station, Harpenden, UK.

Beven, K. (2006) A manifesto for the equifinality thesis. *Journal of Hydrology* **320**: 18–36.

Beven, K.J. & Binley, A. (1992) The future of distributed models – model calibration and uncertainty prediction. *Hydrological Processes* **6**: 279–98.

Beven, K.J. & Freer. J. (2001) Equifinality, data assimilation, and uncertainty estimation in mechanistic modelling of complex environmental systems using the GLUE methodology. *Journal of Hydrology* **249**: 11–29.

Borselli, L., Sanchis, P.S., Cassi, P., *et al.* (2008) EUROSEM (2008): a re-engineered and restructured software tool for soil erosion scenario analysis and sediment connectivity assessment. In de Oliveira Alves Coelho, C. (ed.), *On- and off-site environmental impacts of runoff and erosion*. Universidade de Aveiro, Aveiro, Portugal.

Brazier, R.E., Beven, K.J., Freer, J. & Rowan, J.S. (2000) Equifinality and uncertainty in physically based soil erosion models: Application of the GLUE methodology to WEPP–the water erosion prediction project–for sites in the UK and USA. *Earth Surface Processes and Landforms* **25**: 825–45.

Cai, Q.G., Wang, H., Curtin, D. & Zhu, Y. (2005) Evaluation of the EUROSEM model with single event data on steeplands in the three Gorges reservoir areas, China. *Catena* **59**: 19–33.

Chisci, G. & Morgan, R.P.C. (1986) Modelling soil erosion by water: why and how. In Morgan, R.P.C. & Rickson, R.J. (eds), *Erosion assessment and modelling*. Commission of European Communities Report No. EUR 10860 EN, Office for Official Publications of the European Community, Luxembourg: 237–53.

Clark, M.P., Slater, A.G., Rupp, D.E., *et al.* (2008) Framework for Understanding Structural Errors (FUSE): A modular framework to diagnose differences between hydrological models. *Water Resources Research* **44**: W00B02.

Deasy, C., Quinton, J.N., Silgram, M., *et al.* (2008) *Defra PE0206 Field testing of mitigation options*. Lancaster University, Lancaster, UK.

De Roo, A.P.J., Wesseling, C.G. & Ritsema, C.J. (1996) LISEM: A single-event physically based hydrological and soil erosion model for drainage basins. 1. Theory, input and output. *Hydrological Processes* **10**: 1107–17.

Folly, A., Quinton, J.N. & Smith, R.E. (1999) Evaluation of the EUROSEM model using data from the Catsop watershed, The Netherlands. *Catena* **37**: 507–19.

Freer, J.E., Beven, K.J. & Ambroise, B. (1996) Bayesian estimation of uncertainty in runoff prediction and the value of data: an application of the GLUE approach. *Water Resources Research* **32**: 2161–73.

Hantush, M.M. & Kalin, L. (2005) Uncertainty and sensitivity analysis of runoff and sediment yield in a small agricultural watershed with KINEROS2. *Hydrological Sciences Journal* **50**: 1151–71.

Intermap Technologies. (2007) NEXTMap Britain: Digital terrain mapping of the UK, NERC *Earth Observ. Data Cent.*, Didcot, UK (Available at http://badc.nerc.ac.uk/view/neodc.nerc.ac.uk_AToM_dataen_116583834442118).

Krueger, T., Freer, J., Quinton, J.N. & Macleod, C.J.A. (2007) Processes affecting transfer of sediment and colloids, with associated phosphorus, from intensively farmed grasslands: a critical note on modelling of phosphorus transfers. *Hydrological Processes* **21**: 557–62.

Krueger, T., Quinton, J.N., Freer, J., *et al.* (2009) Uncertainties in data and models to describe agricultural sediment and phosphorus transfer event dynamics. *Journal of Environmental Quality* **38**: 1137–48.

Krueger, T., Freer, J., Quinton, J.N., *et al.* (2010) Ensemble evaluation of hydrological model hypotheses. *Water Resources Research*, 46, W07516, DOI: 10.1029/2009WR007845.

Mati, B.M., Morgan, R.P.C. & Quinton, J.N. (2006) Soil erosion modelling with EUROSEM at Embori and Mukogodo catchments, Kenya. *Earth Surface Processes and Landforms* **31**: 579–88.

Morgan, R.P.C., Quinton, J.N., Smith, R.E., *et al.* (1998a) The European Soil Erosion Model (EUROSEM): A dynamic approach for predicting sediment transport from fields and small catchments. *Earth Surface Processes and Landforms* **23**: 527–44.

Morgan, R.P.C., Quinton, J.N., Smith, R.E., *et al.* (1998b) *The European soil erosion model (EUROSEM): documentation and user guide*. Silsoe College, Cranfield University, Silsoe, Bedford MK45 4DT.

Page, T., Haygarth, P.M., Beven, K.J., *et al.* (2005) Spatial variability of soil phosphorus in relation to the topographic index and critical source areas: sampling for assessing risk to water quality. *Journal of Environmental Quality* **34**: 2263–77.

Pappenberger, F., Matgen, P., Beven, K.J., *et al.* (2006) Influence of uncertain boundary conditions and model structure on flood inundation predictions. *Advances in Water Resources* **29**: 1430–49.

Quinton, J.N. (1994) *The validation of physically-based erosion models – with particular reference to EUROSEM*. PhD Thesis, Cranfield University, Silsoe, Bedford, UK.

Quinton, J.N. (1997) Reducing predictive uncertainty in model simulations: a comparison of two methods using the European Soil Erosion Model (EUROSEM). *Catena* **30**: 101–17.

Quinton, J.N. & Morgan, R.P.C. (1998) EUROSEM: An evaluation with single event data from the C5 watershed, Oklahoma, USA. In Boardman, J. & Favis-Mortlock, D. (eds), *Modelling Soil Erosion by Water*. NATO ASI Series, Series 1: Global Climatic Change, Vol. 55. Springer-Verlag, Berlin: 65–74.

Quinton, J.N. & Catt, J.A. (2004) The effects of minimal tillage and contour cultivation on surface runoff, soil loss and crop yields in the long-term Woburn Erosion Reference Experiment on sandy soil at Woburn, England. *Soil Use and Management* **20**: 343–9.

Rustomji, P. & Prosser, I. (2001) Spatial patterns of sediment delivery to valley floors: sensitivity to sediment transport capacity and hillslope hydrology relations. *Hydrological Processes* **15**: 1003–18.

Smith, R.E., Goodrich, D.C. & Quinton, J.N. (1995) Dynamic, distributed simulation of watershed erosion - the Kineros2 and Eurosem Models. *Journal of Soil and Water Conservation* **50**: 517–20.

Tayfur, G., Ozdemir, S. & Singh, V.P. (2003) Fuzzy logic algorithm for runoff-induced sediment transport from bare soil surfaces. *Advances in Water Resources* **26**: 1249–56.

USDA (1975) *Soil Taxonomy: a basic system for soil classification for making and interpreting soil surveys*. USDA Soil Conservation Service. John Wiley & Sons, New York.

Veihe, A. & Quinton, J. (2000) Sensitivity analysis of EUROSEM using Monte Carlo simulation I: hydrological, soil and vegetation parameters. *Hydrological Processes* **14**: 915–26.

Veihe, A., Quinton, J. & Poesen, J. (2000) Sensitivity analysis of EUROSEM using Monte Carlo simulation II: the effect of rills and rock fragments. *Hydrological Processes* **14**: 927–39.

Veihe, A., Rey, J., Quinton, J.N., *et al.* (2001) Modelling of event-based soil erosion in Costa Rica, Nicaragua and Mexico: evaluation of the EUROSEM model. *Catena* **44**: 187–203.

Wei, H., Nearing, M.A., Stone, J.J. & Breshears, D.D. (2008) A Dual Monte Carlo approach to estimate model uncertainty and its application to the Rangeland Hydrology and Erosion Model. *Transactions of the American Society of Agricultural and Biological Engineering* **51**: 515–20.

6 Scaling Soil Erosion Models in Space and Time

R.E. BRAZIER[1], C.J. HUTTON[1], A.J. PARSONS[2]
AND J. WAINWRIGHT[2]

[1]*School of Geography, University of Exeter, Exeter, UK*
[2]*Department of Geography, University of Sheffield, Sheffield, United Kingdom*

6.1 Introduction

Consider a soil particle sitting on a slope. During a storm event and its immediate aftermath, the particle is subjected to a range of forces that act to detach and transport it both within the slope and ultimately through the catchment in which it is located. These forces will cause the particle to move at a range of velocities and for a range of distances, although for the majority of the time the forces will be insufficient to exceed thresholds for movement and the particle will remain stationary. Soil erosion – detachment, transport and deposition – is made up of very large numbers of these irregular stationary periods and steps. Characterization and prediction of erosion is thus a matter of integrating these very large numbers of individual movements to produce fluxes of sediment movement. Were erosion processes and their direct controls constant, in space and time, the central limit theorem should make this task very simple. Unfortunately they are not, and the problems in characterizing erosion rates are further compounded by spatial and temporal variability in slope and catchment characteristics.

Direct measurement of erosion rates employs two broad techniques. The first considers the change in surface elevation directly using methods such as microtopography meters/erosion bridges

(e.g. Rendell, 1982), or more recently, direct photogrammetry or laser scanning of the surface (e.g. Chandler, 1999). The resulting measurements are thus in units of height/depth of ground gained or lost per unit time $[L\ T^{-1}]$. The second – far more commonly used – employs the direct capture or sampling of sediment passing a point in the landscape. Whereas the first set of techniques comprise point measurements, the second are spatially averaged, albeit not in a straightforward way as will be seen. Direct capture may employ artificial stores (e.g. Gerlach, 1967) or reservoirs and lakes (McManus & Duck, 1985; Rowan *et al.*, 1995; Verstraeten & Poesen, 2000), while sampling may use slot- or wheel-based designs (Brakensiek *et al.*, 1979) or, increasingly, pump samplers originally designed for water samples (Bilotta *et al.*, 2008). A range of different measurement units is typically produced. In artificial sediment traps, the weight of sediment per unit time $[M\ T^{-1}]$ is used, while reservoir or lake studies yield net sedimentation rates $[L\ T^{-1}]$. Depending on the method of sediment capture, slot- or wheel-based samplers may produce weights per unit time, or combined water and sediment samples, as with the pump sampling approach. In the latter, it is common to calculate a sediment concentration $[M\ L^{-3}]$, even though this is based on a direct measurement of the sediment weight, and despite the fact that the suspension is an artificial consequence of the sampling technique. Comparison of these different methodologies shows that quite different measurements are being made, and a number of assumptions are generally necessary to

Handbook of Erosion Modelling, 1st edition. Edited by R.P.C. Morgan and M.A. Nearing. © 2011 Blackwell Publishing Ltd.

convert any of them into the sediment fluxes (usually as mass per unit width per unit time: $M \ L^{-1} \ T^{-1}$) which characterize erosion. However, standard practice has been to convert these measurements *not* into fluxes, but into specific yields. A specific yield calculates erosion as a function of upslope contributing area [$M \ L^{-2} \ T^{-1}$]. At first glance, this approach should account for scale, in that it directly incorporates an upslope area. However, consider again the soil particle moving on a slope. Its movement is controlled by forces generated by local conditions, some of which are independent of upslope conditions (e.g. raindrop detachment in the absence of flow), while others are complex functions of the upslope conditions (e.g. raindrop detachment with flow, flow hydraulics depending on surface microtopography and shape of the upslope area). Under homogeneous conditions, these functions may be reasonably predictable analytically, as discussed below. On slopes and in catchments with varying lithologies, soils, vegetation and land-use practices, not to mention spatially variable rainfall intensities, the relationships tend to be more complicated, and other approaches are required to address them.

Depending on whether measurements are made on a hillslope or at a catchment outlet, it must also be recognized that process domains change. On shorter hillslopes, raindrop-driven processes such as splash and unconcentrated overland-flow erosion are dominant. As slopes get longer, concentrated overland-flow erosion becomes more important, and depending on soil type and land use, subsurface erosion may also occur. Landslides will become more important on steeper slopes, and in steep catchments, undercutting by the channel can mean that landslides are a significant source of sediment reaching the channel system (Ergenzinger, 1992; Korup, 2005). Catchment-based measurements will also incorporate the effect of channel processes such as lateral erosion of banks and floodplains, as well as storage in channels and floodplains, and remobilization of alluvial sediments. For these process-related reasons, catchment-based measurements may be a poor reflection of erosion flux on and from slopes. From the 1950s, and largely developing from empirical modelling approaches to estimate erosion rates on slopes based on plot

data, there was a suggestion that catchment-based measurements tended to underestimate hillslope erosion rates systematically (e.g. Maner and Barnes, 1953; Glymph, 1954). The sediment delivery ratio (SDR) was consequently defined as a way of scaling between 'gross erosion' on slopes and sediment yield at a point in the catchment, and subsequently there has been a significant amount of research on methods for predicting SDRs as a function of catchment conditions (e.g. Roehl, 1962; Walling, 1983). However, as will be discussed below, the SDR is an artefact of the ways in which erosion has been measured, and leads to erroneous understandings of sediment transfers in catchments.

The use of SDRs demonstrates a fundamental point: that data are not independent of the (conceptual) models used in research designs. The focus on measurement techniques and resulting units above shows that not all measurements relate to the same thing – even if they are generally considered to be identical. The failure to consider how models and data relate to each other has produced significant problems in estimating erosion rates at different scales.

What exactly do we mean when we talk about scale and scaling erosion rates and erosion models? As suggested above, a simple consideration that scale relates to different areas of measurement (or different lengths of time) is too simplistic. A lack of explicit consideration of what is meant by the terms 'scale' and 'scaling' often leads to further confusion (see reviews in Blöschl & Sivapalan, 1995; Zhang *et al.*, 2002). Going beyond the cartographic or geographical scales, we need to consider operational or process scales (i.e. the scales over which different processes are important – see above), measurement or observational scale, and modelling scale. Modelling scale relates to the representation of space and time in the model. Different models represent different elements of the landscape (e.g. slope segments, entire slopes, catchment areas) more or less closely, and may integrate over a range of timescales (e.g. event, annual, geological). Measurement scales are characterized by triplets of information (Blöschl, 1996): the extent of the data (total area covered), data spacing or resolution (the number and thus spacing of samples across the extent), and data support (the extent to which a sample is averaged in space or time). The discussion above suggests that all of these uses of scale are intricately linked, and thus require an integrated conceptual framework in modelling studies. Scaling is the movement between different scales of information. We differentiate between upscaling and downscaling. The former refers to the extrapolation of smaller to larger data extents, spacings or supports in terms of measurement scales, and to extrapolation from smaller to larger geographical areas in terms of model scale. Changes in process scale mean that upscaling is not necessarily a straightforward statistical procedure, further emphasizing the need for an integrated conceptual framework. Downscaling is the interpolation of larger to smaller data extents, spacings or supports in terms of measurement scale, or the interpolation of subgrid- or element-scale patterns or sub-timestep results in terms of modelling scale. While upscaling and downscaling are generally considered to be issues of model parameterization, the confounding effect of process scaling again suggests that in reality the case is not so simple. Model evaluations as well as the presentation of results to different audiences will depend on the appropriate up- and downscaling of model results.

This lengthy introduction to the problems and questions of scale and scaling in erosion models has demonstrated a number of common problems with the ways in which erosion rates are measured and interpreted. Consideration of process, measurement and modelling scales requires an integrated approach that allows the consistent movement between these different domains. Scaling erosion models requires integrated process-based *and* empirical (statistical) approaches. In the rest of this chapter, we evaluate the scaling of erosion processes, measurements and models within such an integrated framework.

6.2   Process-Based Scaling in Simple Conditions

A significant issue with all erosion models is that they are non-linear, often highly so. Beyond making them difficult to work with numerically, this issue has implications for the scaling of erosion properties and parameters. Straightforward algebraic manipulation demonstrates why erosion at larger scales is not simply a matter of summing the predicted erosion at smaller scales (and vice versa for downscaling). One approach to scaling is linearization of the underlying equations as a way of estimating the different scaling effects, but this approach introduces errors that propagate through the model domain, and affect the results both quantitatively and qualitatively (e.g. Mokrech *et al.*, 2003). In this section we take an alternative approach, to use process characteristics of an erosional system to evaluate the patterns of erosion relationships with changing scale in a very simple set of conditions.

Parsons *et al.* (2004) investigated the patterns that should emerge on homogeneous slopes in single process domains. They used the measurement of particle travel distance as the underlying concept for characterizing scale-related differences (see also Kirkby, 1991, 1992), and avoided process-related scaling issues as noted above by considering sediment flux as the basic characterization of erosion rate. For particles of a given diameter d [L], the flux $\varphi_d(x)$ [M T^{-1}] will vary with distance downslope x as a direct function of the rate of entrainment ($E_d(x)$ [M T^{-1}]) and the rate of deposition ($D_d(x)$ [M T^{-1}]) at each point on the slope:

$$\frac{d\varphi_d}{dx}(x) = E_d(x) - D_d(x) \qquad (6.1)$$

Making the assumption that all particles of a given size travel the same distance, L_d [L], then the deposition at a specific point is the same as the entrainment rate that distance upslope, so Equation (6.1) can be rewritten as:

$$\frac{d\varphi_d}{dx}(x) = E_d(x) - E_d(x - L_d) \qquad (6.2)$$

It will be seen below that this assumption of uniform distance of movement, which is unrealistic in all but the most exceptionally uniform conditions, is not critical for the results of this argument. In inter-rill (unconcentrated) flows, it has been suggested that the entrainment rate is a function of the square of rainfall energy (e.g. Meyer, 1981) modulated by a function of depth of overland flow once generated. Torri *et al.* (1987) suggested that this function is a negative exponential of flow depth to a power. The median travel distance of particles in shallow flows can be considered to be a non-linear function of rainfall energy and flow energy as well as particle mass (Parsons & Stromberg, 1998). On a planar, uniform slope with a uniform infiltration rate and uniform roughness, flow hydraulics can be considered to be simple functions of distance downslope (discharge increases linearly, so that depth increases with $x^{2/3}$, flow velocity with $x^{1/3}$ and flow energy with $x^{5/3}$). Under these assumptions, Equation (6.2) can be rewritten as:

$$\varphi_d\left(x + L_d\right) = \int_{x}^{x + L_d\,(x)} K_1\, e^{-2\,u^{4/9}}\, du \qquad (6.3)$$

(see Parsons *et al.*, 2004, for full details of the steps in producing Equation (6.3)) where K_1 is a parameter relating to rainfall intensity and energy and flow roughness, and u is an integration variable [L]. Equation (6.3) must be solved numerically, and examples of its form for different particle sizes are shown in Fig. 6.1(a). Typically, at short distances, flux increases with distance downslope, until reaching a maximum several metres from the divide, after which the flux decreases. Parsons *et al.* (2006a), Rejman *et al.* (1999) and Wilcox *et al.* (1997) have demonstrated that this pattern is observed in field settings where there are measurements of flux at different distances from the divide.

In concentrated flows in rills and gullies, different relationships have been found for travel distance of particles, entrainment rates and flow hydraulics. Travel distance is approximated as a function of excess stream power and particle size (Hassan *et al.*, 1992), entrainment can be considered a function of flow shear stress to the 1.5 power (Yalin, 1977), flow discharge will increase with catchment area, or as a function of $x^{1.67}$ (Hack, 1957), and flow depth is equal to discharge to the power 0.4 (Abrahams *et al.*,

1996). Under these assumptions, Equation (6.2) now becomes:

$$\varphi_d\left(x + L\right) = \int_{x}^{x + L_d\,(x)} K_3\, u \, du \qquad (6.4)$$

where K_3 is a parameter reflecting slope and soil (and thus flow) conditions. Equation (6.4) is non-linear because L_d is proportional to $x^{2.18}$, but it is analytically solvable, producing flux as a function of slope length as shown in Fig. 6.1(b). Thus, until sediment exhaustion becomes an issue, erosion rates in rills and gullies increase with scale, albeit at a decreasing rate with distance.

On homogeneous slopes such as the ones considered thus far, the occurrence of multiple process domains will produce scale-erosion relationships with different forms depending on the location where concentrated flows start, so will be a function of initial conditions and storm characteristics, *ceteris paribus*. The difference in the order of magnitude of the fluxes under the different process regimes may make it empirically difficult to differentiate the curves where concentrated flows are initiated close to the divide.

Wainwright *et al.* (2001) relaxed the assumption that all particles travel the same distance under the same flow conditions. It has long been recognized that particles travel a range of different distances as a result of flow variability (especially turbulent bursts where transitional or turbulent flows occur), microtopographic variations and other factors (Wainwright & Thornes, 1991; Hassan *et al.*, 1992; Parsons & Stromberg, 1998). Travel distance of particles can thus be considered to follow a distribution function, with individual particles moving different distances with a given probabilistic form. If F_{L_d} is the cumulative probability distribution function of travel distances of particles of size d, then the flux at any point on the slope is:

$$\varphi_d\left(x\right) = \int_{0}^{x} E_d\left(x - l\right)\left[1 - F_{L_d}\left(l\right)\right] dl, \qquad (6.5)$$

or in other words, the total number of particles entrained from a given distance upslope (*l*) that travel that threshold distance or further. Although

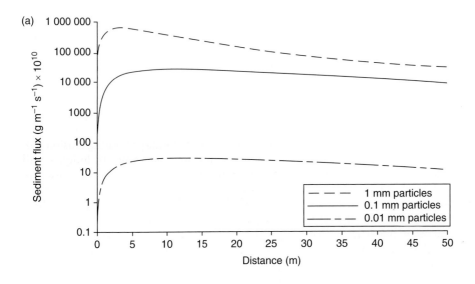

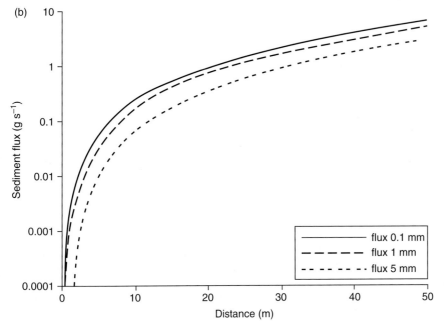

Fig. 6.1 Analytical model of (a) inter-rill erosion and (b) rill erosion as a function of scale, showing the relationship between sediment flux and distance downslope on hillslopes of uniform gradient, assuming spatially uniform rainfall and infiltration, and unlimited sediment supply (after Parsons *et al.*, 2004).

the exact shapes of the curves will vary according to the distribution function used, Equation (6.5) predicts scaling characteristics very similar to those of equations (6.3) and (6.4). The simplest probability distribution, which is very commonly used to describe travel distances, is the exponential distribution. Advantages of this function are that it has a single parameter, which is easily measured (as the reciprocal of the mean observed distance of movement), and that it produces some useful analytical results. For discussions of alternative distribution functions, see Yang and Sayre (1971), Grigg (1970), Kirkby (1991), Wainwright & Thornes (1991), and Wainwright *et al.* (1995). In portions of the slope where the entrainment rate changes slowly with distance (as it does some distance from the divide), Equation (6.5) with an exponential distribution function implies that erosion scales as a negative exponential function of distance.

As Wainwright *et al.* (2001) pointed out, this result implies that the *apparent* sediment delivery ratio should approximate an exponential form, a result that has been obtained empirically by a number of authors (e.g. Williams, 1977; Onstad & Bowie, 1977; Ferro & Minacapilli, 1995). It is also consistent with observations in loess catchments that sediment delivery approximates unity (Walling, 1983), in that the travel distance for very fine particles is high. However, there are difficulties in assuming that this result justifies the use of the SDR approach to scaling. Parsons *et al.* (2004) noted that the SDR is problematic in that it implies that landscapes are far younger than they are known to be geologically, and that it cannot account for differences in landscape appearance where land-use change has occurred on different timescales. Subsequently, they note that this means that Playfair's Law (whereby a valley is proportional in size to its stream, and therefore the sediment load carried past a point on the stream is proportional to the channel length upstream of that point) is violated, and that storage on slopes and floodplains must occur indefinitely, although such stores are not observed (Parsons *et al.*, 2006b). They suggest that this problem has arisen precisely because of the ways in which erosion estimates are made, and in particular, the assumption that fluxes can be converted to specific yields simply by dividing through by the catchment area. The problem with this assumption has already been covered in the introduction. Thus, we believe that the use of SDRs for scaling erosion rates should be avoided, because it produces results that *must be incorrect* in one or more parts of the landscape, even if results are 'correct' at the catchment outlet.

Scaling properties covered in this section have made a series of very simplistic assumptions about the way the landscape works, and have only considered characteristics in single, simple rainfall events. For single events such as these, we suggest that the sorts of relationships presented above *could* be used for downscaling erosion rates. However, in most landscapes the surface conditions and rainstorms generating erosion will be more complicated, and other techniques will also need to be employed. Some of these are addressed in the next two sections. Upscaling erosion estimates requires a recognition that due to the simplifications used, as spatial patterns (and process domains) change, the analytical (or semi-analytical) approaches presented will break down. Erosion rates between events are also not independent, so upscaling in time requires recognition that the starting conditions for the next event are a function of the conditions at the end of the previous event, as modified by other processes (e.g. bioturbation, or human land management). Spatial and temporal scaling thus require a unified framework, and some of the themes required to develop this framework are covered in the next three sections.

6.3   Statistical Scaling in Simple Conditions

Spatial representation of data was discussed above in terms of scale triplets: data extent, spacing or resolution, and support. Sampling strategies for representing erosion variability as a means of scaling rates must account for the characteristics of all of these elements. For example, the sediment-budget approach (e.g. Walling & Collins, 2008) may cover a large extent, but has relatively poor support in that the values obtained are averaged across the whole catchment. As discussed above, such spatial averaging is not straightforward to downscale for process-based reasons. Plot-based studies will tend to have a poor resolution, even if they have a large extent in some cases, because they are time-consuming to set up, maintain and monitor. Studies using erosion pins tend to have an even poorer resolution, and poor temporal support. Erosion measurements emphasize the high degree of spatial variability (see discussion in Brazier, 2004), but for the reasons previously discussed, simple statistical approaches are not valid.

The most widespread use of a statistical scaling approach is that of the slope length and angle factor (LS) in the Universal Soil Loss Equation (Wischmeier & Smith, 1978). The LS factor has been defined by statistical analysis of the USLE erosion plot database, by comparing erosion rates measured on a 'standard' plot length of 22.13 m and slope of 9%, with observations at different plot lengths and gradients. One problem with the statistical underpinning of this approach is that the vast majority of plots used to calculate the USLE relationships were of the standard size. The slope length component (L) is defined as:

$$L = \left(\frac{\lambda}{22.13}\right)^{m} \tag{6.6}$$

where λ is the length of the plot or slope segment (m) and m is a coefficient. Wischmeier & Smith (1978) gave values for m as a function of slope steepness, with $m = 0.5$ for slopes $\geq 5\%$, 0.4 for slopes of 3.5 to 4.5%, 0.3 for slopes of 1 to 3%, and 0.2 for slopes <1%. Foster *et al.* (1977) defined the exponent as:

$$m = \frac{\beta}{1 + \beta} \tag{6.7}$$

where β (dimensionless) is the ratio of rill to inter-rill erosion. In the revised USLE (RUSLE), the value of β is calculated as a function of slope angle for soils that are "moderately susceptible to both rill and inter-rill erosion" (McCool *et al.*, 1997: 105):

$$\beta = \frac{\sin \theta / 0.0896}{3 (\sin \theta)^{0.8} + 0.56} \tag{6.8}$$

where θ is the slope angle (°), which predicts values of β from 0.22 to 0.71 for slopes of 1 to 30°. The effect of the L factor is thus to predict continuously increasing erosion rates with increasing slope length (Plate 1). The observation is contrary to both empirical data (Wilcox *et al.*, 1997; Rejman *et al.*, 1999; Parsons *et al.*, 2006a) and the process-based characterization of changes of rate with scale, which in particular suggests that there should be a qualitative difference in the way in which inter-rill erosion scales compared with rill erosion. This example can also be used to demonstrate the problem with empirical equations if they are used beyond their measured range of applicability. In cases where there is no rill erosion, β and thus m in Equation (6.7) would equal zero, and the approach would predict erosion that was independent of scale, which is clearly contrary to all observations.

Other empirically based methods of regression (Walling, 1983), or 'semi-quantitative models' (de Vente & Poesen, 2005) have, of course, been employed to predict catchment sediment yield, the latter seeking to predict sediment yields from distributed catchment properties. Whilst these models may provide an initial evaluation of the key catchment properties that contribute towards sediment yield, they are often subjective in construction and provide no distributed output to evaluate sediment sources for potential erosion mitigation, or indeed to evaluate the quality of predictions in terms of the spatial variability of erosion processes. Thus, the empirical relationships that underlie such models do not characterize the changing erosion rates observed with scale any more than the aforementioned models.

A further issue relating to scale of data is in the use of field measurements for the testing and calibration of models. Notwithstanding the usual caveats about model calibration (see Wainwright *et al.*, 2009), if a model is tested or calibrated with data that have a different data support from that of the model resolution, then major errors will be produced. Unfortunately, most erosion models include some form of this calibration or are tested against data where the support is different from that which the model produces, at least implicitly (see Chapter 3 for a discussion of different data sources for calibration). For example, Licciardello *et al.* (2009) confused model calibration with validation, and then went on to assume that large-scale model estimates can be validated against average values for areas, without any consideration of scale dependency of either the model estimates or the observed values. Until there are accepted standards and appropriate measurements for testing erosion models at different scales, all erosion predictions should be treated with extreme caution.

6.4 Combined Scaling Approaches and More Complex Conditions

Most hillslopes are more variable than the uniform, planar slopes that have been largely covered above. As the extent of a study area increases, the likelihood that conditions are spatially variable increases, and variability in topography, soils and rainfall are all significant factors in further complicating the evaluation of erosion rates at different scales. When modelling environmental systems and interpreting model results, explicit consideration must be given to the validity of process descriptions and their influence on model output. Process representation should be evaluated in light of the purpose of the model application and against all available data – both qualitative and quantitative and, ideally, spatially explicit data (see above and also Chapter 4 where the impacts on model uncertainty are discussed). The first methodological problem that soil-erosion modellers may encounter is that of scale-appropriate parameterization. In order to measure (or more likely estimate) a parameter for input to a model, a measurement scale has to be chosen. Some thought may be given to the scale at which each parameter is represented, but more often, the modeller is constrained by available data and so chooses the 'best available' or 'cheapest available' data with which to parameterize the model. As an example, the use of a digital elevation model (DEM) is common in many soil erosion model applications, but the choice about the resolution of DEM is often arbitrary or simply reflects the highest resolution of data that are available. Such a choice is pragmatic, but it also

neglects to consider the influence that the measurement scale may have on the performance of the model. It may well be the case that the highest resolution data provide the best results at the plot scale, but not at the catchment scale, as topographic characteristics may scale in a non-linear fashion (Zhang *et al.*, 2002: see below). Clearly, when applying models over larger spatial scales, for computational reasons it is often necessary to reduce the spatial resolution of the model, that is, increase the cell size over which process representations are applied. Applying models at different cell sizes assumes that process descriptions are not affected by a change in resolution, and does not consider how this change may influence the output from the model (see also Section 2.5). Increasing the cell size assumes that a larger area may be represented in a homogeneous manner. This approach may be flawed (unless some stochastic representation of the within-cell properties is included). For example, Kalin *et al.* (2003) showed that as overland flow routing is often derived from surface slope (approximating energy slope), which in turn is extracted from topography, different cell resolutions will produce different model results. In addition, for similar reasons Brazier *et al.* (2005) demonstrated that the resolution of DEM that is employed exerts a strong control over the quality of the predictions that are made in modelling sediment and phosphorus delivery at increasingly coarse scales. Thus, when process descriptions are dependent upon the resolution of the data used to parameterize the model, explicit consideration of the appropriate scale of data required to represent the dominant processes is needed *a priori* and is often not available without further work.

Zhang *et al.* (2002) investigated the scaling characteristics of a simple erosion model in a range of conditions. They evaluated the Musgrave-type erosion model as developed by Thornes (1985) for variable vegetation conditions, focusing on a series of parameters that could be obtained from a range of data sources, with extents up to continental. As pixel size increased, they found that mean estimated erosion rates decreased. Sensitivity analyses demonstrated that

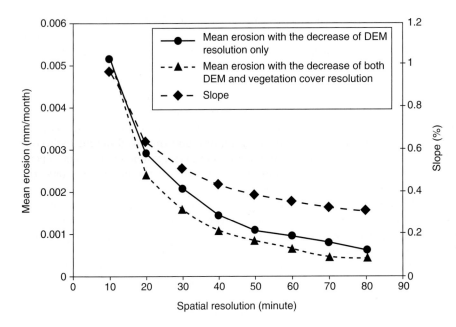

Fig. 6.2 Sensitivity analysis of a simple Musgrave-type erosion model to scale of parameterization (after Zhang *et al.*, 2002).

this decrease was due both to mean slopes decreasing with pixel size, but also to an averaging effect of estimated vegetation cover (Fig. 6.2). Zhang *et al.* used an approach to scale the results that considered the reasons why different parameters change with data resolution. In the case of slope angle, slope decreases with pixel size because extreme values are successively less likely to be sampled. The extent to which this occurs is related to the local fractal dimension of the topography, so that it is possible to define slopes at one scale as a function of slopes at another scale using the fractal dimension as a physically meaningful scaling parameter (see also Zhang *et al.*, 1999). To scale vegetation cover, it was observed that frequency distributions of vegetation cover within a pixel undergo large but predictable changes as the pixel size changes. These changes can be approximated using a Polya distribution function, whose parameters can be estimated by building databases of vegetation at different resolutions for different conditions. Finally, as total erosion is strongly dependent upon individual events, but large-scale data on events are not generally available, Zhang *et al.* (2002) used a statistical model of runoff events as the basis of scaling the estimated overland flow at a finer temporal resolution. This case study illustrates the importance of process-based representations of parameters and their changes with scale, with even very simple representations of the erosion process as described in the Musgrave-type model.

A further methodological scaling problem is that all process parameterization is to some degree spatially lumped (at the measurement scale) in a way that may fail to reflect the spatial variability within a cell, and may have a significant impact for modelling soil erosion (Canfield & Goodrich, 2006). Consequently, process parameterizations that are used in high-resolution models at the plot scale (Wainwright *et al.*, 2008b) may not be applicable for larger cell sizes (but see Section 14.7.1). In certain environments this assumption may hold when cell sizes are of a similar value to the widths of overland-flow paths. However, as cell sizes increase, the topographic information that determines overland-flow depth distribution is lost; flow depth is assumed homogeneous over a cell, which is contrary to evidence for hillslopes that exhibit patchy flow concentration in few threads (i.e. some concentrated flow within broader unconcentrated flows). Given that the relationship between sediment transport models and flow properties is likely to be strong, this approximation is likely to have a non-linear effect on erosion prediction. For example, Abrahams *et al.* (1989) looked at the impact of using mean flow depths as compared to distributions of flow depths across a slope in predicting erosion. They found that mean depths led to significant underpredictions of erosion. Subsequently, Parsons and Wainwright (2006) have argued that the occurrence of distributions of depths and the form of those distributions are critical in the estimation of the onset of concentrated overland flows.

Temporal lumping is a related issue. For process-based erosion models, the most common case where this occurs is in the assumption, largely for consideration of simplicity or numerical stability, that topography is invariant through time, despite the fact that erosion and sedimentation must cause changes. Attempts have been made to update topography in erosion models such as RillGrow (Favis-Mortlock *et al.*, 2000) which consider the evolution of topography explicitly and operate on a very high resolution to allow rills to self-organize. In practice, such updating of feedbacks between erosion process and hillslope form often proves difficult, both to simulate and to evaluate, but if the goal of the erosion modeller is to simulate hillslope evolution on temporal scales that are longer than single events, some updating of topographic evolution will be necessary (see also Section 2.5). In addition, temporal lumping of parameters also occurs in terms of soil conditions, such as antecedent soil moisture status and vegetation cover, both of which may change within or between rainfall events, leading to further misrepresentation of the temporal scale at which key processes operate.

Soil-erosion models are typically driven by some representation of the hydrological system. Process-based erosion models have a further layer of problems in evaluating the complexity of scaling, in that they are by necessity coupled with hydrological models, which themselves may reproduce the flow hydraulics that fundamentally control erosion processes to a greater or lesser extent (Merritt *et al.*, 2003; Wainwright *et al.*, 2008a). For example, Wainwright and Parsons (1998) demonstrated that different forms of erosion model were sensitive in different ways to flow hydraulics, and thus to different hydrology submodels. Because erosion predictions depend non-linearly on the hydraulics predictions, the erosion models in these contexts are ill-conditioned, leading to significant amounts of error propagation (see Chapter 4). Therefore, error propagation in erosion models is *always* worse than that in the underlying hydrological models because of model coupling, and hydrology models often produce poor estimates of even basic runoff properties (Beven, 2004).

As flow parameters are derived from hydrological routing, on all but the simplest topographies, increasing the extent of a study will lead to errors which are not only quantitative but also qualitative, inasmuch as DEM resolution affects aspect changes and thus patterns of flow accumulation (e.g. Zhang & Montgomery, 1994; Holmes *et al.*, 2000). At larger scales, flow routing is often used directly to drive erosion, typically by assuming a relation between discharge and local slope and upslope contributing area (e.g. Peeters *et al.*, 2008). Therefore, the approach taken and the scale over which fundamental equations are applied needs to be taken into consideration for the environment of interest. For example, the assumption that flow may be parameterized as a function of upslope contributing area is inappropriate in semi-arid environments, where localized storms are mismatched with the area of the catchment. In addition, the assumption that erosion is driven by a single, representative discharge is flawed; numerous studies have demonstrated that assumptions of a single discharge produce different results compared with a range of discharges (Wainwright & Parsons, 1998; Zhang *et al.*, 2002; Huang & Niemann, 2006).

Finally, the error in representing spatial variability in erosion predictions (often with lumped or constant parameters) is also cumulative, as not only are key hydrological parameters such as hydraulic conductivity highly heterogeneous (Beven, 2004), but they also control infiltration and therefore overland flow rates, which then interact with equally heterogeneous erodibility parameters. If neither the spatial variability of fundamental hydraulic parameters nor their soil structural counterparts are well represented in erosion models, and therefore models represent process as spatially lumped, then it is unlikely that meaningful (and scaleable) predictions will be forthcoming.

The consequences of the conceptual and methodological problems highlighted above are that improved representations of subgrid scale variability of the parameters that control soil erosion are required in erosion models. Representation of variability can then be used to model the processes that dominate at each scale of interest, in theory allowing the erosion model to 'scale' its predictions appropriately. Understanding scale is thus more than just representing and (hopefully) understanding variability; critically, it must have an underlying process basis in order to support the possibility of being transferable between different areas and environments.

6.5   A Travel-distance Approach to Scaling Erosion Predictions

As has been shown, in complex modelling situations, numerical approaches are required to account for the variability in initial and boundary conditions, and for the high degree of variability in real-world applications. Wainwright *et al.* (2008a,b,c) have developed the Model for Assessing Hillslope-Landscape Erosion Runoff And Nutrients (MAHLERAN) methodology, as a numerical, expanded version of the analytical approaches to scaling erosion rates discussed above. MAHLERAN employs an explicit process basis for scaling in that it uses the travel-distance approach to provide an explicit scaling of deposition rates, and estimates fluxes of sediment using the continuity equation:

$$\frac{\partial h_{s,d}}{\partial t} + \frac{\partial v_{s,d}\,h_{s,d}}{\partial x} - \varepsilon_d + d_d = 0 \qquad (6.9)$$

where $h_{s,d}$ is the equivalent depth of sediment of size d in transport (m), $v_{s,d}$ is the virtual velocity of sediment in movement (m s^{-1}), ε_d is the rate of particle entrainment of the surface (m s^{-1}), d_d is the rate of deposition (m s^{-1}), t is time (s) and x distance (m) along the slope. MAHLERAN is coupled with a pseudo-2D flow-routing model so that x can represent distance in both dimensions on the surface (see Scoging, 1992; Parsons *et al.*, 1996; Wainwright *et al.*, 1999). In this form the model scales erosion explicitly in two ways: firstly, d_d at any point on the slope is a function of entrainment at all points upslope and the probability function of deposition of that sediment, and secondly, $v_{s,d}$ scales the speed of the movement of the sediment. Transport by splash, unconcentrated overland flows and concentrated overland flows are considered separately, and movement can be by bedload or suspension, or a mixture of mechanisms. For simple slopes, the patterns of sediment flux are similar to the analytical results presented above, with the most significant change downslope occurring where the process regime changes to include concentrated flows, and thus erosion rates start to increase again (Fig. 6.3). The location of this change was found to be most strongly dependent on rainfall intensity, and thus will vary both during and between events.

Wainwright *et al.* (2008b) tested MAHLERAN on a 35-m downslope × 18-m across-slope runoff plot which had been subjected to rainfall simulation events with a constant intensity of 80 mm h^{-1}. Sediment fluxes had been measured both at the plot outlet, and at two cross-sections located 12.5 m and 21 m from the upper plot boundary. Rill heads were located between the lower cross-section and the plot outflow. The results of this test demonstrate that MAHLERAN was able to represent the plot sedigraph well (NRMSE = 18.74%, Nash-Sutcliffe index = 0.79), as well as the total fluxes at the cross-sections and the particle-size characteristics of the observed eroded sediment (Plate 2). In a further test, the model was applied

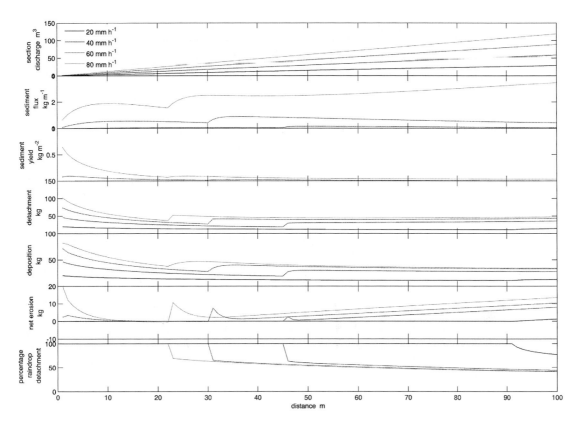

Fig. 6.3 Spatial variability of both unconcentrated and concentrated erosion combined along a uniform, planar, 100-m long × 30-m wide slope according to differing rainfall intensity as predicted by MAHLERAN (after Wainwright *et al.*, 2008b).

to a series of plots of different lengths in the same environment, with similar slopes and vegetation covers (Wainwright *et al.*, 2008c). These plots had been monitored for a period of three years for run-off and erosion responses to natural rainfall events. While the results of this test showed a poorer goodness-of-fit, they produced results that predicted the correct ordering of events, even if the absolute magnitudes were not always well simulated (Fig. 6.4). Further assessment of these tests suggested that there are also a number of limitations of subcomponents of submodels, not least because research has tended not to concentrate on parameterizing transport distances and virtual velocities for different particle sizes in different transporting mechanisms, in part because

typical erosion models make no use of these data, but also because of the technical difficulties in obtaining them. In this case, modelling across different scales has demonstrated the weaknesses of existing empirical and conceptual support for understanding erosion processes, and the link between data collection and theoretical underpinning, as discussed in the introduction.

6.6 Erosion and Landscape Evolution

As discussed above, Zhang *et al.* (2002) demonstrated that scaling from event to monthly and annual timescales in process-based models is not straightforward because of non-linearities in

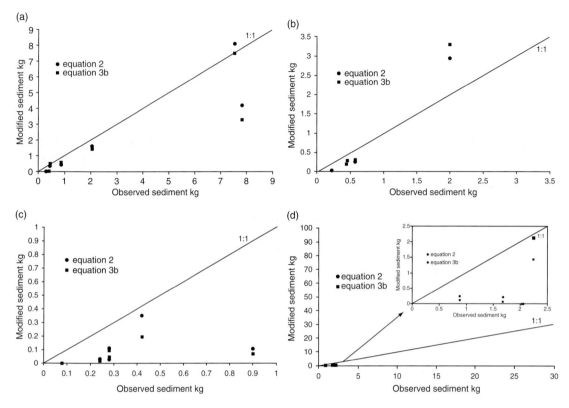

Fig. 6.4 Summary of erosion-model results versus observed sediment yield for the monitored plot events at Walnut Gulch using optimized sediment-transport parameters for hydraulics relationships (equation (2) and equation (3b) in Wainwright *et al.*, 2008b). The four plots are (a) Abbott – 14.48 m long; (b) Dud – 18.95 m long; (c) Laurel – 4.12 m long; and (d) Wise – 22.78 m long. The inset shows details of events with small yields.

erosion models. As system dynamics and feedbacks between components change (even assuming an invariant climate, which is demonstrably not the case), measurements of erosion made over short timescales may not be valid when upscaled (or may show problems with different bases of measurement, as demonstrated by Parsons *et al.*, 2004). Extrapolating from short to long term erosion rates thus needs to be approached with care for a number of reasons.

Over longer timescales, changes in system state (or the pattern of erosion processes within a hillslope or catchment) may occur because of internal system feedbacks that change process domination, or changes in system variables that are dependent on other extrinsic factors

(Le Bissonnais *et al.*, 2005). Modelling studies have allowed us to analyze the relationship between contributing area and sediment yield (Birkenshaw & Bathurst, 2006) for different system states. If we are considering decadal or longer time-periods, the operation of these processes will give rise to long-term deficits *or* gains in sediment yield with increasing area. This may alter system state (via changes in topography, vegetation or surface state) to feed back with process operation and to change dominant erosion processes, thus altering catchment sediment yield relationships. In erosion models, land use, crops, vegetation or even topography may be assumed static for short time-periods (although at the expense of greater model complexity and number of parameters, some erosion models, e.g. WEPP (Nearing *et al.*, 1989), do incorporate vegetation-growth models). However, over longer time-periods important landscape feedbacks may necessitate considering process feedbacks with the changes to such static variables that may have been driven by increased erosion (see Wainwright, 2006, for a dynamic model of interactions between climate, vegetation, erosion and slope-channel coupling). For example, over a single event, it may be argued that a hillslope has not evolved 'enough' to represent those changes in an altered digital elevation model (DEM) of that hillslope as the starting point for the next event. However, over time, after several erosion events have occurred, the hillslope will clearly have evolved even to the naked eye, so surely a representation of this evolution is necessary and an updated DEM should be used. At the other end of the spectrum are landscape-evolution models that dynamically update the DEM, but which often have a poor process representation of hillslope changes (e.g. typically assuming all erosion is diffusive in nature: see Coulthard (2001) for a review of such models).

6.7 Discussion – the Research Frontier in Scaling Erosion Models

In the soil-erosion literature there is often a lack of reporting of the time and the precise space over which soil-erosion rates or sediment yields are measured or modelled. However, Lu *et al.* (2005) illustrated that the spatial variation of erosion and the consequent sediment yield from a catchment is strongly controlled by storm durations and the residence times of sediment on hillslopes or in channels. Thus, good erosion science requires explicit descriptions of spatial and temporal dynamics that can be used to test the quality of erosion model predictions. Averaging (over time or space) may hide important variability in sediment production both within events and between events (Nichols, 2006). For example, during a three-year monitoring period at the Walnut Gulch Experimental Watershed (Parsons *et al.*,

2006a; Brazier *et al.*, 2007), 11 observed erosion events occurred at plots sited within a 0.5 km² catchment – however, only three of these events occurred over all eight of the plots being monitored. Furthermore, none of the rainfall events (durations, intensities) were identical between each plot, illustrating that rainfall/runoff and therefore erosion events are often highly unique, even when observed on plots that may be adjacent to each other and would be expected to respond in a very similar manner. This finding is supported by the work of Nearing (2000) who demonstrated that even 'replicated' hillslope plots subject to the same rainfall conditions were poor predictors of each other and were subject to high levels of variability in erosion rates. Therefore, the assumption that time and space parameters in models can be represented homogeneously, across all but the smallest grid cells, is questionable, particularly if the goal of the modelling exercise is to identify important sources of sediment within the hillslope or catchment, and indicate the potential impacts of important land-use or climatic changes on erosion rates. Conceptually, as well as practically, erosion models need to describe the wide ranges of variability that are associated with our understanding of erosion processes and that occur in the real world, in order to provide meaningful model structures with which to predict erosion. At present, as most erosion models are in some way based on previously collected plot data, they are unlikely to describe the variability associated with the larger fields or catchments that they are then applied to.

There is an additional problem in scaling erosion predictions – that of spatial equifinality. Erosion rates may be predicted at larger scales, due to a misrepresentation of the dominant processes, or the patterns of those processes, as the system becomes inherently more complex with increasing scale (see Chapter 4 for a full explanation of the equifinality problem). The implications of equifinality are all too common in the erosion model evaluation literature that has grown in recent years. Model testing, post-calibration, to demonstrate that adequate predictions can be made of erosion rates, without any consideration of the spatial pattern of erosion, may look impressive (see Peeters *et al.*, 2008 for a recent example), but it does not tell us anything about how well the model represents reality – far from it, such modelling exercises mislead the reader into thinking that the model is better than it actually is, which does not improve the representation of scale within erosion models, or indeed improve the predictions that are made by such models. What is needed is an evaluation of the pattern of erosion as well as the total erosion that has occurred. We will not be able to evaluate the ability of erosion models to predict across scales without datasets that describe the change in erosion rates across each scale of interest. Some progress has been made in this respect using nested-catchment studies (Van Oost *et al.*, 2005; Deasy, 2007); however, very few datasets exist that permit the evaluation of erosion predictions for the right spatial reasons.

Consequently, a sound understanding of all the processes operating in the system is necessary. Modelling tools are required that provide distributed output to evaluate the prediction of sediment sources and sinks within the catchment system, and furthermore, datasets are also required that describe the spatial variability of soil erosion fluxes within a catchment (or plot) in order to evaluate these models. If we continue simply to collect data at plot or catchment outlets and report these data as sediment yields per unit area, thus not describing the spatial pattern of erosion with our area of interest, we have little chance of making erosion predictions for the correct process (and form) reasons.

Specific issues arise when we seek to take our understanding of soil erosion, and representation of processes involved in soil erosion, and apply them over larger spatial and temporal scales to answer different questions about the environment than the models were originally developed to answer. As most 'questions' posed by soil-erosion modellers are represented by applications of models to sites on which they were not originally parameterized/calibrated (and often sites that are larger in scale than the datasets that underpin the model development in the first

place), a common problem is that models are over-extended or simply applied at scales which render their output meaningless.

Finally, the representation of changing dominant processes with scale has led some authors to suggest that multiple scales of erosion model are required (Kirkby *et al.*, 1996; Poesen *et al.*, 1996). Theoretically, coupling plot- or hillslope-scale models that represent detailed processes with larger scale models that can predict soil erosion over appropriately large areas for management decisions, should yield useful and scale-sensitive predictions. However, implementation of such an approach has proven to be very difficult, as *a priori* knowledge of which processes dominate at any given modelling scale is not available (although see the analysis using discriminant functions by Howard, 1995). Moreover, observed data held at different scales very rarely exist with which to evaluate such multi-scale models, although some exceptions are starting to enter the literature (Parsons *et al.*, 2006a; Deasy *et al.*, 2007) and may be useful for model evaluation in the future, as exemplified by Wainwright *et al.* (2008c).

6.8   Conclusion

This chapter has highlighted a number of challenges that environmental modellers face in making scaleable predictions of soil erosion. Although a large number of erosion models exists, very few have been explicitly developed to make robust predictions across the range of scales (plot, hillslope, catchment and basin) that will make results useful to stakeholders who are interested in reducing erosion or mitigating soil loss. In part, this situation has arisen due to the difficulties of observing erosion at multiple scales and therefore understanding how dominant processes change with scale, in order to represent changing process dominance within model structures. There has been an emphasis on using what empirical data we have collected (which is certainly not insignificant) to underpin the development of increasingly complex numerical models of erosion. This approach is understandable, but it is suggested here that it will not provide the necessary interaction between improved understanding of how erosion changes with scale and representation of the effects of scale within erosion model structures. What is required is a recourse to data collection, in a scale-explicit manner that is driven by the requirements of the models that we would like to develop. Model development and data collection must then progress in an iterative sense, such that model evaluation exercises (ideally adopting the sort of approaches to uncertainty analysis that are discussed in Chapter 4 and illustrated in Chapter 5) highlight the type of data that are required, and data collection provides more scale-appropriate descriptions of the real world with which to develop and evaluate improved model structures or hypotheses.

Furthermore, we would suggest that empirical data need to describe the spatial and temporal variability of erosion and the characteristics of the system that control erosion at each scale of interest. As predicting erosion is a complex and highly non-linear problem, describing the 'noise' in the system and studying how this changes with scale will be an important step towards improved predictions of erosion across scales.

Finally, we advocate a particle-based approach to the problem of scaling erosion predictions. Such an approach may provide a means by which improved fundamental understanding of the physics of erosion can be incorporated into erosion models, within a framework that is explicit about changes in scale and can account for the changes in dominant process that occur with scale.

References

Abrahams, A.D., Parsons, A.J. & Luk, S.H. (1989) Distribution of depth of overland flow on desert hillslopes and its implications for modelling soil erosion. *Journal of Hydrology* 106: 177–84.

Abrahams, A.D., Li, G. & Parsons, A.J. (1996) Rill hydraulics on a semiarid hillslope, southern Arizona. *Earth Surface Processes and Landforms* 21: 35–47.

Beven, K.J. 2004. *Rainfall-Runoff Modelling: the Primer*. John Wiley & Sons, Chichester.

Bilotta, G.S., Brazier, R.E., Haygarth, P.M., *et al.* (2008) Rethinking the contribution of drained and undrained grasslands to sediment related water quality problems. *Journal of Environmental Quality* 37: 906–14. DOI: 10.2134/jeq2007.0457

Birkenshaw, S.J. & Bathurst, J.C. (2006) Model study of the relationship between sediment yield and river basin area. *Earth Surface Processes and Landforms* 31: 750–61.

Blöschl, G. (1996) Scale and scaling in hydrology. In *Habilitationsschrift, Wiener Mitteilungen, Wasser-Abwasser-Gewässer* Vol. 132, Institut für Hydraulik, Gewässerkunde und Wasserwirtschaft, Technical University of Vienna, Vienna, Austria.

Blösch, G. & Sivapalan, M. (1995) Scale issues in hydrological modelling – a review. *Hydrological Processes* 9: 251–90.

Brazier, R.E. (2004). Quantifying soil erosion by water in the UK; a review of monitoring and modelling approaches. *Progress in Physical Geography* 28(3): 1–26.

Brazier, R.E., Heathwaite, A.L. & Liu, S. (2005) Scaling issues relating to P transfer from land to water. *Journal of Hydrology* 304: 330–42.

Brazier, R.E., Parsons, A.J., Wainwright, J., *et al.* (2007) Upscaling understanding of nitrogen dynamics associated with overland flow in a semi-arid environment. *Biogeochemistry* 82: 265–78. DOI: 10.1007/s10533-007-9070

Brakensiek, L.D., Osborn, H.B. & Rawls, W.J. (1979) *Field Manual for Research in Agricultural Hydrology*. USDA Agriculture Handbook 224.

Canfield, H.E. & Goodrich, D.C. (2006) The impact of parameter lumping and geometric simplification in modelling runoff and erosion in the shrublands of southeast Arizona. *Hydrological Processes* 20: 17–35.

Chandler, J.H. (1999) Effective application of automated digital photogrammetry for geomorphological research. *Earth Surface Processes and Landforms* 24: 51–63.

Coulthard, T.J. (2001) Landscape evolution models: a software review. *Hydrological Processes* 5: 165–73.

De Vente, J. & Poesen, J. (2005) Predicting soil erosion and sediment yield at the basin scale: scale issues and semi-quantitative models. *Earth-Science Reviews* 71: 95–125.

Deasy, C. (2007) *Effects of scale on phosphorus transfer in small agricultural catchments*. Unpublished PhD thesis, University of Sheffield.

Deasy, C., Brazier R.E., Heathwaite, A.L. & Hodgkinson, R. (2007) Scale-related sediment and phosphorus transfers in small agricultural catchments. *International Association for Hydrological Sciences* Publication No. 314: 79–89.

Ergenzinger, P. (1992) A conceptual geomorphological model for the development of a Mediterranean river basin under neotectonic stress (Buonamico basin, Calabria, Italy). In Walling, D.E., Davies, T.R. & Hasholt, B. (eds), *Erosion, Debris Flows and Environment in Mountain Regions.* IAHS Publication no. 209, IAHS Press, Wallingford: 51–60.

Favis-Mortlock, D., Boardman, J., Parsons, A.J. & Lascelles, B. (2000) Emergence and erosion: a model for rill initiation and development. *Hydrological Processes* 14: 2173–205.

Ferro, V. & Minacapilli, M. (1995) Sediment delivery processes at basin scale. *Hydrological Sciences Journal* 40: 703–17.

Foster, G.R., Meyer, L.D. & Onstad, C.A. (1977) A runoff erosivity factor and variable slope length exponents for soil loss estimates. *Transactions of the American Society of Agricultural Engineers* 20: 683–7.

Gerlach, T. (1967) Hillslope troughs for measuring sediment movement. *Revue de Géomorphologie Dynamique* 17: 173.

Glymph, L.M. (1954) Studies of sediment yields from watersheds. *International Association for Hydrological Sciences Publication* 36: 173–91.

Grigg, N.S. (1970) Motion of single particles in alluvial channels. *Proceedings of the American Society of Civil Engineers, Journal of the Hydraulics Division* 96(HY12): 2501–18.

Hack, J.T. (1957) Studies of longitudinal stream profiles in Virginia and Maryland. *US Geological Survey Professional Paper* 294-B.

Hassan, M.A., Church, M. & Ashworth, P.J. (1992) Virtual rate and mean distance of travel of individual clasts in gravel-bed channels. *Earth Surface Processes and Landforms* 17: 617–27.

Holmes, K.W., Chadwick, O.A. & Kyriakidis, P.C. (2000) Error in a USGS 30-meter digital elevation model and its impact on terrain modeling. *Journal of Hydrology* 233: 154–73.

Howard, A.D. (1995) Simulation modeling and statistical classification of escarpment planforms. *Geomorphology* 12: 187–214.

Huang, X. & Niemann, D. (2006) An evaluation of the geomorphically effective event for fluvial processes over long periods. *Journal of Geophysical Research*, 111(F03015 DOI: 10.1029/2006JF000477).

Kalin, L., Govindarju, R.S. & Hantush, M.M. (2003) Effect of geomorphic resolution on modelling of run-off hydrograph and sedimentograph over small watersheds. *Journal of Hydrology* 276: 89–111.

Kirkby, M.J. (1991) Sediment travel distance as an experimental and model variable in particulate movement. *Catena Supplement* 19: 111–28.

Kirkby, M.J. (1992) An erosion-limited hillslope evolution model. *Catena Supplement* 23: 157–87.

Kirkby, M.J., Imeson, A.C., Bergkamp, G. & Cameraat, L.H. (1996) Scaling up processes and models from the field to the plot to the watershed and regional area. *Journal of Soil and Water Conservation* 51: 391–6.

Korup, O. (2005) Large landslides and their effect on alpine sediment flux: South Westland, New Zealand. *Earth Surface Processes and Landforms* 30: 305–23.

Le Bissonnais, Y., Cerdan, O., Lecomte, V., et al. (2005) Variability of soil surface characteristics influencing runoff and interrill erosion. *Catena* 62: 111–24.

Licciardello, F., Govers, G., Cerdan, O., et al. (2009) Evaluation of the PESERA model in two contrasting environments. *Earth Surface Processes and Landforms* 34: 629–40.

Lu, H., Moran, C.J. & Sivapalan, M. (2005) A theoretical exploration of catchment-scale sediment delivery. *Water Resources Research* 41: W09415. DOI: 10.1029/2005WR004018

Maner, S.B. & Barnes, L.H. (1953) *Suggested criteria for estimating gross sheet erosion and sediment delivery rates for the Blackland Prairies Problem Area in soil conservation.* US Department of Agriculture, Soil Conservation Service, Forth Worth.

McCool, D.K., Foster, G.R. & Weesies, G.S. (1997) Slope length and steepness factors. In Renard, K.G., Foster, G.R., Weesies, G.A., et al. (eds), *Predicting soil erosion by water: a guide to conservation planning with the Revised Universal Soil Loss Equation (RUSLE).* USDA Agricultural Handbook 703.

McManus, J. & Duck, R.W. (1985) Sediment yield estimated from reservoir siltation in the Ochil hills, Scotland. *Earth Surface Processes and Landforms* 10: 193–200.

Merritt, W.S., Letcher, R.A. & Jakeman, A.J. (2003) A review of erosion and sediment transport models. *Environmental Modelling and Software* 18: 761–99.

Meyer, L.D. (1981) How rain intensity affects interrill erosion. *Transactions of the American Society of Agricultural Engineers* 24: 1472–5.

Mokrech, M., Drake, N. & Wainwright, J. (2003) Uncertainty modelling and error propagation in GIS based soil-erosion model. In *GISRUK Conference Proceedings*, London, 9–11 April.

Nearing, M.A. (2000) Evaluating soil erosion models using measured plot data: accounting for variability in the data. *Earth Surface Processes and Landforms* **25**: 1035–43.

Nearing, M.A., Foster, G.R., Lane, L.J. & Finkner, S.C. (1989) A process-based soil erosion model for USDA Water Erosion Prediction Project technology. *Transactions of the American Society of Agricultural Engineers* **32**: 1587–93.

Nichols, M.H. (2006) Measured sediment yield rates from semiarid rangeland watersheds. *Rangeland Ecology and Management* **59**: 55–62.

Onstad, C.A. & Bowie, A.J. (1977) Basin sediment yield modelling using hydrological variables. In *Erosion and Solid Matter Transport in Inland Waters*. IAHS Publication No. 122: pp. 191–202.

Parsons, A.J., Abrahams, A.D. & Wainwright, J. (1996) Responses of interrill runoff and erosion rates to vegetation change in southern Arizona. *Geomorphology* **14**: 311–7.

Parsons, A.J. & Stromberg, S.G.L. (1998) Experimental analysis of size and distance of travel of unconstrained particles in interrill flow. *Water Resources Research* **34**: 2377–81.

Parsons, A.J., Wainwright, J., Powell, D.M., *et al.* (2004) A conceptual model for determining soil erosion by water. *Earth Surface Processes and Landforms* **29**: 1293–1302.

Parsons, A.J. & Wainwright, J. (2006) Depth distribution of interrill overland flow and the formation of rills. *Hydrological Processes* **20**: 1511–23.

Parsons, A.J., Brazier, R.E., Wainwright, J. & Powell, D.M. (2006a) Scale relationships in hillslope runoff and erosion. *Earth Surface Processes and Landforms* **31**: 1384–93.

Parsons, A.J., Wainwright, J., Brazier, R.E. & Powell, D.M. (2006b) Is sediment delivery a fallacy? *Earth Surface Processes and Landforms* **31**: 1325–8.

Peeters, I., van Oost, K., Govers, G., *et al.* (2008) The compatibility of erosion data at different temporal scales. *Earth and Planetary Science Letters* **265**: 138–52.

Poesen, J., Boardman, J., Wilcox, B. & Valentin, C. (1996) Water erosion monitoring and experimentation for global change studies. *Journal of Soil and Water Conservation* **51**: 386–90.

Rejman, J., Usowicz, B. & Debicki, R. (1999) Source of errors in predicting silt soil erodibility with USLE. *Polish Journal of Soil Science* **32**: 13–22.

Rendell, H.M. (1982) Clay hillslope erosion rates in the Basento Valley, S. Italy. *Geografisker Annaler* **64A**: 141–7.

Roehl, J.E. (1962) Sediment source areas, delivery ratios and influencing morphological factors. *International Association of Hydrological Sciences Publication* **59**: 202–13.

Rowan, J.S., Goodwill, P. & Greco, M. (1995) Temporal variability in catchment sediment yield determined from repeated bathymetric surveys: Abbeystead Reservoir, UK. *Physics and Chemistry of the Earth* **20**: 199–206.

Scoging, H. (1992) Modelling overland flow hydrology for dynamic hydraulics. In Parsons, A.J. & Abrahams, A.D. (eds), *Overland Flow Hydraulics and Erosion Mechanics*. UCL Press, London: 105–45.

Thornes J.B. (1985) The ecology of erosion. *Geography* **70**: 222–35.

Torri, D., Sfalanga, M. & Del Sette, M. (1987) Splash detachment: runoff depth and soil cohesion. *Catena* **14**: 149–55.

Van Oost, K., Govers, G., Cerdan, O., *et al.* (2005) Spatially distributed data for erosion model calibration and validation: The Ganspoel and Kinderveld datasets. *Catena* **61**: 105–21

Verstraeten, G. & Poesen, J. (2000) Estimating trap efficiency of small reservoirs and ponds: Methods and implications for the assessment of sediment yield. *Progress in Physical Geography* **24**: 219–51.

Wainwright, J. (2006) Degrees of separation: hillslope-channel coupling and the limits of palaeohydrological reconstruction. *Catena* **66**: 93–106.

Wainwright, J. and Thornes, J.B. (1991) Computer and hardware modelling of archaeological sediment transport on hillslopes. In Lockyear, K. & Rahtz, S. (eds), *Computer Applications and Quantitative Methods in Archaeology 1990*. BAR International Series 565, Oxford: 183–94.

Wainwright, J. & Parsons, A.J. (1998) Sensitivity of sediment-transport equations to errors in hydraulic models of overland flow. In Boardman, J. & Favis-Mortlock, D. (eds), *Modelling Soil Erosion by Water*. Springer Verlag, Berlin: 271–84.

Wainwright, J., Parsons, A.J. & Abrahams, A.D. (1995) Simulation of raindrop erosion and the development of desert pavements. *Earth Surface Processes and Landforms* **20**: 277–91.

Wainwright, J., Parsons, A.J. & Abrahams, A.D. (1999) Field and computer simulation experiments on the formation of desert pavement. *Earth Surface Processes and Landforms* **24**: 1025–37.

Wainwright, J., Parsons, A.J., Powell, D.M. & Brazier, R.E. (2001) A new conceptual framework for understanding and predicting erosion by water from hillslopes and catchments. In Ascough II, J.C. & Flanagan D.C. (eds), *Soil Erosion Research for the 21st Century*. American Society of Agricultural Engineers, St. Joseph, MI: 607–10.

Wainwright, J., Parsons, A.J., Müller, E.N., *et al.* (2008a) A transport distance approach to scaling erosion rates: 1. Background and model development. *Earth Surface Processes and Landforms* **33**: 813–26. DOI: 10.1002/esp.1624

Wainwright, J., Parsons, A.J., Müller, E.N., *et al.* (2008b) A transport-distance approach to scaling erosion rates: 2. Sensitivity and evaluation of MAHLERAN. *Earth Surface Processes and Landforms* **33**: 962–84. DOI: 10.1002/esp.1623

Wainwright, J., Parsons, A.J., Müller, E.N., *et al.* (2008c) A transport distance approach to scaling erosion rates: 3. Evaluating scaling characteristics of MAHLERAN. *Earth Surface Processes and Landforms* **33**: 1113–28. DOI: 10.1002/esp.1622

Wainwright, J., Parsons, A.J., Müller, E.N., *et al.* (2009) Responses to Hairsine's and Sander's "Comment on 'A transport-distance based approach to scaling erosion rates:' Parts 1, 2 and 3 by Wainwright *et al.*" *Earth Surface Processes and Landforms* **34**: 886–90. DOI: 10.1002/esp.1781

Walling, D.E. (1983) The sediment delivery problem. *Journal of Hydrology* **65**: 209–37.

Walling, D.E. & Collins, A.L. (2008) The catchment sediment budget as a management tool. *Environmental Science and Policy* **11**: 136–43.

Wilcox, B.P., Newman, B.D., Brandes, D., *et al.* (1997) Runoff from a semi-arid ponderosa pine hillslope in New Mexico. *Water Resources Research* **33**: 2301–14.

Williams, J.R. (1977) Sediment delivery ratios determined with sediment runoff models. *International Association of Hydrological Sciences Publication* **122**: 168–79.

Wischmeier, W.H. & Smith, D.D. (1978) *Predicting rainfall erosion losses – a guide for conservation planning*. USDA Agricultural Research Service Handbook No. 537, Washington, DC.

Yalin, M.S. (1977) *Mechanics of Sediment Transport*. Pergamon.

Yang, C.T. & Sayre, W.W. (1971) Stochastic model for sand dispersion. *Proceedings of the American Society of Civil Engineers, Journal of the Hydraulics Division* **97**(HY2): 265–88.

Zhang, W. & Montgomery, D.R. (1994) Digital elevation model grid size, landscape representation and hydrologic simulations. *Water Resources Research* **30**: 1019–28.

Zhang, X., Drake, N.A., Wainwright, J. & Mulligan, M. (1999) Comparison of slope estimates from low resolution DEMs: scaling issues and a fractal method for their solution. *Earth Surface Processes and Landforms* **24**: 763–79.

Zhang, X., Drake, N.A. & Wainwright, J. (2002) Scaling land-surface parameters for global scale soil-erosion estimation. *Water Resources Research* **38**: 1180. DOI: 10.1029/2001WR000356

7 Misapplications and Misconceptions of Erosion Models

G. GOVERS

Physical and Regional Geography Research Group, Department of Earth and Environmental Sciences, Katholieke Universiteit Leuven, GEO-Institute, Celestijnenlaan, Heverlee, Belgium

7.1 Introduction

Soil erosion modelling has a long history. The need for accurate soil loss prediction emerged in the 1930s in the United States, after the Dust Bowl. The first models, predecessors of the USLE (Universal Soil Loss Equation), began to appear in the 1940s (e.g. Zingg, 1940), and models have been evolving ever since, albeit not continuously. Erosion model development saw a major paradigm shift from so-called statistical or empirical models to so-called process-based models in the 1980s, mainly stimulated by seminal papers by Foster and Meyer from 1972 onwards (Foster & Meyer, 1972). Present-day process-based erosion models such as WEPP (Nearing *et al.*, 1989) and PESERA (Kirkby *et al.*, 2008) often are sophisticated tools, describing erosion processes in great detail. Some of them are now able to carry out continuous simulations instead of event-based simulations; others just provide long-term average estimates. One might therefore expect that, by now, we have a clear view on what we may achieve with soil erosion models and how we should apply them. As the title of this chapter suggests, this may not always be the case.

In this chapter we try to understand how and why misconceptions about soil erosion models might arise, and what kind of misapplications of models continue to exist. We will take a somewhat provocative stance, not because we think that most of the work that has been carried out hitherto was substandard or ill-conceived, but because we want to stimulate the discussion. It should also be made clear that both issues have been discussed before: Wischmeier (1976) has already warned against the misuse of the USLE, and since then several critical papers on soil erosion model applications have been published (Boardman, 2006; Jetten & Favis-Mortlock, 2006; Jetten *et al.*, 2003). Furthermore, critical considerations on the use of spatial models in related fields such as geomorphology and hydrology also exist (e.g. Grayson *et al.*, 1992). Here, we try to provide the reader with a general overview of model misconceptions and misapplications with respect to soil erosion models: logically, some of the material covered has already been discussed in this earlier work. We also take a rather practical, empirical viewpoint. We thereby mostly forego more philosophical/conceptual issues related to spatial modelling in the Earth Sciences: see Bras *et al.* (2003) for a more conceptual discussion of the issue.

Evidently, misapplication and misconceptions about erosion models are strongly related. Misapplications of erosion models are, in principle, always due to a misconception. On the other hand, not all misconceptions necessarily lead to misapplications. In this chapter we will first discuss misapplications: we will deal with what might be called true misapplications: model

Handbook of Erosion Modelling, 1st edition. Edited by R.P.C. Morgan and M.A. Nearing. © 2011 Blackwell Publishing Ltd.

applications that will lead to erroneous results because there are fundamental reasons why the model is not applicable to the conditions/environment under study. Thereafter we will give attention to other misconceptions about erosion models that may not necessarily lead to erroneous model applications. Such misconceptions may, for instance, hamper a correct interpretation of application results and/or may have important implications in directing future research on soil erosion modelling.

7.2   Misapplications of Soil Erosion Models

The issue of misapplications of erosion models can, in principle, be dealt with at a very conceptual level: models can only work when they are applied to conditions for which they have been calibrated and, if possible, validated. The term 'conditions' has, in this context, a very broad meaning: the model should be applied at the correct temporal and spatial scales, and care should be taken to ensure that the model accounts for the erosion processes under consideration. Only model applications where these conditions are met can be considered to be scientifically valid: if these conditions are not met, model applications may lead to the wrong conclusions about the model's performance, the erosion risk in a certain area and/or the efficiency of erosion control measures (e.g. Nyssen *et al.*, 2006).

7.2.1   The time-scale

The classic example of model misapplication due to an inappropriate time scale is the use of the USLE at time-scales for which this statistical model was not intended. In their original paper, Wischmeier and Smith (1965) clearly stated that a measurement period of over 20 years was necessary in order to establish a true mean soil erosion rate that could be compared with the value predicted by the USLE. As discussed by Wischmeier and Smith, one of the basic reasons for this is that statistically based models such as the USLE are not capable of accounting for the considerable temporal variation that occurs in soil erosion

rates: the model is conceived to represent an 'average annual' situation based upon average annual values for the temporally varying input parameters (most notably rainfall erosivity, soil erodibility and soil cover). Using such a model to simulate what will happen over a short time period or during a single event will inevitably lead to poor results if the system under consideration is characterized by a high temporal variability.

Soil erosion rates do indeed vary significantly over short time-scales, even if all controlling factors except climate (crop type, tillage techniques, soil type, etc.) are kept as constant as possible. Figure 7.1 shows the variation of mean soil erosion rates as measured on three plots under continuous corn at the USDA-ARS research station of Clarinda, illustrating the enormous variability in annual erosion rates. This variability results in a large uncertainty in average erosion rates at a given location, even when measurements are maintained for a long period of time. Coefficients of variation of the mean value decline somewhat with average soil losses but are generally between 0.2 and 0.5, indicating that, even when measurements are maintained for as long as 10–27 years, the 90% confidence interval on the mean is between 40 and 100% of the measured average value (Fig. 7.2).

Evidently, this variation is partly due to yearly variations in climatic conditions at the station's location, and this variation might be accounted for by using annual rainfall erosivity values rather than an overall average annual value. However, climatic variability is only one component of overall variability: variability is also due to variations in plot conditions that cannot be accounted for by the factors included in the original USLE, and one therefore cannot expect that this variation can be simulated or predicted using a model that was designed to predict correctly average values only, such as the USLE.

The implications of this simple observation are perhaps more profound than we usually think. First of all it means that a meaningful calibration of a model of the USLE-type using field data requires the collection of erosion rates under standardized conditions over a sufficiently long

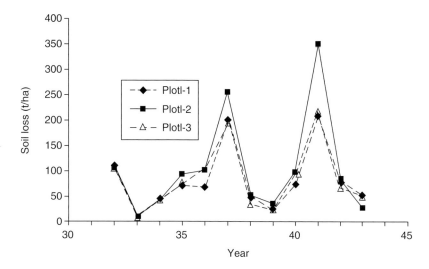

Fig. 7.1 Variations of soil losses under continuous corn as measured by the SCS on three replicate plots between 1932 and 1943. Annual soil losses range from 6 to 355 t ha^{-1} (data source: National Soil Erosion Research Laboratory USLE data repository, Purdue).

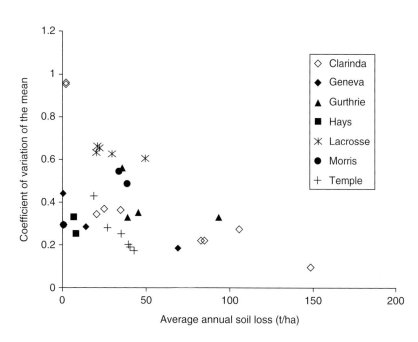

Fig. 7.2 Coefficient of variation of the mean as a function of average annual soil loss rates measured in USDA-ARS research stations under different land uses/crop rotations. Measurement periods ranged between 10 (Morris) and 27 years (Guthrie) (data source: National Soil Erosion Research Laboratory USLE data repository, Purdue).

time period: at least 10 years seems reasonable, but Wischmeier and Smith (1978) considered 20–22 years to be desirable, as measurements should cover a full solar cycle. Very often data are not available for such a long time period. Plot replication cannot entirely solve this issue, as

measuring for a shorter time period may imply that the climatic and vegetation conditions during the measurement period are not representative of the 'true mean'. Alternatively, integrative techniques, assessing soil redistribution rates over longer time spans through a single measurement, may be used, but they need to be applied with care (see below). Logically, the decadal time span requirement does not only apply to model calibration, but also to model evaluation/validation: even then, only the average annual soil erosion value should be compared. These considerations are not only relevant for the USLE: a more recent model such as PESERA (Kirkby *et al.*, 2008) uses a process-based approach to model long-term average water erosion. Yet, it relies on average climatic characteristics (basically the number of rain days per month and the amount of rain per rain day) to calculate the distribution of runoff events. Hence, long-term data on average annual erosion rates are in principle necessary to evaluate its performance.

One might argue that a large part of the investments made in the development of the Revised Universal Soil Loss Equation (RUSLE), the follow-up model of the USLE, as well as the development of other, dynamic, process-based models, were precisely directed towards a better accounting of the temporal variability in factors controlling erosion. This is, for instance, clearly apparent in the new procedure for the calculation of the soil erodibility factor that was incorporated in the RUSLE (Renard *et al.*, 1997); in the USLE, soil erodibility was characterized by a single number, while in the RUSLE a detailed procedure is implemented allowing for temporal variability (see Chapter 8). Experimental and field work had indeed demonstrated that soil erodibility could be highly variable (e.g. Mutchler & Carter, 1983). The developers of the RUSLE used the available experimental data to develop seasonal soil erodibility factors so that the model could in principle account for seasonal changes such as variations in soil moisture, freezing and thawing and soil consolidation. Similarly, the calculation of the cover factor (C) was refined in an attempt to improve the model's capability of accounting for temporal variations in soil cover as well as of the role of the various components of soil cover (canopy and surface). These modifications make it in principle possible to apply the RUSLE to much shorter time spans in comparison with the USLE, as temporal variability is explicitly accounted for.

Process-based models incorporated in a continuous soil-plant simulation model, such as the current WEPP model, are, in principle, applicable to even finer temporal scales, since such models simulate erosion using time-steps of seconds. This temporal resolution is large enough to simulate the variation of erosion within a single event. In the WEPP system, variations in soil conditions (moisture, roughness, density), plant development (growth stage, vegetation cover) and surface cover that may affect erosion rates are updated between each event (Flanagan & Nearing, 1995).

Given the increased time resolution of these models, one would expect them to be better able to predict erosion rates on a seasonal or yearly basis than the USLE, even when the latter is applied using temporally adjusted input values for soil cover and rainfall erosivity. This is in principle easy to test by (i) comparing average annual values as predicted by the USLE and a more sophisticated model such as the RUSLE or WEPP with measured averages; and (ii) by comparing individual annual or seasonal values as predicted by the USLE and a more recent, sophisticated model with measured annual or seasonal values. Yet, there are very few studies empirically investigating these issues. The author is not aware of any literature that empirically substantiates the proposition that RUSLE is indeed capable of better prediction of seasonal variations in erosion rates than the USLE when the latter is applied using seasonally varying input data. However, some intercomparisons of process-based models, the USLE and/or RUSLE have been carried out. Such studies have consistently shown (i) that dynamic, process-based models are often not capable of simulating individual events adequately (Jetten *et al.*, 1999; Favis-Mortlock *et al.*, 2001); and (ii) that, when predictions of average annual values or individual annual values of soil erosion rates are considered, dynamic, process-based models often perform similarly or only marginally better than statistical models. Using 1600 plot-years of data, Tiwari *et al.* (2000) compared predictions using the WEPP, USLE and RUSLE models. WEPP recorded a model efficiency of 0.71 compared with 0.80 and 0.72 for USLE and RUSLE respectively, when average annual values were compared, showing that, in terms of predicting long-term average values, the USLE outperformed more recent, sophisticated models. When looking at individual annual erosion values, one should expect the difference between the different models to reverse. However, Tiwari *et al.* (2000) found that this was not the case, as model efficiencies were 0.58, 0.60 and 0.40 for USLE, RUSLE and WEPP respectively. Thus, at higher temporal resolution the USLE performed similarly to the RUSLE and somewhat better than WEPP.

With respect to event-based erosion modelling, nearly all studies that have been carried out hitherto demonstrate that the application of erosion models to single events often results in very large errors (e.g. De Roo & Jetten, 1999), suggesting that even dynamic, process-based models should not be used or evaluated at the time-scale of a single event. However, they offer a key advantage in comparison with statistical models when it comes to predicting the dynamics of erosion processes, in that they can generate distributions of individual events; it is then not only possible to compare the predicted, average values, but also the simulated distribution of events, so that insight can be gained into what type of event causes most erosion. We may then anticipate how this distribution may shift with a changing climate or a changing agricultural system (Baffaut *et al.*, 1998).

7.2.2 The processes

Obviously, a model should only be used to simulate those processes that it describes. Most water erosion models describe so-called 'sheet and rill' erosion, thereby excluding the effects of soil redistribution by other processes. A valid test of such a model can therefore only be carried out using data on sheet and rill erosion rates. The latter appears to be obvious, yet misapplications occur mainly when direct data are not available and sheet and rill erosion rates are equated to total soil redistribution rates which are measured by looking at an integrative measure of soil redistribution. The latter may, for instance, be the degree of soil redistribution or soil erosion/ deposition rates calculated from radionuclide inventories (mostly ^{137}Cs) or soil truncation.

Integrative measures of soil redistribution provide valuable information about the intensity of soil redistribution rates, but both rates and patterns may be affected by other processes than sheet and rill erosion. On arable land, tillage erosion, ephemeral gully erosion and soil loss by root crop harvesting can all significantly contribute to soil redistribution, and their impact may even dominate the overall soil redistribution pattern. Data from integrative tracers can therefore not be used directly to validate a model that was specifically developed to simulate sheet and rill erosion such as the USLE or WEPP. While in past studies, ^{137}Cs-derived erosion rates and water erosion model predictions were sometimes directly compared since researchers were not aware of the importance of tillage erosion or root crop harvesting (e.g. Busacca *et al.*, 1993), recent literature shows that researchers have become increasingly aware of the fact that, given the importance of other soil redistribution processes, all processes contributing to soil redistribution need to be accounted for when interpreting the results of tracer and/or soil truncation studies (e.g. Belyaev *et al.*, 2005).

7.2.3 The spatial scale

There are several reasons why spatial scale is important for erosion model applications. Firstly, different processes dominate erosion at different scales. For instance, in Europe tillage erosion is often the dominant erosion process at the field scale (Govers *et al.*, 1994). Yet, at the catchment scale tillage does not cause any net soil movement as all soil that is eroded by tillage is redeposited within the same field. A possible solution to this is to view the landscape as a hierarchically structured set of response units, each with their own dominant processes described by relevant submodels (Cammeraat, 2002). A second reason why scale issues are important is that model parameters are often obtained from measurements at a scale that is different from the one at which the model is finally used: model parameters, such as saturated hydraulic conductivity, are often estimated using rainfall simulation experiments at a scale of ca. $1\,m^2$, while models are applied at the field or even the catchment scale.

The above does not imply that erosion models can only be applied at the scale that was used for parameter estimation or erosion rate assessment during model development. What is necessary, though, is that scaling is carried out appropriately and that the results obtained at different scales are interpreted correctly.

Making sure that scaling is appropriate is of tremendous importance when average erosion rates over large surface areas need to be estimated. The advent of GIS software has made the manipulation of large spatial datasets much easier, and over recent years several global estimates of soil erosion have been produced (Yang *et al.*, 2003; Ito, 2007; Van Oost *et al.*, 2007); the accuracy of such estimates depends critically on the use of correct scaling factors. For simple, topographically-driven models the length or area factor that is present within the model can be considered as a scaling factor. If the erosion rate per unit surface area increases with increasing slope length, average erosion rates measured on small plots will be lower than average regional soil erosion rates. If, on the other hand, the erosion rate decreases with increasing slope length/contributing areas, average regional erosion rates will be lower than average soil erosion rates. The slope length effect may vary considerably, depending on local conditions and the type of erosion process. In RUSLE calculations it is assumed that erosion scaling with slope length depends on the ratio of rill to inter-rill erosion, which is primarily controlled by slope gradient but may also depend on other factors such as soil type and soil conditions (Loch, 1996; Renard *et al.*, 1997). If inter-rill erosion is dominant, the length factor can be small (the length exponent is close to zero), which implies that erosion rates are more or less constant as a function of slope length. If rill erosion is dominant, the length exponent is assumed to be much higher (approaching 1), which implies that erosion rates increase with increasing slope length/ contributing areas.

There is a significant, albeit relatively small, amount of experimental data backing up the slope length factors that are generally used for arable land, especially for cases where rill erosion is dominant (Govers *et al.*, 2007). However, the situation may well be different on hillslopes under natural vegetation: under these conditions, erosion rates per unit surface area may well decrease with increasing slope length, the fundamental reason being that total runoff amounts do not increase, or increase only slowly with increasing slope length, as the spatial variability of hydraulic conductivity is very high and increasing water depths promote increasing infiltration (Dunne *et al.*, 1991). Net erosion rates on such slopes will then become limited by the amount of sediment that can be transported by overland flow off the slope. If time-averaged transport capacity does not increase at least linearly with slope length, erosion rates per unit area will decrease with increasing slope length as proposed by Parsons *et al.* (2006). Slope length factors used in (R)USLE applications on non-agricultural land should be adjusted accordingly, but experimental data on the variation of sediment fluxes with slope length on surfaces with natural vegetation is at present very scarce.

Process-based models generally model runoff generation and transfer explicitly: they are therefore in principle capable of simulating erosion at larger scales without an *a priori* specification of a length or area factor. However, this evidently requires that the mechanisms controlling the scaling of runoff generation and transfer are explicitly accounted for in the model. If, for instance, a model is applied in an environment where infiltration rates are increasing with increasing water depth while the model description of infiltration and runoff generation does not allow for this, the model may be expected to perform badly when applied to slopes of greater length than that of typical erosion plots (1–20 m).

Another issue that may cause misapplication at large scales is that the erosion response of a larger catchment is not only controlled by the amount of rill and inter-rill erosion, but also by other erosion processes such as (ephemeral) gully erosion, landsliding and river incision and/or bank erosion and, importantly, also by sediment deposition. There is therefore no direct relationship between simulated or predicted erosion rates and river sediment yield. Accounting for sediment deposition is extremely important, given that in many cases up to 80% of the eroded sediment is (re)deposited within a short distance (<10 km) of the sediment source (Van Rompaey et al., 2001; Wilkinson & McElroy, 2007). Relationships between catchment sediment yields and soil erosion rates are further complicated by the fact that, at the catchment scale, internal mechanisms lead to compensatory effects: a decreased sediment supply from the slopes due to the implementation of soil conservation measures may lead to increased mobilization by the river of previously deposited sediment through bank erosion and incision, so that sediment yield is maintained over a long time period (Trimble, 1999). The poor relationship between soil erosion rates and river sediment yields is well illustrated by the fact that the massive increase of soil erosion due to agricultural activities has only led to an increase of global river sediment fluxes from the land to the ocean by ca. 20% (Syvitski et al., 2005). One may therefore expect that using an inverse approach (i.e. estimating soil erosion rates from total river sediment fluxes) may lead to errors; a substantial overestimation of soil erosion rates on agricultural land in large catchments may result if sediment yields are back-converted to erosion rates using a simple sediment delivery approach (Lal, 2003).

The above does not imply that coupling river sediment yields to erosion rates is never possible. In areas where soil erosion by overland flow is the dominant sediment production process, river sediment yields can indeed be coupled to soil erosion rates if a spatially distributed approach is used and sediment deposition is adequately accounted for, and if the model allows for the two-dimensional nature of true landscapes (e.g. Van Rompaey et al., 2001).

Accounting for the two-dimensional nature of landscapes is to some extent possible by adapting a hillslope model so that it can account for flux divergence (on convexities) and convergence (in concavities). By doing so it is implicitly assumed that erosion rates and patterns display continuity over various landscape elements, and that gullying in hollows can be described using the same process descriptions as for rill erosion on hillslopes. There is some empirical support for this in the case of ephemeral gullying (Desmet & Govers, 1997). Clearly, such an approach is not possible when sediment production in concavities is controlled by processes other than those described by the model, such as permanent gullying driven by headcut retreat due to surface runoff and/or groundwater seepage.

Thus, erosion models can be applied at larger scales than the hillslope provided that appropriate scaling functions are used and the results are correctly interpreted: water erosion models may well be used to predict spatial variations in soil erosion rates by sheet and rill erosion and to some extent by ephemeral gullying, but not due to other processes. Coupling simulated soil erosion rates to temporal variations in sediment yield is only possible under well-defined conditions: soil erosion should (by far) be the dominant sediment production process in the basin, and the model used needs to be able to simulate sediment transfer and deposition within the catchment. A model such as WATEM/SEDEM (Van Rompaey et al., 2001) allows this using a steady-state approach for areas where soil erosion by water is the only important sediment production process. A promising approach to develop models incorporating a wider range of processes is the use of parsimonious algorithms, as is done in the SEDNET model (Wilkinson et al., 2004).

7.3 Misconceptions About Erosion Models

Soil erosion modelling is an ongoing scientific activity: there is not only an increasing literature on the application of soil erosion models, but new approaches to hillslope erosion modelling continue to be developed (Wainwright et al., 2008; Heng et al., 2009; Wei et al., 2009). It is therefore not only important to reflect upon misapplications of models, but also on (mis)conceptions that may direct which models we will develop in the future and how we will test them. Here, we discuss a number of such misconceptions that we consider to be important. We explore to what extent commonly accepted conceptions about erosion models are indeed tenable, considering the available empirical evidence and what the possible implications of these conceptions are for future model development and applications.

7.3.1 Misconception: 'Better' models lead to more accurate predictions (of soil erosion rates)

The continuous development of soil erosion models has been driven, to a large extent, by the desire to improve predictions. Although not always explicitly stated, it was assumed that statistical 'average' models were inadequate to describe the complexity of soil erosion processes and did not allow sufficiently for the impact of temporal and spatial variability as well as the role of different subfactors in model input variables. While evidence of the importance of temporal variability and the role of soil and vegetation subfactors gradually accumulated in the literature from 1960 onwards, the seminal papers published by Foster and Meyer (1972, 1975) in conjunction with the advent of cheap computer power caused the development and testing of process-based models to progress relatively quickly from 1980 onwards.

There is no doubt that erosion models have been 'improving' over time: rather than being purely empirical and simulating average conditions only, many models are now capable of accounting for temporal variability and often contain some form of process description. Research on erosion processes is still ongoing and may lead to a further improvement of the process descriptions in the future.

One might therefore expect that the predictive capability of such newer models would exceed that of old models such as the USLE. Over recent years, several publications have appeared where the performance of statistical, average models

and dynamic, process-based models has been compared. In most cases these publications focus on predictions of soil erosion rates only, i.e. the amount of soil leaving a pre-defined area subject to erosion, divided by the surface area considered and the amount of time over which measurements took place. Most often, data are collected on areas subject to net erosion over the whole surface, so the plots rarely contain zones which are characterized by net deposition.

The outcome of these comparison exercises is somewhat surprising. Overall, average annual and even annual values of soil erosion rates as measured on erosion plots are not better predicted by using a dynamic, process-based model rather than a statistical approach. As already mentioned, Tiwari et al. (2000) found that, using the data used to develop the USLE, the overall performance of the WEPP model was somewhat worse than that of the USLE. The RUSLE performed similarly to the WEPP model for average annual values, and similarly to the USLE for annual values. Other studies show similar results. Bathurst and Lukey (1998) used the SHETRAN model to simulate sediment fluxes in experimental catchments strongly affected by soil erosion in the French Alps, and found that SHETRAN performance was very mixed, with good agreement for some events while predictions were out by more than an order of magnitude for other events; overall the correlation between predicted and measured data was not significant. Using the same data and a simple, empirical model based on rainfall characteristics only as an input, Brochot and Meunier (1995) arrived at good predictions over a wide range of events ($R^2 = 0.72$–0.78). Risse et al. (1993) extensively tested the USLE and arrived at R^2 values of 0.58 for annual and 0.75 for average annual values, respectively, while a similar study by Zhang et al. (1996) using the WEPP model obtained values of 0.50 and 0.65 for annual and annual average erosion rates before optimization, and values of 0.54 and 0.68 after optimization; these results are thus in line with those of Tiwari et al. (2000). Rapp et al. (2001) did not find any improvement in soil loss predictions from 206 natural runoff plots when the RUSLE was used instead of the

USLE: model efficiencies were 0.73 for average annual and 0.58 for annual values for both models. Klik and Zartl (2001) reported that WEPP reasonably simulated soil losses for individual storms as measured on plots in Austria, but only after increasing standard inter-rill erodibility values by a factor of 9–10 and rill erodibility by a factor of 1–3. With respect to annual values, the performance of the calibrated WEPP and uncalibrated RUSLE was similar. Stolpe (2005) found that measured erosion rates in south-central Chile were equally well predicted using the USLE and WEPP (R² = 0.86), but predictions using the RUSLE were weaker (R² = 0.50). Spaeth *et al.* (2003) compared the performance of the USLE and RUSLE using rainfall simulation data and found that both models performed poorly: the greater flexibility of the RUSLE did not offer any advantage over the USLE. Results are broadly similar for catchment models; again, spatially distributed, dynamic models do not significantly outperform lumped models when sediment yields are compared. Im *et al.* (2007) compared the HSPF model, which uses detailed process descriptions but a rather rough spatial discretization, with the spatially distributed SWAT model with respect to the prediction of runoff and sediment yield for a 12,000 ha watershed in Virginia, and found that both models performed similarly with slightly better results for HSPF. Parajuli *et al.* (2009) found the process-based SWAT and the AnnAGNPS model to perform similarly with respect to runoff and sediment yield. SWAT did perform better, however, with respect to phosphorus export. Shen *et al.* (2009) reported that WEPP outperformed SWAT with respect to runoff prediction in the Zhangjiachong Watershed, but both models performed similarly with respect to sediment yield, with model efficiencies (after calibration) of 0.83 and 0.82 for WEPP and SWAT respectively.

Thus, the development of more sophisticated models does not seem to lead to better predictions. This situation is not at all unique with respect to soil erosion; similar observations have been made in hydrology, where the feasibility of deterministic hydrological modelling has been questioned for a long time (Grayson *et al.*, 1992)

and no detectable progress was made in flood prediction over two decades, despite increasing model sophistication (Welles *et al.*, 2007).

There are at least two basic reasons for this rather uneasy situation: firstly, process-based models require much more input data, and with every input value that is required there is necessarily some uncertainty associated. As an example, WEPP uses an inter-rill erodibility factor (K_i) that is calculated as the product of a base inter-rill erosion factor and seven subfactors (Alberts *et al.*, 1995). Each of these eight factors is calculated using an empirical relationship to account for canopy effects, ground cover, roots, and so on. Assuming, as proposed by Alberts *et al.* (1995), a baseline value of 5.3×10^6 (kg s m^{-4}) for the base inter-rill erodibility factor and assuming, for the sake of simplicity, that each of the adjustment factors has a value of 1 and that each of the factors has an uncertainty (error standard deviation) of 10%, the resulting inter-rill erodibility value will already have an uncertainty of around 28%. An increase in the estimation uncertainty of the subfactors to 20% would lead to an uncertainty on the final value of around 60%. Thus, as models become more sophisticated, the uncertainty associated with estimating input values inevitably increases. Even if one would, for a given point in the landscape, be able to estimate K_i with a very high degree of accuracy, we would face a problem of uncertainty as the values for different subfactors vary significantly over space. Basically, further sophistication of a model will only result in improved predictive capabilities if the increase in prediction error resulting from an additional input data is smaller than the reduction in prediction error to a better model prediction (Van Rompaey & Govers, 2002).

Secondly, even the most sophisticated model will not be able to capture all the variability in soil conditions that lead to variations in erosion response. This can be shown by comparing erosion measurements from replicate plots. Many erosion study sites of the USDA contained replicate plots: one might consider replicate plots as being physical models of each other (i.e. 'identical twins'). It is therefore unreasonable to expect that a simulation model can make a better prediction or simulation of the erosion rate observed on a single plot than that measured on the replicate plot. Indeed, the idea behind replicate plots is that they are maintained in exactly the same condition: input values used to simulate replicate plots are therefore the same. Nearing et al. (1999) clearly showed that replicates do show a high degree of variation, especially for low erosion rates. Consequently, the predictive ability of erosion models, which in principle cannot outperform the physical model (the replicate plot), is fundamentally limited. Nearing (2006) found that the upper limit that may be reached is an R^2 of 0.77 for erosion rates > 75 t/ha. For smaller values the maximum achievable R^2 is lower, as the inherent variability of erosion becomes relatively more important. It is worthwhile noting that R^2 values reported in model comparison studies for average annual erosion rates are similar in magnitude (see above): this implies that the upper limit of erosion predictability appears to have been achieved, even by a statistical model such as (R)USLE.

7.3.2   Misconception: A model is well calibrated and validated if the field/catchment erosion rate/sediment yield is well simulated

Most model evaluations use data collected at the outlet of the system (field, catchment) under study for calibration and validation purposes. Yet, this approach poses an important problem: it is perfectly possible that a model reproduces the right answer for the wrong reasons. Van Oost et al. (2005a) used a simple example to illustrate this: suppose that we have a catchment consisting of only two fields, A and B, of equal magnitude, each making up exactly 50% of the catchment and situated on both sides of a single valley that cuts the catchment in half. It is obvious then that the catchment response will be the same when the input data for field A and field B are interchanged. Thus, a model test will not be able to show which set of input data/parameter values is associated with field A and which one is associated with field B.

Although this example is overly simplistic, it clearly illustrates the general problem. Spatially distributed, process-based erosion models have an almost infinite number of degrees of freedom because of the many variable inputs and parameter values that need to be provided as spatial fields. Equifinality is a necessary consequence of this situation: (very) different sets of parameter values/input data lead to similar model outcomes, even over a relatively wide range of events (Beven, 2006; see also Chapter 4).

Although we often realize the above problem, we continue to compare the performance of process-based, spatial models based on 'output' at the field or catchment scale only, as these are the data that are generally available. Although the latter may be meaningful in a practical application where it needs to be decided which model is going to be applied to solve a given problem (e.g. Nasr et al., 2007), it is unlikely that such output-based model comparisons will further our scientific understanding of what is happening within the catchment. If we focus on outlet response only, we simply have too little information to optimize unequivocally a spatially distributed, process-based erosion model. Assuming that the model that is best reproducing catchment output in terms of total sediment quantities is also best at representing the processes within the catchment is, at the least, risky. Takken et al. (1999) showed that, when a spatially distributed model is optimized based on total catchment response, internal patterns of sediment production, transfer and deposition may be completely misrepresented.

We can then also expect that, when the performance of models is compared, there will be no model consistently outperforming the other models; the performance of a model in a particular situation will depend as much on its ability to accommodate variations in the catchment response through variations in the input data and the parameter values chosen for optimization, as on its 'correctness' in describing the processes occurring within the catchment. An important point here is that models having more degrees of freedom should be expected to perform better, and that this improved performance should be compared with the increased amount of information needed to run the model, something that is almost never accounted for in model comparison studies.

The above suggests that the problem may be (partly) solved by complementing output-based model evaluation with other information. Jetten *et al.* (2003) have already discussed this issue and proposed that more attention should be given to what may be called internal model validation, i.e. the degree to which the model is indeed capable of representing the spatial and temporal variations in soil (and water) redistribution within a catchment (see also Chapter 3). While data on internal sediment redistribution are usually not available on an event basis, tracer studies and soil truncation studies (Govers *et al.*, 1996; Quine *et al.*, 1997; Polyakov *et al.*, 2009) can provide information on sediment redistribution patterns over the long term. Other spatial data may also be useful, such as the spatial distribution of ephemeral gullies (Nachtergaele *et al.*, 2001) or the spatial distribution of deposition zones after an important event.

However, we should realize that there may be fundamental limits with respect to what may be achieved using spatial data; the spatial and temporal resolution of the spatial data required that we may collect will in general be at least an order of magnitude smaller than that of the input data for a spatially distributed, dynamic model, as well as of the simulated model response. Furthermore, these data will be characterized by important uncertainty. For instance, correctly estimating deposition volumes on a field is hampered by the variation in field microtopography. The uncertainty on such estimates may therefore well exceed values typical for reservoir sedimentation (i.e. 30–35%; Verstraeten & Poesen, 2002). Thus, spatially distributed data on sediment distribution may help to identify major shortcomings in model results, but will not allow solution of the equifinality problem entirely: the use of spatial data may help to constrain the range of 'possible' parameter values and to compare those values with physically meaningful values. By doing so we may be able to investigate to what extent a pro-

posed model is indeed capable of describing the observed response. If the observed response cannot be described with reasonable parameter values, then this might be an indication that the model is structurally flawed, and is not capable of describing the processes controlling the field response.

7.3.3 Misconception: The major advantage of process-based models is that they can be applied without a priori calibration

The development and testing of process-based models has often been promoted through the idea that statistical models can be applied only to conditions for which they have been developed, while a process-based model may be applied to conditions different from those for which the model was developed and tested. The implicit idea behind this is that process-based models describe the basic processes leading to runoff and sediment detachment, transport and deposition. Hence, such a model should, in principle, be applicable without extensive *a priori* model calibration and it should be possible to use such models for *ex ante* evaluations.

This assumption is wrong for several reasons. One of the most important is that, while process-based models may use deterministic, well-tested process descriptions, the inputs that are necessary to run them (e.g. erodibility, soil hydrological characteristics) are usually estimated through statistical procedures. For instance, the effective hydraulic conductivity (K_{erange} (mm h^{-1})) for rangeland in WEPP is estimated using the following equation when the surface has >45% rills (Alberts *et al.*, 1995):

$$K_{erange} = -14.29 - 3.40\ln(ROOT10) + 37.83s \, and$$
$$+ 208.86orgmat + 298.64RR$$
$$- 27.39RESI + 64.14BASI$$

$$(7.1)$$

where *ROOT10* is the root biomass in the top 0.1 m of the soil (kg m^{-2}), *sand* is the fraction of sand in the top 0.1 m of the soil, *RR* is the random roughness (m), *RESI* is the fraction of litter surface

128

G. GOVERS

cover in inter-rill areas, and *BASI* is the production of the fraction of basal surface cover in inter-rill areas and total basal surface cover.

These procedures have been developed using a calibration and validation dataset and are therefore subject to the same limitations as statistical erosion models. As process-based models are very sensitive to input data and/or parameter values, one should therefore only apply them for these conditions where the statistical input value estimation procedures were properly validated. If input values need to be estimated for conditions that are different from those for which input value estimation procedures were developed, erroneous results may be expected.

Even if the model is used in conditions similar to those for which input value estimation procedures were developed, there will be considerable uncertainty associated with the parameter values obtained from the estimation procedures, and the resulting uncertainty on the final model result may well be bigger than that for a statistical model. Quinton (1997) convincingly showed that, in a simple application of the EUROSEM model, the impact of input value uncertainty on the final model result was such that a meaningful model evaluation became impossible. The range of possible output simulations was such that the measured outcome always fell within the lower and upper possible limits, and it was therefore impossible to evaluate the model properly; given the uncertainty in parameter estimates, it was possible that the model properly simulated the measured events but poor simulations were in fact equally probable (see also Chapter 5).

The uncertainties in input values are not only due to the statistical nature of parameter estimation techniques. There is also a considerable error associated with the actual measurement or estimation of the data necessary for input value estimation. We often have to rely on very few and very small samples or a limited number of point measurements to estimate soil bulk density, moisture content, roughness and vegetation cover. The loss of prediction accuracy due to uncertainty in both input data and in input value estimation may outweigh possible gains in accuracy due to a better process description, thereby leading to a decrease in model performance with increasing model complexity (Van Rompaey & Govers, 2002).

This does not imply that we should not apply process-based, spatially distributed models to ungauged fields and catchments, but we should not expect their predictions to be quantitatively superior to those obtained with simpler, statistical or black-box model structures. *A priori* calibration/evaluation is as necessary for spatially distributed process-based models as for statistical models; neglecting this prerequisite may lead to erroneous evaluations which may have important practical implications (Nyssen *et al.*, 2006; see also Chapter 3).

7.3.4 Misconception: Erosion models now simulate what is needed by science and society: further model development and testing is therefore unnecessary

Given the fact that a wide range of soil erosion models exists, one might indeed expect there to be an erosion model for every potential application either from a scientific or a societal viewpoint. We would argue that this is not (yet) the case. Below we focus on two 'missing links' which are relatively poorly represented in most erosion models: the modelling of sediment size-selectivity and the modelling of erosion–soil interactions.

Current needs with respect to soil erosion modelling are no longer limited to the simulation of the total amount of soil redistribution within a landscape. Recent literature devotes a considerable amount of attention to the movement of sediment-associated nutrients, pollutants and organic matter. As the latter are mainly bound to the fine fraction, understanding how the fine soil fraction is mobilized and transported through the landscape becomes crucial. Many current models have the ability to deal with different size fractions and are capable of simulating the redistribution of sediment-associated soil constituents. Various approaches have been proposed, ranging from the use of simple enrichment ratios in association with statistical models (Foster *et al.*, 2003), to process-based approaches describing the dynamics of the various size fractions in detail (Hairsine & Rose, 1992). However, compared with the efforts that have historically been made to collect data for model testing focusing on total soil erosion, there are relatively few datasets available that allow the testing of models describing size-selectivity. One of the reasons is that traditional field plots almost exclusively focused on areas where net erosion occurs; while size-selectivity may occur when sediment is detached, depositional processes are potentially even more selective than detachment processes (Beuselinck *et al.*, 1999). Understanding deposition is also important because deposition may lead to temporary storage and remobilization of sediment (Krueger *et al.*, 2009). As most field experiments on erosion were (and probably still are) located on areas without net deposition, the amount of data that is available to test various approaches to selective deposition modelling is relatively limited, and only limited model testing has been carried out (e.g. Beuselinck *et al.*, 1999). Sometimes the potential of a model to predict the movement of fine sediment and associated soil constituents is evaluated using data collected at the catchment outlet (Nasr *et al.*, 2007). It is evident that this approach holds potential, but that the resulting model calibration may be very specific for the conditions tested due to the issues we discussed earlier (see Chapter 13 as an example).

One of the major difficulties in deposition modelling is the prediction of the actual sediment size distribution of the transported sediment. It has been known for a long time that eroded sediment is often transported in the form of (micro-) aggregates, a fact that has profound implications. The fact that colluvial soils often have a grain-size composition similar to the eroding soils from which they were derived can only be explained by the fact that most of the colluvium was indeed deposited in the form of (micro-)aggregates (Beuselinck *et al.*, 2000). Understanding the dynamics of aggregation during an erosion event is key to understanding size selectivity and enrichment. Recently developed measurement techniques offer some perspective here: it is now for instance possible to measure the size distribution of transported sediment on the fly using laser diffractometry

(Williams *et al.*, 2007). An additional complication is that the (micro-)topography of depositional surfaces and therefore the runoff hydraulics may change very rapidly once deposition occurs. Often, one can observe depositional surfaces being re-incised by consecutive events or even in the last phase of the event causing the deposition in the first place; the mobilization of previously deposited material is possible when the sediment load of the runoff water reaching the depositional zone is well below sediment-transporting capacity (Plate 3). Models accounting for these interactions between sediment deposition, runoff hydraulics and sediment load are at present non-existent.

At this point one might wonder why we should develop our models further in order to include these complicated processes. It is indeed unlikely that further refinement of the description of depositional processes in process-based, dynamic models will greatly improve our predictive capabilities with respect to erosion rates and/or gross sediment yield. Yet, the objective of models is not only to predict the impact of future events or to simulate past data correctly; models are also scientific tools to enhance our understanding of how landscapes function. Evaluating models for sediment deposition and particle size-selectivity using appropriate field and laboratory data will allow us to judge to what extent we really understand the processes we are interested in, which is a valid scientific objective on its own. Also, better understanding does not necessarily have to lead to more complex process models, which are often more difficult to calibrate and/or validate. In some cases better understanding may also help to decide where and how a model can be simplified without losing predictive power, thereby allowing for more efficient model calibration. Van Oost *et al.* (2004) were able to show that sediment re-entrainment was not important to describe sediment deposition on, and export from, eroding fields in Central Belgium: hence, a simpler model structure not including re-entrainment could also be used successfully. However, the effect of such simplifications may strongly depend on local conditions and they therefore need to be applied with care. A better understanding of the factors controlling deposition rates and sediment enrichment has important practical applications as well, since it will, for instance, help us to decide which measures in a catchment will be most effective in reducing not only total sediment yield, but also the movement of different sediment size classes, even if we cannot always make a correct quantitative prediction.

The focus of erosion modelling on simulating correctly the total quantity of soil movement has also led to neglect of what is actually eroding – the soil. Soils are an important part of ecosystems, providing services to human society. The potential of soils to provide these services depends on their quality, which is undoubtedly affected by erosion. The most important service provided by agricultural soils is the production of crops, and one would therefore expect that over the years a large observational database would have been established on the relationship between erosion and crop productivity. Surprisingly, this is not the case: although some data have been collected, many of them were obtained using rapid desurfacing, a method that is not capable of simulating the gradual removal of soil by erosion and leads to strong overestimation of erosion effects on agricultural productivity (Bakker et al., 2004). Other data were collected under more realistic conditions, but do not account for the confounding effect of landscape position: eroded soils are located in specific landscape positions (steep slopes, convexities) that may be inherently less productive than concavities and footslopes (Stone et al., 1985; Bakker et al., 2004).

This important hiatus can be filled by setting up specific activities whereby the erosion-productivity link is studied using an integrated approach under various agro-ecological conditions. The latter is a necessity, as the effect of erosion on crop productivity is known to depend on soil properties (Dercon et al., 2006), but will also depend on the agricultural system. Under high-input agriculture the negative effects of erosion on nutrient supply are compensated for by fertilization, so that the effect of erosion on crop yield is ultimately controlled by the effect of erosion on the ability of the soil to store water and make it available for plant growth (Bakker et al., 2005).

However, in low-input agriculture, the effects of nutrient losses may indeed have strong effects on crop productivity (Dercon et al., 2006).

Erosion–soil interactions are not limited to soil productivity. Soil properties can dramatically change due to erosion effects: for instance, if rock fragments are present, erosion rates may be dramatically reduced on a decadal time-scale due to the concentration of protective rock fragments at the surface. As the reduction of erosion rates will be greatest in locations where initial erosion rates are highest, this will also lead to significant changes in erosion patterns in the landscape (Govers et al., 2006). Erosion will also lead to the redistribution of soil organic matter which has profound implications for the carbon exchange between the soil and the atmosphere (Van Oost et al., 2005b). Such long-term interactions need to be explicitly accounted for in models designed to simulate erosion over longer time spans.

Further model development efforts will not result in the ultimate universal soil erosion model. Depending on the research objectives, different approaches are necessary. Modelling within-field carbon or nutrient redistribution by erosion over long time-scales does not require a dynamic model, but can be studied using a steady-state approach (Van Oost et al., 2003). On the other hand, studying sediment delivery to watercourses where event size may be crucial requires the use of a high temporal resolution model (Van Oost et al., 2004; Fiener et al., 2008). Thus, several types of models will also co-exist in the future in order to address different issues.

7.4 Conclusions

Misapplications and misconceptions of erosion models can be discussed from different perspectives. Here we took a rather pragmatic viewpoint of this issue and discussed how misapplications of models may be related to applications of models at inappropriate spatial and temporal scales and/ or to a spatial domain where processes other than those represented in the model contribute significantly to sediment redistribution. Upscaling a model to large areas requires not only that all processes that are relevant over a larger spatial scale are included in the model description. It is also necessary to use appropriate scaling functions; understanding how soil erosion rates depend on slope length or unit contributing area over the domain to be modelled is critical in this respect.

Misconceptions about soil erosion models do not always lead to misapplications. Even if this is not the case, however, misconceptions may have important consequences as they may (mis)guide scientific efforts. A particularly important misconception is that increasing model quality will be expressed in improved predictions of absolute erosion rates at increasingly smaller temporal scales. This will not happen as there is a fundamental upper limit to prediction accuracy, and it appears that even the performance of older, statistically based models, such as the (R)USLE, is close to this limit.

The latter does not imply, however, that sophisticated, spatially distributed, process-based models are useless. Rather, they are tools that allow us to improve our understanding of how erosion and deposition processes work at different scales, allowing us to move away from the plot to real landscapes. The development of such models requires a continuous confrontation of field and laboratory data with model predictions. Temporally and spatially detailed data are necessary to test thoroughly the last generation of erosion models. Confrontation of model performance with field or laboratory observations should be designed in such a way that it allows assessment of which of any competing model concepts is 'better' in terms of describing the data and why this is the case. As the necessary field data or laboratory data to do this are often not available, there is a risk that efforts are uniquely directed towards the improvement of model performance through parameter optimization. By doing so, we miss the opportunity that models give us to evaluate to what extent they are really capturing the complex interplay between hydrology, erosion and (size-selective) deposition that takes place in real landscapes. Erosion models will also need further development in order to make them suit-

able tools to answer scientifically and societally relevant questions that even the most sophisticated, present-day models cannot fully address, such as erosion–soil interactions at different scales and the size-selective transfer of sediment and associated nutrients, pollutants and organic matter through a landscape.

References

Alberts, E.E., Nearing, M.A., Weltz, M.A., *et al.* (1995) Soil component. In Flanagan, D.C. & Nearing, M.A. (eds), *USDA – Water Erosion Prediction Project: Hillslope profile and watershed model documentation.* NSERL Report No. 10. USDA-ARS National Soil Erosion Laboratory, West Lafayette, Indiana, pp. 7.1–7.46.

Baffaut, C., Nearing, M.A. & Govers, G. (1998) Statistical distributions of soil loss from runoff plots and WEPP model simulations. *Soil Science Society of America Journal* **62**: 756–63.

Bakker, M.M., Govers, G. & Rounsevell, M.D.A. (2004) The crop productivity-erosion relationship: an analysis based on experimental work. *Catena* **57**: 55–76.

Bakker, M.M., Govers, G., Ewert, F., *et al.* (2005) Variability in regional wheat yields as a function of climate, soil and economic variables: assessing the risk of confounding. *Agriculture Ecosystems & Environment* **110**: 195–209.

Bathurst, J.C. & Lukey, B. (1998) Modelling badlands erosion with SHETRAN at Draix, southeast France. In Summer, W., Klaghofer, E. & Zhang, W. (eds), *Modelling Soil Erosion, Sediment Transport and Closely Related Hydrological Processes.* IAHS Publication 249, pp. 129–36.

Belyaev, V.R., Wallbrink, P.J., Golosov, V.N., *et al.* (2005) A comparison of methods for evaluating soil redistribution in the severely eroded Stavropol region, southern European Russia. *Geomorphology* **65**: 173–93.

Beuselinck, L., Govers, G., Steegen, A., *et al.* (1999) Evaluation of the simple settling theory for predicting sediment deposition by overland flow. *Earth Surface Processes and Landforms* **24**: 993–1007.

Beuselinck, L., Steegen, A., Govers, G., *et al.* (2000) Characteristics of sediment deposits formed by intense rainfall events in small catchments in the Belgian Loam Belt. *Geomorphology* **32**: 69–82.

Beven, K. (2006) A manifesto for the equifinality thesis. *Journal of Hydrology* **320**: 18–36.

Boardman, J. (2006) Soil erosion science: reflections on the limitations of current approaches. *Catena* **68**: 73–86.

Bras, R.L., Tucker, G.E. & Teles, V. (2003) Six myths about mathematical modelling in geomorphology. *Geophysical Monograph* **135**: 63–79.

Brochot, S. & Meunier, M. (1995) Erosion de badlands dans les Alpes du sud: synthése. In Meunier, M. (ed.), *Compte-rendu de recherches n°3: BVRE de Draix. Equipements pour l'eau et l'environnement n°21.* Cemagref Editions, pp. 141–74.

Busacca, A.J., Cook, C.A. & Mulla, D.J. (1993) Comparing landscape-scale estimation of soil-erosion in the Palouse using Cs-137 and RUSLE. *Journal of Soil and Water Conservation* **48**: 361–7.

Cammeraat, L.H. (2002) A review of two strongly contrasting geomorphological systems within the context of scale. *Earth Surface Processes and Landforms* **27**: 1201–22.

De Roo, A.P.J. & Jetten, V.G. (1999) Calibrating and validating the LISEM model for two data sets from The Netherlands and South Africa. *Catena* **37**: 477–93.

Dercon, G., Deckers, J., Poesen, J., et al. (2006) Spatial variability in crop response under contour hedgerow systems in the Andes region of Ecuador. *Soil & Tillage Research* **86**: 15–26.

Desmet, P.J.J. & Govers, G. (1997) Two-dimensional modelling of the within-field variation in rill and gully geometry and location related to topography. *Catena* **29**: 283–306.

Dunne, T., Zhang, W.H. & Aubry, B.F. (1991) Effects of rainfall, vegetation, and microtopography on infiltration and runoff. *Water Resources Research* **27**: 2271–85.

Favis-Mortlock, D.T. Boardman, J. & MacMillan, V.J. (2001) The limits of erosion modeling: why we should proceed with care. In Harmon, R.S. & Doe, W.W. III (eds), *Landscape Erosion and Evolution Modeling.* Kluwer Academic, New York: 477–516.

Fiener, P., Govers, G. & Van Oost, K. (2008) Evaluation of a dynamic multi-class sediment transport model in a catchment under soil-conservation agriculture. *Earth Surface Processes and Landforms* **33**: 1639–60.

Flanagan, D.C. & Nearing, M.A. (eds) (1995) *USDA Water Erosion Prediction Project: hillslope profile and watershed model documentation.* NSERL Report 10, USDA-ARS National Soil Erosion Laboratory, West Lafayette, Indiana.

Foster, G.R. & Meyer, L.D. (1972) A closed-form equation for upland areas. In Shen, H. (ed.), *Sedimentation.* Symposium to Honor Prof. H.A. Einstein, Fort Collins, pp. 12.1–12.17.

Foster, G.R. & Meyer, L.D. (1975) Mathematical simulation of upland erosion by fundamental erosion mechanics. In *Present and prospective technology for predicting sediment yields and sources.* Agricultural Research Service Report ARS-S-40, pp. 190–207.

Foster, G.R., Toy, T.E. & Renard, K.G. (2003) Comparison of the USLE, RUSLE 1.06c and RUSLE2 for application to highly disturbed lands. In Renard, K.G., McElroy, S., Gburek, W., et al. (eds), *Proceedings, 1st Interagency Conference on Research in the watersheds.* Benson, AZ, pp. 154–60.

Govers, G., Vandaele, K., Desmet, P., et al. (1994) The role of tillage in soil redistribution on hillslopes. *European Journal of Soil Science* **45**: 469–78.

Govers, G., Quine, T.A., Desmet, P.J.J. & Walling, D.E. (1996) The relative contribution of soil tillage and overland flow erosion to soil redistribution on agricultural land. *Earth Surface Processes and Landforms* **21**: 929–46.

Govers, G., Van Oost, K. & Poesen, J. (2006) Responses of a semi-arid landscape to human disturbance: a simulation study of the interaction between rock fragment cover, soil erosion and land use change. *Geoderma* **133**: 19–31.

Govers, G., Gimenez, R. & Van Oost, K. (2007) Rill erosion: exploring the relationship between experiments, modelling and field observations. *Earth-Science Reviews* **84**: 87–102.

Grayson, R.B., Moore, I.D. & McMahon, T.A. (1992) Physically based hydrologic modeling. 2. Is the concept realistic? *Water Resources Research* **28**: 2659–66.

Hairsine, P.B. & Rose, C.W. (1992) Modeling water erosion due to overland-flow using physical principles. 2. Rill flow. *Water Resources Research* **28**: 245–50.

Heng, B.C.P., Sander, G.C. & Scott, C.F. (2009) Modeling overland flow and soil erosion on nonuniform hillslopes: A finite volume scheme. *Water Resources Research* **45**: W05423.

Im, S.J., Brannan, K.M., Mostaghimi, S. & Kim, S.M. (2007) Comparison of HSPF and SWAT models performance for runoff and sediment yield prediction. *Journal of Environmental Science and Health Part A: Toxic/Hazardous Substances & Environmental Engineering* **42**: 1561–70.

Ito, A. (2007) Simulated impacts of climate and landcover change on soil erosion and implication for the carbon cycle, 1901 to 2100. *Geophysical Research Letters* **34**: [Note(s): L09403.1-L09403.5].

Jetten, V., de Roo, A. & Favis-Mortlock, D. (1999) Evaluation of field-scale and catchment-scale soil erosion models. *Catena* **37**: 521–41.

Jetten, V., Govers, G. & Hessel, R. (2003) Erosion models: quality of spatial predictions. *Hydrological Processes* **17**: 887–900.

Jetten, V. & Favis Mortlock, D. (2006) Modelling soil erosion in Europe. In Boardman, J. & Poesen, J. (eds), *Soil Erosion in Europe*. John Wiley & Sons, Chichester: 695–716.

Kirkby, M.J., Irvine, B.J., Jones, R.J.A., *et al.* (2008) The PESERA coarse scale erosion model for Europe. I. Model rationale and implementation. *European Journal of Soil Science* **59**: 1293–1306.

Klik, A. & Zartl, A.S. (2001) Comparison of soil erosion simulations using WEPP and RUSLE with field measurements. In Ascough, J.C. II & Flanagan, D.C. (eds), *Soil Erosion Research for the 21st Century*. American Society of Agricultural Engineers, St. Joseph, MI: 350–53.

Krueger, T., Quinton, J.N., Freer, J., *et al.* (2009) Uncertainties in data and models to describe event dynamics of agricultural sediment and phosphorus transfer. *Journal of Environmental Quality* **38**: 1137–48.

Lal, R. (2003) Soil erosion and the global carbon budget. *Environment International* **29**: 437–50.

Loch, R.J. (1996) Using rill/interrill comparisons to infer likely responses of erosion to slope length: implications for land management. *Australian Journal of Soil Research* **34**: 489–502.

Mutchler, C.K. & Carter, C.E. (1983) Soil erodibility variation during the year. *Transactions of the American Society of Agricultural Engineers* **26**: 1102–4.

Nachtergaele, J., Poesen, J., Steegen, A., *et al.* (2001) The value of a physically based model versus an empirical approach in the prediction of ephemeral gully erosion for loess-derived soils. *Geomorphology* **40**: 237–52.

Nasr, A., Bruen, M., Jordan, P., *et al.* (2007) A comparison of SWAT, HSPF and SHETRAN/GOPC for modelling phosphorus export from three catchments in Ireland. *Water Research* **41**: 1065–73.

Nearing, M.A. (2006) Can soil erosion be predicted? In Owens, P.N. & Collins, A.J. (eds), *Soil Erosion and Sediment Redistribution in River Catchments*. CAB International, Wallingford: 145–52.

Nearing, M.A., Foster, G.R., Lane, L.J. & Finkner, S.C. (1989) A process-based soil-erosion model for USDA-Water Erosion Prediction Project Technology. *Transactions of the American Society of Agricultural Engineers* **32**: 1587–93.

Nearing, M.A., Govers, G. & Norton, L.D. (1999) Variability in soil erosion data from replicated plots. *Soil Science Society of America Journal* **63**: 1829–35.

Nyssen, J., Haregeweyn, N., Descheemaeker, K., *et al.* (2006) Comment on "Modelling the effect of soil and water conservation practices in Tigray, Ethiopia" [Agric. Ecosyst. Environ. 105 (2005) 29–40]. *Agriculture Ecosystems and Environment* **114**: 407–11.

Parajuli, P.B., Nelson, N.O., Frees, L.D. & Mankin, K.R. (2009) Comparison of AnnAGNPS and SWAT model simulation results in USDA-CEAP agricultural watersheds in south-central Kansas. *Hydrological Processes* **23**: 748–63.

Parsons, A.J., Brazier, R.E., Wainwright, J. & Powell, D.M. (2006) Scale relationships in hillslope runoff and erosion. *Earth Surface Processes and Landforms* **31**: 1384–93.

Polyakov, V.O., Kimoto, A., Nearing, M.A. & Nichols, M.H. (2009) Tracing sediment movement on a semi-arid watershed using rare earth elements. *Soil Science Society of America Journal* **73**: 1559–65.

Quine, T.A., Govers, G., Walling, D.E., *et al.* (1997) Erosion processes and landform evolution on agricultural land – new perspectives from caesium-137 measurements and topographic-based erosion modelling. *Earth Surface Processes and Landforms* **22**: 799–816.

Quinton, J.N. (1997) Reducing predictive uncertainty in model simulations: a comparison of two methods using the European Soil Erosion Model (EUROSEM). *Catena* **30**: 101–17.

Rapp, J.F., Lopes, V.L. & Renard, K.G. (2001) Comparing soil erosion estimates from the USLE and RUSLE on natural runoff plots. In Ascough, J.C. & Flanagan, D.C. (eds), *Soil Erosion Research for the 21st Century*. American Society of Agricultural Engineers, St Joseph, MI: 24–7.

Renard, K.G., Foster, G.R., Weesies, G.A., *et al.* (1997) *Predicting soil erosion by water: a guide to conservation planning with the Revised Universal Soil Loss Equation (RUSLE)*. Agricultural Handbook 703, US Department of Agriculture.

Risse, L.M., Nearing, M.A., Nicks, A.D. & Laflen, J.M. (1993) Error assessment in the Universal Soil Loss Equation. *Soil Science Society of America Journal* **57**: 825–33.

Shen, Z.Y., Gong, Y.W., Li, Y.H., *et al.* (2009) A comparison of WEPP and SWAT for modeling soil erosion of the Zhangjiachong Watershed in the Three Gorges Reservoir Area. *Agricultural Water Management* **96**: 1435–42.

Spaeth, K.E., Pierson, F.B., Weltz, M.A. & Blackburn, W.H. (2003) Evaluation of USLE and RUSLE estimated soil loss on rangeland. *Journal of Range Management* **56**: 234–46.

Stolpe, N.B. (2005) A comparison of the RUSLE, EPIC and WEPP erosion models as calibrated to climate and soil of south-central Chile. *Acta Agriculturae Scandinavica Section B: Soil and Plant Science* **55**: 2–8.

Stone, J.R., Gilliam, J.W., Cassel, D.K., *et al.* (1985) Effect of erosion and landscape position on the productivity of Piedmont soils. *Soil Science Society of America Journal* **49**: 987–91.

Syvitski, J.P.M., Vorosmarty, C.J., Kettner, A.J. & Green, P. (2005) Impact of humans on the flux of terrestrial sediment to the global coastal ocean. *Science* **308**: 376–80.

Takken, I., Beuselinck, L., Nachtergaele, J., *et al.* (1999) Spatial evaluation of a physically-based distributed erosion model (LISEM). *Catena* **37**: 431–47.

Tiwari, A.K., Risse, L.M. & Nearing, M.A. (2000) Evaluation of WEPP and its comparison with USLE and RUSLE. *Transactions of the American Society of Agricultural Engineers* **43**: 1129–35.

Trimble, S.W. (1999) Decreased rates of alluvial sediment storage in the Coon Creek Basin, Wisconsin, 1975–93. *Science* **285**: 1244–6.

Van Oost, K., van Muysen, W., Govers, G., *et al.* (2003) Simulation of the redistribution of soil by tillage on complex topographies. *European Journal of Soil Science* **54**: 63–76.

Van Oost, K., Beuselinck, L., Hairsine, P.B. & Govers, G. (2004) Spatial evaluation of a multi-class sediment transport and deposition model. *Earth Surface Processes and Landforms* **29**: 1027–44.

Van Oost, K., Govers, G., Cerdan, O., *et al.* (2005a) Spatially distributed data for erosion model calibration and validation: The Ganspoel and Kinderveld datasets. *Catena* **61**: 105–21.

Van Oost, K., Govers, G., Quine, T.A., *et al.* (2005b) Landscape-scale modeling of carbon cycling under the impact of soil redistribution: the role of tillage erosion. *Global Biogeochemical Cycles* **19**: GB4014. DOI: 10.1029/2005GB002471

Van Oost, K.. Quine, T.A., Govers, G., *et al.* (2007) The impact of agricultural soil erosion on the global carbon cycle. *Science* **318**: 626–9.

Van Rompaey, A.J.J. & Govers, G. (2002) Data quality and model complexity for regional scale soil erosion prediction. *International Journal of Geographical Information Science* **16**: 663–80.

Van Rompaey, A.J.J., Verstraeten, G., Van Oost, K., *et al.* (2001) Modelling mean annual sediment yield using a distributed approach. *Earth Surface Processes and Landforms* **26**: 1221–36.

Verstraeten, G. & Poesen, J. (2002) Using sediment deposits in small ponds to quantify sediment yield from small catchments: possibilities and limitations. *Earth Surface Processes and Landforms* **27**: 1425–39.

Wainwright, J., Parsons, A.J., Müller, E.N., *et al.* (2008) A transport-distance approach to scaling erosion rates: I. Background and model development. *Earth Surface Processes and Landforms* **33**: 813–26.

Wei, H., Nearing, M.A., Stone, J.J., *et al.* (2009) A new splash and sheet erosion equation for rangelands. *Soil Science Society of America Journal* **73**: 1386–92.

Welles, E., Sorooshian, S., Carter, G. & Olsen, B. (2007) Hydrologic verification – a call for action and collaboration. *Bulletin of the American Meteorological Society* **88**: 503–11.

Wilkinson, B.H. & McElroy, B.J. (2007) The impact of humans on continental erosion and sedimentation. *Geological Society of America Bulletin* **119**: 140–56.

Wilkinson, S., Henderson, A. & Chen, Y. (2004) *SedNet User Guide*, Client Report. CSIRO Land and Water, Canberra.

Williams, N.D., Walling, D.E. & Leeks, G.J.L. (2007) High temporal resolution in situ measurement of the effective particle size characteristics of fluvial suspended sediment. *Water Research* **41**: 1081–93.

Wischmeier, W.H. (1976) Use and misuse of Universal Soil Loss Equation. *Journal of Soil and Water Conservation* **31**: 5–9.

Wischmeier, W.H. & Smith, D.D. (1965) *Predicting rainfall erosion losses from cropland east of the Rocky Mountains*. Agricultural Handbook 282, US Department of Agriculture.

Wischmeier, W.H. & Smith, D.D. (1978) *Predicting rainfall erosion losses – a guide to conservation planning*. Agricultural Handbook 537, US Department of Agriculture.

Yang, D.W., Kanae, S., Oki, T., *et al.* (2003) Global potential soil erosion with reference to land use and climate changes. *Hydrological Processes* **17**: 2913–28.

Zhang, X.C., Nearing, M.A., Risse, L.M. & McGregor, K.C. (1996) Evaluation of WEPP runoff and soil loss predictions using natural runoff plot data. *Transactions of the American Society of Agricultural Engineers* **39**: 855–63.

Zingg, A.W. (1940) Degree and length of land slope as it affects soil loss in runoff. *Agricultural Engineering* **21**: 59–64.

Part 2
Model Applications

8 Universal Soil Loss Equation and Revised Universal Soil Loss Equation

K.G. RENARD[1], D.C. YODER[2], D.T. LIGHTLE[3] AND S.M. DABNEY[4]

[1]USDA-ARS, Southwest Watershed Research Center, Tucson, AZ, USA
[2]Biosystems Engineering and Soil Science, University of Tennessee, Knoxville, TN, USA
[3]USDA-NRCS, National Soil Survey Center, Lincoln, NE, USA
[4]USDA-ARS, National Sedimentation Laboratory, Oxford, MS, USA

8.1 Introduction

8.1.1 History of the USLE (Universal Soil Loss Equation)

Conservation of soil and water requires both knowledge of the factors affecting these resources, and methods for controlling those factors to preserve those resources. Over the years, field, plot and small watershed studies have provided much valuable information regarding the complex factors and interactions involved in the environmental operations of land use and farming. These studies are the basis of the Universal Soil Loss Equation (USLE), which is a conservation planning tool that has been demonstrated to do a reasonably good job of estimating erosion for many disturbed-land uses. Predicting soil loss associated with modern land use is based on guidelines developed from research information in combination with additional experience from many sources. Information from empirical experiments and physically-based principles both assist in effective conservation planning.

The process of pulling together research results and experiences from agricultural practices began with Hugh Hammond Bennett (Helms, 2008),

who was undoubtedly the most influential soil conservationist in the US. His early efforts led to his recognition as the 'father of soil conservation'. Bennett's early preaching against the menace of soil erosion led to Congressional action in 1929 establishing ten experimental stations, primarily in the cultivated agricultural areas of the US (Meyer & Moldenhauer, 1985; Renard, 1985). Later expansion of the research programmes included a large number of plots, crops, and management conditions that ultimately resulted in over 10,000 plot-years of data, collected over seven decades. Most of the plots involved the familiar dimensions 6.0 ft (1.8 m) wide by 72.6 ft (22.1 m) long, or a plot 35 ft (10.7 m) long used for some rainfall simulator studies. These plots simplified the computing of runoff and erosion on a per unit area basis (0.01 acre for the 6×72.6 ft or nominally $40 \, m^2$ for the 1.8×22.1 m). Typical plot configurations were described in Brakensiek et al. (1979) and Laflen and Moldenhauer (2003).

In 1954, the National Runoff and Soil Loss Data Center was established by the US Department of Agriculture – Agricultural Research Service (USDA-ARS) at Purdue University in West Lafayette, Indiana. The Center was established to provide a central location for compiling and analysing soil erosion data collected from studies throughout the US. The Center, under the direction of W.H. Wischmeier, was responsible

Handbook of Erosion Modelling, 1st edition. Edited by R.P.C. Morgan and M.A. Nearing. © 2011 Blackwell Publishing Ltd.

for summarizing and analysing the more than 10,000 plot-years of soil erosion and runoff data mentioned above, which resulted in the USLE (Wischmeier & Smith, 1965, 1978).

It has now been more than 50 years since the first releases of erosion prediction technology based on what have become widely known as the factors affecting sheet and rill erosion and, ultimately combining those in the USLE. Table 1 in Laflen and Moldenhauer (2003) gives an excellent synopsis of the published chronology of soil erosion prediction technology in the US.

The USLE and its predecessors were meant as field-level conservation planning rather than research tools, and were therefore structured to be 'user friendly' for USDA programmes in the Soil Conservation Service (SCS) (now the Natural Resources Conservation Service (NRCS)), and designed for tailoring erosion-control practices to the needs of specific fields and farms. The USLE was a 'paper-based' model where factors were found in printed tables and charts, and calculations were done by hand.

"Had digital computers been available in the 1940s when erosion became recognized as a national problem, current prediction methods might more closely mimic the theory contained in Ellison's classic paper (1947) than the current empiricisms of the USLE." (Renard, 1985: 5)

What follows is a description of the evolution of the USLE–RUSLE effort, beginning with the improvements over the USLE leading to the RUSLE1 computer program and publication of the USDA Agriculture Handbook No. 703 (Renard *et al.*, 1997). We will then describe the development of RUSLE2, leading to its release in 2004 and its continuing documentation. The final section of this chapter will examine continuing and possible future developments of the technology.

8.1.2 USLE/RUSLE factor values

The fundamental concept in establishing factor values in the USLE was the Unit Plot. This conceptual plot was composed of a land parcel 72.6 feet (22.1 m) in length with a 9% slope, maintained in a continuous, regularly tilled fallow condition

with up-and-down hill tillage, thereby representing a condition very near the worst-case management. Such a plot was used as a base condition to which all other topographic, cropping, management and conservation practices were compared. Data from plots with different slopes, lengths and crops were adjusted to the unit plot, and compared across locations to establish reliable factor values. Benchmark soil erodibility and other terms (rainfall, slope length, slope steepness, cover-management and the support practice factors) used in the USLE/RUSLE have evolved over the years from data derived for varied conditions. Few if any unit plots were ever actually developed, but the concept was used to determine how the conditions of actual plots related to the unit plot.

The USLE soil loss equation is:

$$A = R\,K\,L\,S\,C\,P \qquad (8.1)$$

where A is the computed soil loss per unit area, expressed in the units selected for K and for the period selected for R (in common practice these are usually selected such that they compute A, soil loss in US tons per acre per year); R, the rainfall and runoff factor, is the number of rainfall erosion index units, plus a factor for runoff from snowmelt or applied water where such runoff is significant; K, the soil erodibility factor, is the soil loss rate per rainfall erosion index unit for the specified soil under Unit Plot conditions; L and S are the slope length and steepness factors in relation to the conditions on a unit plot; C, the cover and management factor, is the ratio of soil loss from an area with specified cover and management to that from an identical area under the tilled continuous fallow Unit Plot conditions (C thus ranges from a value of zero for completely non-erodible conditions, to a value of 1.0 for the worst-case Unit Plot conditions); and P, the support practice factor, is the ratio of soil loss with a support practice like contouring, stripcropping, or terracing to that with straight-row farming up and down slope.

Because the USLE was based on empirical erosion data collected from relatively small plots or subwatersheds on relatively uniform hillslopes, the resulting erosion estimates were limited to similar situations. In essence, these results did not include any impact (either erosion or deposition) of the concentrated flow channels that form in the natural swales at the bottom of the roughly planar hillslopes, and certainly did not address classical gullying processes that often occur at steep boundaries such as headcuts and sidewall sloughing.

Use of the plot data to establish values for the factors above began with an analysis of rainfall erosivity by correlating the erosion measured under Unit Plot conditions with a whole series of measured rainfall values. A very strong correlation was found between this worst-case erosion and a combination of two rainfall factors, namely the total storm energy E and the maximum storm 30-minute intensity, or I_{30} (Wischmeier, 1959). The R factor was then calculated by summing over the calendar year the $E \cdot I_{30}$ values for all storms of over 12 mm (0.5 in.) or with more than 6.5 mm (0.25 in.) falling in 15 minutes, and taking the average of those annual values over all years of record. The soil erodibility (K) values were then determined for Unit Plot conditions ($C = P = LS = 1.0$) solving for K using measured A and R values. With the K values in hand, the values for C, P and LS could be determined by replicated plot studies on similar soils using different management practices or topographies.

Techniques for determining factor values to insert in the USLE (Equation (8.1)) were first presented for general use in the USDA's Agriculture Handbook No. 282 (Wischmeier & Smith, 1965). As use of this technology expanded and new studies were carried out to fill gaps and address weaknesses, new data were incorporated into the USLE, resulting in the second and most widely known release of the USLE technology in the USDA's Agriculture Handbook No. 537 (AH537) (Wischmeier & Smith, 1978). The values for the USLE factors as presented in AH537 were generally created to represent an average annual basis, although the form of the relationship does not demand that. The exception to this was the C factor, which was recognized as changing substantially through the year, leading to the cropping-period approach presented in AH537.

Following the release of AH537, the USLE became very widely used, both within the US and internationally. Perhaps its most common use was as one of the primary tools of the USDA Soil Conservation Service for conservation planning on agricultural lands. As use of the USLE expanded and it was applied in other situations, like disturbed forest lands (Dissmeyer & Foster, 1981, 1984), limitations of the technology became apparent. At the same time, continuing soil erosion research on both natural plots and under simulated rainfall led to improved understanding of the physical processes involved in hillslope sheet and rill erosion. Recognized limitations and advancements in erosion science pointed to the need for updating the USLE.

8.2   RUSLE

8.2.1   RUSLE1 development

In 1985, scientists and engineers from the USDA-ARS and the USDA Soil Conservation Service and affiliated academics with expertise in soil erosion assembled in West Lafayette, Indiana. At that workshop, two important decisions evolved, including the need to (1) develop technology to replace the USLE with a physically-based model (subsequently called the Water Erosion Prediction Project or WEPP); and (2) to computerize and update the 1978 version of the USLE with an improved model, subsequently called the Revised USLE or RUSLE. All subsequent material in this chapter is directed to a description and analysis of the various portions of the RUSLE effort, including both RUSLE1 and RUSLE2.

The first version of RUSLE1, a software program designed to operate in a DOS-based computer environment, was released in 1997. RUSLE1 was supported by USDA-ARS through Agriculture Handbook No. 703 (AH703) (Renard *et al.*, 1997). The computer system soil erosion model described therein was a major conversion of the factor approach presented in AH537. Perhaps the most significant change was the subfactor approach to the calculation of the cover-management factor C, thereby allowing use of RUSLE1 for any land use that could be adequately addressed by these subfactors. This broke the previous bonds of the

RUSLE
Soil loss estimation

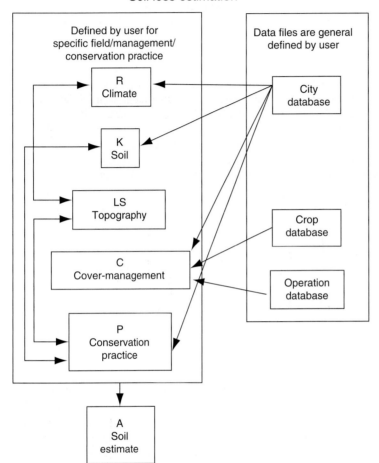

Fig. 8.1 RUSLE1 software flow chart (from AH703).

USLE to agricultural settings, as described in some detail below. The new AH703 development took an appreciable amount of effort involving scientists and engineers with experience in areas of knowledge representing each factor. It also involved a significant amount of testing by SCS-NRCS personnel in various locations, each having specific expertise in the crops, soils and climates involved. The DOS program developed for the 1997 RUSLE1 version permitted English unit calculations only. To input or output metric units required hand conversions of individual factor values using conversion factors (Foster *et al.*, 1981) which were included in Appendix B of AH703.

Another change in the RUSLE1 approach was to begin grouping the expanded list of required user inputs into a crude database, defined as shown in Fig. 8.1. This allowed for saving and re-use of sets of inputs corresponding to, for example, a specific location. In addition to the computerization of the model, every USLE factor underwent significant changes in moving to RUSLE1. These changes are generally described in the paragraphs that follow.

(i) Rainfall erosivity factor (R)   The most common way of presenting rainfall erosivity information in the US has been through the use of isoerodent maps, allowing the reader to interpolate the corresponding R value for a specific location. The isoerodent maps in AH703 were calculated using the same criteria as in AH537, namely summing the storm kinetic energy times the maximum 30-minute intensity for storms larger than 0.5 in (12 mm), unless at least 0.25 in (6.5 mm) fell in 15 min. These calculations were computed for all non-snow storms within a period of N years. Normally at least 22 years of storms were included for the calculations (see AH537 for details), but longer periods are advisable when the coefficient of variation of annual precipitation is large. A total of 181 key precipitation locations with 15-min data were used for the map in AH537, and a few additional locations to fill in gaps were added to produce Figure 2.1 in AH703. AH537 included very little erosivity information for the western US, with only 11 western station isoerodent values used to estimate the two-year 6-h precipitation amount. A power relationship developed by Wischmeier (1974) to fit those values provided some measure of the expected erosivity, but the results were not thought to be very accurate, or to reflect adequately the known intermontane climate variability. Through an agreement between Oregon State University, USDA-ARS, USDA-SCS and the National Weather Service, data from 713 stations with 15-min measurement intervals were used to calculate EI values, and thereby to construct new isoerodent maps for the western US in AH703, although all storms were included in the western erosivity calculations (excluding snow). Analysis of these records showed that 225 precipitation-measuring locations had records longer than 12 years and precipitation resolutions of 0.01 in. (0.25 mm). Values of the coefficient of determination (R²) in excess of 0.8 were obtained with the model $EI_{15} = b(EI_{60})$. Values of the regression parameter b ranged from 1.08 to 3.16, varying widely among climate zones.

To supplement this work, 1082 hourly stations were used to calculate EI_{60}. Of these stations, 790 had record lengths of 20 years or longer. These data were adjusted to a 15-min measurement interval using the cited correction. R factors were also adjusted to equivalent break-point data using the Weiss (1964) relationship $R = 1.0667 (R_{15})$. The isoerodent map was prepared by hand contouring on large-scale maps, reflecting the major topographic influences in mountain and range topography. The newer isoerodent maps (Figures 2-2, 2-3 and 2-4 of AH703) were thus felt to be a significant improvement over those in AH537.

In addition, seasonal EI distributions were developed for 84 climate zones in the western US (Fig. 2-7, AH703). The distributions were developed for calculating the time-varying C factor in RUSLE1, building on the crop growth stage approach found in AH537.

City database files were then developed in RUSLE1 to provide the climatic data needed for erosion calculations. This included the R-factor value, the EI distribution values for 24 bimonthly periods, and the 10-yr frequency storm maximum EI that was needed for calculating the P factor credit for contour farming. Maps of these values were calculated for precipitation gauge locations and are presented in Figures 2-9 to 2-12 in AH703.

Two additional modifications to the classical USLE R-factor approach were included in RUSLE1 to address specific geographical needs. In areas with very low relief and high rainfall intensities (such as in the Mississippi River delta), research has found that runoff ponds to substantial depths before running off, and that this ponded water absorbs some of the raindrop impact that could cause detachment (Mutchler, 1970). Based on these data, RUSLE1 included a term to adjust downwards the erosivity experienced by the soil, based on slope steepness and rainfall erosivity (taken as a surrogate for intensity). The other modification to the R factor was for frozen and thawing soils, encountered in the Pacific Northwest (Northwest Wheat and Range Region (Austin, 1981)), and in some of the southern plains of Canada. In these cases, a soil with much weakened structure exposed to even a low-erosivity event will experience high erosion rates, so an alternative means of selecting an equivalent R value for these conditions was included in RUSLE1.

(ii) Soil erodibility factor (K) The soil erodibility factor (K) represents the effect of soil properties and soil profile characteristics on soil loss (see Chapter 2 in Renard *et al.*, 1997). In a practical sense, K is a lumped parameter representing an integrated annual average of the soil and profile reaction to erosion and hydrological processes. The processes consist of soil detachment and transport by raindrop impact and surface flow, deposition due to topography and tillage roughness and rain infiltration into the soil profile.

The best erodibility factors are obtained from long-term direct soil-loss measurement on natural plots. Rainfall simulation data has also been used, but is recognized as being less accurate (Römkens, 1985). Only inherent soil properties are considered determinants of the USLE soil erodibility factor, which means that soil erodibility must be measured under the Unit Plot conditions described earlier. The minimum adequacy of the observation period for soil erodibility was usually taken as two years, but longer periods provide better results due to the likelihood of experiencing a broader range of climatic and soil conditions. Most of the plots used in measuring soil erodibility were in the Midwestern cropping areas of the US (see Table 3.1 in Renard *et al.*, 1997).

In most cases, US RUSLE1 users will have little trouble in selecting specific K values, because NRCS has identified values for most major soil mapping units. Site-specific values can be obtained from the widely available NRCS soil surveys, or directly from USDA soil databases. If such data are not available, the erodibility nomograph (Fig. 8.2), based on a relationship fitting the data as described above, is the most commonly used tool to estimate K, although there are some soils where it does not apply, and one of the site-specific relationships for specific soils (Renard *et al.*, 1997: 75) may be a better choice in the US. Users should contact their NRCS state soil scientist or other local soil specialist to certify the value to be used for their location. In other areas of the world, users may have to resort to soil sampling and the use of Fig. 8.2.

(iii) Topography factors (LS) There are more questions and concerns about the LS topographic factor than for any other term in RUSLE. The primary reason for these concerns is that the choice of slope length involves substantial judgment; different users choose different slope lengths for similar situations. The two primary questions here are what hillslope (downslope runoff path) to use to represent an area, and how then to define that hillslope in terms of specific length and steepness values. The first question is really one of policy rather than science (do we choose the worst-case hillslope, or the median slope, or some other?), while the second question is a more technical yet qualitative one of how to define where runoff begins, the path it takes down the slope and when it reaches a concentrated flow channel, thus ending the hillslope. The attention given to slope length is not always warranted because soil loss is often less sensitive to slope length than to any other USLE/RUSLE factor. For typical slope conditions, a 10% error in slope length results in a 5% error in computed soil loss. In contrast, soil loss is much more sensitive to changes in slope steepness than to errors in slope length. In the USLE, for example, a 10% error in slope steepness will usually give about a 20% error in computed soil loss. RUSLE has a more linear slope steepness relationship than did the USLE. Improvements in the relationship for steep slopes mean that computed soil loss for slopes less than 20% are similar in RUSLE and USLE, but on steep slopes, computed soil loss in RUSLE is just over half that predicted by the USLE, whose relationship did not include data for steep slopes. In addition, RUSLE makes more explicit the reliance of the length relationship on the susceptibility of the soil to rilling, which may be influenced by the slope steepness, soil characteristics, and management impacts. Finally, RUSLE includes a slope relationship specifically for the frozen soil region of the Northwest Wheat and Range Region (Austin

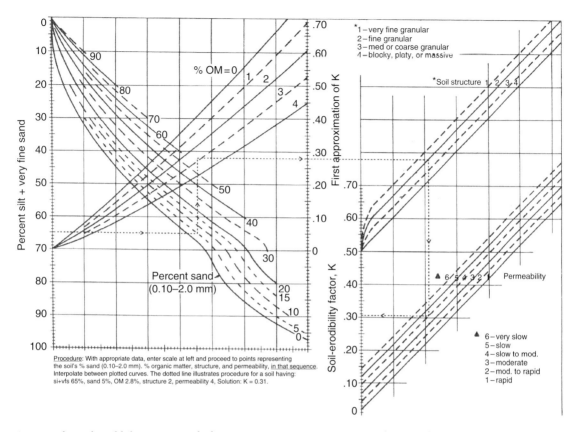

Fig. 8.2   The soil erodibility nomograph that gives *K* in US Customary Units (see Wischmeier & Smith, 1978).

1981). Detailed information on the selection of slope calculations is given in AH703 Chapter 4.

The difficulty in defining slope length is, however, substantial enough to have served as the primary impediment for employing GIS-based systems in using RUSLE. The topographic data available to populate GIS databases generally does not have the spatial resolution necessary to pick out the small concentrated flow channels commonly found at the bottom of a USLE/RUSLE hillslope. As a result, slope lengths computed using these data are almost always far too long. In fact, most attempts to use GIS with USLE/RUSLE recognize this and simply cut off the slope lengths at some arbitrary value. This poor resolution also causes the GIS system to miss the flat floodplains often found at the bottom of a hillslope, where

substantial deposition may occur. This may change as higher-resolution topographic data (such as those collected using Lidar) become available, although how best to use these extensive datasets must still be decided.

In using the USLE, the slope length was defined as beginning at the top of the hillslope where runoff starts, and extending down to where the sheet and rill flow reaches either a concentrated flow channel or a depositional area. This limit of the depositional area was required because such deposition rarely occurred on the plots used to collect USLE data. Deposition can be caused by anything that slows the runoff and causes sediment to deposit, such as an increase in roughness caused by a management change (e.g. a strip of dense vegetation), or a decrease in slope grade.

In RUSLE1, this definition was expanded slightly to include areas of deposition caused by management changes on the hillslope, which was accomplished by including some of the more process-based routines used in CREAMS (Foster *et al.*, 1980).

Slope length factor (L) Plot data used to derive slope length (L) show that erosion for slope length λ (ft) varies as:

$$L = (\lambda/72.6)^m \qquad (8.2)$$

where 72.6 = the RUSLE unit plot length (ft) and *m* is a variable slope length exponent. The slope length λ is the horizontal projection. The value for *m* can be found from $m = \beta/(1 + \beta)$, where the slope-length exponent β is related to the ratio of rill erosion (caused by overland flow) to inter-rill erosion (principally caused by raindrop impact). The ratio of rill to inter-rill erosion when the soil is susceptible to both rill and inter-rill erosion is:

$$\beta = (\sin \theta/0.0896)/[3.0(\sin \theta)^{0.8} + 0.56] \qquad (8.3)$$

where θ is the slope angle. For a value of β, the slope-length exponent *m* is calculated using the relation above. When runoff, soil, cover, and management conditions indicate that the soil is highly susceptible to rill erosion, the exponent should be increased (see AH703, Chapter 4). These conditions are expected, for example, for steep, freshly prepared construction slopes. In such cases where the soil is highly susceptible to rilling, AH703 recommended doubling the value of β resulting from Equation (8.3). When conditions favour more inter-rill and less rill erosion, as in cases of consolidated soils like those found in no-till agriculture, *m* should be decreased by halving the β value. A low rill to inter-rill erosion ratio is typical of conditions on rangelands. With thawing, and cultivated soils dominated by surface flow, a constant value of 0.5 should be used (McCool *et al.*, 1989, 1993). In RUSLE1 the choice between these alternatives was made by selecting a general land-use category; in RUSLE2 the pro-

gram automatically and continuously adjusts the *m* value based on slope steepness, soil type and management impacts.

Slope steepness factor (S) Soil loss increases more with steepness than with slope length. In RUSLE, the slope steepness has changed from that used by the USLE, and is evaluated with the relationship (McCool *et al.*, 1987):

$$S = 10.8 \sin \theta + 0.03 \quad S<9\% \qquad (8.4)$$

$$S = 16.8 \sin \theta - 0.50 \quad S>9\% \qquad (8.5)$$

The relationship is based on the assumption that runoff is not a function of slope steepness for slopes greater than 9%. Slope effect on runoff and erosion as a result of mechanical disturbance, cover and vegetation is considered in the cover-management (C) or support practice factor (P). For slopes shorter than 4.6 m (15 ft), use:

$$S = 3.0 (\sin \theta)^{0.8} + 0.56 \qquad (8.6)$$

Equation (8.6) applies to conditions where the water drains freely from the slope end. For the slope steepness factor above, it is assumed that rill erosion is insignificant on slopes shorter than 4.6 m (15 ft), and that inter-rill erosion is independent of slope length.

When freshly tilled soil is thawing, in a weakened state and primarily subjected to surface flow, use the following (McCool *et al.*, 1993):

$$S = 10.8 \sin \theta + 0.03 \quad S<9\% \qquad (8.7)$$

$$S = (\sin \theta/0.0896)^{0.6} \quad S>9\% \qquad (8.8)$$

In most practical applications, a single plane or uniform slope can be a poor representation of the hillslope topography, and erosion can vary greatly between concave or convex slopes of equal average steepness. Users are cautioned and encouraged to use the complex slope calculations, because differences can be significant when contrasted with a uniform plane.

Actual selection of the hillslope used to represent a field can be a complicated choice, and is best done through examples rather than verbiage. Additional detail and guidance for field measurement of the *LS* factor for varying field scenarios is given in AH703.

Cover-management Factor (C) The cover-management factor, *C*, is possibly the most important of the RUSLE/USLE factors because it represents the most readily managed condition for reducing erosion. In the USLE, the *C* factor was described as providing a measure of how erosion from the current condition compares with that for the Unit Plot condition, which is considered as nearly worst-case. The individual values of *C* vary between 0 for a completely non-erodible condition, to a value somewhat greater than 1.0. Values greater than 1.0 imply conditions more erodible than those normally experienced under Unit Plot conditions, which can occur for conditions with very extensive tillage (e.g. roto-tilling), leaving a very smooth surface that produces much runoff and makes the soil especially susceptible to erosion. *C* values are weighted average soil-loss ratios (SLRs), each of which represents the ratio of soil loss under current conditions for a short period of time to the expected soil loss under Unit Plot conditions during that same period. The SLRs vary throughout the year as soil and cover conditions change with soil disturbance and plant growth. The *C* value then represents the average of the time-varying SLR values, each weighted by the portion of rainfall erosivity during that same time period.

In contrast to the tables of *C* factors presented in AH282 and AH537, RUSLE1 uses a subfactor method to compute SLRs as a function of five factors:

$$C = PLU \cdot CC \cdot SC \cdot SR \cdot SM \qquad (8.9)$$

where *C* is the overall cover-management factor, *PLU* is the prior landuse subfactor, *CC* is the canopy cover subfactor, *SC* is the surface cover subfactor, *SR* is the surface roughness subfactor, and *SM* is the soil moisture subfactor (used only in

the Northwest Wheat and Range Region area (Austin, 1981), otherwise unity). Expanded details for evaluating *C* factors are presented in AH703.

Although ground cover is known to affect erosion more than the other subfactors, it is wrong to give it exclusive attention without considering within-soil effects such as those associated with root mass and tillage. A 30% surface cover after planting is the criterion frequently used for conservation tillage; the USLE relationships predict that this 30% cover will reduce soil loss by about 72%. By comparison, the soil loss for a freshly ploughed meadow is reduced by about 75% from that for Unit Plot conditions, showing that within-soil effects can have a substantial impact. Although the effects are not as pronounced, the impacts of canopy cover and surface roughness can also provide substantial benefits, especially in the absence of surface cover.

The structure of Equation (8.9) implies that the effects of subfactors in reducing erosion are multiplicative. For example, if there is a canopy cover that reduces erosion by 45% from Unit Plot conditions, this means that $CC = 0.55$. If there is also enough surface cover to reduce erosion by 60% from Unit Plot conditions ($SC = 0.4$), then assuming all other factors are under Unit Plot conditions ($PLU = SR = SM = 1.0$), the overall factor value would be $C = 0.55 \times 0.4 = 0.22$, or a 78% reduction in erosion from Unit Plot conditions.

The subfactor approach in RUSLE1 was designed to break the dependence of the USLE structure on specific land-use data. Without this break, calculations would require separate complete and expensive datasets for each possible combination of land uses. The subfactor analytical approach was carried out under the basic assumption that the erosion impact of various factors such as surface cover and roughness is really independent of the type of land management controlling that factor. For example, the impact of covering the surface with straw mulch and of growing grass should be relatively independent of whether this is done as part of normal agricultural field operations or to control erosion on construction sites. Under this assumption, we start with the relationships to estimate erosion based on the parameters that control the subfactors for surface cover, biomass and roots in the soil, surface roughness, vegetative canopy cover, and soil moisture. Once those relationships are developed using field data, if the RUSLE1 program can model the effect of any field management operation on those parameters (soil, vegetation, biomass), it should be able to model the resulting erosion.

The subfactor approach and the equations controlling it are described in great detail in AH703. What follows is a brief introduction to the subfactors included in RUSLE1.

Prior land-use subfactor (PLU) The *PLU* subfactor is calculated in RUSLE as the product of soil consolidation and soil biomass effects:

$$PLU = C_f \cdot C_b \cdot \exp\text{-}[(c_{ur} \cdot B_{ur}) + (c_{us} \cdot B_{us}/C_f^{cuf})] \quad (8.10)$$

where *PLU* is the prior land-use subfactor (ranging from 0 to 1), C_f is a surface-soil-consolidation factor, C_b represents the relative effectiveness of subsurface residue in consolidation, B_{ur} is the mass density of live and dead roots in the upper 100 mm (lb acre^{-1} in^{-1}), B_{us} is the mass density of incorporated surface residue in the upper 100 mm of soil (lb acre^{-1} in^{-1}), C_{uf} represents the soil consolidation impact on the effectiveness of incorporated residue, and c_{ur} and c_{us} are calibration coefficients indicating subsurface residue impacts.

The B_u variables calculate the impact on erosion rates of live and dead roots and incorporated residue. The effectiveness of such materials can take two forms. Firstly, roots and residue can control erosion directly by physically binding soil particles together and acting as mechanical barriers to soil and water movement. Secondly, roots and residue exude binding agents and serve as a food source for micro-organisms that produce other organic binding agents. These serve to increase soil aggregation and thereby reduce susceptibility to erosion. The RUSLE software keeps track of the biomass in each layer, continuously adjusting the rootmass and subsurface residue to

account for residue additions or losses by decomposition.

Canopy cover subfactor (CC) The canopy-cover subfactor indicates the effectiveness of the vegetative canopy in reducing the energy of rainfall striking the soil surface. Although most of the rainfall intercepted by canopy eventually reaches the soil surface, it usually does so with much less energy than rainfall directly striking the ground. The intercepted drops fracture into smaller drops, or drip from leaf edges, or travel down crop stems to the ground. The canopy-cover effect is given as:

$$CC = 1 - F_c \cdot \exp(-0.1 \cdot H) \quad (8.11)$$

where *CC* is the canopy-cover subfactor ranging from 0 to 1, F_c is the fraction of land surface covered by canopy, and *H* (ft) is the distance that raindrops fall after striking the canopy.

Surface cover subfactor (SC) Surface cover affects erosion by reducing the transport capacity of runoff, by causing deposition in ponded areas, and by decreasing the surface area susceptible to raindrop impact. This is perhaps the single most important factor in lowering SLR values. Surface cover includes crop residue, rocks, cryptogams, and other non-erodible and non-mobile material in direct contact with the soil surface. The effect of surface cover on soil erosion is given as:

$$SC = \exp[-b \cdot S_p (0.24 / R_u)^{0.08}] \quad (8.12)$$

where *SC* is the surface cover subfactor, *b* is a coefficient, S_p is the percentage of land area covered by surface cover, and R_u is the surface roughness, as will be defined later.

Land area percentage covered by residue can be estimated from residue weight by the relationship of Gregory (1982):

$$S_p = [1 - \exp(-\alpha \cdot B_s)] \cdot 100 \quad (8.13)$$

where S_p is percentage residue cover, α is the ratio of the area covered by a piece of residue to its mass (acre lb^{-1}), and B_s is the dry weight of crop residue on the surface (lb ac^{-1}). If more than one type of residue is present, the resulting total surface area cover is calculated as:

$$S_p = \{1 - \exp[-\Sigma \, (\alpha_i \cdot B_{si})]\} \cdot 100 \qquad (8.14)$$

where α_i is the ratio of area covered to the mass of that residue for each type encountered. The summation is for each type of residue, as each residue type may have a unique α_i value.

Surface roughness subfactor (SR)   Surface roughness has been shown to affect soil erosion directly, and also to affect it indirectly through the impact of residue effectiveness as controlled by the b value in Equation (8.12). The surface roughness subfactor is a function of the surface random roughness, which is defined as the standard deviation of surface elevations across the slope, when changes due to land slope or non-random tillage marks (such as dead furrows, traffic marks, and/or disk marks) are removed from consideration. A rough surface has many depressions and barriers. During a precipitation event, these trap water and sediment, causing rough surfaces to erode at lower rates than do smooth surfaces under similar conditions. Increasing surface roughness decreases transport capacity and runoff detachment by reducing flow velocity.

Roughness and cloddiness of soils also affect the degree and rate of soil sealing from raindrop impact. Soils that are left rough and cloddy typically have greater infiltration rates. Soils that are finely pulverized are usually smooth, seal rapidly, and have low infiltration rates. RUSLE assumes that roughness decreases with the time since tillage by the relationship:

$$D_r = \exp[\; \tfrac{1}{2}(-0.14 \, P_t) + 1/2(-0.012 \cdot EI_t)] \quad (8.15)$$

where D_r is the dimensionless roughness decay coefficient, P_t is the total inches of rainfall since the most recent soil-disturbing surface operation, and EI_t is the total EI amount since that operation.

If the initial roughness is defined as R_i, then surface roughness just before a new tillage operation (R_u) can be defined as:

$$R_u = 0.24 + [D_r \, (R_i - 0.24)] \qquad (8.16)$$

where R_u is in inches. Since many field operations affect only a portion of the surface, R_u is also the roughness of that field portion left undisturbed by the current operation.

For that surface portion affected by the field operation, the resulting roughness has been found to be a function of subsurface biomass present in the top 4 in. of soil. The relationship is:

$$R_a = 0.24 + (R_t - 0.24)$$
$$\{0.8 \, [1 - \exp \, (-0.0012 \, B_u)] + 0.2\} \quad (8.17)$$

where R_a is the roughness after biomass adjustment (in.), R_t is the original roughness based on the assumption of ample subsurface biomass such as is found with high-yielding US-type corn, and B_u is total subsurface biomass density in the top inch of soil (lb ac^{-1} in^{-1}), with $B_u = B_{ur} + B_{us}$ as used in Equation (8.10).

The adjusted tillage roughness is then combined with that of the undisturbed portion of the surface as follows:

$$R_n = R_a \, F_d + R_u \, F_u \qquad (8.18)$$

where R_n is the net roughness following the field operation (in.) and F_d and F_u are respectively the fractions of the surface disturbed and undisturbed, such that their sum equals one.

Similarly, the roughness decay coefficient must be adjusted to reflect that only a portion of the field is disturbed using the relation:

$$D_e = D_r \, F_u + 1.0 \, F_d \qquad (8.19)$$

where D_e is the equivalent roughness decay coefficient. RUSLE then reorganizes the relationships described above to calculate the R_t, P_t and EI_t values corresponding to the equivalent roughness decay coefficients, under the assumption of a constant EI_t/P_t ratio. If a site is clean-tilled and left without human intervention, two things will happen: (1) the tillage roughness will decrease as defined previously; and (2) as time passes, vegetation will tend towards its climax community, with attendant roughness caused by protruding roots, soil pushing up around old basal areas, rocks, and so on. RUSLE assumes that the formation of this vegetative roughness follows a typical sigmoidal growth curve, increasing from the minimum roughness (r_{min} with a default of 0.24 in.) to the total roughness when soil is consolidated (r_{max}) over the time required for consolidation (t_{con}).

Once the current roughness R_u has been defined based on the tillage roughness and all the roughness decay calculations described above, the surface roughness subfactor for this time period is then:

$$SR = \exp\left[-0.66\left(R_u - 0.24\right)\right] \qquad (8.20)$$

Soil moisture subfactor (SM) In non-irrigated portions of the Northwest Wheat and Range Region (NWRR; Austin, 1981), soil moisture during critical crop periods depends upon crop rotation and management. In such cases, the addition of a soil-moisture subfactor (*SM*) is suggested. *SM* reflects dry fall conditions and increasing soil moisture over winter. The soil moisture decrease during the growing season depends upon crop rooting depth and soil depth, and the soil moisture replenishment during the winter and spring depends upon precipitation amount and soil depth. Research to make such a correction is needed. In most instances this factor is assumed to be unity, which means that there is no substantial impact of soil moisture extraction by the vegetation on erosion. This assumption of *SM* = 1.0 is probably valid for all areas except those experiencing erosion caused by light rains on frozen-thawing soils.

(iv) Conservation practice factor (*P*) It is not always clear how the conservation practice factor (*P*) differs from the cover management factor (*C*), because both are meant to indicate the impact of management practices on erosion. In general terms, the basic difference is that the *C* factor reflects the positive impact over the larger portion of the management area, through factors like vegetation, biomass on the surface or within the

soil, and roughness. The *P* factor is generally seen as reflecting the positive impacts of management through the control of runoff, with special emphasis on how the management changes the direction and speed of that runoff, but also reflecting to some degree management practices that control the amount of runoff. Traditionally the *P* factor has been used to reflect the impact of agricultural practices such as the various forms of strip-cropping (buffer strips, filter strips, rotational strip-cropping), terraces, contour tillage, and subsurface drainage. In other land uses, *P* would reflect the impact of analogous practices, such as filter strips for water quality control, or the use of diversions on construction sites. RUSLE1 brought to the USLE structure a subfactor approach for the *P* factor as well as the *C* factor, with separate subfactors for contouring, strips, terraces, and subsurface drainage. As with the *C* factor, these subfactor values are multiplied together to give the overall *P* factor.

Contouring subfactor Data on the effect of contouring show a tremendous amount of scatter, but there are some trends, as shown in Figure 6-2 of AH703. These indicate that higher ridges give more benefit than lower ridges, that contouring is more effective for areas with lower rainfall intensities, and that the effectiveness reaches a peak at about 9% slope, losing effectiveness at lower slopes due to less inherent erosion, and at higher slopes due to potential breakover of the ridges by ponded runoff. In addition, contouring is most effective when the ridges are perfectly on the contour, with its impact decreasing rapidly as the furrows have more grade.

RUSLE1 fits the scattered contouring data with a series of equations used to describe the base contouring *P* value for different slope steepnesses. It then adjusts these for climate and storm intensity using a runoff scaling factor based on the 10-year storm *EI* compared with a value for the central part of the US, and finally adjusts the results based on the contour furrow grade, using the relationship (AH703 eqn. 6-11):

$$P_g = P_o + (1 - P_o)(s_f / s_l)^{1/2} \qquad (8.21)$$

where P_g is the P factor for off-grade contouring, P_o is the P factor for on-grade contouring calculated using the sequence described above, s_f is the grade along the contour furrow, and s_l is the slope grade.

As reflected in the data summarized in tables in AH537, contouring tends to lose effectiveness on very long slopes, as runoff tends to build up behind the contour ridges and cause breakover of the ridges, which can be assumed to make the lower contour ridges ineffective. RUSLE estimates the maximum slope length over which contouring is effective (called the 'critical slope length' in AH703) using a variation on a relationship developed by Foster *et al.* (1982) for mulch stability. Once again, this relationship depends on slope steepness and runoff, and it is calibrated against the critical slope lengths shown in the tables in AH537. RUSLE then gives P-factor credit (i.e. reduces the erosion estimate) for the area upslope of the critical length, but not for the downslope area.

Strip-cropping subfactor The impact of management on runoff and its ability to carry sediment is probably the single factor that has changed most in the USLE/RUSLE evolutionary process. As described above, this has included substantial changes in how the hillslope is defined. RUSLE1 included a process-based approach to estimating the amount of deposition caused by changes in management and the resulting slowing of runoff. This started with the definition of a slope segment as being a portion of the topography with constant soil, management, and steepness. The approach taken was a simplified version of the CREAMS approach (Foster *et al.*, 1980), which looks at four possible cases for each slope segment, where a segment is defined: (1) where there is no runoff leaving the segment, so all incoming sediment is deposited; (2) where there is erosion throughout the segment; (3) where there is deposition throughout the segment; and (4) where deposition occurs at the top of the segment and erosion at the bottom. These four cases are examined by calculating the increase in transport capacity within the segment, and comparing that

with the amount of additional sediment added by erosion within the segment. This requires estimation of a runoff rate, which in RUSLE1 is based on the ten-year EI storm erosivity.

The impact of the deposited sediment on the P factor is somewhat subjective, as the P factor is meant primarily as a measure of soil resource conservation, while the primary effect of deposition is on sediment delivery. Because sediment deposition does not preserve the soil resource as much as preventing erosion in the first place, RUSLE1 does not give as much conservation credit for practices that cause sediment deposition as for practices that prevent soil erosion. RUSLE1 gives credit for deposition that occurs based on its location on the slope, using the relationship:

$$B = M \left(1 - x^{1.5}\right) \qquad (8.22)$$

where B is the benefit, M is the mass of sediment deposited, and x is the location of the deposition as a fraction of the total distance downslope. This benefit is calculated into the P factor as:

$$P_s = (g_p - B)/g_p \qquad (8.23)$$

where P_s is the P factor for strip-cropping, and g_p is the potential sediment load that would occur if there was no deposition.

Terracing subfactor Within RUSLE, terraces (or diversions on construction sites) provide two benefits: (1) they break the hillslope profile into a combination of multiple shorter profiles, thereby reducing erosion; and (2) they cause some deposition to occur up on the hillslope, thereby providing some benefit in conserving the soil resource. The first of these benefits is taken into account through the LS topographic factor described above. For the second benefit, RUSLE uses sediment yield data collected on watersheds with terraces to estimate the amount of sediment deposition that will take place, then gives that a credit benefit identical to that described above for the benefit of deposition in strip-cropping.

Subsurface drainage subfactor There are some data that suggest subsurface drainage can be effective in reducing erosion, presumably by reducing soil moisture and thereby decreasing runoff during a storm event (Formanek *et al.*, 1987; Bengtson & Sabbage, 1988). These data show substantial scatter, but indicate an average erosion reduction of about 40% for a subfactor $P = 0.6$.

Sediment delivery estimate As the first step in the evolution from the USLE, RUSLE1 is still primarily geared towards planning based on soil conservation. In spite of this, using the techniques described above for strip-cropping, it does provide a crude sediment delivery estimate. This is done by using a value of $B = M$ in Equation (8.22), providing the ratio of sediment delivered over sediment eroded, or essentially a sediment delivery ratio. Multiplying this P factor for sediment delivery by the other factors then provides the hillslope sediment delivery for evaluating off-site impacts.

8.2.2   *RUSLE1 program implementation*

The RUSLE1 computer code was written in the C programming language. Chapter 7 in AH703 included a fairly detailed description of the RUSLE1 program layout and operation. This was deemed necessary because as dissemination of and training in RUSLE1 proceeded, it quickly became apparent that a program with this level of complexity could not be assumed to be intuitive to a first-time user. Part of this complexity was inherent in the level of input information required from the user, while an additional portion was due to the program structure. This structure was based on the USLE 'paper' implementation, and hid nothing from the user.

8.2.3   *RUSLE1 implementation history and experience*

Perhaps the two primary lessons learned from the USDA-NRCS implementation of RUSLE1 were: (1) the importance of an iterative feedback process in developing the program; and (2) the sheer scale of the effort necessary to implement such a model on the national scale. Although

these lessons were partially due to the specifics of the situation, they are also broad enough to be instructive to other individuals and groups within or outside the US who are implementing a program like RUSLE. One of the key elements in the development of the RUSLE1 computer program was the close contact between the program developers and a variety of user representatives. Although the development began with defined user requirements, these underwent substantial changes as the program was presented to users through a variety of feedback and training sessions, involving a mixture of skilled and novice users. Only through that iterative feedback process did the program begin to meet the true user needs, as these needs often only became apparent when users were exposed to the program. Based on the RUSLE experience, it simply does not work to introduce a new model under the presumed process of setting initial user requirements and declaring success once those are met.

Although the RUSLE1 computer program itself was first deemed ready for full review and delivery in 1991, the process of developing the database information necessary to allow full implementation took an additional 4–5 years. This included a strong collaborative research effort sponsored by USDA-NRCS and carried out by researchers at North Carolina A&T University, Alcorn State University, and Alabama A&M University to collect the data required for the vegetation descriptions, including especially time-varying data on vegetative canopy cover, rootmass and biomass.

Substantial effort also went into determining exactly how the program would be implemented in the USDA-NRCS field offices, with special attention paid to consistency of results across political boundaries, and consistency of use patterns. One of the important concepts developed during this period was the development of *C*-Factor Zones, which recognized that climatic differences rather than political boundaries controlled the possible management scenarios, leading to shared management descriptions across state lines. National, state and regional NRCS personnel developed the required management descriptions to describe the bulk of schemes used in these areas, which in turn defined the operation and vegetation descriptions that had to be developed. In other words, broad implementation required both local expertise and substantial cooperation and oversight.

In spite of all the work that needed to be done and all the decisions that needed to be made (and remade!), full NRCS implementation of RUSLE1 began in 1993, using version 1.04 of the program, which is the version represented and documented in far more detail in AH703. This was actively used for conservation planning throughout the US, and for the Conservation Compliance portion of NRCS responsibilities associated with the 1985 and 1990 US Farm Bills.

8.2.4   Science problems with RUSLE1

As the USLE came into general use, it quickly became apparent that the impact of management on erosion could vary greatly among periods within a year or among years within a rotation. This was recognized by the later USLE methods, with AH537 using a time-varying SLR based on cropping periods. RUSLE1 carried this further, using a daily time-step for the *C*-factor calculations, and also for some of the *P*-factor calculations. However, due to user requirements that the structure of RUSLE1 reflect that of a 'paper implementation' of the USLE, this was not carried to its logical extreme. The time-varying values of each of the individual factors were aggregated over the year, and the resulting annual values were multiplied as shown in Equation (8.1). Unfortunately, this aggregated approach is not correct, as the sum of products is not equal to the product of sums, which can be seen in the simple calculation $(2 + 3) \times (4 + 5) = 45 \neq (2 \times 4) + (3 \times 5) = 23$. Clearly, the proper approach was to take any time-varying values and multiply these for each day or period, then add the daily products to get the total erosion. This was recognized as a problem early in RUSLE1 development, but it could not be dealt with while retaining the 'paper implementation' capability. Calculations showed

that the erosion results could vary by up to 30% between the two approaches.

8.2.5   RUSLE1 program weaknesses

In addition to the weaknesses inherent in the RUSLE1 science development, some more general weaknesses in the program operation became apparent during implementation.

The first program weakness was that the RUSLE1 structure was based on science rather than on how the user saw things. For example, one parameter used in the *LS* calculations is the soil texture, which affects the susceptibility of the soil to develop rills, thereby impacting the *LS* β value (Equation (8.2)). In spite of this, the user will clearly think of texture as a soil property, and not as something related to topography. This is one of many examples in RUSLE1 where there was a need to approach things more from the user's viewpoint, and not from the modelling viewpoint.

Another weakness of the RUSLE1 approach is that any user could change any database value. Although NRCS had put substantial effort into developing specific databases for climates and vegetations, any user could change the values, either intentionally or by accident. This resulted in many implementation headaches, such that two users using the same inputs would get very different results because one of the underlying database files had been modified.

Finally, the DOS-based interface used in RUSLE1 was already dated at the time of its delivery, and users repeatedly asked for a Windows®-based or similar graphical user interface, with which they were becoming increasingly familiar.

8.3   RUSLE2

As RUSLE1 developed, it quickly became apparent that there were some scientific weaknesses with the approach taken that were primarily caused by its close linkage to the methodology used in the USLE. In addition, through the training and implementation process, some lessons were learned about how the general program could be improved. Work began to address these issues in 1996, culminating in the release of a first RUSLE2 version in 2001 and the beginning of a US-wide NRCS implementation with actual distribution of the program to the field offices beginning in 2004. Based on some of the lessons learned in RUSLE1 implementation, this included a much earlier push to begin establishing the required databases, as well as to begin the iterative process of developing and modifying the program based on user feedback.

A primary change in the RUSLE2 implementation of the USLE relationships could be described as downplaying the importance of the individual factors. In the original USLE concept, these factors (except perhaps for *C* and *P*) were generally considered to be independent. This was clearly no longer the case in RUSLE1, as exemplified by the dependence of the β term in the *LS* relationship (Equation (8.3)) on various soil and management factors, which also impact the *K* and *LS* values. The factor-based RUSLE1 implementation caused both science and user problems. Inconsistent values could be entered in the various factors (a science error) and the program required the user to jump back and forth between factors in order to enter relevant and related data.

Implementation of RUSLE1 also made clear many places where the science was not specifically in error, but could be greatly enhanced. The most obvious of these was in full implementation within RUSLE2 of the CREAMS (Foster *et al.*, 1980) sediment transport and deposition approach, allowing the definition of the RUSLE2 hillslope to include depositional zones all the way down to the concentrated flow channel, and making RUSLE2 much more applicable to water quality problems. Other places where it was thought that the science could be enhanced by smaller improvements were many, especially in reducing the need for user selection of values by developing ways for the program to calculate needed values from information already available in databases. For example, in RUSLE1 the user needed to describe in several places the susceptibility of the soil to rilling; but this would vary

with time, and can be estimated using parameters such as the soil texture, slope steepness, and management parameters already calculated within RUSLE2.

As with RUSLE1, many scientists and engineers were involved in producing and delivering the RUSLE2 technology, including those involved in data collection and preparation for analysis. Most of these individuals are acknowledged in the references for the corresponding documents.

8.3.1 General approach to RUSLE2 science problems

The following summary brings up to date earlier and more extensive summaries of science improvements in RUSLE2 (Foster *et al.* 2000, 2003; USDA-ARS, 2008a,b). In this treatment, distinctions are drawn between RUSLE2 and RUSLE1 version 1.04, as documented in AH703. Some of the science enhancements in RUSLE2 exist in later versions of RUSLE1, specifically in RUSLE1.05 and RUSLE1.06, which were developed with the support of the US Department of Interior Office of Surface Mining (Foster *et al.*, 2003).

RUSLE2 retains the conceptual use of the USLE factors, makes computations that are based on soil loss estimates referenced to unit plot conditions, and uses ratios to adjust predictions to other conditions. However, RUSLE2 goes beyond the USLE. It uses process-based equations derived from fundamental erosion science and professional judgment to make RUSLE2 applicable to situations beyond the scope of USLE or RUSLE1. As scientific approaches improved, RUSLE2 was calibrated to reproduce the core SLRs for different cropping systems and crop growth stages listed in Table 5 of Agriculture Handbook 537 (Wischmeier & Smith, 1978). This calibration ensured that RUSLE2 erosion estimates for common situations would be similar to the established and accepted values that have been used for decades in the US for conservation compliance assessment.

A major change in RUSLE2 was the de-emphasis of the USLE factors, and the organization of information into 'objects'. This object-oriented organization applies to both the computer programming and the way that data are input by the user. The RUSLE2 developers made an effort to group and consolidate information needed by RUSLE2 into objects or descriptions that reflected how users think about the USLE factors. In the example mentioned above, with RUSLE1 the user had to use soil-related information not only in determining the K factor, but also in determining the LS factor, where the user chose among soil classes differing in their relative susceptibility to rill or inter-rill erosion (Table 4-5, Renard *et al.*, 1997). In RUSLE2, all soil-related information is included in a soil description, and all management information is contained in a management description. RUSLE2 combines these descriptions with the topographic description to define another description, that of a hillslope profile object, and extracts the information it needs from the descriptions to make erosion computations based on climate information contained in a location description.

Databases are maintained at the object level. Objects may contain other objects and sub-objects. For example, a management object is composed of the dates of occurrence of operation objects (like tillage, planting, or other soil-disturbing operations) and vegetation objects. Vegetation objects contain descriptions of growth patterns, and canopy and residue characteristics needed by RUSLE2 to compute the vegetation's influence on erosion. RUSLE2 does not simulate the growth of vegetation, but rather takes the information contained in the vegetation description and accounts for its effect on the L, C and P factors through numerous influences on variables tracked or calculated internally by RUSLE2, including soil biomass, surface residue cover, surface roughness, canopy cover, Manning's roughness, and the runoff curve number. In the USLE, all the factors were independent of each other; the K, L, S and P factors were annual constants, while the R and C factors were broken down into crop growth phases. In RUSLE1, the R, K and C factors varied among 24 half-month periods but remained largely independent of each other, although the LS and

ground-cover effects varied with the ratio of rill to inter-rill erosion, which in turn varied with soil texture, slope steepness and cover-management variables. In RUSLE2, all factors except S vary on a daily basis, and there are numerous interactions among the factors (USDA-ARS, 2008a). Annual averages of the RUSLE2 factors can be calculated, but the products of these averages will not equal the average annual erosion predicted by RUSLE2.

A major improvement in RUSLE2 is that the user can now define any number of steepness, soil, or management breaks along the slope, and the program will accordingly break the slope into segments representing each combination, and complete the calculations on those. RUSLE2 overcame limitations in describing complex hillslopes that existed in USLE and RUSLE1 by conceiving of hillslopes as being composed of three layers: topography, soil, and management. Each of these layers can be segmented independently to represent any complex one-dimensional hillslope situation. RUSLE2 then defines slope segments as each unique combination of topography, soil, and management layers. Because of the inclusion of deposition routines that were not part of the USLE or RUSLE1, RUSLE2 applies to hillslopes that include concave areas where sediment deposition occurs. Also, channels at the slope bottom, terraces with channels within hillslopes, impoundments, and sediment basins may all be described. These features allow RUSLE2 to compute sediment deposition and fine-particle enrichment of delivered sediment using process-based equations. Currently RUSLE2 does not simulate erosion in channels.

This ability to consider slope segments has also enabled RUSLE2 to deal nicely with the application of terraces or diversions as a management alternative. From a USLE/RUSLE perspective, the terrace channel becomes the concentrated flow channel defining the bottom of an upper hillslope profile, while the top of the terrace itself defines the beginning of a new lower profile. Within RUSLE2 this is handled automatically, defining not only the profiles, but allowing the user to specify the type of concentrated flow

channel transferring water down from the ter-
race/diversion to the hillslope bottom. This channel can currently be modelled to cause deposition,
but cannot currently be modelled as experiencing
erosion. The ability to easily add or remove terraces for the hillslope description is important
because it allows these to be approached as
another management alternative, rather than
requiring redefinition by the user of the hillslope
profile itself.

(i) Changes in the climate description The climate data required to calculate soil loss in
RUSLE2 are monthly averages for precipitation,
temperature and erosivity, plus the desired location's ten-year 24-h precipitation amount ($P_{10y,24h}$).
Climate description changes from RUSLE1 to
RUSLE2 include: specification of $P_{10y,24h}$ rather
than the ten-year *EI* event; updating the underlying record to the period from 1960 to 1989 (1960
to 1999 in many cases); and development of the
erosivity density concept. Specification of
monthly average precipitation and monthly average erosivity density is the preferred way of
describing monthly erosivity in RUSLE2, and
these values are contained in all the NRCS location climate files (USDA-NRCS, 2008). Erosivity
density is defined as the amount of rainfall erosivity per unit of precipitation. Erosivity
density has units of energy per unit area per unit time
(e.g. MJ ha^{-1} h^{-1}), and when multiplied by the
depth of precipitation over an interval (event, day,
month, year) yields the appropriate average erosivity value. Using erosivity density has several
advantages over directly calculated rainfall erosivity: (1) because it is the ratio of storm erosivity
to storm precipitation, missing data have less
impact on monthly means; (2) a shorter period of
record is needed to arrive at a stable value of this
ratio than a stable absolute value of erosivity;
(3) because erosivity density was found to be relatively independent of elevation up to 3000 m, it
was possible to interpolate a smoothly-varying
erosivity density surface for the entire nation,
making it possible to calculate erosivity for each
county (common use in the US) or each precipitation zone (USDA-ARS, 2008a,b). The effect of

elevation on erosivity was reflected by defining
precipitation zones within counties of 11 mountainous western US states. The erosivity density
approach allows geographically consistent erosion predictions needed for a conservation/ero-
sion planning tool, and maximizes information
that can be extracted from available 15-min precipitation data.

(ii) Changes in the soil description Changes in
the soil description and *K*-factor computations
include the development of a modified nomograph for highly disturbed soils, the development
of new routines to describe time-variation in the
K factor based on location temperature and precipitation data, and the ability to reflect the
impact of subsurface drainage by specifying a soil
hydrological class. RUSLE2 contains equations
representing both the standard nomograph (Fig.
8.2) and a modified nomograph that applies to
disturbed soils such as construction sites or
reclaimed mine soils. The modified nomograph
is the same as the standard nomograph for fine
granular soils ($S = 2$), but the structural trend in
erodibility is reversed in the modified nomograph, so that erodibility decreases as structure
varies from very fine granular to massive. In the
modified nomograph, the labels for class 1 and 3
structures would be exchanged and the line for
class 4 structure would be to the left of all structure lines shown in Fig. 8.2. The modified nomograph is recommended for highly disturbed lands
such as reclaimed mined land and construction
sites, whereas the standard nomograph is recommended for agricultural soils because of its
empirical support. For equivalent soil properties,
both the standard and modified nomograph
return a base *K* factor for Columbia, MO, which
is a reference location and the centre of the
RUSLE2 domain.

RUSLE1 included a time-varying *K* factor
that was based on a few data points collected in
the central US that indicated a time-varying
change in Unit Plot erosion from storms with
similar erosivity. New relationships in RUSLE2
capture the effect of temperature and precipitation on the likelihood of runoff and hence the *K*

factor. For example, during cool and wet periods, higher antecedent soil water is likely to increase runoff and soil erosion, thus K should be higher. Similarly, increased temperature is expected to increase evapotranspiration, leading to lower antecedent soil moisture, lower runoff, and reduced K values. Relationships in RUSLE2 capture the main effects of seasonal variation in K at each location based on the ratio of temperature and precipitation values at each location to the average annual values at the reference location (Columbia, MO). For identical soil descriptions, these adjustments will increase the annual effective K at locations that are cooler and wetter than Columbia, MO, while average K values will be lower than the nomograph value at locations that are hotter and drier than Columbia, MO.

In RUSLE2, inclusion of the CREAMS (Foster *et al.*, 1980) sediment transport and deposition relationships requires knowledge of the sediment size distribution at the point of detachment, so the diameter, specific gravity, and primary particle composition of each of five size classes is calculated as a function of soil clay using equations similar to those in CREAMS (Foster *et al.*, 1985). The effect of drainage on runoff and sediment transport is discussed below with regard to the P factor.

(iii) Changes to the topographic description
Whereas the rill to inter-rill erosion ratio in RUSLE1 was selected by the user, in RUSLE2 this ratio is calculated internally based on soil texture, prior land use (soil biomass and soil consolidation) effects, ground cover and slope steepness. This ratio determines the slope length exponent, m, in Equation (8.2), which controls the sensitivity of sheet and rill erosion to slope length. Instead of using Equation (8.3), the ratio of rill to inter-rill erosion in RUSLE2 is computed from (USDA-ARS, 2008a):

$$\beta = \left(\frac{K_r}{K_i}\right)\left(\frac{c_{pr}}{c_{pi}}\right)\left(\frac{\exp(-b_r f_g)}{\exp(-0.025 f_g)}\right)\left(\frac{s/0.0896}{3s^{0.8} + 0.56}\right)$$

$$(8.24)$$

where the ratio K_r/K_i is the inherent rill to inter-rill soil erodibility ratio computed as a function of soil texture (as discussed in the text following Equation (8.3)); the term c_{pr}/c_{pi} reflects the effect of prior land use on the rill to inter-rill erosion ratio; the ratio $\exp(-b_r f_g)/\exp(-0.025 f_g)$ reflects how ground cover affects rill erosion more than it affects inter-rill erosion, b_r and 0.025 are coefficients (%$^{-1}$) that express the relative effectiveness of ground cover for reducing rill erosion and inter-rill erosion, and f_g is ground cover expressed as a percentage. The last term is the same as Equation (8.3). Equation (8.24) shows how RUSLE2 takes the information stored in the topographic, management, and soil objects and uses it to calculate needed coefficients, thus reducing the need for users to specify unfamiliar parameters. The fact that the rill to inter-rill erosion ratio, as calculated from Equation (8.24), is independent of slope length (when it really is not) illustrates the price that RUSLE2 pays for the ability to retain the simple and familiar USLE equation structure.

Complex slopes can be represented in RUSLE2 to provide a better approximation of topography. A broad range of process-based routines allows for calculation of deposition caused by either management or topographic changes. This means that, for RUSLE2, the hillslope is defined as from where runoff begins until it enters a concentrated flow channel, which is the same definition as for WEPP.

(iv) Changes to the management description
One significant change from RUSLE1 to RUSLE2 was the grouping of field operations and vegetations into a separate management object or description. Management objects comprise descriptions of field operations (their dates of occurrence, and their effects on surface cover and surface roughness) with vegetation descriptions whose growth is begun by the operation (if any) and the yield expected for that vegetation, and the amount and type of external residue added to the surface if a mulching operation. Management descriptions result in daily tracking of an extensive suite of variables that affect sheet and rill erosion, including canopy cover, standing residue,

surface residue, surface roughness, ridge height, and the depth distribution of buried residue and soil biomass. Some of these, like standing stubble and ridge height, are variables that did not exist as USLE or RUSLE1 subfactors, but even the more familiar variables have received new and more detailed treatment in RUSLE2. In addition to surface and standing residue, RUSLE2 tracks dead biomass in 24 2.5-cm-thick soil layers in the soil profile. By default, standing residue decays at a rate that is a fraction of that of the surface residue, buried residue, or dead roots, which all decay at a rate controlled by climatic and residue variables using the same relationships as in RUSLE1. Mechanical tillage operations are described much more fully in RUSLE2 database files than in RUSLE1, in terms of the impact they have on flattening standing residues, disturbing the soil, or affecting the growth of vegetation. Soil disturbance is described in terms of the fraction of the soil disturbed, the intensity and depth of soil disturbance, the creation of ridges and random roughness, and the effect on burying, redistributing, or re-surfacing residues.

In a vegetation description, users define the base crop yield, the time course of canopy and root mass development (a 'growth chart'), and the characteristics of the residue produced when the crop dies. RUSLE2 uses this information once a 'begin growth' operation in a management description calls for that vegetation. The growth of the vegetation in RUSLE2 is independent of the location's climate data, so it must be properly described by the user for the situation being analysed. Several 'wizards' are available in the RUSLE2 interface to help users to develop vegetation descriptions, to define canopy/biomass relationships, canopy shape and intercepted raindrop fall height, and yield/flow retardance relationships. A new portion of the program specifically designed to help the database developer and program user properly to account for residue and root production in perennial vegetation systems is being developed, and is discussed subsequently.

One key feature added to the vegetation/operation/management descriptions in RUSLE2 is the ability of the user to vary crop yield. Vegetation is described for a specific assumed base yield, but when the vegetation is actually used within a management regime, the user can specify a higher or lower yield value. The vegetation description includes how the biomass varies with yield, allowing adjustment of all of the vegetation parameters by the program.

(v) Changes to the support practice factor
Whereas the RUSLE1 user selects a cover management condition that, together with the soil hydrological group, defines a 'runoff index' analogous to the runoff curve number (CN), RUSLE2 calculates a CN internally as a function of soil hydraulic class, soil biomass, soil consolidation, soil roughness, and soil residue cover, thus reflecting the combined effects of soil, management and climate. RUSLE2 calculates runoff for the $P_{10y,24h}$ rainfall event every day. It also calculates sheet and rill erosion for this index event, and uses process-based equations to determine sediment transport, deposition, and fines enrichment. 'Infiltration' is calculated on slope segments with a low CN as the difference between $P_{10y,24h}$ precipitation depth and the 'initial abstraction', taken as 0.2 times the 'maximum retention' parameter, a transform of the CN (USDA-ARS, 2008a). The RUSLE2 equations for sediment transport capacity and deposition, and robust simplifications of the equations used in CREAMS, give RUSLE2 the ability to reflect the effects of spatial variation of soil erodibility, slope steepness, and cover management along a slope on detachment, transport and deposition. This approach results in estimates of the long-term average sediment production, erosion rate, transport capacity, deposition, and sediment characteristics along the slope, as well as the sediment amount and characteristics of sediment leaving the slope (Foster *et al.*, 2000). In fact, RUSLE2 goes further than other 'process-based' models, in that it approximates backwater effects when it determines the effectiveness of dense narrow vegetative buffers on sediment trapping (USDA-ARS, 2008a). RUSLE2 also includes the ability to approximate the effect of simple impoundments and channels on sediment delivery and fines enrichment.

8.3.2 RUSLE2 implementation and lessons learned

(i) RUSLE2 websites There are two 'official' RUSLE2 web sites: an ARS site, http://www.ars. usda.gov/Research/docs.htm?docid=6010, and an NRCS site, http://fargo.nserl.purdue.edu/rusle2_dataweb/RUSLE2_Index.htm. Both sites offer the same model, but with different databases, permission (access) levels, and templates. The ARS site provides a minimal database and access levels that allow scientists and engineers to see and change more parameters. The NRCS site includes much more extensive databases, and templates including a wide variety of additional tools, but the permissions for database manipulation are more restricted. The USDA-NRCS website is the single national point of delivery for the NRCS-approved RUSLE2 management templates and database components. Both websites contain documentation and training materials.

The NRCS website is remotely maintained and kept current by the NRCS database manager, who posts frequent database updates, revised soils data in RUSLE2 format, and updates to the 24,000+ management templates. Although the current version installer is posted for downloading and installation by private sector users, the NRCS has recently begun using an automatic software installation process for new releases of RUSLE2. This minimizes the amount of support time necessary to remove and install RUSLE2 on NRCS field office computers.

(ii) RUSLE2 interface: plasticity and security An internal NRCS oversight and evaluation review of RUSLE1 implementation uncovered significant differences in soil loss estimates from RUSLE1 across county lines in adjacent states and regions due to a lack of consistent RUSLE1 databases within NRCS. With this past experience in mind, RUSLE2 was implemented with a hierarchical approach that allows users to see and change only those factors they fully understand. In RUSLE1, any user could change any parameter, sometimes leading to a very unlikely combination of inputs that gave them the output they

desired. In RUSLE2 inappropriate changes are controlled by three mechanisms. The first is the user interface, which is very user-configurable. This allows more complicated inputs or outputs to be removed from the visible set, simplifying the model to a degree matching the user's interests and abilities. Since in the RUSLE2 calculation engine a parameter that is not needed is not calculated, removing unnecessary parameters also accelerates calculations. The second control mechanism is called access control, which limits what the user is allowed to see or edit. Access must be granted to the user by a higher-level user, providing a very flexible control structure that can be modified as a user is trained and needs greater control over the program. The third control mechanism is protection of specific records or groups of records within the database. For example, records created by a user with a high access level can only be edited and re-saved by a user with that access level or a higher one in the same access chain. Other users can edit the record, but can only save it as another record, over which they can exert control. As a result, once NRCS creates and locks a record, they can distribute it with the confidence that it cannot be modified by less knowledgeable users.

(iii) RUSLE2 database development and management RUSLE2 is supported by databases that store factor data and data entered by users. The climatic data are held in a location/climate description stored in the database, as are the soil data in their own separate description. These can then be accessed for re-use simply by calling for them by name. The most extreme example of this approach is in the management descriptions. A management description is a list of the field operations and associated dates, including what vegetation is planted or residue added (if any). These field operation, vegetation, and residue descriptions are each stored in their own named database descriptions for potential re-use by other managements, which in turn can also be stored.

Database development began in early 2000 with the designation of a USDA-NRCS National Database Manager or 'czar' who was given the task of expanding the initial minimal core vegetation and operation descriptions for NRCS use on cropland and pastureland, as well as assisting with the development of detailed climate descriptions, and directing and managing the importing of soils data for all available soil surveys. Working with many colleagues, the database czar populated a single nationally-coordinated database of climate, soils, operations, vegetations, residue and support practice descriptions. For consistency, field office users were 'locked out' of editing the data in these parts of the database.

Because the national database was vast, it was organized into sections that could be downloaded from the NRCS website for use in local conservation planning. Soils data were organized by state and county or soil survey area. Thus, only the soils data that a particular field office or user needed would be contained in the local database, although another soils description could be imported as needed. Climate data were organized by state for use in the same way. Management records were organized by Crop Management Zones (CMZs), 75 regions of the country with similar crops and tillage systems.

Climate records. Climate data were populated for the entire US, including Alaska, Hawaii and the Pacific islands, Puerto Rico and the Virgin Islands. The effort included extraction of the monthly parameters from the national 1960–1989 dataset (1960–1999 in some cases), with calculation of monthly EI values for stations with recording intervals of 15 min or shorter. These data were smoothed using several routines and visual inspection to provide a relatively smooth erosivity density 'surface', which was then used to provide point values or, more commonly, an average value over a county.

Soil records. Creation of soil descriptions in RUSLE2 was eased by making direct use of the NRCS NASIS/SURGO soil database and tools, available online at http://soils.usda.gov/technical/nasis/. This is based on an NRCS soils expert (usually the State Agronomist) downloading from NASIS all of the necessary soil descriptions, then running those through a RUSLE2 utility that extracts the necessary information, tests it for con-

sistency, and puts it into the required RUSLE2 format. Most RUSLE2 soils databases include some generic soil descriptions based on soil texture, and these are often more appropriate for use with highly disturbed and mixed soils like those on construction sites and mine reclamation projects.

Management records. The RUSLE1 experience used the approach of organizing the US by C factor or EI distribution zones in order to develop and coordinate the issuance of C-factor sets for common single crop and crop rotation scenarios. With RUSLE2 implementation, this cropping region concept was built upon with the creation of 75 Crop Management Zones (CMZs), in which common crops and tillage systems were described in detail and saved as 'locked' RUSLE2 management templates. CMZs are zones in which the climate and other factors thought to control management are assumed to be constant and unaffected by political boundaries. In other words, within a CMZ the crops are likely to be grown with very similar planting and harvest dates, as well as similar tillage systems, and so on. For example, one CMZ representing the central Corn Belt stretches east from the southeastern corner of Nebraska and northwestern corner of Kansas through northern Missouri, and across central Illinois, Indiana and Ohio. Another CMZ stretches south along the eastern side of the Appalachian range from Maryland into Alabama.

With national coordination, this effort involved significant coordination among NRCS state agronomists in setting typical dates of operations and creating these management templates to represent the typical tillage systems used in growing the important crops in each CMZ. Once a set of crop management template descriptions was created by a CMZ coordinator, it was submitted to the database manager for inclusion in the national NRCS RUSLE2 database. Each CMZ set was contained in a separate RUSLE2 export file so it could be imported into the local RUSLE2 database in each field office located within the boundaries of that CMZ. This provided a starting point for field offices as they implemented RUSLE2, and also provided consistency in the use of RUSLE2 since the locked management templates were based on typical dates and typical tillage systems used in the CMZ. Internally, the vegetations, operations and residues used by these managements all came from the same national NRCS RUSLE2 database. Users could copy templates into a local management file and change or edit the tillage system details, yields, crops and dates of operations to tailor them for use locally in specific runs, whereas the locked templates remained unedited for future use.

(iv) RUSLE2 database status   Although the data developed for RUSLE1 (especially the vegetation and operation descriptions) proved invaluable in the NRCS database development efforts, development of the RUSLE2 database was still a tremendous effort. The USDA-NRCS, with initial guidance from ARS and aided by university and other cooperators, has compiled a database that includes (as of 21 July 2008): (1) 105 residues, describing how much cover each provides and how fast it decomposes; (2) 917 vegetations, from asparagus to zucchini, with each describing how the vegetation grows in terms of providing canopy cover and biomass; (3) 438 field operations, describing what happens to the soil, residue and vegetation as a result of the operation; (4) 10,976 climate descriptions; (5) 1,048,659 soil component descriptions, representing 649,032 soil map units in 3100 soil survey areas; (6) 467 special descriptions describing saved descriptions of strip-cropping, contouring, terracing, and sediment control basin practices; and (7) 26,361 managements for 75 CMZs, describing how the field operations, vegetations and residues fit together into management schemes.

8.3.3   *Implementation needs and training requirements*

(i) Preliminary training   Initial RUSLE2 training was conducted by NRCS with assistance from the RUSLE2 development team in regional testing sessions during the period from 1999 to 2000. A minimal database, which included generic soils and only a few major crops and operations, was used for testing RUSLE2. As the training sessions progressed, it became clear that there were sev-

eral background requirements that would be required of the trainees prior to full-scale training, including:
• enough background in the underlying USLE/RUSLE science to allow the trainee to understand the conceptual approach, the use of the inputs, and the meaning of the results; and
• a general understanding of how the computer program organizes information and reflects the 'conceptual model' behind RUSLE. This was enhanced by the flexibility of the program in developing very simple user interface templates, which allowed the program to be introduced at a rather basic level.

In addition, a fuller database adequately reflecting the broad range of situations that users would need to address was required for full-scale training.

(ii) NRCS RUSLE2 training   Beginning in the summer of 2001, USDA-NRCS conducted regional 'train the trainer' sessions for NRCS state and area agronomists and others with erosion prediction responsibilities. These sessions were conducted by the NRCS Water Erosion cooperating scientist, national database manager, and the RUSLE2 development team. Training focused on the erosion science on which RUSLE2 is based, how to navigate the user interface, how the database structure is organized, the content of records in the various parts of the database, and hands-on experience in creating management scenarios and making simple RUSLE2 runs. One or two individuals from each state attended and began learning the model as well as learning how to train field office employees within their states. Each of these trainers then went back and conducted a series of 1–2 day RUSLE2 model training sessions to allow field office staff to develop sufficient skills such that they could make soil loss estimates using a relatively simple user template.

The regional 'train the trainer' sessions proved very valuable not only to the NRCS state personnel but also to the RUSLE2 development team, in that several NRCS user needs were identified that eventually led to enhancements and modifications to the RUSLE2 user interface. One enhancement was the development of a rotation builder module to allow creation of multi-year and multi-crop rotations from concatenation of single-year management scenarios. This also allows rapid substitution of individual tillage system years as treatment alternatives are explored during conservation planning with producers. NRCS users also expressed a need to group RUSLE2 runs for these different alternative treatments into a 'worksheet screen', and to assemble the worksheets for multiple fields within RUSLE2, thereby representing an entire farm. This need was addressed through the development of the 'plan view' in the interface. As states began to conduct field office training sessions, an NRCS User Guide for the RUSLE2 interface was developed and distributed. Additional 'how to' guides and references were prepared for specific tasks, such as importing and exporting database components, importing soils data from NASIS soils descriptions, installing new versions, and performing database updates.

As implementation and use of RUSLE2 expanded, and as NRCS and private sector users gained experience in the model, regional advanced RUSLE2 training sessions were conducted in all regions of the US. These sessions built on the initial training and provided more in-depth training, resulting in a deeper understanding of operations, vegetations, and support practice records, of modelling erosion and sediment deposition on complex slopes, of database management, of more complex screen views, and of organizing outputs and dealing with complex management scenarios. Additionally, as users became more sophisticated, more complicated screen views and printing templates that included more detailed outputs and analysis were developed and released.

(iii) Day-to-day support A significant amount of day-to-day support was provided to states and field offices by the NRCS water erosion cooperating scientist and the national database manager during the implementation years of 2001 through to the present. This was provided through a combination of telephone and e-mail support, and direct computer-to-computer sharing of software applications. Several 1–3 hour training sessions were conducted via this latter method to provide training to multiple states on new enhancements and timely topics. Support personnel also processed hundreds or even thousands of individual requests for additional vegetations, operations and support practices, as RUSLE2 use expanded across the US.

As various other applications were being developed, access to the 24,000 + RUSLE2 crop management templates became necessary. A common file exchange format was used so that these files could be exported and utilized by the Wind Erosion Prediction System (WEPS) model (Hagen, 1996). Several other applications have used this file exchange format to utilize all subsets of the RUSLE2 management templates.

8.3.4 *Most significant application enhancements*

RUSLE2 includes a user convenience called the 'worksheet view' that allows comparisons or combinations of a series of hillslope profiles, each of which represents a single RUSLE2 erosion/delivery calculation. For example, the management alternative worksheet uses a single climate, soil and topographic description, but below that shows a table of management alternatives and resulting erosion and sediment delivery (Plate 4). Each line in the table represents a single RUSLE2 calculation, and all lines share the common climate, soil and topography. The idea is that each worksheet in this case represents a field, with a list of likely management options and resulting erosion values. A group of worksheets can then be combined/compared in a 'plan', which can represent a farm, with each worksheet representing a field within that farm or land parcel. Within the worksheet, the user can control which management alternatives are brought into the plan for each field, allowing for comparison of all the alternatives. Once the planning decision is made, yes/no toggles can be set to display only the scenarios representing the 'before' and 'after' management alternatives. Thus they provide documentation of the 'before' and 'after' soil losses.

As indicated earlier, a feature added to the RUSLE2 program that greatly eased the database development effort was the inclusion of the Rotation Builder. This allowed the management scenarios to be developed as single crops, allowing the user within the program to 'paste' these together into the desired sequence. For example, the long growing seasons in the southeast US allows for multiple vegetable crops to be grown in sequence, resulting in a huge number of permutations requiring a large number of database descriptions. The Rotation Builder allows for limiting the descriptions to the single vegetations, which the RUSLE2 user can then combine within the program run as desired. If a specific combination is used frequently, the program allows users to save the combination as a single management, allowing for easy re-use.

Another RUSLE2 enhancement, developed with the support of the Wisconsin Department of Natural Resources and Dane County Land Conservation, is that the erosion and sediment delivery no longer need to be summed for the year. Conservationists working on construction sites are often interested in what happens only during some accounting period, over which the site operator is liable for erosion and sediment control. For example, the Wisconsin project defined a successful management plan as that which would keep the total sediment delivered from the site under 11 Mg ha^{-1} (5 US t ac^{-1}) during the period from the time of first soil disturbance until either placement of some non-erodible cover, or 60 days of growth of a permanent perennial vegetation. RUSLE2 allows for flexible definition of the accounting period, and of the target that must be achieved.

RUSLE2 also allows for printing a report describing the inputs and outputs of a RUSLE2 calculation. The form of the report is user-configurable, allowing users to define what they would like to see and in what form in a Microsoft Word® document. The resulting document can then be locked so that the user cannot change the results, and the associated RUSLE2 inputs can be saved into the document, allowing for a regulator to inspect the underlying information.

8.3.5 NRCS tools added to the NRCS RUSLE2 interface

Several additional calculations have been added to the NRCS RUSLE2 interface. The most prominent of these is calculation of the Soil Conditioning Index (SCI) (USDA-NRCS, 2002, examined by Zobeck *et al.*, 2007), which provides a rough estimate of whether a specific location/management/soil combination will tend to cause an increase or decrease in soil organic matter. One component of the SCI that has also proved useful is the Soil Tillage Intensity Rating (STIR), which makes use of the tillage type, tillage depth, operation speed, and percentage surface disturbance as a rough estimate of the soil disturbance cause by the operation. The STIR value is used as a criterion for NRCS's National Residue Management Practice Standards (available for download at http://www.nrcs.usda.gov/Technical/Standards/nhcp.html, accessed 3 September 2008). The STIR and SCI calculations have no impact on the RUSLE2 erosion/delivery calculations, but make use of the management operation and erosion results.

NRCS has added a calculation of management fuel usage based on the sequence of operations to the RUSLE2 interface. Several state phosphorus index calculations have also been added to the interface. Other tools were added to the NRCS RUSLE2 interface to compute a Nitrogen Leaching Index and an Energy and Fuel Use Calculator based on the tillage operations. Examples of the SCI and the Fuel Use Calculator results are shown in Plate 4.

8.3.6 Future of the technology

The science supporting RUSLE2 continues to advance and will be incorporated into future releases of the model. Two active areas of research include (1) residue production in perennial systems, and (2) ephemeral gully erosion estimation.

In most USDA erosion models (e.g. WEPP, WEPS, RUSLE1), residue production occurs only during senescence of a crop and is calculated from the decline in live biomass. This is equivalent to the assumption that there is no dead biomass production during periods of increasing biomass, and no additional growth after the peak biomass is reached. This is probably a reasonable and acceptable assumption for the treatment of annual crops. However, in perennial vegetation and in mixed stands where different components mature at different times, death and growth usually occur simultaneously. A new type of vegetation description is being developed for RUSLE2 in which residue production is more continuous, based on the assumption that live biomass has an effective life span. In the absence of forage harvest or biomass removal, the daily change in live biomass amount is calculated as the difference between new growth and the death of old growth. Live biomass that is not harvested is added to a dead biomass pool after its lifespan is reached, thereby providing the soil the benefits of additional residue cover. Users input monthly potential growth patterns and shoot and root life-spans, and RUSLE2 calculates corresponding residue production patterns. Growth patterns are altered in response to management operations involving biomass removal. Daily changes in residue biomass are then calculated as the difference between death and decomposition or residue harvest. RUSLE2's new routines will simplify the creation of vegetation descriptions for perennial systems, providing more realistic estimates of residue creation throughout the year, and thereby improving runoff and erosion estimation for pastures, hay fields, and other systems dominated by perennial vegetation.

To predict average annual ephemeral gully erosion is challenging because there is no existing long-term database of ephemeral gully erosion rates comparable to the plot database underlying the USLE, which in turn underlies RUSLE. Ephemeral gully erosion is a process inherently driven by larger-than-average runoff events (see Chapter 19). Many process-based models have developed climate-generators (e.g. CREAMS:

Knisel, 1980) that reproduce the stochasticity of weather. Applying these long-term weather records to an ideal runoff and erosion model would create a distribution of runoff and erosion events. Taking the monthly means of this population of ephemeral gully erosion events would represent the long-term average values needed to complement RUSLE2 sheet and rill erosion estimates and to estimate long-term average ephemeral gully erosion. The RUSLE2 developers proposed that modelling the correct storm amount and sequence of storms could reproduce the mean values. Toward this end, techniques to predict a sequence of index storms for any combination of soil and management anywhere within the continental US (and elsewhere, with appropriate calibration) have been developed, and require only RUSLE2 climate and profile-level information. The results approximate the mean monthly runoff, annual runoff event frequency, and a gamma distribution function scale parameter that characterizes 30-year stochastic runoff predictions generated using the AnnAGNPS (annualized Agricultural Non-Point Source) model (Bingner & Theurer, 2001).

By taking the largest in a series of runoff events as a 6-month return period event, and scaling the magnitudes of the periodic runoff events proportional to the long-term average disaggregated daily runoff amounts on event days, these parameters allow estimation of the date and size of a series of index runoff events that are proposed as the basis for an ephemeral gully calculation capability within RUSLE2. Index event RUSLE2 hillslope runoff, sediment yield, and sediment size distribution will be coupled with a physically-based ephemeral gully erosion model, possibly that used in CREAMS, to predict annual average ephemeral gully erosion.

8.3.7 *RUSLE2 examples*

RUSLE2 is so flexible that it is very difficult to decide which capabilities to show in a few examples, and in which form to display those. In narrowing the possibilities, it was decided to concentrate on three examples. The first example is included especially to dispel the notion that RUSLE2 is difficult and complicated to use. The other examples represent two common uses of RUSLE2. The second example compares management alternatives on a single field, as would probably be done by a conservation planner working with an agricultural producer. The third example demonstrates planning to meet a specific sediment delivery target on a construction site. In both of these latter cases the figure will appear relatively complex, but this complexity was added so that specific features could be highlighted.

(i) Example 1. Very simple view One of the complaints sometimes lodged against RUSLE2 is that it is too complex and difficult for a novice user. As described in the sections above, the complexity the user sees in RUSLE2 is totally controlled by what the user asks to see and how they ask to see it. The calculations are exactly the same for a simple view (Plate 5) as for a more complex one, except that fewer calculations may be needed because fewer outputs are requested. The user views are completely user-configurable, so there is an infinite number of possible views, not just some pre-specified simple, medium and complex views. The RUSLE2 screen capture (Plate 5) shows one of the simpler views, which could be used by someone with a minimal understanding of soil erosion. In order to get an erosion and sediment delivery result, the user need only select a location (climate), a soil, and a management from pre-existing lists in the database. They then enter a slope length and steepness (which assumes a uniform slope), and can immediately see the resulting erosion and sediment delivery. If desired, this view also allows the user to select from pre-defined contour or cross-slope tillage systems, to put in pre-defined vegetated barriers on or at the bottom of the slope, or to see what happens if pre-defined terrace systems are installed. If the trainer or program supplier believes that even these few conservation practices will not be understood by the target user, even these entries associated with Step 5 in the view above may be easily removed.

In this access level and view, the user has no way of directly modifying any of the inputs except slope length and steepness. Everything else must be selected from pre-defined descriptions in the database, presumably placed there by someone with the training and knowledge to do so. Most users are not long satisfied with so little flexibility. For example, they may want to be able to see the impact of a complex slope shape rather than being forced to assume a uniform slope. This increased power comes at the cost of increased complexity, as the user must now be faced with a user template allowing them to enter length and steepness values for the slope segments. This constant desire for more power and flexibility results in what the RUSLE2 development team calls 'template creep', which is the tendency of user templates to become increasingly complex over time in order to provide additional power. The RUSLE2 complexity that some users complain of is not built into the RUSLE2 program, but rather exists because other users who developed that user template thought those entries and outputs were necessary.

Finally, notice that Plate 5 shows the inputs and results in metric units, while the values shown in Plate 4 were in Imperial units. This demonstrates some of the additional flexibility of the RUSLE2 interface, which allows for any desired mixing and matching of units, and also for selecting the desired units within a system (e.g. cm or mm for height).

(ii) Example 2. Agricultural conservation planning The RUSLE2 screen capture shown earlier in Plate 4 presents the results of conservation planning on a hypothetical field. In this view the field is defined as having a single climate, soil, and uniform slope. Each line in the table then represents a single RUSLE2 erosion calculation, using the climate, soil and topography defined above, and combining it with a unique combination of contouring, terraces and cropping sequence to yield erosion, fuel use, and Soil Conditioning Index (SCI) results. In order from the top of the table, the lines represent: (1) corn with moldboard ploughing in the fall and disking in the spring, tilled and planted up-and-down the slope; (2) the same tillage, but close to the contour; (3) the same as (2), but with a single terrace in the middle of the slope; (4) fall chisel ploughing up-and-down the slope; (5) fall chisel ploughing, but close to the contour; and (6) fall strip tillage, where in the fall only a narrow strip is disturbed in knifing in nitrogen. The results for each line show the planner not only the erosion and sediment yield associated with each alternative, but also the estimated fuel cost and the SCI value for that option, with values > 0 indicating a net increase in soil organic carbon over time. These generally show the expected results, with the reduced tillage option resulting in the lowest erosion, fuel cost, and highest SCI values.

The graphs shown in Plate 4 indicate some of RUSLE2's capability in graphically representing results. In this case the graphs are of the percentage of soil surface covered by crop residue, with the graph on the left for the fall moldboard plough scenario, and that on the right for the strip till management. In addition, although it is not displayed here, the crop yields for each of the management alternatives can be set by the user, if it is thought that the management sequence has an effect on those.

(iii) Example 3. Construction site sediment control As described above, although the RUSLE2 calculations for estimating erosion and sediment yield for construction sites are no different from those for agricultural settings, the RUSLE2 flexibility allows for a substantially different look and feel, which makes it easier to use in construction settings. Several of these differences are shown in Plate 6.

One primary difference seen here is that for construction sites the primary output of interest is not the soil erosion on the hillslope, but rather the sediment delivery to the receiving channel, representing the off-site impact. In fact, it is often comparison of this value to some defined standard rather than comparison of average annual soil loss with the soil loss tolerance (T) (Johnson, 1987) that indicates the success or failure of a construction plan.

Another difference is the look and feel of the screen itself, including especially the visible icons and the text. These can be things as trivial as using a bulldozer icon instead of a tractor to represent field operations, or as substantial as completely different text shown on the screen for the same parameter, reflecting differences in terminology. For example, in agricultural settings we generally speak of crops and of crop residues added to the surface, while in construction settings we would use the more generic vegetation and surface cover materials, including synthetic blankets and added mulches as well as residues from the vegetation grown on the site.

Another difference mentioned above is the ability of RUSLE2 to aggregate results not only on an average annual basis, but over a user-defined accounting period. For example, in the situation shown here, the accounting period is defined as beginning from the time of the first soil disturbance until either the application of some non-erodible permanent material (e.g. pavement, sod, or landscaping materials) or 60 days of growth of perennial vegetation, with days whose average temperature falls below 35°F not counted. In Plate 6, the two bottommost results in the lower left-hand corner indicate whether the system meets the definition of the accounting period, and a green or red colour in the rectangle indicates whether the system did or did not meet the allowable sediment delivery threshold, in this case set by the regulatory body as a total of no more than 5 Mg ha^{-1} (2 US t ac^{-1}) over the entire accounting period.

Users indicated that for construction site use – unlike for agricultural use – there would be little need for the capability to save and re-use management descriptions, as the timing of field operations would vary tremendously due to many factors. Because of this, the view in Plate 6 shows the management scenario description (dates and descriptions of field operations) directly within the general RUSLE2 profile view, rather than named and stored as a separate database record.

These users also indicated a need to define complex slope topography, as they wanted to be able to account for the deposition occurring on the flatter portion at the bottom of an S-shaped slope: thus the complexity of Step 3 in Plate 6, which in the previous views was shown simply as single uniform slope length and steepness. The slope schematic in the upper-right corner of the view displays this complexity. The upper (management) layer of this schematic shows a management break about 55 ft (16.8 m) down the slope. This is caused by the selection in Step 5 of a pre-defined strip-barrier system, which in this case puts a single 20-ft strip of poor stand cool-season grass at the bottom of the slope. In addition, Step 5 sets that the runoff from the bottom of the slope feeds to a sediment basin, which is pre-defined as having an 80% settling efficiency for a silt loam soil that has not experienced previous deposition. This last clause is important because any deposition occurring before the runoff hits the sediment basin will cause the coarse material to settle out, thereby reducing the actual efficiency of the basin. In the specific case shown here, there will be deposition at the bottom of the slope caused by both the decreased slope steepness and the grass strip, so the basin will not provide 80% efficiency. If they had so desired, the users could have added additional complexity to the view to show where the deposition actually occurred, but this was not deemed worthwhile.

8.4   Summary

Soil erosion has long been recognized as a serious problem. Considerable efforts have been expended to address this problem, beginning in Missouri in 1923 and supported by the US Congress in a 1929 appropriation that initiated intensive soil erosion research. Early efforts to preserve soil and prevent erosion through the work of pioneers like H.H. Bennett led to an early period of plot scale conservation research at sites representing the ten major farming regions in the US. The 6 ft (1.8 m) wide by 72.6 ft (22.1 m) long (0.01 acre, 40 m²) research plots were constructed to represent various crops and rotations. Primary measurements included precipitation, runoff and soil

loss (erosion). The results from this research, in combination with additional crops and cultural practices data, ultimately provided a repository of data widely used by engineers and scientists to evaluate conservation practices. These data were the foundation of the empirical erosion prediction technologies and ultimately the Universal Soil Loss Equation (USLE).

The USLE was developed at Purdue University under the direction of Walter Wischmeier, with able assistance from Dwight Smith, and was published in 1965 and 1978 in two handbooks (AH282 and AH537). The handbooks became widely accepted for conservation farming (and especially soil erosion by water) in the US. In the early 1980s a program to develop technology to replace the USLE was initiated. The computer-based RUSLE (Revised Universal Soil Loss Equation) model was published in 1997. RUSLE incorporated significant advances over the USLE and permitted application of soil erosion estimation for a greater variety of crops and management practices beyond those in the original USLE database.

RUSLE was subsequently revised to include advanced scientific and interface technology and subsequently delivered as RUSLE2, along with expanded databases and more control over the parameters that specific users could see and change. The USDA-NRCS has accepted responsibility for the underlying databases within the US, which include descriptions of climates, vegetations and soils, along with extensive files describing common management practices. RUSLE2 is widely recognized as a major advance in erosion prediction and conservation technology, and provides a very flexible tool allowing resource conservationists, managers and developers to compare a broad range of management alternatives in deciding on an optimum resource use.

References

Austin, M.E. (1981) Land resource regions and major land resource areas of the United States. *USDA Agricultural Handbook No. 296.* 156 pp.

Bengtson, R.L. & Sabbage, G. (1988) *USLE P-factors for subsurface drainage in a hot, humid climate.* ASAE

Paper 88-2122, American Society of Agricultural Engineers, St. Joseph, MI.

Bingner, R.L. & Theurer, F.D. (2001) AnnAGNPS: estimating sediment yield by particle size for sheet and rill erosion. *Proc. 7th Federal Interagency Sedimentation Conference*, Reno, NV, 25–29 March 2001. I-1–I-7.

Brakensiek, D.L., Osborn, H.B. & Rawls, W.J. (Coordinators) (1979) Field manual for research in agricultural hydrology. *USDA Agriculture Handbook* No. 224. 550 pp.

Dissmeyer, G.E. & Foster, G.R. (1981) Estimating the cover-management factor (C) in the Universal Soil Loss Equation for forest conditions. *Journal of Soil and Water Conservation* **36**: 235–40.

Dissmeyer, G.E. & Foster, G.R. (1984) *A Guide for Predicting Sheet and Rill Erosion on Forest Land.* USDA-Forest Service Technical Publication R8-TP.

Ellison, W.D. (1947) Soil erosion studies. *Agricultural Engineering* **28**: 145–6, 197–201, 245–8, 297–300, 349–51, 402–5, 442–4.

Formanek, G.E., Ross, E. & Istok, J. (1987) Subsurface drainage for erosion reduction on croplands in northwestern Oregon. In *Irrigation Systems for the 21st Century*. Proceedings of the Irrigation and Drainage Division Special Conference, American Society of Civil Engineers, New York: 25–31.

Foster, G.R., Lane, L.J., Nowlin, J.D., *et al.* (1980) A model to estimate sediment from field-sized areas. In *CREAMS, A field scale model for chemicals, runoff, and erosion from agricultural management systems*. USDA, Conservation Research Report No. 26, pp. 36–64.

Foster, G.R., McCool, D.K., Renard, K.G. & Moldenhauer, W.C. (1981) Conversion of the USLE to SI metric units. *Journal of Soil and Water Conservation* **36**: 355–9.

Foster, G.R., Johnson, C.B. & Moldenhauer, W.C. (1982) Hydraulic failure of unanchored cornstalk and wheat straw mulches for erosion control. *Transactions of the American Society of Agricultural Engineers* **25**: 940–47.

Foster, G.R., Young, R.A. & Neibling, W.H. (1985) Sediment composition for nonpoint source pollution analyses. *Transactions of the American Society of Agricultural Engineers* **28**: 133–9, 146.

Foster, G.R., Yoder, D.C., McCool, D.K., *et al.* (2000) *Improvements in science in RUSLE2*. 2000 ASAE Annual International Meeting, Technical Papers; Engineering Solutions for a New Century 2: 2871–89 (paper no. 00-2147).

Foster, G.R., Toy, T.J. & Renard, K.G. (2003) Comparison of the USLE, RUSLE1.06c, and RUSLE2, for application to highly disturbed land. In *First Interagency Conference on Research in the Watersheds*. USDA-ARS Agricultural Research Service. Washington, DC: 154–60.

Gregory, J.M. (1982) Soil surface prediction with various amounts and types of crop residue. *Transactions of the American Society of Agricultural Engineers* **25**: 1333–7.

Hagen, L.J. (ed.) (1996) *Wind Erosion Prediction System technical documentation*. USDA-ARS. Available at http://www.weru.ksu.edu/weps/docs/weps_tech.pdf [accessed 4 October 2008].

Helms, D. (2008) *Hugh Hammond Bennett and the Creation of the Soil Erosion Service*. USDA Natural Resources Conservation Service. Historical Insights No. 8. 13 pp.

Knisel, W.G. (1980) A field scale model for chemicals, runoff and erosion from agricultural management systems. *USDA Conservation Research Report No.* 26. USDA, Washington, DC. 643 pp.

Johnson, L.C. (1987) Soil loss tolerance: fact or myth? *Journal of Soil and Water Conservation* **42**: 155–60.

Laflen, J.M. & Moldenhauer, W.C. (2003) *Pioneering soil erosion prediction: the USLE story*. Special Publication No. 1, World Association of Soil & Water Conservation. Beijing, China. 54 pp.

McCool, D.K., Foster, G.R., Mutchler, C.K. & Meyer, L.D. (1987) Revised slope steepness factor for the Universal Soil Loss Equation. *Transactions of the American Society of Agricultural Engineers* **30**: 1387–96.

McCool, D.K., Foster, G.R., Mutchler, C.K. & Meyer, L.D. (1989) Revised slope length factor for the Universal Soil Loss Equation. *Transactions of the American Society of Agricultural Engineers* **32**: 1571–6.

McCool, D.K., George, G.E., Freckleton, M., *et al.* (1993) Topographic effect of erosion from cropland in the Northwestern Wheat Region. *Transactions of the American Society of Agricultural Engineers* **36**: 771–5.

Meyer, L.D. & Moldenhauer, W.C. (1985) Soil erosion research: a historical perspective. In *Agricultural History*. University of California Press, Berkeley, CA: 192–204.

Mutchler, C.K. (1970) Splash of a waterdrop at terminal velocity. *Science* **169**: 1311–12.

Renard, K.G. (1985) Rainfall simulators and USDA erosion Research; history, perspective, and future.

In Lane, L.J. (ed.), *Proceedings of the Rainfall Simulator Workshop*, Tucson, AZ. Society for Range Management, Denver, CO: 3–6.

Renard, K.G., Foster, G.R., Weesies G.A., *et al.* (Coordinators). (1997) *Predicting Soil Erosion by Water: A guide to conservation planning with the Revised Universal Soil Loss Equation (RUSLE)*. USDA Agricultural Handbook No. 703, 404 pp. Available at http://www.ars.usda.gov/SP2UserFiles/Place/64080530/RUSLE/AH_703.pdf [accessed 3 September 2008].

Römkens, M.J.M. (1985) The soil erodibility factor: a perspective. In El-Swaify, S.A., Moldenhauer, W.C. & Lo, A. (eds), *Soil Erosion and Conservation*. Soil and Water Conservation Society, Ankeny, IA: 445–61.

USDA-ARS (2008a) Draft science documentation, Revised Universal Soil Loss Equation, Version 2. Available at http://www/ars.usda.gov/sp2UserFiles/Place/64080510/RUSLE/RUSLE2_Science_Doc.pdf [accessed 3 September 2008].

USDA-ARS (2008b) Draft user's reference guide, Revised Universal Soil Loss Equation, Version 2. Available at http://www.ars.usda.gov/sp2UserFiles/Place/64080510/RUSLE/RUSLE2_User_Ref_guide.pdf [accessed 3 September 2008].

USDA-NRCS (2002) National agronomy manual, section 508. Available at http://www.info.usda.gov/media/pdf/M_190_NAM.pdf [accessed 3 September 2008].

USDA-NRCS (2008) ftp://fargo.nserl.purdue.edu/pub/RUSLE2/NRCS Base Database/ [accessed 3 September 2008].

Weiss, L.L. (1964) Ratio of true to fixed-interval maximum rainfall. *Journal of the Hydraulics Division ASCE* **90**(HY1): 77–82.

Wischmeier, W.H. (1959) A rainfall erosion index for a universal soil-loss equation. *Soil Science Society of America Proceedings* **23**: 246–9.

Wischmeier, W.H. (1974) *New developments in estimating water erosion*. Proceedings, 29th Annual Meeting of the Soil Science Society of America, Madison, WI: 179–86.

Wischmeier, W.H. & Smith, D.D. (1965) *Predicting rainfall-erosion losses from cropland east of the Rocky Mountains: a guide for selecting practices for soil and water conservation*. USDA Agricultural Handbook No. 282. Available at http://www.ars.usda.gov/SP2UserFiles/Place/64080530/RUSLE/AH_282.pdf [accessed 3 September 2008].

Wischmeier, W.H. & Smith, D.D. (1978) *Predicting rainfall-erosion losses – a guide to conservation farming*. USDA Agricultural Handbook No. 537. Available in scanned pdf format at http://www.ars.usda.gov/SP2UserFiles/Place/64080530/RUSLE/AH_537.pdf [accessed 3 September 2008].

Zobeck, T.M., Crownover, J., Dollar, M., *et al.* (2007) Investigation of Soil Conditioning Index values for Southern High Plains agroecosystems. *Journal of Soil and Water Conservation* **62**: 433–43.

9 Application of WEPP to Sustainable Management of a Small Catchment in Southwest Missouri, US, Under Present Land Use and with Climatic Change

J.M. LAFLEN

USDA-ARS (retired), Buffalo Center, IA, USA

The WEPP (Water Erosion Prediction Project) is a daily simulation model developed to predict soil erosion on all lands. It is based on hydrology and erosion processes – detachment of soil by raindrops and flowing water, sediment transport by flowing water, and sediment deposition in flowing water and impoundments. It maintains a daily accounting of the condition of the soil and above- and below-ground biomass. It computes an estimate of the disposition of all water that falls on the land surface, whether from rain, snow or irrigation. It computes rates and volumes of runoff and estimates hydraulic shear, the detaching force occurring in rills and channels. It also computes the size distribution of detached sediment and the size distribution of sediment discharged from the area of interest.

It can be used on individual hillslopes (Hillslope version), and on small watersheds (Watershed version) as shown in Fig. 9.1. It includes soil, climate and management databases for the United States. It is widely used on public lands in the US by the US Forest Service and Bureau of Land Management. It is available on a Forest Service website (http://forest.moscowfsl.wsu.edu/fswepp/), on a US Department of Agriculture

Handbook of Erosion Modelling, 1st edition. Edited by R.P.C. Morgan and M.A. Nearing. © 2011 Blackwell Publishing Ltd.

website (http://milford.nserl.purdue.edu/), and it is available for download at http://www.ars.usda.gov/Research/docs.htm?docid=10621. The WEPP watershed version can also be used in a GIS environment (http://www.geog.buffalo.edu/~rensch/geowepp/). Technical documentation, tutorials, databases, and other WEPP information are also available at http://www.ars.usda.gov/Research/docs.htm?docid=10621. It has been extensively evaluated for a very broad range of conditions (Laflen *et al.*, 2004).

WEPP can be applied in other countries. The biggest obstacle is developing a suitable climate file. Methods have been prepared for developing the climate files using existing climate records (http://www.ars.usda.gov/Research/docs.htm?docid=10621). Plant parameters will probably need adjustment, some farm operations may need parameter adjustment, and soil files will need to be developed. Most information needed to develop such parameters and files is contained in the WEPP Help (part of the WEPP program), or in the technical documentation. WEPP contains a graphing capability that is very helpful for evaluating plant parameters, managements, and operations.

WEPP has been applied to various problems in the US, the most significant being its application to clean-up of the Rocky Flats nuclear weapons production plant near Denver, CO. The plant made

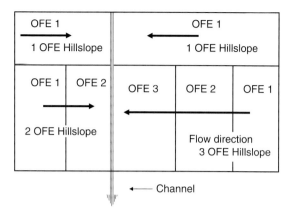

Fig. 9.1 The WEPP Hillslope version models one hillslope only, but a hillslope can have one or more overland flow elements (OFEs) which are homogeneous areas. Water flows downhill in a rill, passing from one OFE to the next lower OFE, with the lowest OFE on a hillslope discharging into a channel. The WEPP Hillslope version models the processes on a hillslope. The WEPP Watershed version models the processes that occur in the channels and impoundments which receive flow from hillslopes, other channels and impoundments.

components for the US nuclear weapons arsenal using various radioactive and hazardous materials, including plutonium and uranium and toxic metals such as beryllium and other hazardous chemicals (http://arq.lanl.gov/source/orgs/nmt/nmtdo/AQarchive/06springsummer/index.shtml).

WEPP was used in closure of a problem landfill near Las Vegas, Nevada (Smith *et al.*, 2002), and has been used in estimating the movement of radioactive materials in an urban area (unpublished work by Laflen). It is used to estimate daily erosion for the state of Iowa (Cruse *et al.*, 2006), with results published each day on the Internet (http://wepp.mesonet.agron.iastate.edu/).

The US has a total land area of nearly 9.4 billion km². Major land uses are about 264 million ha of forest, 238 million ha of grassland pasture and range land, 199 million ha of cropland, 120 million ha of special use (mostly parks and wildlife areas), 92.3 million ha of miscellaneous use, and about 24.3 million ha of urban lands. WEPP has been applied to all of these uses, perhaps most extensively on forest and range lands.

In this paper the focus will be on the application of WEPP to management of a small agricultural watershed, being a portion of a much larger watershed of great value because of the uses of the streams and reservoirs for power generation, cooling for power plants, water supply for municipalities, and as a major recreation area in the state of Missouri. The application to the small watershed is one in which the Universal Soil Loss Equation (USDA, 1961) or its derivatives (Williams, 1975) or revisions (Wischmeier & Smith, 1978; Renard *et al.*, 1997) might be applied to fields within the small watershed.

Soil erosion modelling needs related to agriculture have dramatically increased in the half-century since the release of the USLE in 1961. Soil erosion modelling is used in evaluating fertilizer application and manure management to achieve environmental goals (Iowa NRCS, 2004). Soil erosion models are a part of every large-scale model dealing with land use and management effects on sediment delivery to downstream locations. In the preface to the *User Requirements* for WEPP (Foster & Lane, 1987), the need for enhanced capabilities in erosion modelling was given as the justification for its development.

9.1 Description of Problem Area

The example agricultural problem is the application of WEPP to a small piece of land in southwest Missouri, US (Fig. 9.2). The land is typical of much of the area in western Missouri and eastern Kansas. The area chosen is located about 5 km south of the Marmaton river, a tributary to the Little Osage river, which flows into the Osage river, and, after receiving flow from several tributaries, and passing through the Lake of the Ozarks reservoir in Central Missouri, flows into the Missouri river. The Osage river, its tributaries and their reservoirs are major recreational areas. They include several reservoirs that are important recreational waters and that provide water to a number of municipalities, generate power, supply water for cooling power plants, and provide for flood control. The last reservoir (and the first to be

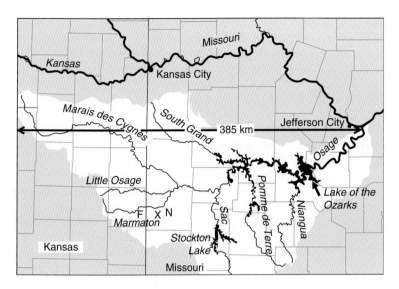

Fig. 9.2 Location of Osage River Watershed and the small agricultural watershed (X), west of Nevada, MO (N), and east of Fort Scott, KS (F).

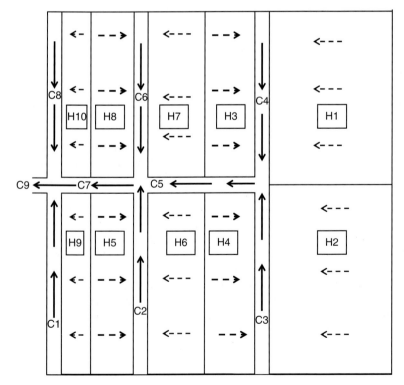

Fig. 9.3 Layout of small watershed. Individual hillslopes are numbered as H1, H2, ... H10. Individual channels are numbered as C1, C2, ... C9. Dashed lines show overland flow directions, and solid lines show channel flow directions.

constructed) on the Osage river is the Lake of the Ozarks impounded by the Bagnell Dam, originally constructed in 1931 to provide hydropower to much of eastern Missouri – mainly St. Louis. Reducing sediment loads to these rivers and reservoirs will ensure a sustainable water resource for these multiple purposes for generations to come. To help ensure that sustainability, the potential impact of global warming should be evaluated.

An additional problem for the entire Midwest agricultural area of the US is the contribution to the annual creation of an area of hypoxia in the Gulf of Mexico. The hypoxia area is partially caused by soil erosion and the ensuing delivery of nutrients to the Gulf of Mexico. One way of reducing the delivery of nutrients is to reduce soil erosion and sediment delivery, and this may well be required to reduce the hypoxia area in the Gulf; in fact, nutrient reduction goals for a number of states are already under discussion (UMRSHNC, 2005).

It is also important in selecting farming practices to evaluate the long-term implications of global climate change on sediment lost from this watershed. While the exact nature of the climate change in particular regions is still unknown, tools are already being developed and some are in place (Southwest Watershed Research Center, 2008) to estimate the impact of climate change when this change is known.

The problem is to select for this small watershed a suitable, sustainable land-management system that will reduce soil erosion, runoff volume, and sediment and nutrient delivery, and allow for intense use of this land in an area with a soil that is suitable for row crops, small grains and permanent grass production. The watershed layout as it will be modelled is shown in Fig. 9.3. While sustainability can be defined in a number of ways, in this case the focus will be on soil erosion and sediment delivery.

9.2 Model Criteria

The factors needed to evaluate for this watershed the impact of land management on sediment yield and water quality, and to design facilities that could mitigate damages, are:

(1) an average annual soil loss and sediment delivery estimate for an extended period;
(2) sediment characteristics;
(3) return period estimates for single storms for design purposes, and for estimation of onsite and offsite damages from rare storms;
(4) an identification of areas on the watershed hillslopes or channels where remedial measures might be needed; and
(5) an estimate of the impact of global climate change on soil erosion and sediment delivery.

The annual soil loss and sediment delivery estimates, as well as sediment characteristics, are important for the long-term viability of the streams and lakes in the Osage river watershed. Return period estimates for single storms are more related to the design of nearby structures and water conveyance systems, perhaps for zoning to minimize damage to high-value real estate and to minimize danger to humans and livestock. Siting and design of roadways, waterways, impoundments and conservation practices could be based on the results from such an analysis.

Global climate change estimates are being made, and these estimates could be included in such an evaluation. While there are many unknowns, there seems reasonable consensus that rainfall intensities and amounts will increase (Nearing *et al.*, 2005). With easily used tools available to develop scenarios for climate change, it seems prudent to use models and data that could at least give some reasonable estimate of the potential impact of global change. Such information could result in improved selection and design of systems, and in lowered future costs and damages.

9.3 Models that Might be Applicable

There are several models that could potentially be used (http://soilerosion.net/doc/models_menu.html). Not all have all the abilities needed to provide the information necessary to meet the criteria above. Several widely used models include:
• RUSLE (http://fargo.nserl.purdue.edu/rusle2_dataweb/RUSLE2_Index.htm)
• SWAT (http://www.brc.tamus.edu/swat/)

- AGNPS (http://www.wsi.nrcs.usda.gov/produ cts/w2q/h&h/tools_models/agnps/)
- EUROSEM (http://www.cranfield.ac.uk/sas/ nsri/index.jsp)
- GeoWEPP (http://www.geog.buffalo.edu/~ren sch/geowepp/)

Each of these models has its strengths and weaknesses. A major weakness for several of the models listed above is the use of USLE technology for erosion prediction. A major strength of several of the models above is their application to large watersheds. WEPP could not be applied to the entire Osage river watershed, but several of the above models might be suitable for that purpose; however, they would not capture the detail of the particular watershed studied here. RUSLE is not a watershed model, but it could make reasonable estimates of average annual soil erosion on each of the hillslopes in the small watershed here, but not sediment delivery or daily soil erosion values. GeoWEPP (Renschler *et al.*, 2002) is a geospatial interface for WEPP, using available digital geo-referenced information as model input to model watersheds. GeoWEPP can provide all the information that the WEPP Watershed version supplies but not the detailed analysis of hillslopes.

WEPP was chosen because it could work at the level of detail for this watershed to give information about the individual hillslopes and channels, the erosion from each and the deposition in each. These are the levels at which individual farmers operate, and where they make their decisions. Also, it has the necessary tools for easy use in the US, and particularly easily accessible databases and management files. Additionally, it was able to make estimates of the impact of climate change, and provide return period information useful in design for every scenario.

9.4   WEPP Setup and Implementation

The particular small watershed chosen for this example is in western Missouri, US (longitude 94.4° W, latitude 37.9° N). It is modelled in its state of about 30 years ago before extensive conservation work involving terraces and water-ways. The Watershed version of WEPP was used. The Watershed version includes the Hillslope version for hillslope erosion and sediment delivery estimates to channels, and then, the channels that detach, transport and deposit sediment are modelled in the Watershed version. The watershed could have been modelled using GeoWEPP, but the watershed as modelled is simple (Fig. 9.3) and could easily be parameterized. For more complex watersheds, GeoWEPP provides the necessary information to run the WEPP Watershed version. This information includes hillslope delineation, channel location, and slopes and dimensions of channels and hillslopes.

There are four kinds of information used by WEPP. These are topography, climate, soil, and management.

9.4.1   Topography

The small square watershed is about 0.8 km on each side. The individual hillslopes are modelled in either three or four slope length segments, flatter at the upper and lower ends than in the midsection. The slopes and slope lengths are typical of those found in the upland areas of eastern Kansas and western Missouri. All hillslopes had slope segments of 2% and 4%. All channels were modelled as having 2% slopes, fairly typical of these upland areas. The hillslopes were modelled as being uniform of soil and management (only one overland flow element per hillslope), and channels were assumed to be planted through, rather than set aside in a permanent waterway, a typical practice in this region when channels are not severely eroded. The entire watershed was farmed as a single field, typical of the larger farms and the larger equipment used in this region.

Except for the two upper hillslopes, H1 and H2, the area between channels drains to adjacent channels to the east or west. H1 and H2 drain to the west. Delivery of runoff and eroded material to the channels is assumed to be uniform along the channel. Topographic information for all hillslopes and channels is given in Table 9.1. This information includes the channel and hillslope dimensions, average slope, and the channels

Table 9.1 Topographic data for hillslopes and channels.

Hillslopes (H) or channels (C)	Area (ha)	Drains to	Average slope (%)	Length (m)	Width (m)
H1	12.22	C4	1.7	325	376
H2	13.78	C3	1.7	325	424
H3	4.70	C4	2.8	125	376
H4	5.30	C3	2.8	125	424
H5	5.30	C2	2.8	125	424
H6	6.36	C2	2.8	150	424
H7	5.64	C6	2.8	150	376
H8	4.70	C6	2.8	125	376
H9	3.19	C1	2	75	425
H10	2.81	C8	2	75	375
C1	0.04	C9	2	425	1
C2	0.09	C7	2	425	2
C3	0.09	C5	2	425	2
C4	0.08	C5	2	375	2
C5	0.06	C7	2	275	2
C6	0.08	C7	2	375	2
C7	0.06	C9	2	200	3
C8	0.08	C9	2	375	2
C9	0.02	Outlet	2	81	2
Total watershed area	64.57 ha				

that hillslopes and channels discharge into. No topographic surveys were available, so slopes, lengths and channel dimensions were estimated.

For this example application, hillslopes and channels were laid out on a rectangular grid, making it fairly simple to enter the various dimensions. In WEPP, a watershed can be constructed using an aerial photograph or a contour map as a background, locating the channels and hillslopes using the visible features on the background as a guide to draw in the various features. There are tools to draw and scale the various features. While WEPP internally computes every hillslope as a rectangle, the watershed can be constructed using a series of polygons, and WEPP will convert the polygon to a rectangle, preserving the hillslope area.

9.4.2 Climate

The climate data for WEPP was generated using the software program Cligen. Cligen (http://www.ars.usda.gov/Research/docs.htm?docid=18094) is a stochastic weather generator which, for a single point, provides a daily estimate of solar radiation, wind, dewpoint and temperature, using monthly parameters derived from historical measurements. It produces the information needed for the soil erosion model as well as other information to estimate the status of the land, biomass and water. The Cligen database was developed to have at least one Cligen station for most of the approximately 3000 counties in the US. The nearest Cligen station is Fort Scott, KS, about 25 km west of the area to be modelled. The Cligen database and the program to generate a climate is part of the WEPP package. Additionally, methods and data are available to develop climate files from international data (http://www.ars.usda.gov/Research/docs.htm?docid=18094). WEPP can also use historical data or breakpoint precipitation data.

The Cligen parameters can be adjusted to compensate for orographic effects and climate change. WEPPCAT was developed to adjust Cligen data for climate change (http://typhoon.tucson.ars.ag.gov/weppcat/index.php). It includes PRISM

Table 9.2 Return period daily precipitation amounts (mm) for precipitation intensification and increase in precipitation. Also shown is average annual precipitation for each scenario.

Return period (y)	Climate scenarios						
	Base (mm)	5%I (mm)	10%I (mm)	20%I (mm)	5%P (mm)	10%P (mm)	20%P (mm)
2	81	84	86	92	86	90	98
5	103	106	110	117	108	114	123
10	123	127	131	140	129	136	147
20	146	152	155	166	153	161	175
25	156	161	167	179	164	173	187
50	189	195	202	216	199	208	226
100	206	213	220	235	216	225	247
200	222	230	238	255	234	246	266
500	268	278	288	308	282	296	321
Average annual precipitation (mm)	1028	1006	1012	1029	1081	1129	1229

(http://www.prism.oregonstate.edu/) to adjust for orographic effects. PRISM is also available on the Forest Service website (http://forest.moscowfsl.wsu.edu/fswepp/).

Cligen was used to generate seven climate files for Fort Scott. The files were the baseline (unadjusted for climate change or orographic effects – no orographic adjustments were needed), three files with rainfall intensification of 5, 10 and 20%, and three files with precipitation increases of 5, 10 and 20%. The precipitation intensification and precipitation increase climate files were developed using WEPPCAT to adjust Cligen parameters for climate change.

The potential effect of climate change on precipitation for various return periods is shown in Table 9.2. Each of the seven climate files was generated for a period of 1000 years, so that precipitation, runoff volume, peak runoff rates and sediment delivery could be estimated for very long return periods.

9.4.3   Soil

The dominant soil on the area is the Parsons silt loam, with a surface layer of silt loam and a subsoil of clay to silty clay loam. It is moderately well to poorly drained. There may be perched water tables in wet seasons because of the clay subsoil. Slopes are quite moderate, while runoff may be high because of high precipitation, low available soil moisture storage, and very low infiltration and deep percolation rates.

For the cropped area in the Osage river watershed, the Parsons soil is a reasonable representation. Most of the cropped area is underlain by a clay subsoil. This clay pan drastically reduces water movement below about 0.5 m, which causes increased surface runoff for the cropped areas. Additionally, crop yields can be drastically reduced by short periods of drought.

While there is considerable cropping in this area of the Osage river watershed, towards the east, timber begins to increase and cropland decreases. A high percentage of the Osage river watershed is in grass.

The soil characteristics used in these model runs for the Parson silt loam are shown in Table 9.3.

9.4.4   Management

The major crops in this region are corn, soybeans and wheat. There is also considerable grassland for beef cattle production. These form the basis for the modelling of four management scenarios – corn, soybeans and wheat with

Table 9.3 Soil parameter values used in WEPP runs.

Layer	Depth (mm)	Sand (%)	Clay (%)	OM (%)	CEC (meq/ 10)	Rock (%)
1	305	6.4	24.5	0.75	16	1.1
2	2032	23.3	47.5	0.25	285	1.2
Interrill erodibility		$4702760\,\text{kg s m}^{-4}$				
Rill erodibility		$0.0074\,\text{s m}^{-1}$				
Critical hydraulic shear		$3.5\,\text{Pa}$				
Effective hydraulic conductivity		$1.44\,\text{mm h}^{-1}$				
Albedo		0.23				
Initial saturation		0.75				

Table 9.4 Dates and times of operations on watershed for 2-year corn–soybean–wheat rotation with three levels of tillage.

Date		Moldboard plough	Chisel plough	No till
4/1/1	Tillage	Tandem disk	Tandem disk	–
4/15/1	Plant corn	Planter, dbl disk	Planter, dbl disk	No-till planter
9/30/1	Harvest corn	–	–	–
10/1/1	Tillage	Moldboard plough	Chisel plough	–
10/15/1	Tillage	Tandem disk	Tandem disk	–
10/16/1	Drill wheat	Drill, dbl disk	Drill, dbl disk	No-till drill
6/15/2	Harvest wheat	–	–	–
6/20/2	Tillage	Moldboard plough	Chisel plough	–
6/25/2	Tillage	Tandem disk	Tandem disk	–
7/1/2	Plant soybeans	Planter	Dbl disk	No-till planter
10/15/2	Harvest soybeans	–	–	–
11/1/2	Tillage	Moldboard plough	Chisel plough	–

(1) moldboard plough, (2) chisel plough, (3) no-till, and (4) alfalfa production. The particular management (Table 9.4) was corn planted in the spring followed by a fall harvest, tillage was performed after harvest (for the moldboard plough and chisel plough managements), and winter wheat was planted. Winter wheat was harvested in June, tillage was performed (but not for no-till management) and soybeans planted. Soybeans were harvested in the fall, followed by tillage (but not for no-till management). The grass was cut for hay on June 1, July 15 and September 1. The managements modelled are all practised in the Osage river watershed.

The plants selected (corn, soybean, wheat and bromegrass) were those in the WEPP database for high yields. Dates of operation were based on discussions with a local farmer. When compared with published alfalfa, corn, wheat and soybean yields for Missouri (USDA, 1997), average yields for grass, corn and wheat were quite satisfactory, while soybean yields were low. The soybean yield when wheat is harvested and the land then planted to soybeans in the same year is frequently low in this area, and the yield results from the WEPP plant growth reflected this. The combination of planting before the soil moisture has any opportunity for recharge, the existence of a clay-pan, and high transpiration during a period when rainfall is low, does not produce many opportunities for high yields for double-cropped soybeans in this region. Additionally, average yield for the last hay cutting was low, reflecting the fact that in many years drought impacted upon hay yield late in the growing season.

There are many variations of tillage sequences that could have been used, but the sequences and equipment chosen in these four scenarios encompass the broad range of management practices that might occur in this area, from the most erodible to the least erodible. The types of machinery used in the model are frequently used in this area and were selected from the WEPP database.

WEPP allows the user to develop a user database for local conditions. This includes the parameterization of plants, machinery, and initial conditions for the local situation. WEPP also includes graphical and text outputs helpful in evaluating existing or user-developed plants, machinery and local conditions, as well as dates of various operations.

9.5   Model Results

Model results needed to make decisions about management of this small agricultural watershed include average annual soil loss and sediment delivery, sediment characteristics, return period estimates, location of erosion and deposition, and an estimate of the impact of global climate change. Additionally, information about performance of potential management systems during very unusual erosion events might be helpful in determining best management systems.

9.5.1   *Soil loss and sediment delivery*

The standard summary watershed model output in its standard format is shown in Table 9.5 for the chisel plough management system. Values given are average annual values of precipitation volume, runoff volume, soil loss and sediment yield for each hillslope, channel, and for the entire watershed. Dimensions of fields and channels are also given, as well as the overall sediment delivery ratio (ratio of sediment delivered from watershed to that detached on the watershed – including erosion on hillslopes and detachment in channels). Sediment and water yields for impoundments would also be available if impoundments were present.

A one-year portion (Year 170) of the 1000 years of daily output is shown in Table 9.6. The WEPP daily output is a text file that can easily be transferred to a spreadsheet for extensive analyses. Most WEPP output files are text files that can also be imported into spreadsheets.

Similar outputs, but for months and years, can also be selected. The year 170 was selected for presentation because it contained the largest one-day sediment yield, 77 t ha^{-1} for the moldboard plough system on 12 October. Table 9.7 shows the maximum daily sediment yield for each management system, and the sediment yield for all systems for the day, month and year in which a maximum occurred. These were gleaned from the daily output for each of the 1000 year runs for the baseline climate condition for each management system.

The maximum sediment yields never occurred on the same day (or even the same year) for any of the management systems. Also, while for the moldboard and chisel plough systems maximum loss occurred towards the end of a wet period, for the no-till and grass systems the maximum losses occurred as relatively isolated events.

Generally, a few storms produce the majority of soil erosion and sediment yield (Edwards & Owens, 1991; Ghidey & Alberts, 1996). This is also true for this application (Fig. 9.4). Fewer than 10% of the storms having sediment yield (about one storm a year) produced about 50% of the total sediment yield. A major contributing factor to the high erosion rates for the moldboard and chisel plough systems was that corn harvest was on 30 September in all even-numbered years, with moldboard or chisel ploughing occurring on 1 October in preparation for planting wheat on 16 October. The severe erosion occurred after that date.

The severe erosion for the moldboard plough occurred in the period with the highest 6-day rainfall in the 1000-year period, a total of 534.1 mm of rainfall on 8–13 October. This Cligen-generated rainfall amount was nearly identical in amount and timing to a 6-day rainfall amount on 29 September to 4 October 1986, when 533 mm of rainfall was recorded at Fort Scott. This event caused major flooding on the Marmaton river in Fort Scott, only a few miles from the location of this watershed. This was an extensive storm, with

Table 9.5 Model output for the chisel plough management system for a 1000-year period.

Contributing area (ha)	Discharge volume (m³ y⁻¹)	Sediment yield (t y⁻¹)	Sediment yield (t ha⁻¹y⁻¹)	Sediment Delivery Ratio	Precipitation volume in contributing area (m³ y⁻¹)
64.61	107,239	1549.6	24	0.843	663,677

Hillslopes	Runoff volume (m³ y⁻¹)	Soil loss (kg y⁻¹)	Sediment deposition (kg y⁻¹)	Sediment yield (kg y⁻¹)	Field area (ha)	Length (m)	Width (m)
H1	17,900	398,487	24,557	373,929	12.2	325	376
H2	20,188	449,379	27,693	421,686	13.8	325	424
H3	8435	134,267	509	133,758	4.7	125	376
H4	9511	151,390	574	150,817	5.3	125	376
H5	9512	151,407	574	150,834	5.3	125	424
H6	11,053	163,001	737	162,265	6.4	150	424
H7	9802	144,549	654	143,895	5.6	150	376
H8	8435	134,267	509	133,758	4.7	125	376
H9	6101	57,309	23	57,287	3.2	75	425
H10	5383	50,567	20	50,547	2.8	75	375

Channels	Discharge volume (m³ y⁻¹)	Sediment yield (t y⁻¹)	Length (m)	Width (m)
C1	6215	55	425	2
C2	20,690	281	425	2
C3	29,819	507	425	2
C4	26,447	448	375	2
C5	56,342	936	275	2
C6	18,353	249	375	2
C7	95,468	1451	200	3
C8	5483	49	375	2
C9	107,239	1550	81	2

many cities in Oklahoma, Kansas and Missouri experiencing similar rainfall amounts and major flooding. Flows into three of the reservoirs on the Osage river during runoff from that storm period were the highest on record.

Monthly and yearly outputs from hillslopes and channels are also available. Individual hillslopes may be modelled separately from the Watershed version and daily delivery of sediment and water to the various channels can be calculated. For hillslopes, a graphical output that allows plotting of nearly every input or output is available.

The average annual sediment yield from individual hillslopes (Table 9.8) and channels was used to compute where detachment and deposi-

tion occur in the channel system (Table 9.9). As shown in Table 9.9, on average, deposition occurred in the channels when erosion rates and sediment yields from hillslopes were high. When erosion rates and sediment yields from hillslopes were low, detachment occurred in these channels. Deposition rates in the channels averaged nearly 900 t ha⁻¹ of channel area (total channel area = 0.6 ha) for the moldboard plough system, and nearly 400 t ha⁻¹ for the chisel plough system. Detachment rates were small, averaging 7.3 t ha⁻¹ of channel area for the permanent grass, and 7.8 t ha⁻¹ for the no-till system.

Based on these results, there should be less maintenance of the channels for the no-till or

Table 9.6 Watershed daily storm output for all runoff events in Year 170 for moldboard plough management.

Day	Month	Year	Precipitation depth (mm)	Runoff volume (m³)	Peak runoff (m³ s⁻¹)	Sediment yield (kg)
9	4	170	30	784	0.4	7614
16	4	170	55	15,072	4.7	595,893
17	4	170	57	25,489	6.1	1,014,696
2	5	170	21	912	0.7	21,764
12	5	170	26	5538	1.6	133,525
21	5	170	26	6866	1.7	146,379
27	5	170	19	2665	1.3	13,962
17	6	170	36	1143	0.6	2863
23	6	170	24	504	0.3	891
25	6	170	24	797	0.4	1130
26	6	170	78	35,729	9.7	999,330
28	6	170	19	1223	0.5	1132
25	9	170	22	1574	1.2	5529
26	9	170	16	1265	0.6	2222
27	9	170	14	1339	0.6	2019
11	10	170	74	35,350	8.8	1,214,482
12	10	170	222	135,796	21.3	4,996,518
13	10	170	199	122,286	22.8	4,447,701
22	10	170	219	118,833	19.5	4,735,493
6	11	170	33	11,840	3.0	279,951
8	11	170	14	57	0.0	794
10	11	170	25	7429	1.2	30,832
9	12	170	16	668	0.3	1720

Table 9.7 Sediment yield for the day, month and year when maximum daily sediment yield occurred, for each management.

Sediment yield period	Sediment yields for managements			
	Moldboard plough (t ha⁻¹)	Chisel plough (t ha⁻¹)	No-till (t ha⁻¹)	Grass (t ha⁻¹)
Sediment yield on:				
21 October, year 60	74	60*	3.4	3.0
12 October, year 170	77*	33	2.5	3.4
26 August, year 212	20	20	6.3*	2.7
13 October, year 650	65	32	5.0	4.1*
Sediment yield in:				
October, Year 60	166	134	10.4	9.1
October, Year 170	238	125	7.5	7.9
August, Year 212	23	23	6.9	3.2
October, Year 650	65	33	5.0	4.1
Sediment yield in:				
Year 60	228	195	18.8	13.4
Year 170	288	170	9.7	9.1
Year 212	99	88	13.0	6.3
Year 650	91	57	6.7	5.1

* Maximum daily sediment yield over the 1000-year simulation.

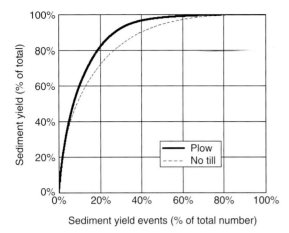

Fig. 9.4 Total sediment yield over a 1000-year simulation versus the average number of sediment yield events per year.

Table 9.8 Average annual sediment yield from hillslopes for the base climate.

Hillslope	Sediment yield (t ha⁻¹)			
	Plough	Chisel	No-till	Grass
H1	45.9	30.6	2.3	1.4
H2	45.9	30.6	2.3	1.4
H3	42.6	28.5	2.6	1.9
H4	42.6	28.5	2.6	1.9
H5	42.6	28.5	2.6	1.9
H6	39.0	25.5	2.4	1.8
H7	39.0	25.5	2.4	1.8
H8	42.6	28.5	2.6	1.9
H9	26.7	18.0	2.8	2.1
H10	26.7	18.0	2.8	2.1

permanent grass systems as compared to the moldboard or chisel plough systems. If the hillslopes are to be farmed with the moldboard or chisel plough systems, additional practices should be evaluated that would control erosion on the hillslopes and sediment delivery to channels. Some of these practices might include terracing, buffer strips, and contour farming. These can be evaluated using WEPP. While channel erosion is predicted to be low for the no-till and permanent grass systems, it

Table 9.9 Average annual detachment and deposition in the watershed channels for the various management systems for the base climate.

Channels	Sediment detached (+) or deposited (−) (t y⁻¹)			
	Moldboard plough	Chisel plough	No-till	Alfalfa
C1	−5.5	−2.0	0.3	0.3
C2	−72.0	−31.8	0.6	0.5
C3	−157.1	−65.9	0.6	0.5
C4	−141.5	−59.6	0.5	0.4
C5	−43.3	−18.8	0.7	0.7
C6	−65.3	−29.0	0.5	0.5
C7	−35.9	−15.4	0.6	0.6
C8	−4.2	−1.8	0.2	0.3
C9	−14.1	−4.9	0.6	0.6
Total	−539.0	−229.2	4.7	4.4

could be reduced further by constructing wider channels on lower slopes. Sediment delivery from the watershed could be reduced further by the use of impoundments as sedimentation basins. These too could be modelled in WEPP.

The results shown in Table 9.9 are average annual results. It is likely that there will be some storms for the moldboard and chisel plough systems where detachment occurs in channels, and there may be deposition for some storms in no-till and grass systems. These conditions will depend on storm, surface, subsurface and biomass conditions.

9.5.2 Sediment characteristics

Information about sediment characteristics estimated by WEPP is shown in Table 9.10, which gives average particle information for all sediment generated in the 1000-year run for the moldboard plough management. A summary for all managements that includes sediment characteristics is shown in Table 9.11. For this example watershed, it is apparent that the material discharged from the watershed is quite similar to the surface soil.

While the information in Tables 9.10 and 9.11 is for the total period and represents information about the material exiting the watershed, this information can also be generated for yearly and monthly periods when using the Watershed version

Table 9.10 Particle information for sediment leaving the watershed.

		Particle composition				
Diameter (mm)	Specific gravity (g cm^{-3})	Sand (%)	Silt (%)	Clay (%)	Organic matter (%)	Fraction in flow exiting
0.002	2.60	0	0	100	3.1	0.086
0.010	2.65	0	100	0	0	0.280
0.030	1.80	0	73.8	26.2	0.8	0.509
0.490	1.60	21.0	50.3	28.7	0.9	0.110
0.200	2.65	100.0	0	0	0	0.150

Primary particle type	Fraction
Clay	0.251
Silt	0.711
Sand	0.038
Organic matter	0.008

Index of specific area = 57.03 m^2 g^{-1}
Enrichment ratio of specific area = 1.02

Table 9.11 Sediment delivery and sediment characteristics predicted by 1000-year long WEPP runs.

Management systems	Total sediment delivery from hillslopes (t y^{-1})	Total channel sediment detachment(+) or deposition (t y^{-1})	Total sediment delivery from watershed (t y^{-1})	Size distribution and organic matter of eroded soil and surface soil				Enrichment ratio	Surface area (m^2 g^{-1})
				Clay (%)	Silt (%)	Sand (%)	Organic matter (%)		
Moldboard plough	2672	−539	2133	25.1	71.1	3.8	0.8	1.03	57.2
Chisel plough	1779	−229	1550	24.9	70.7	4.4	0.8	1.02	56.6
No-till	158	5	163	25.0	69.1	6.4	0.8	1.00	55.7
Grass	109	4	114	24.6	69.3	6.1	0.8	1.00	56.0
Surface soil	-----------	-----------	-----------	24.5	69.0	6.5	0.8	-----------	-----------

of WEPP. Very similar information can be generated for individual hillslopes by using the WEPP Hillslope version; in that case, information for individual storms can be produced, as well as the monthly, yearly, and total simulation information.

9.5.3 Return period

The output from the return period analysis is shown in Table 9.12, an output option available in both the Hillslope and Watershed versions of WEPP. If the return period output is selected, WEPP will produce return period lengths of up to 100 years for 2, 5, 10, 20, 25, 50 and 100 years. There is also a provision for the user to set desired return period outputs, but no lengths can be longer than half of the run length. The return period analyses in WEPP use the Weibull plotting formula (Haan, 1977).

The daily precipitation and the published precipitation amounts (US Weather Bureau, 1961) for the same return periods are shown in

Table 9.12   Return period information from 1000-year WEPP run for the plough management system. The comparable published precipitation return period information is also shown.

Return period (y)	Runoff volume (mm)	Sediment leaving (t ha⁻¹)	Peak runoff rate (m³ s⁻¹)	Daily precipitation (mm)	Published 24-h precipitation for return periods (mm)
2	53	14.4	8.9	81	97 (+20%)
5	74	20.7	12.3	103	127 (+23%)
10	93	26.4	15.1	123	145 (+18%)
25	120	34.1	17.9	156	170 (+9%)
50	147	44.0	19.5	189	190 (+0.5%)
100	175	51.9	21.3	206	213 (+3%)
200	194	68.9	22.7	222	–
500	210	73.9	23.6	268	–

Table 9.12. In all cases, published values were higher than those derived in this study, with the largest differences occurring at shorter return periods. The two- to ten-year published return period estimates were about 20% greater than those from this study, with differences at the 25–100 year return periods being much smaller.

9.5.4   Identification of problem areas

The information shown in Table 9.11 indicates that the hillslopes generate most of the sediment, and that those management practices that have considerable tillage have high soil losses. While average slopes are small, slopes are quite long, and all hillslopes have areas of appreciable slope. Additionally, both annual precipitation and rainfall intensities are quite high.

The channels were susceptible to some deposition for the plough and chisel systems, and a minor amount of erosion was predicted in the channels when no-till and perennial grass was used. Reducing the slope of the channel, and/or widening it, would be expected to increase deposition and reduce erosion in the channels. Using grass in the channels for the corn–wheat–soybean rotation would be expected to increase deposition and reduce sediment delivery.

No-till has been shown to reduce the need for conservation practices, and it is a very acceptable practice in this region. Additionally, much of the area in this region is in permanent perennial grass. Hence, additional practices were not evaluated. Practices that could have been evaluated using WEPP include terracing, grassed waterways, buffer strips and sedimentation basins. Additionally, variations in the management systems used could have been evaluated.

A frequent use of erosion prediction on agricultural lands is the evaluation of different management systems. WEPP has a 'project set' feature that allows for the evaluation of a number of managements simultaneously. It has been used in the Iowa Daily Erosion Project (Cruse *et al.*, 2006) to compute soil erosion every day from 17,848 different locations in Iowa, each with a unique combination of climate, topography, soil and management, in one setup. It is particularly useful when evaluating a number of standard management practices for a particular piece of land.

9.5.5   Climate change

The effect of climate change on average runoff volume and sediment yield is shown in Fig. 9.5. These were all evaluated for the Fort Scott climate, using the WEPPCAT tool (http://typhoon. tucson.ars.ag.gov/weppcat/index.php) to adjust the Cligen parameter files for rainfall intensification and precipitation increases of 5%, 10% and 20%.

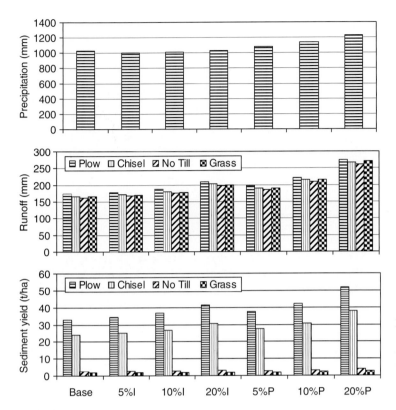

Fig. 9.5 Effect of climate change on runoff and sediment yield. Scenarios are intensification of precipitation of 5%, 10% and 20% (5%I, 10%I and 20%I) and increases in precipitation amount of 5%, 10% and 20% (5%P, 10%P and 20%P) as computed using WEPPCAT for Cligen parameters.

The generated climates were used to run WEPP for 1000 years for the small watershed for each of the four management scenarios. Results are shown in Figs 9.5 and 9.6.

WEPP was responsive to both intensification of precipitation and to increases in precipitation amount. This was reflected in both runoff and sediment yield for all managements as shown in Fig. 9.5. While increases in average annual precipitation were fairly small, the largest being of the order of a 20% increase in average annual precipitation, increases in runoff and sediment yield were much greater, ranging to over a 50% increase in average annual runoff and sediment yield.

The effect of precipitation intensification and precipitation increase due to climate change seemed to have a smaller effect on sediment yields for various return periods than its effect on

average annual sediment yields as shown in Fig. 9.6. Based on these results, while climate change would increase sediment yields in rare storms, it would increase average annual sediment yields by a greater percentage.

9.6   Summary

The WEPP model was applied to a small watershed about 5 km from the Marmaton river in southwest Missouri, and about 25 km east of Fort Scott, KS. The Marmaton River runoff eventually flows into the Osage river, a major river system that contains important reservoirs and channels for power generation, recreation and water supply for the state of Missouri. Reducing sediment yield to the river system from small upstream watersheds like the one modelled in

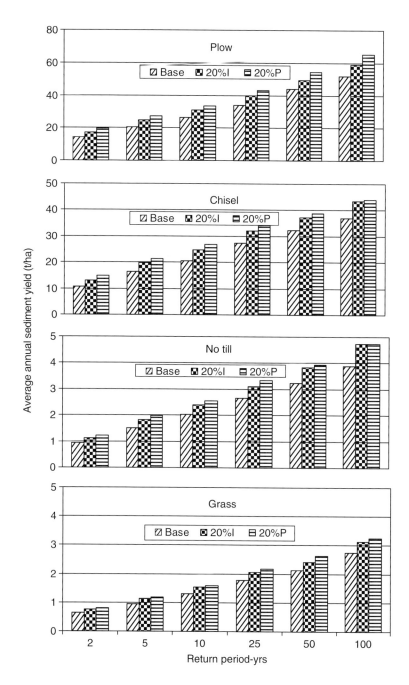

Fig. 9.6 Effect of climate change on sediment yield for various return periods. Climate change is represented by a 20% increase in rainfall intensity and precipitation.

this paper will maintain the capacities of this system, and reduce downstream effects that extend well beyond the borders of Missouri. Additionally, in this important agricultural area with shallow soils, maintaining the productive capacity of the land while controlling sediment yield is an important objective of the land management system.

WEPP has important capabilities in evaluating various scenarios for small watersheds. The ability to determine where and when erosion occurs in such a small watershed can be important in targeting treatment. The ability to study rare storms (Mutel, 2010) to select special measures to control runoff and sediment yield from them can be important in preventing damages on and off site, and can also be important for protecting public investment in roads, bridges and other structures, and in protecting human life.

In this case, WEPP was used to evaluate treatments on farmed agricultural areas to determine not only what benefits might be expected on this watershed, but also what benefits might be expected on tilled watersheds that occur in the western half of the Osage river watershed.

WEPP seemed to be well suited for this type of project. While it was not used on the forested watershed and the soils on the eastern half of the Osage river watershed, it could have been. WEPP is well supported by soils, climate and management databases for the US, and is applied to nearly all areas and all erosion and sediment yield problems in the US. It is supported by tools that can be applied to develop climate files to model climate change, and these were applied here.

WEPP has a number of different default managements and default field operations databases. The necessary tools to allow WEPP to be used in other countries can easily be developed. Details of field operations in a country can be added to the databases and management files and saved for later use within or outside that country. Plants can be parameterized easily to match plant characteristics in another country, or even new plants that are not in the original database. WEPP has tools that can be used to apply it to most of the world, including the development of the necessary local climate databases.

References

Cruse, R., Flanagan, D., Frankenberger, J., *et al.* (2006) Daily estimates of rainfall, water runoff, and soil erosion in Iowa. *Journal of Soil and Water Conservation* **61**: 191–9.

Edwards, W.M. & Owens, L.B. (1991) Large storm effects on total soil erosion. *Journal of Soil and Water Conservation* **46**: 75–8.

Foster, G.R. & Lane, L.J. (1987) *User Requirements: USDA-Water Erosion Prediction Project (WEPP)*. NSERL Report No. 1, USDA-ARS National Soil Erosion Research Laboratory, West Lafayette, IN: 43 pp.

Ghidey, F. & Alberts, E.E. (1996) Comparison of measured and WEPP predicted runoff and soil loss for Midwest claypan soil. *Transactions of the American Society of Agricultural Engineers* **39**: 1395–1402.

Haan, C.T. (1977) *Statistical Methods in Hydrology*. Iowa State University Press, Ames, IA.

Iowa NRCS (2004) Iowa Technical Note 25. Iowa Phosphorus Index. Available at http://www.ia.nrcs. usda.gov/technical/Phosphorus/phosphorusstandard. html

Laflen, J.M., Flanagan, D.C. & Engel, B.A. (2004) Soil erosion and sediment yield prediction accuracy using WEPP. *Journal of the American Water Resources Association* **40**: 289–97.

Mutel, C.F. (ed.) (2010) *A Watershed Year: Anatomy of the Iowa floods of 2008*. University of Iowa Press, Iowa City.

Nearing, M.A., Jetten, V., Baffaut, C., *et al.* (2005) Modeling response of soil erosion and runoff to changes in precipitation and cover. *Catena* **61**: 131–54.

Renard, K.G., Foster, G.R., Weesies, G.A., *et al.* (1997) *Predicting Soil Erosion by Water – a guide to conservation planning with the Revised Universal Soil Loss Equation (RUSLE)*. USDA Agricultural Handbook No. 703.

Renschler, C.S., Flanagan, D.C., Engel, B.A. & Frankenberger, J.R. (2002) *GeoWEPP – the geo-spatial interface for the Water Erosion Prediction Project*. American Society of Agricultural Engineers Meeting, Paper No. 022171. Available from ASAE, St. Joseph MI, USA.

Smith, S.B., Mezzacappa, D.J., Gaddy, A. & Laflen, J.M. (2002) *Evaluating landfill final cover soil loss (USLE, RUSLE, and WEPP)*. SWANA 39th Annual Conference, Baltimore, MD.

Southwest Watershed Research Center (2008) WEPP CAT – Water Erosion Prediction Project Climate Assessment Tool. Available at http://typhoon.tucson. ars.ag.gov/weppcat/index.php

UMRSHNC (2005) Final Report: Gulf hypoxia and local water quality concerns workshop. Upper Mississippi River Sub-basin Hypoxia Nutrient Committee, Ames, IA, September 2005. Available from American

Society of Agricultural and Biological Engineers, St. Joseph, MI.

USDA (1997) *Agricultural Statistics*. US Department of Agriculture, National Agricultural Statistics Service. US Govt. Printing Office, Washington, DC.

USDA Agricultural Research Service (1961) *A universal equation for predicting rainfall-erosion losses*. ARS 22–66.

US Weather Bureau (1961) *Rainfall Frequency Atlas of the United States*. US Department of Commerce, Weather Bureau Technical Paper 40, January 1961, Washington, DC.

Williams, J.R. (1975) Sediment yield prediction with universal equation using runoff energy factor. In *Present and Prospective Technology for Predicting Sediment Yield and Sources*, ARS-S-40, United States Department of Agriculture: 244–52.

Wischmeier, W.H. & Smith, D.D. (1978) Predicting rainfall-erosion losses – a guide to conservation farming. *USDA Agricultural Handbook No. 537*.

10 Predicting Soil Loss and Runoff from Forest Roads and Seasonal Cropping Systems in Brazil using WEPP

A.J.T. GUERRA[1] AND A. SOARES DA SILVA[2]

[1]*Department of Geography, Institute of Geosciences, Federal University of Rio de Janeiro, Rio de Janeiro, Brazil*
[2]*Federal University of Rio de Janeiro, Rio de Janeiro, Brazil*

10.1 Introduction

There have been many papers, MSc dissertations and PhD theses authored on WEPP applications in Brazil in the last two decades (Table 10.1). Favis-Mortlock and Guerra (1999) used WEPP in Mato Grosso State and noted that simulated average annual sediment yield increased in one of the scenarios tested and decreased in the other two, reflecting the range of uncertainty in predictions of future rainfall. In another study in Mato Grosso State, Favis-Morlock and Guerra (2000) found that WEPP-CO_2 underestimated current-climate mean annual sediment yield. The model calculated a value of 6.86t ha^{-1} y^{-1}, while the value measured during four years of monitoring was 12t ha^{-1} y^{-1}.

Martins Filho *et al.* (2003) validated the model to predict inter-rill erosion for three Oxisols. The initial proposed model for the dimensionless slope factor (S_f) did not provide a good relationship between the actual and predicted soil loss. The slope factor was better expressed using the equation $S_{fa} = 1.061 - 1.037\ e^{-4\ \sin\theta}$, than by the equation $S_f = 1.050 - 0.850\ e^{-4\ \sin\theta}$ where θ is the slope angle (degrees). The model $D_i = R\ I\ K_i\ S_{fa}$

predicts erosion well on the studied Oxisols where D_i is the inter-rill soil loss (kg m^{-2} s^{-1}), R is the rate of runoff (m s^{-1}), I is the rainfall intensity (m s^{-1}), K_i is the inter-rill soil erodibility (kg s m^{-4}) and S_{fa} is the modified slope factor.

Martins Filho *et al.* (2004) studied the cover-management subfactor (C_{iII}) and concluded that the equations $C_{iII} = e^{-2.50\ CS/100}$ and $C_{iII} = e^{-2.238\ CS/100}$ allow good estimates for the cover-management subfactor, for slopes between 5.3% and 15.3%, where CS is the percentage ground cover on the soil surface. For slopes ~36.3% the equation $C_{iII} = e^{-0.795\ CS/100}$ is recommended. Nevertheless, the authors suggested that more research is needed to improve the calibration for steep slopes.

Many of the dissertations and theses in Brazil using WEPP have been completed in Agricultural Engineering departments, reflecting the considerable concern over soil loss due to agricultural activities in Brazil. We have selected examples from two PhD theses (Garcia, 2001; Gonçalves, 2007) where the authors have applied WEPP in real situations and considered the associated limitations and potential. We describe in detail how the model was established and run, and the necessary data obtained for each application. Through their methodology and results, it is expected that WEPP users, in different parts of the world, will be able to use these two case studies as examples for subsequent model use, development and

Handbook of Erosion Modelling, 1st edition. Edited by R.P.C. Morgan and M.A. Nearing. © 2011 Blackwell Publishing Ltd.

Table 10.1　WEPP applications in Brazil (papers and PhD theses).

Year	Authors	Title of paper/PhD	Journal/University
1990	Chaves, H.M.L.	Uncertainty analysis of a steady state erosion model.	Purdue University
1994a	Chaves, H.M.L.	Novidades sobre o Water Erosion Prediction Project: WEPP.	*Solos Altamente Suscetíveis à Erosão*
1994b	Chaves, H.M.L.	Adaptação do modelo WEPP para as condições brasileiras.	*Solos Altamente Suscetíveis à Erosão*
1998	Guerra, A.J.T. & Favis-Mortlock, D.	Land degradation in Brazil: the present and the future.	*Geography Review*
1999	Favis-Mortlock, D. & Guerra, A.J.T.	The implications of general circulation model estimates of rainfall for future erosion: a case study from Brazil.	*Catena*
2000	Favis-Mortlock, D. & Guerra, A.J.T.	The influence of global greenhouse-gas emissions on future rates of soil erosion: a case study from Brazil, using WEPP-CO_2.	*Soil Erosion- Application of Physically Based Models*
2001	Garcia, A.R.	Uso do modelo WEPP (Water Erosion Prediction Project) modificado para estimar taxas de erosão em estradas florestais.	Universidade Federal de Viçosa
2003	Martins Filho, M.V., *et al.*	Modelagem do processo de erosão entressulcos para latossolos de Jaboticabal – SP.	*Engenharia Agrícola Jaboticabal*
2004	Martins Filho, M.V., *et al.*	Modelos para estimativa do sub-fator cobertura-manejo relativo à erosão entressulcos.	*Engenharia Agrícola Jaboticabal*
2007	Gonçalves, F. A.	Validação do modelo WEPP na predição de erosão hídrica para condição edafoclimática na região de Viçosa-MG.	Universidade Federal de Viçosa

validation. Our intention is to help WEPP users with the practical use of the model.

10.2　Case Study 1: Estimation of Erosion Rates on Forest Roads in Brazil

The main objective of this case study (Garcia, 2001) was to determine the total runoff volume and sediment yield from different forest road segments under normal rainfall conditions. The measured data were compared with the predicted data generated by the WEPP model. The research site is located in Agudos Municipality, São Paulo State, on a reforestation project with *Pinus caribaea* and *Pinus oocarpa*.

Roads are the main means of transport for agro-industrial production within Brazil. The non-paved rural roads in areas with steep slope angles make it difficult to carry goods, especially during the rainy season. This situation causes several problems for the state and private sectors. The roads need major modifications in order to reduce overflow volumes. Rainwater represents the main cause of erosion, and so it is important to capture and control these waters, so as to mitigate their erosional effects.

The environmental impacts on the rural roads affect water quality, because it increases turbidity and the consequent silting in the channels. Compaction by heavy vehicles decreases infiltration rates and thus increases overland flow and river discharge. Furthermore, the increased turbidity and associated eutrophication of rivers damages water flora and fauna, and this significantly decreases fish diversity, population and health.

Erosion on roads caused by rainfall and runoff occurs mainly because of inadequate construction and maintenance, including problems with the terrain surface, steep slope angles and inadequate drainage. The erosion process on roads starts with the earthwork of the road itself, and carries on with the cut and fill talus, the gutters and the pavement. Erosion rates on the roads are influenced by water flow, hydraulic conductivity and soil erodibility (Garcia, 2001).

The study was carried out in a reforestation project, where the main relief characteristics are a flat surface, on a sandy Oxisol. Total runoff and sediment yield were monitored under forest road conditions. Four replicates were made for each of four treatment combinations consisting of two slope angles (1% and 7%) and two lengths (20 m and 40 m), for a total of 16 road segments. Fourteen rainfall events were measured, ranging between 7 and 79 mm of measured rainfall depth. In each event, total rainfall, runoff and soil loss were monitored. The monitoring period was from July 2000 to March 2001, with most runoff and sediment yield occurring during the rainy season (December to March).

To measure the amount of eroded material and runoff, four barrels were placed on the lower part of each road segment. Furthermore, data regarding meteorological conditions, rainfall, soil properties, slope angle and segment length were included. This procedure aimed to test the model, in order to adapt it to the Brazilian forest environment.

Monitoring always took place one day after the rain event. Soil data were obtained by IAC (Agronomic Institute of Campinas) and rainfall data were obtained for all of São Paulo State. The rainfall range between 7 and 19 mm did not cause significant differences in the total amount of runoff, when the slope length ranged between 20 and 40 m and the slope steepness ranged between 1% and 7%. On the other hand, the slope angle significantly influenced the runoff amount for the 20 m slope length for rainfall amounts of 22 and 28 mm, and on the 40 m length for 60 mm of rainfall. The variation in slope length did not significantly influence runoff amount for the 7% slope at 28 mm rainfall, or for the 1% slope with 79 mm rainfall.

There is a direct relationship between soil loss and both slope steepness and slope length. If we take a 7 mm rainfall as an example, for a 20 m long slope with 7% grade, the total amount of soil loss was 900% higher than on the 1% slope. On the 40 m slope segments, total soil losses from the 7% slopes were 487.5% higher than the values found for the 1% slope segments.

In all rainfall events, total runoff was higher from the 40 m slope length segment. The proportional differences were higher when total rainfall amounts were smaller. The influence of slope angle on erosion was more evident for smaller rainfall events. When rainfall was >19 mm, the slope angle effect was weak or non-existent for the 20 m slope. For the 40 m slope, the slope angle influence occurred for rainfall events ≤9 mm.

When total rainfall was >9 mm, there was no significant difference between slope angle and slope length in terms of runoff amount. In some cases, there was even more runoff from the 1% slope than from the 7% slope. Garcia (2001) suggested that this probably happened because of differences in rainfall intensity, which was not measured. During low rainfall intensity events there was little runoff, because most water infiltrated.

The correlation was positive between total soil loss and total rainfall amount. In all rainfall events, there was an increase in total soil loss on slopes between 1% and 7%, except for the 10 mm rainfall event. According to Garcia (2001), it is not possible to explain the higher amount of soil loss for this rainfall event for the smaller slope gradient, because rainfall intensity was not measured.

For the 8 mm rainfall event and 1% slope, total soil loss for the 40 m slope was 159.4% higher than for the 20 m segment. For the 20 m slope, total soil loss on the 7% slope was 275% higher than on the 1% slope.

In terms of total runoff, the highest values occurred during the 91.8 mm rainfall event on the 20 m long slope at 1% slope. The highest total soil loss occurred during a 79 mm rainfall event, for the same slope length and steepness. For the 20 m and 7% slope, most runoff was measured from the 61.1 mm rainfall event. For the same slope length and slope steepness the highest soil loss took place on the 69.7 mm event. On the

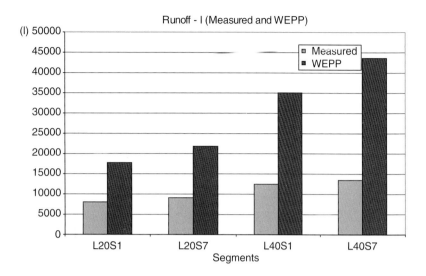

Fig. 10.1 Runoff (l): comparison of measured values and those predicted by WEPP.

40 m slope, most runoff occurred during the 62.9 mm event on the 1% slope, and during the 68.5 mm event on the 7% slope. On the 40 m slope, the highest soil loss occurred during the 67.2 mm rainfall event on the 1% slope, and during the 61.0 mm rainfall event on the 7% slope.

As runoff amount increases, runoff velocity also increases, favouring sediment transport. Therefore, the amount of soil loss augments exponentially, due to runoff enlargement. This occurred in all treatments, although Garcia (2001) noticed that slope steepness had more influence on soil loss than slope length. However, slope length variations had more influence on runoff amount than slope steepness.

The total amount of precipitation was used to estimate runoff and soil loss amount using WEPP (Fig. 10.1). Runoff amount predicted by WEPP on the 20 m long and 1% slope, 20 m and 7%, 40 m and 1%, and 40 m and 7% slopes varied by 120.8, 140.0, 181.5 and 224.1%, respectively, in relation to the monitored measurements (Table 10.2).

The total amount of soil loss on the slopes of 20 m length and 1% angle, 20 m and 7%, 40 m and 1%, and 40 m and 7%, varied by −1,125.1, −6.7, −724.6 and 120.8%, respectively, compared with the monitored measurements (Table 10.2;

Fig. 10.2). These differences confirm the need for the calibration of model variables, since the soil and climate files were adapted for the experimental site.

The model did not prove satisfactory, showing soil-loss prediction to be much less than the actual measurements on the experimental site for the 1% slope. Garcia (2001) suggested that on such gentle slopes the model might underestimate soil loss, which agrees with the observations of Martins Filho *et al.* (2003). On the other hand, soil-loss data predicted by WEPP were higher than but also closer to the observed data on the 7% slope, despite the difference of 166.6% between the actual and predicted data (Garcia, 2001).

Figure 10.3 shows the percentage variations between the observed and predicted values of runoff and soil loss. The negative values indicate that the model underestimated measured values only for soil loss on the 1% slope, whereas for the 7% slope, the model overestimated measured values. For runoff, in all situations, the model overestimated measured values. This shows that WEPP cannot always accurately predict soil loss. For runoff, although WEPP overestimated measured values, there was general

Table 10.2 Total runoff and total soil loss, measured (for the whole experiment) and predicted by WEPP (modified from Garcia, 2001).

Segment	Runoff (l)			Soil loss (kg)		
	Measured	WEPP	Variation %	Measured	WEPP	Variation %
L20S1	8052.86	17,780.10	120.8 (+)	6.615	0.540	1125.1 (−)
L20S7	9114.33	21,871.27	140.0 (+)	16.628	17.747	6.7 (+)
L40S1	12,453.02	35,051.46	181.5 (+)	11.519	1.397	724.6 (−)
L40S7	13,467.72	43,644.78	224,1 (+)	31.534	69.453	120.8 (+)

L = Slope length (m); S = Slope steepness (%).

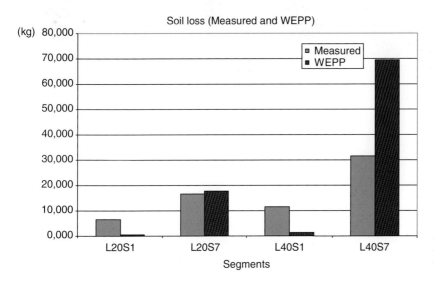

Fig. 10.2 Soil loss (kg): comparison of measured values and those predicted by WEPP.

consistency, because in the four situations there was a very similar trend.

10.3 Case Study 2: Prediction of Water Erosion for Soil and Climatic Conditions in Viçosa Municipality, Minas Gerais State

The aim of this application (Gonçalves, 2007) was to assess soil and water losses on an experimental station under natural rainfall, and to compare these results with data obtained through the equations generated by the WEPP model. The values obtained experimentally for the soil parameters K_i (inter-rill erodibility), K_e (effective hydraulic conductivity), K_r (rill erodibility) and τ_c (critical shear stress) were compared with the data generated by the WEPP model equations. The data obtained from plots were used as input to the WEPP model.

Six plots were established on a Distrofic Haplic Cambisol (Inceptisol) to determine soil loss and runoff, for five seasonal crops. The soil texture is: 51% sand, 8% silt and 41% clay. The model input data were estimated for the same soil properties, soil management and slope steepness conditions of the experimental station.

Three simulations were generated to estimate soil loss and runoff by WEPP. Estimate 1 used the values of K_i, K_r, τ_c and K_e obtained experimentally.

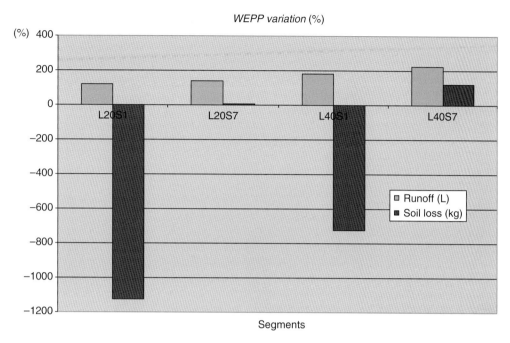

Fig. 10.3 Percentage variation between the observed and predicted runoff and soil loss in different segments.

Estimate 2 used a value for K_i obtained according to the methodology proposed by Kinnell (1993), and the other parameter values were the same as for Estimate 1. Estimate 3 used the parameter values generated by WEPP.

The six plots used for this experiment had the following characteristics: soybean grown along the contour (SCL); soybean grown up-and-down slope (SUD); maize along the contour (MCL); maize up-and-down slope (MUD); bare soil along the contour (BCL) and bare soil up-and-down slope (BUD). The mean effect of vegetation cover (SCL, SUD, MCL and MUD) compared with the bare plots (BCL and BUD) resulted in a mean reduction of soil loss by 84% and runoff by 52%. Maize was more efficient in decreasing soil loss and runoff than soybean.

The combined effect of soil management practices and vegetation cover increased the efficiency of the mean reduction of soil loss by 92% (MCL) and 88% (SCL), compared to the bare plots (BCL).

Runoff was less influenced by soil management practices and vegetation type than soil loss. There was a mean reduction of 73% on the maize on the contour-line plot (MCL), compared with the bare plot (BUD).

Runoff was obtained through simulated rainfall. Ultisol (Red Yellow Argisol) and Inceptisol (Distrofic Haplic Cambisol) presented distinct behaviours. Sediment concentration was higher at the beginning of the simulated rainfall and decreased during the experiment. The Argisol (34% sand, 13% silt and 53% clay) presented higher soil loss concentrations during the whole experiment, whereas the Cambisol only had higher soil loss concentrations at the beginning of the experiment. This situation is associated with the higher proportion of loose particles due to recent soil ploughing.

Inter-rill erosion increased during the simulated rainfall, although rates of inter-rill erosion stabilized on the Cambisol before the Argisol.

Table 10.3   Erodibility parameters: comparison of experimental values and those generated by WEPP.

Parameters	Cambisol			Argisol		
	Experimental	WEPP	%	Experimental	WEPP	%
K_i (kg s m^{-4}) × 10^6	1.26433	3.30430	161.35 (+)	3.06176	3.11220	1.65 (+)
K_r (kg N^{-1} s^{-1})	0.0108	0.0145	34.26 (+)	0.0268	0.0140	47.76 (−)
τ_c (N m^{-2})	3.481	2.950	15.25 (−)	0.495	2.950	495.95 (+)
K_e (mm h^{-1})	13.3	2.5	81.20 (−)	7.3	0.7	90.41 (−)

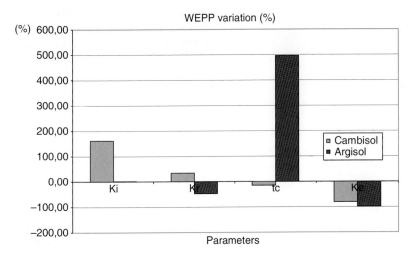

Fig. 10.4   Percentage variation between values obtained by experiment and those generated by WEPP for K_i, K_r, τ_c and K_e.

Inter-rill erodibility was obtained from the maximum detachment rate and from the equation of Foster (1982). The following values were obtained: 3.062 × 10^6 and 1.264 × 10^6 kg s m^{-4}, respectively, for the Argisol and Cambisol. Using the modified equation of Kinnell (1993), the respective inter-rill erodibilities were: 3.786 × 10^6 and 1.926 × 10^6 kg s m^{-4}.

Due to the soil characteristics, the duration of runs was different; for the Argisol it was 360 minutes and for the Cambisol it was 540 minutes. The effective hydraulic conductivity was considered as the stable infiltration rate, which was 7.3 mm h^{-1} for the Argisol and 13.3 mm h^{-1} for the Cambisol. The time taken for stabilization for the Argisol (~90 minutes) was less than for the Cambisol (~270 minutes). The explanation is the concentration of clay in the Bt horizon in the

Argisol, which reduced infiltration capacity and, consequently, increased runoff.

The lowest detachment rates occurred at the beginning of each experimental storm, but then increased rapidly because of the low particle cohesion of the topsoil. On the Argisol the detachment rates increased to a maximum after about 45 minutes, before decreasing again. On the Cambisol the detachment rates continued to increase and did not stabilize until near the end of the storm. The reason for this difference is that at the beginning of runoff, the water does not have the capacity to transport all the sediments, but as runoff continues, this capacity increases such that transportation of the detached sediment occurs, with different particle diameters. Because the Argisol has a higher proportion of aggregates, but also contains loose particles, it resists detachment more at the end of the storm, when the aggregates are dominant, whereas on the Cambisol, with a higher proportion of larger diameter particles, detachment continues to the end of the storm.

The parameter values estimated by WEPP for the Cambisol were higher than those for the Argisol, except for critical shear strength (τ_c), which was the same for the both soils. For the Cambisol, values for K_i and K_r were overestimated and values for τ_c and K_e were underestimated. For the Argisol, K_i and τ_c values were overestimated and the K_r and K_e values were underestimated (Table 10.3; Fig. 10.4). Nevertheless, K_i was slightly overestimated for the Argisol (1.65%) and strongly overestimated for the Cambisol (161%). The factor K_e was underestimated for both the Cambisol (−81.20%) and Argisol (−90.41%). WEPP underestimated the critical shear strength for the Cambisol (−15.25%) but overestimated it for the Argisol (496%). Inter-rill erosion was overestimated for the Cambisol (34.26%) and underestimated for the Argisol (−47.76%).

The hydraulic conductivity underestimation is probably related to soil characteristics, which generated the equations used for the WEPP prediction. The differences are likely to be associated with the high cation exchange capacity of the Argisol and high sand content of the Cambisol.

10.4   Conclusions

In the two case studies it is evident that there are considerable differences between measured and model values. In case study 1, the values for runoff predicted by WEPP were higher than those measured. The values of soil loss measured on the 20 m and 40 m length segments on a 1% slope angle were higher than those predicted by WEPP. The values for soil loss measured on the 20 m and 7% slope were very close to the ones predicted by WEPP, differing by only 6.7%. On the other hand, on the 40 m and 7% slope, the measured values were lower than the ones predicted by WEPP. This demonstrates the need for further model calibration.

In case study 2, we conclude that the Argisol has higher topsoil erodibility than the Cambisol,

because it had higher rill and inter-rill erosion values. Furthermore, the calculated critical shear strength and effective hydraulic conductivity values are less for the Cambisol than for the Argisol. Thus, the WEPP equations overestimated inter-rill and rill erodibility values and underestimated the effective hydraulic conductivity and critical shear strength of the Cambisol. For the Argisol, the inter-rill erodibility and critical shear strength values were overestimated, and rill erodibility and effective hydraulic conductivity values were underestimated. The use of the experimentally-derived soil parameters as input data to WEPP improved the soil loss estimates by 306% and runoff estimates by 135%.

We conclude that there is a great need to use local data to calibrate the values of the input parameters used in WEPP to describe the soil properties in order to achieve better predictions. The lack of accurate data, such as rainfall intensity, may also cause considerable errors when we compare measured and estimated values. We also conclude that for tropical soils, which usually have higher clay contents than European and North American soils, there is a trend of overestimation, both for runoff and soil loss, when we compare measured data with data predicted by WEPP.

Acknowledgements

We are grateful to Professor Michael Fullen (University of Wolverhampton, UK) for reading the first version of this chapter and making useful corrections and suggestions.

References

Chaves, H.M.L. (1990) *Uncertainty analysis of a steady state erosion model.* PhD thesis, Purdue University.

Chaves, H.M.L. (1994a) Novidades sobre o Water Erosion Prediction Project – WEPP. In Pereira, V. de P., Ferreira, M.E. & da Cruz, M.C.P. (eds), *Solos Altamente Suscetíveis à Erosão.* Jaboticabal – SP: UNESP/SBCS: 207–42.

Chaves, H.M.L. (1994b) Adaptação do modelo WEPP para as condições brasileiras. In Pereira, V. de P., Ferreira,

M.E. & da Cruz, M.C.P. (eds), *Solos Altamente Suscetíveis à Erosão*. Jaboticabal – SP: UNESP/SBCS: 213–21.

Favis-Mortlock, D. & Guerra, A.J.T. (1999) The implications of general circulation model estimates of rainfall for future erosion: a case study from Brazil. *Catena* **37**: 329–54.

Favis-Mortlock, D. & Guerra, A.J.T. (2000) The influence of global greenhouse-gas emissions on future rates of soil erosion: a case study from Brazil, using WEPP-CO$_2$. In Schmidt, J. (ed.), *Soil Erosion - Application of Physically Based Models*. Springer-Verlag, Berlin: 3–31.

Foster, G.R. (1982) Modeling erosion processes. In Hann, C.T., Johnson, H.P. & Brakensienk, D.L. (eds), *Hydrologic Modeling of Small Watersheds*. American Society of Agricultural Engineers, Monograph 5, St Joseph, MI: 296–380.

Garcia, A.R. (2001) *Uso do modelo WEPP (Water Erosion Prediction Project) modificado para estimar taxas de erosão em estradas florestais*. Viçosa (MG). Tese de Doutorado. Universidade Federal de Viçosa.

Gonçalves, F.A. (2007) *Validação do modelo WEPP na predição de erosão hídrica para condição edafo-climática na região de Viçosa-MG*. Viçosa (MG). Tese de Doutorado. Universidade Federal de Viçosa.

Guerra, A.J.T. & Favis-Mortlock, D. (1998) Land degradation in Brazil – the present and the future. *Geography Review* **12**: 18–23.

Kinnell, P.I.A. (1993) Runoff as a factor influencing experimentally determined interrill erodibilities. *Australian Journal of Soil Research* **31**: 333–42.

Martins Filho, M.V., Andrade, H., Dias Junior, M.S. & Pereira, V.P. (2003) Modelagem do processo de erosão entressulcos para latossolos de Jaboticabal – SP. *Engenharia Agrícola Jaboticabal* **23**: 9–20.

Martins Filho, M.V., Engler, M.P.C., Izidorio, R., *et al.* (2004) Modelos para estimativa do sub-fator cobertura-manejo relativo à erosão entressulcos. *Engenharia Agrícola Jaboticabal* **24**: 603–11.

11 Use of GUEST Technology to Parameterize a Physically-Based Model for Assessing Soil Erodibility and Evaluating Conservation Practices in Tropical Steeplands

C.W. ROSE[1], B. YU[2], R.K. MISRA[3], K. COUGHLAN[4] AND B. FENTIE[5]

[1]The Griffith School of Environment, Griffith University, Brisbane, Queensland, Australia
[2]School of Engineering, Griffith University, Brisbane, Queensland, Australia
[3]Faculty of Engineering and Surveying, University of Southern Queensland, Toowoomba, Queensland, Australia
[4]Annerley, Queensland, Australia
[5]Queensland Department of Environment and Resource Management, Queensland, Australia

11.1 Introduction

There was more than one motivation that led to the development of the closely related series of computer-implemented mathematical models of soil erosion known as the Griffith University Erosion System Template (GUEST). The dictionary meaning of the word 'template' refers to a guide or means of transferring a design. Choice of this term indicates the original intent of the model, namely to guide the analysis of data from bare-soil runoff erosion plots so as to yield a physically-based measure of the erodibility of the soil. This elusive soil characteristic is sought, for example, in the K factor of the Universal Soil Loss Equation (USLE). The erosion behaviour of bare soil plots is also commonly used as a baseline against which erosion from alternative cropping practices can be compared.

Handbook of Erosion Modelling, 1st edition. Edited by R.P.C. Morgan and M.A. Nearing. © 2011 Blackwell Publishing Ltd.

A general motivation that led to the development of GUEST was to seek a more physically-based soil erodibility measure than that provided by the K factor in the USLE. This motivation was strengthened by the finding that use of the USLE methodology in the extensive wheat-growing belt in Australia, where annual soil loss is very variable, required several decades of experimentation in order to determine the K factor with useful accuracy. This made the methodology inappropriate in such contexts (Edwards, 1987). Furthermore, the technique used to determine soil loss in establishing the USLE, thought to be suspect, was later shown to lead to serious underestimation of soil loss, especially in well-structured soils (Ciesiolka et al., 2006).

A more specific motivation for the development of GUEST was a recognition of the seriousness of both on-site and off-site consequences associated with water erosion in tropical steeplands, both within Australia and in Australia's neighbouring countries in southeast Asia. The opportunity to develop a multi-country project to address these problems in the context of tropical steepland cropping systems was provided by funding from the Australian Centre for International Agricultural Research (ACIAR), in collaboration with a range of institutions in southeast Asia. Two successive collaborative projects were funded in 1985 and 1992, with three general aims covering a number of related soil conservation concerns (Coughlan & Rose, 1997a), which included:

• testing a range of locally-applicable technologies to reduce soil-loss rates to some acceptable level, such as less than 10 t ha^{-1}y^{-1};
• quantifying hydrological and sediment transport processes with a view to matching soil conservation technologies to dominant processes at different sites;
• developing methodologies to predict runoff, soil and nutrient losses, and the consequences of these losses in terms of soil productivity.

These soil conservation concerns were deemed to be sufficiently serious by authorities and institutions in Malaysia, Thailand, the Philippines and Australia that they agreed to provide substantial staff and funding support for the projects, supplemented by ACIAR support for staff, design and technical input, and training in soil erosion processes and research methods.

In the collaborating countries, a substantial fraction of the cultivated area commonly remained bare during crop growth. Thus it was decided that the layout of field experiments would in all cases include a bare soil treatment, sustained during the cropping season by hand-weeding. The results obtained from the bare soil treatment plots then provided a base from which any improvement resulting from a soil conservation measure employed on other plots could be judged. Because of its recognised importance to soil erosion, the runoff rate was a measurement incorporated into the experimental methodology of the ACIAR-supported multi-country experiments. A special issue of *Soil Technology* (Rose, 1995) and chapters in Coughlan and Rose (1997a) describe the methodologies and some of the outputs from these projects.

The GUEST program can be used even if data on runoff rates are not available. For example, prior to the development of GUEST, long-term soil erosion plots had been established in a wide range of countries based on the methodology set out for the USLE. Measurement associated with these USLE plots included rainfall rate, but not runoff rate – a factor recognised in GUEST to be of great importance to erosion. Because of their historical significance, it was decided to apply the GUEST framework of soil erosion analysis to data from 11 USLE sites in the international network known as the ASIALAND Management of Sloping Lands project, co-ordinated by IBSRAM (the International Board for Soil Research and Management). For this to be accomplished, methods were developed to infer runoff rates from the USLE-recorded measurements of total runoff and rainfall rate, as described by Yu (1997), Yu *et al*. (1998, 1999), and Yu and Rose (1999).

The well-documented program WEPP (Water Erosion Prediction Project) uses a range of information which was readily available in the US, but it was not available for the range of countries involved in the multi-country projects referred to. Whilst this precluded the use of WEPP in these projects, there were other more theoretical objections to the use of WEPP, for example, the manner in which the consequences of the size or settling velocity distribution of soil was dealt with, and in its lack of recognition of the role of a deposited layer. Also, instead of using shear stress as adopted in WEPP, advantages (as shown by Nearing *et al*., 1997) were seen in using stream power. In GUEST an effective fraction of stream power is assumed to overcome the cohesion of the original soil in entrainment, and to raise saltating sediment into the water layer against its immersed weight. Despite such differences in conceptual approach, since both conserve mass of sediment and water, there are some similarities in mathematical form between WEPP and approaches expressed in GUEST in steady-state situations (Yu, 2003).

11.2   Short History of the Development of GUEST

A precursor to the development of GUEST was recognition of the utility of the concept of stream power in describing erosion dominated by overland flow (Rose *et al*., 1983a,b). This approach was developed significantly for flow-driven erosion by Hairsine and Rose (1992a,b), and by Hairsine and Rose (1991) for rainfall-driven erosion. This original version of program GUEST, referred to as Type A, was described by Misra and Rose (1996), and incorporated the simultaneous effects of rainfall, runoff, and deposition on sediment concentration. The Type A version of GUEST combined the steady-state solutions of the Hairsine and Rose theories for both rainfall and flow-driven types of erosion processes acting together. It was assumed that the sediment concentration produced by rainfall could simply be added to that produced by overland flow whose stream power exceeded a threshold value.

The theory on which GUEST is based begins with a governing equation for soil erosion, transport and deposition:

$$\frac{\partial (c_i D)}{\partial t} + \frac{\partial (c_i q)}{\partial x} = e_i + e_{ri} + r_i + r_{ri} - d_i \quad (11.1)$$

where D is water depth (m), q is unit discharge (flow rate per unit flow width, m^2 s^{-1}), c_i is the sediment concentration for particle size class i (kg m^{-3}), e_i and e_{ri} are rates of rainfall detachment and re-detachment (kg m^{-2} s^{-1}), r_i and r_{ri} are rates of flow entrainment and re-entrainment (kg m^{-2} s^{-1}), and d_i is the rate of deposition (kg m^{-2} s^{-1}). The detachment and entrainment terms are related to the process of dislodging primary particles and aggregates from the original soil. The dislocated particles and aggregates are continuously being returned to the soil surface under gravity. The processes when these loose, deposited materials are detached once again by the rain or entrained by the flow are called re-detachment and re-entrainment, respectively. Equation (11.1) is based on a mass balance for individual particle size classes. It is necessary to separate sediment into different classes according to their particle size because the associated settling velocity, which characterizes the rate of deposition, is closely related to particle size. While interactions between rainfall-driven and flow-driven erosion processes have been the subject of subsequent investigation (e.g. Asadi *et al.*, 2007), Equation (11.1) assumes that the effects of both erosion

processes can simply be added together. In the practical applications referred to in this chapter, the steady-state situation is assumed, so that the time-variant term in Equation (11.1) is ignored.

The effectiveness of this Type A version of GUEST was much enhanced by its implementation as a Fortran-based computer program (Misra & Rose, 1989, 1992), and this was used to assess parameter sensitivity for both rainfall and runoff-driven processes (Misra & Rose, 1996), and the relationship between erodibility parameters and soil strength (Misra & Rose, 1995). This version of GUEST was also used both to assess the relative importance of the range of factors governing soil loss, and their interaction, and in evaluating the likely effectiveness of soil conservation options. Furthermore, this computer program greatly facilitated extraction of soil erodibility parameters from experimental data collected when erosion was due to either, or both, rainfall-driven or flow-driven processes. Following extensive experimental investigation it was shown that in many situations of significant soil erosion, flow-driven erosion was dominant over that due to rainfall, at least in sediment mass terms. This is not, however, to downgrade the significance of rainfall-driven processes, especially in structural breakdown and chemical enrichment, and in low slope contexts where rainfall-driven processes can dominate.

In soil erosion investigations, a newly designed facility called the Griffith University Tilting Flume Simulated Rainfall Facility (or GUTSR) (Misra & Rose, 1995) provided valuable data on erosion consequences for a wide range of soil types and conditions and erosion contexts (e.g. Proffitt & Rose, 1991a,b; Proffitt *et al.*, 1991, 1993a,b; Hairsine & Rose, 1991, 1992a,b; Misra & Rose, 1995, 1996). Experiments using this facility showed that even during constant flow-driven situations, sediment concentration can fluctuate with time between recognizable upper and lower limits. The upper limit can be associated with the so-called 'transport limit' (Foster, 1982). The lower limit, where the strength of the soil matrix controls sediment concentration, has been termed the 'source limit' (Hairsine & Rose, 1992a,b). The transport limit can be achieved, for example, when rill wall collapse provides a ready supply of weak, eroded sediment.

The theory of Hairsine and Rose described such behaviour as due to the formation, as soil erosion proceeds, of a layer of sediment consisting of previously eroded material, and formed by net deposition. The original soil matrix in general is characterized as having some strength, while the deposited layer is much weaker. This deposited layer can grow in depth during erosion, providing partial or perhaps complete protection from erosion of the underlying soil matrix. Complete protection corresponds to the essentially unstable upper transport limit – unstable since continued erosion of the deposited layer will eventually reduce the completeness of its coverage. At the potentially more stable source limit the sediment concentration is determined by the rate of erosion of sediment sourced from the soil matrix. Such observations made during experiments in the GUTSR facility on fluctuations in sediment concentration over time during erosion provided insight into what may occur in field plot experimentation, where commonly only measurements of soil loss for the entire erosion event are available.

Sensitivity analysis using the Type A GUEST program showed the importance, amongst other factors, of land slope and a sediment characteristic called 'depositability', defined as the mean settling velocity of sediment components in water – a measurable characteristic (Misra & Rose, 1996; Lisle *et al.*, 1995). Data collected in the GUTSR facility with simulated rainfall and overland flow on soil beds prepared with differing soil strengths demonstrated the ability of GUEST to interpret the effect of varied soil strengths on sediment concentration (Misra & Rose, 1995, 1996).

The theory of Hairsine and Rose (1992a,b) for flow-driven erosion introduced an important fundamental parameter, J, related to soil strength, and defined as the energy required, per unit mass of soil eroded, to remove soil from the matrix by overland flow, a process termed 'entrainment'. As noted above, where the rate of entrainment limits the concentration of eroded sediment, the sediment concentration achieved is described as 'source-limited'. This situation is distinct from that at the 'transport limit' when the soil matrix is completely covered by sediment eroded previously in the same event (as, for example, following rill collapse), and when soil strength can be assumed to play no role in determining sediment concentration. Hairsine and Rose (1992a,b) developed analytical expressions for the transport limit and also the source limit (involving J). Figure 11.1 illustrates how, as the soil strength (and so J) increases, sediment concentration c decreases for any particular value of stream power (Ω).

As described earlier, careful flume experimentation showed that even during a flow-driven erosion event in which flow conditions are constant, sediment concentration can fluctuate through time, generally not remaining at either of the two limits described earlier (e.g. Rose, 1993). In order to describe the actual erodibility in this normal complex flow-driven situation, a Type B version of the family of GUEST programs was developed in which an empirical parameter β was introduced, defined so that at the upper or transport limit of flow-driven erosion, β had the value of unity. Thus finding a value of $\beta < 1$ following analysis of experimental data indicates that during the erosion event, soil strength did play some role in reducing sediment concentration to a value below that of the transport limit. Should sediment concentration remain at the source limit during an erosion event, then there is a direct relationship between β and J. This is more likely to be the case where the eroding surface is reasonably stable, without very active rilling, which can involve head cutting and rill wall collapse. However, β quantifies soil erodibility whether or not persistence of sediment concentration at the source limit occurs. The use of β has the advantage of not requiring evaluation of J, which currently depends on measurement of sediment concentration as a function of time – data not commonly available in field experiments.

The Type B program involving β is more versatile than Type A since it is less demanding in data than that required for the determination of J. However the theory developed for J, which describes the fundamental manner in which soil strength can reduce sediment concentration in

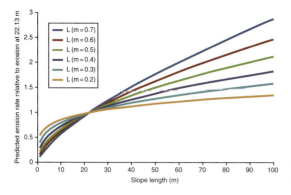

Plate 1 Relative erosion rate as a function of the slope for different slope-length factors in the RUSLE model.

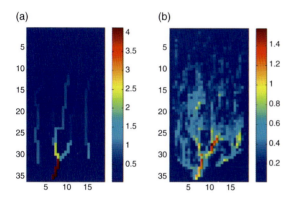

Plate 2 Comparisons of spatial patterns of sediment production for the E2 event on the large shrubland plot at Walnut Gulch: (a) total runoff in litres at a point; (b) total sediment movement in kilograms at a point. Plot dimensions are 20×35 m (after Wainwright *et al.*, 2008b).

Plate 3 Colluvial deposits re-incised during the last phase of an important erosion event at Huldenberg, Central Belgium.

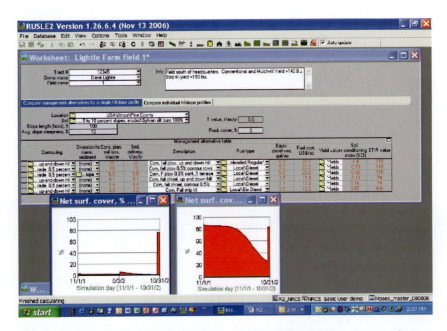

Plate 4 Example of a RUSLE2 worksheet view applied to agricultural conservation planning that compares six management and conservation practice combinations for a single climate, soil and topography. The inset graphs show the soil residue cover for fall ploughing (left) vs strip tillage (right).

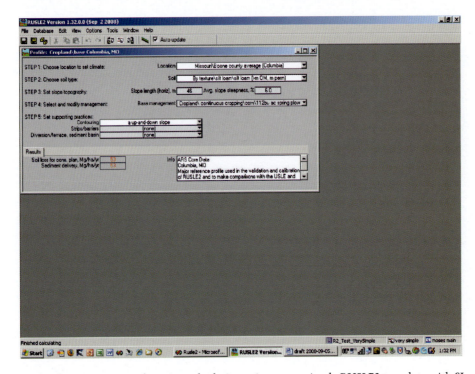

Plate 5 Example of an average annual erosion calculation using a very simple RUSLE2 template with SI units.

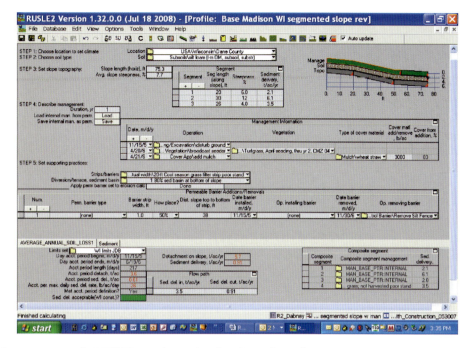

Plate 6 Screen capture of a RUSLE2 template tailored to the analysis of a construction site.

Plate 7 Runoff plots at 70% slope on the VISCA site, Leyte, the Philippines. Bare plot in the foreground shows rill development with eroded sediment deposited in the Gerlach trough. The thatched covers to the left of the photograph provide shelter for each plot's tipping-bucket runoff-rate measuring equipment and suspended load sampler (from Presbitero *et al.*, 2005, with permission).

Plate 8 Typical landscape of the hilly part of the Loess Plateau, northern Shaanxi Province.

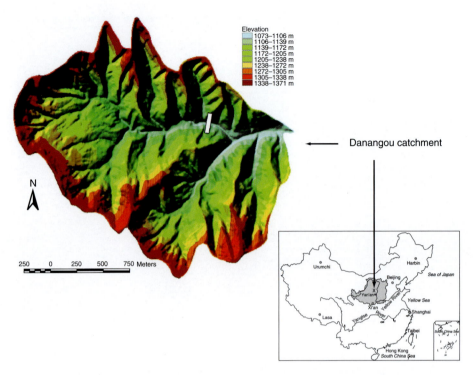

Plate 9 Elevation of Danangou catchment. Position of weir indicated with white bar. Map of China from ReVegIH (2005); the grey area indicates the Loess Plateau.

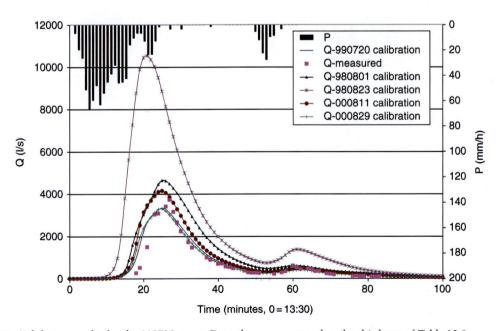

Plate 10 Validation results for the 990720 event. Data shown correspond to the third row of Table 12.8.

Plate 11 Sediment delivered by a gully to the buffer at Site 2.

Plate 12 Grass waterway established along the line of the gully at Site 2.

Plate 13 Astroturf mat placed within the buffer.

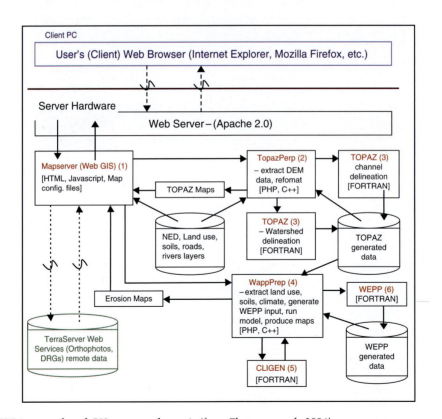

Plate 14 WEPP internet-based GIS system schematic (from Flanagan *et al.*, 2004).

(a)

(b)

Plate 15 A young cap-batter gully developed on a rehabilitated mine site in northern Australia (Willgoose & Loch, 1996). (a) This view is from midway up the gully at its point of maximum depth looking down at the depositional fan in the background. The picture shows the large diameter armour on the floor and the relatively finer substrate in the gully walls. The depositional fan has smothered all vegetation out to where the people are standing in the background, but evidence of deposition exists out beyond the first line of trees. (b) This view is taken from the base of the depositional fan (just beyond the two people in the background of (a)) looking over the depositional fan with the gully in the background. Photo (a) (i.e. the point of maximum depth) is taken about three-quarters of the way up the slope between the person and the top of the slope.

(a)

(b)

Plate 16 (a) Permanent gully (ca. 1.5 m deep and ca. 3 m wide) in rangeland cut into a vertisol (Jimma, South Ethiopia, September 2008). Note mass movement processes (soil fall and soil slumping) following hydraulic erosion, undercutting of sidewalls and tension crack development. (b) Permanent gully (ca. 8 m deep and ca. 15 m wide) in an olive grove cut into marls down to limestone bedrock (Andalucia, Spain, March 2008).

Plate 17 Ephemeral gully (mean depth is ca. 0.10 m and width is ca. 5 m) cut into loess-derived topsoil (Rutten, eastern Belgium, June 2008). Land use is cropland (sugar beet). This ephemeral gully was formed on 29 May 2008 during a rain event in which 60 mm rain depth was recorded in 2 h.

Plate 18 Bank gully or edge-of-field gully (mean depth is ca. 2 m and width is ca. 3 m) cut into a loess-derived soil (Bertem, central Belgium, November 1993). Runoff generated on the sealed and compacted topsoil, left bare after harvesting potatoes, first flowed through a pipe in the sunken lane bank, after which the roof of the pipe collapsed.

Plate 19 Bank gully (ca. 10 m deep and ca. 20 m wide) in an urbanized area cut into deeply weathered, unconsolidated sandy deposits (São Luis, Brazil, January 2007). This gully formed as a consequence of urban sprawl, resulting in vegetation removal and an increased runoff response (Guerra *et al.*, 2007). Note the pisolithic petroplinthite at the soil surface in the bottom right corner.

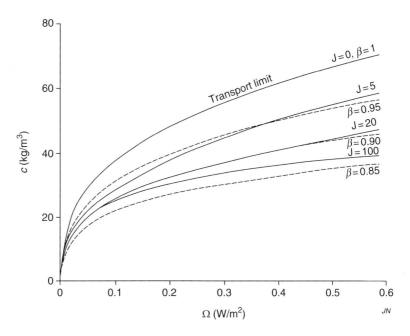

Fig. 11.1 Sediment concentration (*c*) calculated as a function of streampower (Ω) for particular values of parameter *J* and *β* defined in the text. The set of curves for *J* > 0 and *β* < 1 are calculated assuming that the sediment concentration remains at the source limit corresponding to the chosen values of *J*. Reproduced with permission from Rose (1993), p. 324 in *Hydrology and Water Management in the Humid Tropics* by M. Bonell *et al.*, 1993, Cambridge University Press.

the flow-driven erosion of bare soil, is invaluable, not only in itself, but also in providing justification for the form in which the empirical soil erodibility parameter *β* was introduced. As shown in Rose (1993) any decrease in *β* below its theoretical upper limit (of *β* = 1) affects sediment concentration variation with the driving variable stream-power in a manner very similar indeed to the effect of increasing *J* above its lower limit of *J* = 0 (at the transport limit). This is illustrated in Fig. 11.1, and this similarity relieves the empiricism of *β* by giving more confidence to its physical meaning when flow-driven erosion is dominant.

It is the Type B version of GUEST employing *β* which has been widely used in extensive multi-country experiments in the tropics, subtropics and southeast Asia referred to earlier. Hereafter in this chapter this version of the program will be simply referred to as GUEST. In these multi-country studies it was experimentally established that flow-driven erosion was usually the major process involved for steeper slopes and significant erosion events (Rose, 1995), and in such contexts

the soil erodibility parameter *β* has a physical meaning which can be related to *J*. In such analysis the possible contribution of other erosion processes to soil loss, such as rainfall detachment, rolling, or mass movement, is recognized as increasing the value of *β* obtained from the analysis, even possibly to a value greater than unity (which is the upper limit for flow-driven erosion alone). In some experiments, evidence of the expected additional erosive role of rainfall has also been found (Yu *et al.*, 1999).

Monitored erosion events in which *β* was found to modestly exceed unity were such that soil loss by processes that add to flow-driven erosion could be expected to occur, or where mass movement was observed in some steep slope erosion events under intense tropical rainfall (Presbitero *et al.*, 2005).

Explicit recognition of the separate physical role of settling velocity, and of the effects of rilling (if it occurs), both increase the physical relevance of erodibility *β*.

The information on runoff rate required to operate GUEST can be measured directly (e.g. Ciesiolka & Rose, 1998), or else inferred from other data. Yu *et al.* (1999) showed that consistent values of erodibility parameters are obtained if: (i) an effective runoff rate is calculated from a measured hydrograph; (ii) a hydrograph is estimated from rainfall rate and total runoff amount; or (iii) scaling techniques using peak rainfall intensity and a gross runoff coefficient are used.

11.3   Theory Outline for GUEST (Type B)

This theory outline is for the Type B version of GUEST which provides a physically meaningful erodibility parameter (β) provided that runoff-driven processes are dominant, which has been shown to be the case on steeplands (Rose *et al.*, 1997). If other erosion processes contribute significantly to soil loss, or plots are not bare of vegetation, then the physical interpretation of the value of β is less certain, although it is still useful in describing the erodibility of the measurement plot (assumed to be essentially planar, although if rilling is observed, its effects are recognized).

Under steady-state conditions, for a given runoff rate, Q (mm h^{-1}), the unit discharge (or volumetric discharge per unit slope width), q (m^2 s^{-1}), is determined by

$$q = LQ/3600000 \qquad (11.2)$$

where L is the slope length (m). Rill geometry, if rilling occurs, can be defined by its average spacing, W_r (m), bottom rill width, W_b (m) and the side slope of rills, z (defined as the horizontal increment per unit vertical increment).

The continuity equation for the surface runoff is given by:

$$W_r q = A V \qquad (11.3)$$

where A is the cross-sectional area (m^2), and V (m s^{-1}) the mean flow velocity. In effect, an average catchment area of $W_r L$ is assumed for each rill. The velocity is determined using Manning's formula:

$$V = \frac{1}{n} R^{2/3} S^{1/2} \qquad (11.4)$$

where S is the land slope (the sine of the slope angle), and R the hydraulic radius (m). For rills of different shapes, analytical expressions for R are given in Yu and Rose (1999).

Other than for perfect plane geometry of flow, when $R = D$, the flow depth, there is no straightforward analytical solution to determine the water depth for a given unit discharge. Water depth, however, is determined numerically (using Newton's method) to solve the continuity equation (Equation (11.3)):

$$A(D) V(D) - W_r q = 0 \qquad (11.5)$$

Once the water depth is determined, hydraulic radius and flow velocity can be computed, and the stream power per unit area, Ω (W m^{-2}), calculated from:

$$\Omega = \rho_e\, g\, S\, R\, V \qquad (11.6)$$

where ρ_e is the density of water and sediment mixture. Soil erosion due to flow-driven processes occurs only when stream power exceeds a threshold value Ω_0 (about 0.008 W m^{-2} for cultivated soils (Proffitt *et al.*, 1993b)).

A key assumption in the GUEST theory is that a certain fraction, F, of the excess stream power $(\Omega - \Omega_0)$ is involved in maintaining the sediments in suspension. A value of $F = 0.1$ is used in GUEST, although values as high as 0.2 have been measured at low stream powers (Proffitt *et al.*, 1993b).

Supported by direct observation (Heilig *et al.*, 2001), a layer of previously eroded and deposited sediment quickly develops which sits on top of the original uneroded soil matrix. This deposited layer, whilst providing some degree of protection to the soil matrix from entrainment by the eroding agent, also provides a ready source of easily erodible material, due to its low or negligible strength. The sediment concentration would be expected to be at a maximum when this weak deposited layer completely covers the soil matrix, and an equilibrium sediment concentration would be achieved when the rate of re-entrainment of sediment from this complete layer is equal to the oppositely directed rate of sediment deposition caused by gravity. This maximum equilibrium concentration is identified with what Foster (1982) called the 'transport limit', here written as c_t.

The rate of deposition, d, is written as:

$$d = \Sigma v_i c_i \quad \text{(kg m}^{-2}\text{ s}^{-1}\text{)} \qquad (11.7)$$

where sediment is characterized by a distribution of settling velocities (v_i, m s^{-1}) with sediment concentration c_i.

The transport limit is conceived as the steady-state sediment concentration achieved in overland flow when the rate of re-entrainment of recently deposited sediment is equal to the rate of deposition. If this recently deposited sediment is assumed to have negligible strength, then the rate of energy expenditure in re-entrainment is that required to lift an upward flux of sediment against its immersed weight, where the flux rate is equal to d given in Equation (11.7). The stream power (Equation (11.6)) supplies this rate of energy expenditure. The immersed weight of sediment is $\dfrac{(\sigma - \rho_e)}{\sigma}$ multiplied by its un-immersed weight. Then, as shown by Rose and Hairsine (1988) and Hairsine and Rose (1992a) (and in simplified form by Rose, 2004), it follows that, neglecting Ω_0 in comparison to Ω, c_t is given by:

$$c_t = \frac{F \rho_e}{\phi}\left(\frac{\sigma}{\sigma - \rho_e}\right) SV \quad \text{(kg m}^{-3}\text{)} \qquad (11.8)$$

where ϕ is the average settling velocity of all classes (m s^{-1}), σ is the sediment density (kg m^{-3}), and S is the land slope.

The value of the soil depositability, ϕ, depends upon the distribution of the values of the settling velocity v_i across the size range of particles and aggregates which engage in settling during erosion. Perhaps the most relevant method of measuring this distribution is the modified bottom withdrawal tube method, for which the theory and practice is given by Lovell and Rose (1988a,b). However, there is a variety of methods which can be used to obtain sediment depositability, includ-

ing an approximate estimate based on a determination of sediment size distribution and use of an empirical equation relating settling velocity to size (reviewed by Fentie *et al.*, 2004a), such as that of Cheng (1997). The Griffith University Depositability Program (called GUDPRO; Lisle *et al.*, 1995) allows analysis of data from a variety of sources to yield ϕ.

Determination of the fundamental erodibility parameter J requires sediment concentration to be measured as a function of time, which is so difficult to measure in field experiments that it is hardly attempted. Also, as seen earlier, since sediment concentration can fluctuate between source and transport limits (Rose *et al.*, 1990), some overall average measure of erodibility is required. Such problems have been overcome in GUEST by introducing an empirical erodibility parameter β, defined by:

$$c = c_t^{\beta} \quad \text{(kg m}^{-3}\text{)} \qquad (11.9)$$

Figure 11.1 illustrates that the form of the relationship between c and Ω for any particular value of J can be closely approximated by c_t^{β}, where $\beta \leq 1$.

Whilst c and c_t in Equation (11.9) are instantaneous values, and although GUEST allows such calculations of c if β is known, such measurements are not usually made (for reasons mentioned earlier), with only the flow-weighted average concentration on an event basis ($\bar{c}$) being available. Instantaneous values when summed for the event as a whole, so that Equation (11.9) also holds for event average quantities, allow $\bar{c}$ to be calculated by summation which can be expressed as:

$$\bar{c}_t = \frac{\displaystyle\int_0^T c_t\, Q\, dt}{\displaystyle\int_0^T Q\, dt} \qquad (11.10)$$

and then:

$$\beta = \frac{\ln \bar{c}}{\ln \bar{c}_t} \qquad (11.11)$$

where T is the duration of runoff, and Q is recognised to be a function of time $(Q(t))$. Then the total soil loss (SL) during an erosion event is given by:

$$SL = c_t^{\beta} \Sigma Q \Delta t \quad (\text{kg m}^{-2}) \qquad (11.12)$$

As described under experimental methods (Section 11.5), in many applications of GUEST, Q was measured as a function of time (i.e. $Q(t)$). However, GUEST has also been applied when the runoff rate is not measured. Yu et al. (1997a,b, 1999) and Yu and Rose (1999) describe a number of alternative methods of employing GUEST where runoff rate is not measured (as, for example, in experiments using USLE-type methodology). Most of these methods involve use of an 'effective runoff rate' Q_e, which is now derived.

It follows from Equation (11.8) and Manning's equation that:

$$c_t = kQ^{0.4} \quad (\text{kg m}^{-3}) \qquad (11.13)$$

where k is a constant for any given plot characteristic and soil, and is given by:

$$k = \frac{F\sigma S}{(\sigma / \rho - 1)\phi} \left(\frac{L^{2/3}S^{1/2}}{n} \right)^{3/5} \quad (\text{kg m}^{-3.4}\ \text{s}^{0.4}) \qquad (11.14)$$

With sediment concentration at the transport limit, the total soil loss during an erosion event is equal to

$$\Sigma kQ^{0.4}Q\Delta t \quad (\text{kg m}^{-2}) \qquad (11.15)$$

So, since k is a constant within any event, the event average value of c_t is given by:

$$\bar{c}_t = k\frac{\Sigma Q^{1.4}}{\Sigma Q} \quad (\text{kg m}^{-3}) \qquad (11.16)$$

Thus a single effective steady state runoff rate (Q_e), which is required to compute the average sediment concentration at the transport limit during an erosion event, can be invoked such that:

$$\bar{c}_t = kQ_e^{0.4} \quad (\text{kg m}^{-3}) \qquad (11.17)$$

Then from equations (11.16) and (11.17) it follows that

$$Q_e = \left(\frac{\Sigma Q^{1.4}}{\Sigma Q} \right)^{2.5} \quad (\text{m s}^{-1}) \qquad (11.18)$$

Thus, use of Q_e provides an alternative method of calculating $\bar{c}_t$ (and also β). Of course if Q is measured at small time intervals, there is no conceptual advantage in using equations (11.17) and (11.18) rather than Equation (11.10) to calculate $\bar{c}_t$ and β, since the two methods are equivalent. However, the major advantage of using Q_e as a single hydrological driver is in allowing its prediction in situations where $Q(t)$ is not measured. As shown in Yu et al. (1997a,b, 1999), and Yu and Rose (1999), Q_e can be estimated with much less weather data than is required to predict the complete hydrograph of $Q(t)$.

If the erodibility parameter β has been determined or can be estimated, then GUEST can be used to explore predictive scenarios (Yu and Rose, 1997). Yu et al. (1997b) provide and illustrate methodologies for six different soil erosion prediction scenarios with decreasing quality of available data.

11.4 Subsequent Development of GUEST

During experience gained in processing very large bodies of experimental data, some further refinement of GUEST took place, although the effect of such refinement on results obtained is noticeable only in rather unusual experimental conditions. These refinements (described in Yu & Rose, 1999, and in Presbitero et al., 2005) include recognition of the following factors, which are probably of very small or possibly negligible effect in many applications. Firstly, at very high sediment concentrations, sediment can significantly enhance fluid density above that of water (see Chapter 12 for a case study of such situations). Secondly, at shallow depths of flow, some of the larger water-stable aggregates may be inadequately submerged to participate in saltation. Another factor associated with high sediment concentration is that there is a continuous call on the shear stress exerted by overland flow in order to provide the momentum to sediment involved in its removal from the soil surface, this acquired momentum being lost on return to the soil surface. The 'saltation stress' involved in this process reduces the effective shear stress.

The variant of GUEST named GUEPS (Yu & Rose, 1997; Rose & Yu, 1998) incorporates these refinements, together with the option to predict soil loss in any given hydrological event, should the erodibility β be known or can be estimated from other information. In GUEPS the sediment concentration at the transport limit is written (Presbitero *et al.*, 2005; Fentie *et al.*, 1999) as:

$$c_t = \frac{F\rho SV\left(\dfrac{\sigma}{\sigma - \rho}\right)(1 - C)\,Rb}{D\,\phi_e\,K} \qquad (11.19)$$

with

$$K = 1 + \frac{F(1 - C)}{gD}V^2 b \\ - \frac{0.62\,FVS}{D\,\phi_e}\left(\frac{\sigma}{\sigma - \rho}\right)(1 - C)\,Rb \qquad (11.20)$$

where F is the fraction of the stream power effective in erosive processes, ϕ_e is the effective depositability (m s⁻¹) and $(1 - C)$ is the fraction of the soil fully immersed in the flow for a given water depth, b is a shape factor depending on the rill geometry, and σ and ρ are sediment and water density (kg m⁻³), respectively. In Equation (11.19), K can be regarded as a modifying factor of the order of 1. The second term of the modifying factor K (Equation (11.20)) takes into account the effects of the saltation stress, and the third term increases the available stream power due to sediment concentration. Such an increase of the stream power occurs because the density of the water and sediment mixture increases as the sediment concentration increases. These two modifying terms are significant when the sediment concentration is high, say >200 kg m⁻³, and/or the flow depth is low so that the depositability, ϕ, is

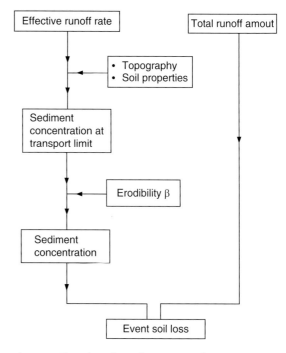

Fig. 11.2 Flowchart for soil erosion prediction using GUEST technology (from Yu *et al.*, 1997a, with permission).

reduced to an effective value, ϕ_e, which can be calculated by program GUDPRO (Lisle *et al.*, 1995).

If a value of β is determined, or can be estimated based on past experience, then it is possible to use program GUEST to predict sediment loss in any selected or given hydrological scenario or context. A flow chart indicating the connection between the various sources of information involved in event soil loss prediction is shown in Fig. 11.2.

11.5 Experimental Methods Commonly Used in GUEST-Based Projects

Event soil loss has often been measured in USLE-based experimentation by collection of runoff in a container, with any overflow subsampled by flow splitting. The sediment collected in containers is subsequently stirred vigorously and subsampled to yield a sediment concentration thought to be typical for the container. It has been shown by Ciesiolka *et al.* (2006) that this type of technique can lead to serious underestimation of soil loss. The reasons for this underestimation include the effect of rapid settling of larger soil particles which escape sampling, and the difficulty in bringing the collected sediment to a spatially uniform concentration by stirring.

The technique for measuring soil loss from runoff plots which was developed in association with GUEST is described in Ciesiolka *et al.* (1995), Coughlan (1997), and Coughlan and Rose (1997b). Sediment leaving the hydrologically-defined runoff plot is collected in a shallow trough of low slope ($\approx$1%). This low slope encourages deposition of the coarser and more rapidly settling fraction of the eroded sediment. This deposited sediment is collected after the erosion event, weighed and subsampled to determine its water content, which is then used to convert the mass of deposited sediment to an oven-dry basis.

The flow of suspended sediment leaving the trough is then passed through a flow measuring device. For larger plots this can be a flume, or for plots of more modest size a 'tipping bucket' device is convenient (Ciesiolka *et al.* 1995; Ciesiolka & Rose, 1998). Using either technique, an average concentration of this suspended sediment can be obtained, yielding a total loss of suspended sediment which is then added to the deposited sediment load collected in the trough to yield the total soil loss for the erosion event. Use of either of these techniques also directly yields information on runoff rate for use in GUEST.

The separate measurement of the fine suspended sediment fraction also allows information on the loss of nutrients and carbon to be obtained, which are commonly enriched ingredients in the suspended load. It is the loss of such components that can subvert the long-term sustainability of productive land uses.

Plate 7 illustrates the described experimental arrangements for the very steep experimental plots at the Visayas College of Agriculture (VISCA), Leyte, the Philippines. The hand-cultivated plots

varied in length at different sites, from 30 m at low slopes to 12 m at higher slopes (reflecting farmer practice), with experimental plots being essentially planar and hydrologically defined by boundaries at the top and sides.

11.6 Results of Some Field Projects Using GUEST Technology

11.6.1 ACIAR projects in southeast Asia and Australia

This section summarizes the results of Projects 8551 and 9201 supported by the Australian Centre for International Research (ACIAR), which used the GUEST program and experimental methodology as described by Rose *et al.* (1997) and Rose and Yu (1998). Figure 11.3 shows the location of the seven ACIAR project sites, together with sites for the IBSRAM-ASIALAND projects to be described later. The two sites in Australia were located on mechanized commercial pineapple farms in a subtropical zone, whilst the tropical southeast Asian plots were cultivated by hand or oxen plough. Some details of the sites, including soil type, and the size and slope of plots are given in Table 11.1. At Kemaman the main treatment plot of cocoa and bananas was accompanied by a small bare plot. Tropical crops at other sites included rozelle at Khon Kaen, maize at Nan, maize and mungbean at Los Banos, and maize and peanuts at VISCA. At VISCA there were both low-slope plots at 10% as well as the quite steep slopes typical of hand-hoe cultivation on the island of Leyte. Plots of different lengths (and therefore area) were investigated in the Australian pineapple plantation sites.

Some information on plot treatment and experimentation at the different sites is given later.

(i) Soil and water loss   Table 11.2 provides a summary of average annual runoff, runoff coefficient (defined as the ratio of average annual runoff to average annual rainfall), soil loss and sediment concentration for the five sites involved in ACIAR Project PN9201. Results for each site are given, firstly for a reference plot kept essentially bare by

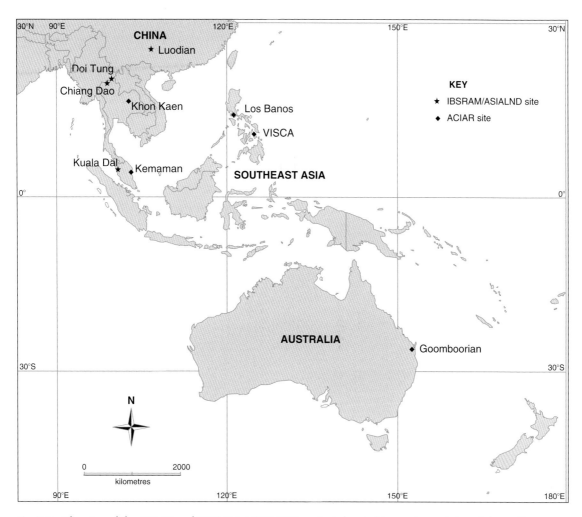

Fig. 11.3 The sites of the ACIAR and IBSRAM-ASIALAND projects located in southeast Asia and Australia.

Table 11.1 Details of the ACIAR soil erosion research sites in SE Asia and Australia.

Site	Parent material	Soil type	Slope (%)	Plot area (m²)
Kemaman, Malaysia	Shale	Orthoxic Tropodult	17	2 (bare plot) 1000 (treatment plots)
Khon Kaen, Thailand	Sandstone	Oxic Paleustult	4	150
Nan, Thailand	Shale	Oxic Paleustult	12–50	288
Los Banos, Philippines	Volcanic tuff	Typic Tropudalf	14–21	72
VISCA, Philippines	Basalt	Oxic Dystropept	10, 50, 60, 70	72
Imbil, Australia	Shale/rhyolite	Lithic Eutropept	38	18–3500
Goomboorian, Australia	Sandstone	Typic Eutropept	5	18–3500

Table 11.2 Average annual runoff, runoff coefficient, soil loss and sediment concentration from experimental plots in ACIAR Project 9201 (from Coughlan & Rose, 1997b).

Site*	Treatments	Runoff (mm)	Runoff co-efficient**	Soil loss (t ha^{-1})	Sediment concentration (kg m^{-3})
Kemaman, Malaysia	Bare plot (sandy clay loam)	2245	0.62	127	5.7
4.5 years	No living ground cover	1287	0.35	90	7.0
17% slope	Grass and legume ground cover	413	0.11	17	4.1
Average annual rainfall = 3638 mm					
Los Banos, Philippines	Bare plot (clay)	393	0.19	184	47
6 years	Clean cultivated farmers practice	387	0.19	119	31
Avg. slope = 18%	Alley cropping and mulching	114	0.06	6	5.3
Average annual rainfall = 2037 mm					
VISCA, Philippines	Bare plot (clay)	55	0.02	69	125
2 years	Clean cultivated furrows up-and-down slope	84	0.03	38	45
50% slope plots	Alley cropping and mulching	16	<0.01	3	19
Average annual rainfall = 2800 mm					
Goomboorian, Gympie, Australia	Bare plot (loamy sand) (landslope = 14%)	286	0.27	216	76
3 years	Conventional plot, no surface contact cover	213	0.20	51	24
Furrow slope <6%	Improved practice – furrow mulching	150	0.14	3	2
Average annual rainfall = 1045 mm					

*Information on length of experimental period, slope, and average annual rainfall over the experimental period is given.

**Runoff coefficient, R_c = average annual runoff/average annual rainfall.

weeding, followed in descending order by results for the local conventional farmer practice, and lastly by results for an 'improved practice' judged likely to be acceptable and also effective in conserving soil. Soil texture is also given.

Unacceptably high bare soil erosion losses of well over 100 t ha^{-1} y^{-1} were measured at three of the five sites. At VISCA (Presbitero *et al.*, 1995), the very high permeability of the soil yielded a very low runoff coefficient (Table 11.2). Conventional farmer practice at the three sites of excessive soil loss did not reduce these high losses to tolerable levels. Improved practices varied with site, but these reduced the rate of soil loss to less than 20 t

ha^{-1} y^{-1} at Kemaman, Malaysia (Hashim *et al.*, 1995), and to less than 10 t ha^{-1} y^{-1} at other sites, a figure which may possibly be tolerable, given the expected high rates of soil formation in the tropics. All improved practices were agronomic in nature, making use of the considerable potential for biomass production in the humid tropics, and involved the use of cover sufficiently close to the ground surface to impede overland flow, referred to as 'surface contact cover'. This type of cover may be contrasted with 'canopy cover' provided by vegetation in less intimate contact with the soil surface.

The effectiveness of surface contact cover in reducing soil loss is illustrated by comparing results for the two non-bare soil treatments at the Kemaman site (Table 11.2), for which canopy cover by the cocoa tree crop and companion shade trees was similar. However, the grass and legume contact cover in the improved practice reduced average annual soil loss by a factor of five, which is a greater reduction than that due to the accompanying reduction in runoff coefficient.

At both Los Banos and VISCA sites in the Philippines, the improved practice for which data are given in Table 11.2 included the introduction of leguminous hedgerows whose trimmings were a component of mulch added to the cropped alley between contour-planted hedgerows (Paningbatan *et al.*, 1995). Although the relative improvement in soil loss due to the presence of the hedgerow or the mulch cannot be dissected from the data, runoff coefficient and sediment concentration are both reduced by this combined practice (Table 11.2). Although hedgerows can fail during extreme typhoon events, causing a dramatic increase in soil loss akin to contour bank failure, hedgerows do reduce soil loss more than would be expected in terms of reduced runoff. This reduction is evidently due to net deposition from sediment-laden water which is slowed down and infiltrates as it moves through the half-metre or so of hedgerow. The reasons for this net deposition have been clarified in separate studies (Rose *et al.*, 2002, 2003; Hussein *et al.*, 2007a,b).

Total runoff from plots with a conventional farmer practice of up-and-down slope cultivation was less than for bare soil except at the VISCA site, where weed control by hand hoe in the bare plot under wet conditions appeared to produce large stable aggregates and enhance infiltration. However, soil loss was reduced by conventional practices compared with bare soil, sometimes considerably, and this was mostly due to a reduction in sediment concentration (Table 11.2).

Results in Table 11.2 for the improved practice (the last listed treatment for each site) show a substantial reduction in runoff, notably at the Kemaman site, although longer period data are needed for the Nan site. For soils of light texture (at Goomboorian in Australia, and Khon Kaen in Thailand), the improved practice had a greater effect on sediment concentration than on runoff.

The runoff coefficient, R_c is shown in Table 11.2 to be very variable. However, the range of R_c (0.27 to 0.62) for soils of lighter texture is higher than for the range for clay soils (0.02 to 0.19). Clay content can aid better water-stable aggregation and associated biotic activity, perhaps lowering R_c, although R_c is affected by rainfall amount and distribution and other factors.

(ii) Soil erodibility *Khon Kaen, Thailand*: The relative importance of rainfall-driven erosion versus erosion driven by overland flow was investigated as follows. Soil from the runoff plots at each site was placed in specially constructed detachment trays as described in Rose (1993) designed to yield the average sediment concentration during an erosion event due to rainfall impact alone. The sediment concentration from these small low-slope trays was compared with the average sediment concentration measured in plot runoff. At the low slope (4%, Table 11.1) plots at Khon Kaen, these sediment concentrations were similar, with that from the detachment tray sometimes exceeding that from the plot, and a mean value of the soil erodibility parameter β (defined in Equation (11.11)) of 1.05 was obtained, with very little storm-to-storm variation (Sombatpanit *et al.*, 1995). Whilst this value of β indicates that overland flow was virtually at the transport limit, the detachment tray data indicate that the sediment concentration at this limit could all be provided by the impact of rainfall. Note, however, that the parameter β incorporates the effects of any process that contributes to erosion. Detachment tray data showed flow-driven erosion to be the dominant erosion mechanism at other ACIAR project sites with higher slopes than those at Khon Kaen. The local concern with soil erosion at Khon Kaen may have had more to do with loss in soil fertility associated with the measured preferential loss of fine soil components and associated nutrients than with excessive soil losses. It is well known that nutrient loss is enriched in eroded sediment (Rose & Dalal, 1988), and this was measured in these projects (Hashim *et al.*, 1997).

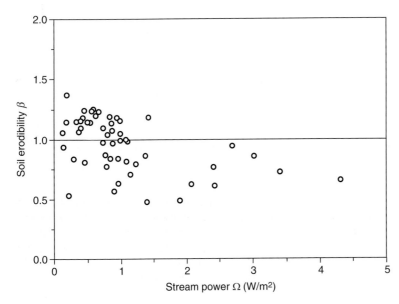

Fig. 11.4 Values of β calculated using data collected from each erosion event on steep plots with observed rilling at the VISCA site, plotted against the mean streampower for the event. Reproduced with permission from Presbitero *et al.* (2005).

Los Banos, the Philippines: At all tropical sites, in order to maintain a bare soil plot for substantial time periods, weed growth had to be controlled by cultivation, usually by hand as at the Los Banos site. Cultivation of the different land management treatments (including current farmer practice and alley cropping systems) used oxen and up-and-down slope cultivation. At Los Banos, the erodibility parameter β was evaluated for the bare plot over the four years 1990 to 1993. As shown in Paningbatan *et al.* (1995) and Rose *et al.* (1997), β was very responsive to the cultivation required to remove weed growth, especially in the first year. At the end of a fallow period, during which soil had consolidated, β was as low as 0.4, but then would rise to a little above 1.0 for the first erosion event following weeding. Variation in β was not as extreme in subsequent years, declining from 0.92 to 0.7 during 1991, and from 1.2 to 0.65 during 1993. Change through time in ϕ (the average sediment settling rate) was monitored, increasing from 0.096 to 0.165 m s⁻¹ during 1990. Soil strength was measured for pre-saturated soil using a torvane apparatus, and strength was found to increase monotonically with time since the last cultivation. Despite some scatter in the value

of β, there is some support for the expected decline in β as soil strength increased (Rose *et al.*, 1997). Also, over a 5-year period of measurement the soil organic matter fell by approximately half from some 4% to 2%, depending on treatment details to some extent. Since commencement of the GUEST program, the significance of interactions between soil erosion and loss of carbon stored in the soil has received much greater recognition.

VISCA, the Philippines: At VISCA there were both low-slope plots at 10%, and plots with quite steep slopes typical for the hand-hoe cultivation carried out on the island of Leyte (Plate 7). All experimental plots at the VISCA site were 6 m wide and 12 m long in the downslope direction. For the low slope (10%) plots, the average value of β for all 13 events during 1991 was 0.87 with a standard deviation of 0.12 (Presbitero *et al.*, 1995). For the four events with obvious rills, the average value of β was a little lower (0.78), perhaps due to rills eroding into somewhat more consolidated soil. On the adjacent steep-slope plots of 50%, 60% and 70% slope (Table 11.2), β was calculated for 18 events where rilling was recorded on all three slopes, yielding 18 × 3 = 54 datasets, with results shown in Fig. 11.4.

Table 11.3 Plot characteristics and soil erodibility for five ACIAR sites in southeast Asia and Australia (after Yu *et al.*, 1999).

Site	Country	Location	Soil texture	Slope (%)	No. of events	Soil erodibility β
Goomboorian	Australia	26°04'S, 152°48'E	Sand	5	79	1.049 ± 0.009
Kemaman	Malaysia	4°18'N, 103°19'E	Sandy loam	17	50	0.319 ± 0.018
Los Banos	Philippines	14°6'N, 121°12'E	Clay	26	19	0.879 ± 0.058
VISCA	Philippines	10°45'N, 124°49'E	Clay	50	4	0.890 ± 0.128
Khon Kaen	Thailand	16°30'N, 102°50'E	Loamy sand	3.6	22	0.928 ± 0.245

In another 14 events rilling may have occurred, but was not obvious following the event, providing some uncertainty in flow geometry, and hence in β. Figure 11.4 shows some decline in β as streampower of the flow increased, a trend in β also observed for a site in Thailand (Yu *et al.*, 1999). Values of β exceeding unity at lower values of streampower may indicate a contribution of rainfall impact to soil erosion. A detailed investigation was made of the effects of weeding on β. This showed that if weeding was carried out whilst the soil was wet, then sediment concentration, soil loss and β were all reduced for a storm event following weeding (which disturbed the soil surface). The hand cultivation of moist soil produced a rough, cloddy and well aggregated soil condition. However, this trend was reversed (i.e. β increased) if weeding was carried out when the soil was drier, and soil attached to the removed weeds was scattered onto the plot.

Imbil and Goomboorian, Australia: Similar experiments to those at other ACIAR sites were carried out on commercial pineapple farms at Imbil and Goomboorian on soils with different parent material and slopes (Tables 11.1 and 11.2). At Imbil, with a gravelly, weak-crumbed loam to clay loam soil, measurements were made on plot lengths of 7, 12 and 22 m (Ciesiolka *et al.*, 1995). Initially, when soil strength was low following cultivation, values of β were high and similar for all three plot lengths. However, as events continued over the period 1989 to 1991, values of β declined following the removal of the finer soil fractions, and as consolidation and armouring of

the soil surface with stone and gravel fractions developed. This decline was more rapid for the 7 and 12 m length plots than the 22 m plots, which developed deep rills. For the 22 m long plots, β fell from 1.24 following cultivation to 0.32 some 28 months and 19 erosion events later, hence the large standard deviation in a mean value of 0.67 (see Table 11.3). Whether or not the runoff rate was sufficient to remove stone armouring from the surface seemed to affect the value of β, small events which left armour in place yielding lower values of β (Ciesiolka *et al.*, 1995). Thus at this site, the dynamics of armouring, combined with general soil consolidation through time, played important roles in modifying the erodibility of the soil.

In contrast, at Goomboorian the sandy loam soil had no stone or gravel component. Preparation for planting the pineapple crop was preceded by intense cultivation, but since weeds were controlled chemically, there was no further cultivation. Over the three-year period of experimentation, the value of β fell steadily from about 1.1 to 1.0, probably due to soil consolidation, with some minor fluctuation which may be due to pulsing of sediment through the 36-m long ridge/furrow system plots (Rose *et al.*, 1997). Since the furrow slope was only 5%, the reason for β being greater than 1.0 initially could well be due to a rainfall-driven contribution to sediment concentration. The gradual limited decline in β with time was accompanied by an increase in soil strength as measured by torvane, presumably associated with gradual soil compaction or consolidation following the initial cultivation.

11.6.2 General discussion of ACIAR project erodibility results

Table 11.3 summarises the plot characteristics and soil erodibility values for the five ACIAR sites. For both clay-textured soils at Los Banos and VISCA, the calculated value of β fluctuated during each year, especially in response to cultivation, which was a necessary practice in order to keep the bare soil plots essentially weed-free (Paningbatan *et al.*, 1995). Thus, as shown in Table 11.3, soil erodibility is not a constant for a particular soil. Rather, at least in the agricultural context represented by the program sites, erodibility can be significantly affected by any land management activity, including cultivation, and the effects of rainfall in structural breakdown and compaction also appear to play an additional role. The value of β obtained using the Type B version of GUEST analysis consistently used in the ACIAR projects includes, but does not explicitly recognize, the contribution that rainfall impact can make to sediment contribution.

Results showed that the value of β calculated from bare plot data using GUEST varied in a manner which depended upon soil type, cultivation, and time since last cultivation (cultivation commonly being the method used for weed control), and evidently upon soil strength, consolidation, selective loss of finer fractions, and related factors. The types of cultivation varied considerably between sites. The bare soil at Kemaman was not cultivated, leading to a decline in β with time as initially loose soil was lost, followed by soil consolidation. These strong effects of initial cultivation followed by consolidation probably blur the dependence of soil erodibility on soil type at this site.

The ability to evaluate soil erodibility β on an event-by-event basis allows investigation of the effects of time, soil management and other factors. For cultivated plots, values of β were commonly in the range 0.7–1.0, tending to be higher after cultivation or weeding. Consolidation (and associated soil strength increase), together with lack of rill development led to lower values of β. Soils which were not cultivated to control weeds showed a substantial decline in β with time, as at Imbil and Kemaman. Thus soil erodibility, at least as indicated by β, appears not to be a general soil-type constant, although soil type has a major role in the likely range of values of β.

11.6.3 Cover effects in soil conservation

The aerial cover dominantly provided by leaves of vegetation can offer some protection against rainfall impact, although damaging drops larger than raindrops can be formed. However, much more effective protection of soil against erosion is provided by surface contact cover; this cover consists of mulch, plant parts, and so on, which are in such intimate contact with the soil surface as to impede overland flow, as well providing protection against raindrop impact. Figure 11.5, based on data from the Los Banos site (Paningbatan *et al.*, 1995), illustrates the commonly-found dramatic effect of increasing surface contact cover in reducing sediment concentration and so soil loss. The relationship between the contact cover fraction (C_f) and sediment concentration c relative to that from bare soil (c_b) has often been found to be of the general functional form:

$$c \, / \, c_b = \exp(-\kappa \, C_f) \qquad (11.21)$$

In Fig. 11.5 the coefficient κ has the approximate value of 10, this rapid exponential decline indicating the strong effectiveness of surface contact cover in this experiment, which was provided by crop residue and the trimmings of contour-planted leguminous hedgerows.

11.6.4 Use of ACIAR project data in developing a three-parameter hydrological model

These ACIAR projects provided a vast body of data on runoff rate, possibly the largest body of data on plot runoff rate measured at one-minute time intervals with infiltration-excess overland flow. This led to a search for the most appropriate and efficient form of infiltration equation to interpret the data. At all sites, apparent infiltration

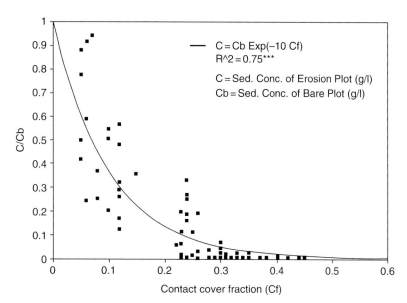

Fig. 11.5 The sediment concentration ratio c/c_b as affected by surface contact cover fraction. Reproduced with permission from Paningbatan *et al.* (1995) Alley cropping for managing soil erosion of hilly lands in the Philippines. *Soil Technology* **8**: 193–204, Elsevier.

rate was closely related to rainfall rate, with much less dependence on cumulative infiltration amount than would be indicated by models such as the Green-Ampt equation (Yu, 1999). The Green-Ampt infiltration equation predicts little sensitivity to depth of ponded water, which implies little sensitivity to rainfall rate, which is not the case in this extensive dataset. The outcome of this extensive examination of alternative infiltration models was the finding that the most efficient and accurate of the range of alternative models investigated (Yu *et al.*, 1997c) for the spatially-averaged infiltration rate $\bar{I}$ is that given by the equation:

$$\bar{I} = I_m(1 - \exp(-P/I_m)) \quad (\text{m s}^{-1}) \quad (11.22)$$

where the parameter I_m is the maximum possible value of the spatial mean infiltration rate for the complete plot area and P is the rainfall rate (m s⁻¹). This model describes in parametric fashion the spatial variation in infiltration rate using a single parameter.

The infiltration equation (11.22) shows that the spatially-averaged infiltration rate increases

(non-linearly) with rainfall rate up to a maximum limiting value, when the entire plot is generating runoff. This behaviour in response to increasing rainfall rate is consistent with an increasing fraction of the plot area generating excess rainfall, together with the remaining unponded area experiencing increased infiltration by its rainfall acceptance. That excess rainfall is generated from only some fraction of the catchment or plot area is commonly described as the 'partial-area' concept of runoff generation (Rose, 2004).

The infiltration equation (11.22) provides one of the three components in a predictive model relating the dynamics of runoff from an area to the time-varying rainfall it receives. The other two components are as follows:

• In general a certain amount of rainfall needs to fall on a given area before any runoff is produced. This threshold amount of rainfall (or infiltration) depends on how wet or dry the land area is prior to the rainfall received, as illustrated by Yu *et al.* (2000a).

• There is also a time lag between the generation of excess rainfall and its appearance as runoff due

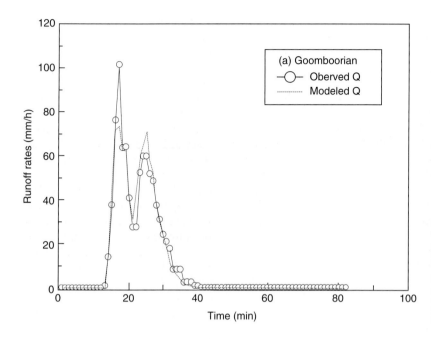

Fig. 11.6 A comparison of measured and modelled runoff rates at the Goomboorian site. Reproduced with permission from Yu *et al.* (1997a).

to the time lag involved in overland flow moving over the land surface to reach the exit from the plot where runoff rate is measured. Yu *et al.* (2000b) provided an analytical expression for this lag time and illustrated its application. The effective value of Manning's *n* involved in this lag can be evaluated from the lag between peaks in rainfall and runoff.

That such a three-parameter model of runoff provides a very good description of runoff rate at one-minute intervals within a storm event is illustrated in Fig. 11.6.

11.6.5 *Some soil conservation outcomes of ACIAR Projects*

Results from the extensive ACIAR experiments given earlier have focused on soil loss from the bare-soil treatments included at each site, and the interpretation of the data in terms of the soil erosion parameter β using program GUEST. With two exceptions, results from the bare soil plots at all ACIAR sites showed unacceptably high erosion losses of over $100\,t\,ha^{-1}\,y^{-1}$. These

results also provided a base against which the soil-conserving effectiveness of alternative management systems was compared. At each site, soil, water and nutrient loss were recorded for plots with different management scenarios, firstly for the common farmer practice for the crop of interest, and then for plots on which a range of soil-conserving practices was implemented. These practices were judged by collaborating scientists in the countries concerned to be potentially effective, realistic in terms of adoption feasibility, socially acceptable by land users in the various regions of Asia and Australia, and economically justifiable. All the soil-conserving technologies investigated were agronomic and not structural in character. At each site, agronomically-based systems of cultivation have been confirmed to yield generally low levels of soil loss, even on steep slopes. Rose (1995) described some of these management systems and the results obtained with these systems in place.

As an example, at the Los Banos site in the Philippines, the traditional farmer practice involved up-and-down slope tillage prior to planting maize. Soil-conserving treatments included alley cropping between contour-planted hedgerows of a leguminous shrub. Whether or not crop residuc and hedge trimmings were returned to the cropped alley between hedgerows were sub-treatments. Substantial reductions in loss of soil and water due to adopted soil-conserving practices were recorded in all but limited extreme events involving typhoons (Rose, 1995).

Acknowledging the importance of decreases in soil loss due to the adoption of more soil-conserving practices, the accompanying reduction in loss of chemical nutrients can also be vital, especially for crop production in the longer term (Hashim *et al.*, 1997). The link between soil erosion and nutrient loss is amplified by nutrient enrichment of eroded soil, which is of most importance in soils of lighter texture (e.g. at the Khon Kaen and Goomboorian sites).

Using cost, production, and soil loss data from the Los Banos site, an investigation was carried out into the long-term effects of alternative land management strategies on soil loss, crop yield, and on-farm economics (Nelson *et al.*, 1997). This investigation, using simulation-based cost-benefit methodology, was based on hedgerow intercropping data from the Los Banos site. The labour input required for the establishment of this system made it economically less attractive to poor farmers with no access to formal credit markets. Less labour-intensive soil conservation methods investigated in ACIAR projects included the return of maize crop residue into contour cultivation, and trash line cultivation. Adoption of improved farming practices with the potential for longer-term benefits is greatly inhibited by the insecurity of land tenure, which is typical in much of the Philippines' rural areas.

11.7   Soil Erosion Research at IBSRAM-ASIALAND Sites

Recognising the widespread and severe problem of soil erosion in southeast Asia, in 1988 the International Board for Soil Research and Manage-

ment (IBSRAM) organised an international network on the management of sloping lands for sustainable agriculture in Asia (ASIALAND Management of Sloping Lands). Previous soil erosion studies in the ASIALAND network countries were based on USLE technology, so that measurements included rainfall rate and event runoff amount, but not runoff rate. The purpose of the work described here was to develop and test methods so that data from USLE-type experiments could be used to evaluate soil erodibility using the physically-based erosion model GUEST. As described earlier, the use of models such as GUEST requires data on either runoff rate or an effective runoff rate (as distinct from runoff amount). Hence in order to apply GUEST to the data collected in ASIALAND projects, methods were investigated by Yu and Rose (1999) on how to infer runoff rates (or an effective runoff rate) for erosion events in which runoff rate was not measured.

When rainfall rate and total event runoff has been measured (as in USLE-type experimentation), a complete hydrograph can be generated using the simple but well-tested infiltration model of Yu *et al.* (1997a,b) given in Equation (11.22) in which P is rainfall rate and I_m is the only model parameter. I_m represents the spatially-averaged maximum rate of infiltration for the plot. Yu (1997) developed a computer program GOSH (Generation Of Synthetic Hydrograph) to facilitate this hydrograph generation process. Once the hydrograph is generated, the erosion model can be used in the same way as if the runoff rates had been measured experimentally. Rather than using the complete hydrograph, an effective event runoff rate Q_e can be computed using Equation (11.18) and then used in GUEST to calculate $\bar{c}_t$ (Equation (11.17)), and so β (Equation (11.11)). Yu and Rose (1999) showed that if only total runoff and rainfall are measured, but peak rainfall rate is known, then Q_e can be estimated using a scaling technique, again yielding β. Data were collected using USLE-type methodology at the four experimental sites in China, Malaysia and Thailand listed in Table 11.4 and shown in Fig. 11.3. Using data from

Table 11.4 Summary description of the four experimental sites from China, Malaysia and Thailand, together with calculated soil erodibility β (after Yu *et al.*, 1999).

Site	China Luodian, Guizhou	Malaysia Kuala Dal, Perak	Thailand Chiang Dao, Chiang Mai	Thailand Doi Tung, Chiang Rai
Soil texture	Silty clay loam	Clayey	Silty or clay loam	Silty clay or clay loam
Median particle size* (mm)	0.327	0.765	0.200**	1.15**
Depositability ϕ (m s^{-1})	0.0942	0.0856	0.0224	0.0738
Slope (%)	40	15	28	30
Plot length (m)	25	25	36	36
Mean soil erodibility β	0.421 ± 0.144	0.421 ± 0.166	0.248 ± 0.165	0.962 ± 0.172

*From wet-sieving analysis
**Measured in 1997

these sites, Yu *et al.* (1999) showed that for any particular runoff event, consistent values of β are obtained using runoff rates inferred from use of program GOSH, or from Q_e inferred using either of the methods just mentioned. This finding provided no direct proof that the values of β obtained using such inferred values of Q_e agree with values of β based on direct measurement of runoff rate, on which Q_e is physically based (Equation (11.18)). Examination of such agreement could be obtained in the future using the ACIAR database. Should such agreement be justified, then this would provide a yet firmer basis for employing GUEST methodology to data lacking the measurement of runoff rate.

Whilst measurement of runoff rate is a significant experimental burden in soil erosion research, as mentioned earlier in Section 11.5 there are advantages in its measurement using techniques outlined in that section and in Ciesiolka *et al.* (1995) and Coughlan and Rose (1997b).

Soil samples taken from the four ASIALAND sites were subject to wet sieving from which soil depositability was inferred using the equation of Cheng (1997). In all but the Malaysian site, weeds on the bare plot were controlled chemically. Rubber tree seedlings were planted into the Malaysian site, probably affecting later results from that site. Plot steepness ranged from 40% at the site in China to 15% at the Malaysian site

(Table 11.4). As in the ACIAR projects referred to earlier in this chapter, values of β were found to vary with successive erosion events at each site. An important feature of this study was to examine statistically the possible role of factors that might have an effect on the values of β obtained.

At the Chinese and Malaysian sites there was sufficient length of record to show a general downward trend with time in β, despite considerable variability. This time trend could indicate a gradual loss of the more erodible components of the soil, or some form of soil consolidation or strengthening, as weed control was achieved without cultivation. This time trend in decreasing erodibility has been observed elsewhere in similar regions. However, the decline in values of β in Malaysia could be partly due to the cover provided through time by rubber tree seedlings established in the second year of the experiment. The value given for β by GUEST is an effective erodibility figure; however, if the soil has some cover, then the value obtained for β is less than that of bare soil.

Although the length of record was only five months at the Chiang Rai site in Thailand, the values of β, calculated assuming an essentially planar land surface, followed a downward trend with peak runoff rate (see Fig. 11.7, and a similar trend in Fig. 11.4). Whilst the explanation of this trend is uncertain, at lower runoff rates higher values of β could be given both by

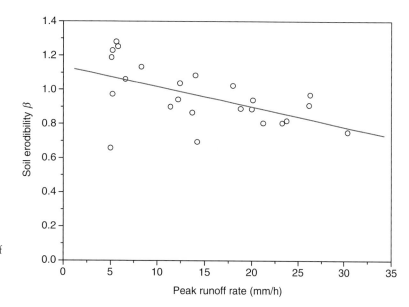

Fig. 11.7 Relationship between soil erodibility β and peak runoff rate for the Chiang Rai site. Reproduced with permission from Yu *et al.* (1999).

flow being more concentrated than assumed, and by the influence of rainfall detachment and re-detachment. This latter possibility is supported by the finding that for three of the four sites, statistical analysis supported a role for peak rainfall intensity in effecting an increase in the values of β. This could point to the purposeful or chosen limitation in the Type B analysis of GUEST that yields a value of β which does not explicitly recognize the role of rainfall in evaluating soil erodibility. Whilst this choice was made partly as an appropriate approximation in the spheres of higher-slope application, and partly on the grounds of simplicity in application compared with Type A analysis, it may limit the physical interpretation of the meaning of β. However, the influence of rainfall in this dataset on β was highly variable from site to site, with no clear pattern emerging.

Table 11.4 gives information on soil texture at the four ASIALAND sites, together with the arithmetic average and standard deviation in calculated values of β. Since soils with abundant fine materials tend to be well structured (increasing depositability and yielding lower sediment concentration), and have greater strength (increasing resistance to removal by rainfall or flow-driven processes, also decreasing sediment concentration), we would expect that soil erodibility would decrease as the proportion of fine material increases. There was some experimental support that increased soil strength was associated with a decrease in the soil erodibility parameter β (Misra & Rose, 1995; Misra & Teixeira, 2001). Furthermore, since size distribution determined using the wet-sieving technique most resembles the actual size distribution of soil particles during rainfall and runoff events, this technique was used to provide an appropriate size distribution. We then took the ratio of percentage of particles (both primary particles and aggregates) > 0.5 mm to that ≤ 0.5 mm, Θ, as a measure of the particle size. Figure 11.8 shows a scatterplot of this ratio of coarse to fine materials against the average soil erodibility. It can be seen that as the particle size from wet-sieving analysis decreases, the average soil erodibility at these sites also tends to decrease. The linear relationship shown as the

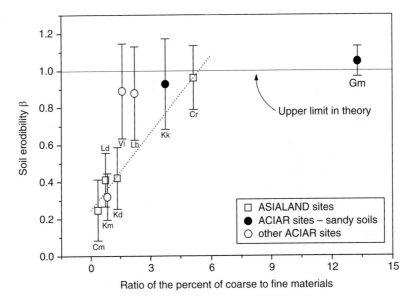

Fig. 11.8   The relationship between the ratio of coarse to fine materials and the average soil erodibility for ASIALAND and ACIAR sites. The dashed line represents the best fit for the four ASIALAND sites. For the ACIAR sandy soils, the ratio of percentage of sand (>0.02 mm) to the sum of percentage of silt and clay from mechanical analysis is used. For all other sites, the ratio of coarse particles (both primary particles and aggregates >0.5 mm) to fine particles (≤0.5 mm) obtained using the wet-sieving technique is used. Other legends are: Ld, Luodian, China; Kd, Kuala Dal, Malaysia; Cm, Chiang Mai, Thailand; Cr, Chiang Rai, Thailand; Gm, Goomboorian, Australia; Km, Kemaman, Malaysia; Lb, Los Banos, the Philippines; Vi, VISCA, the Philippines; Kk, Khon Kaen, Thailand. Reproduced with permission from Yu *et al.* (1999).

dashed line fitted to data from ASIALAND sites in Fig. 11.8 is given by:

$$\bar{\beta} = 0.246 + 0.139\,\Theta \qquad (11.23)$$

Also included in Fig. 11.8 are the results of data from the ACIAR experiments presented earlier in this chapter (see Table 11.3). Whilst the line fit shown in Fig. 11.8 and Equation (11.23) is to the ASIALAND data, the main discrepancies of the data from ACIAR sites from this line are for the two Philippines sites, which suggest a curvilinear relationship. However, as noted earlier, the frequent cultivation required for manual weed control may have led to the high erodibility of these two clay soil sites. The decrease in β with increase in fine soil fraction could be correlated with increased soil strength, and/or with decreased depositability (which increases c_i (Equation (11.8)), thus decreasing β (Equation (11.11)).

There are a number of implications, if not conclusions, from this comparative study of soil erodibility at ASIALAND and ACIAR sites. Firstly, variability in the erodibility parameter is considerable, both site to site and event to event. Secondly, the soil erodibility parameter is positively related to wet-sieved particle sizes, although the nature and reliability of this relationship are uncertain, and further compilation of the calculated soil erodibility values is clearly needed in order to provide greater confidence in such a predictive relationship. Thirdly, for sandy soils, the soil erodibility approaches its theoretical upper limit – unity – independent of the exact particle-size distribution, although this may be a qualified statement, because the two ACIAR sites with sandy soils also happen to be the ones with very gentle slopes.

The results obtained from the ASIALAND sites show that databases from USLE-type of experiments can be readily re-analysed to estimate necessary parameters for the physically-based erosion model GUEST. In other words, the erosion model GUEST can be tested, validated and used at no greater costs than validating the USLE. In fact, validation and use of the GUEST approach is actually easier because no long-term experimentation (perhaps >20 years) is required, as is the case for the USLE (Edwards, 1987).

11.8   Concluding Comments

Probably the largest body of data on soil erosion has been that collected using the USLE methodology, thus lacking measurement of runoff rates. In the ASIALAND section of this chapter it is demonstrated that the GUEST methodology can be applied to data collected for analysis using the USLE methodology. Thus there is the opportunity, in future, to analyse other USLE datasets to yield values of β which would be expected to provide further evidence of how soil erodibility is affected by soil characteristics and soil management (e.g. Fig. 11.8). The advantage of using the erodibility parameter β described in the GUEST methodology is that it has an approximate physical basis in fundamental soil erosion theory. In comparison, the soil erodibility K in the USLE equation is simply a proportionality factor introduced to ensure agreement between measured soil loss and the adopted product-type arrangement of other factors which do not include runoff or runoff rate.

Field applications of the GUEST program, such as those recorded in this chapter for the ACIAR and *ASIALAND* projects, benefited from a more fundamental controlled erosion program using simulated rainfall equipment

(e.g. Proffitt *et al.*, 1993b; Misra & Rose, 1995, 1996). These controlled studies supported the theory used in program GUEST, and its recognition of the pertinence of soil structural stability and soil strength as two factors affecting soil erodibility. In GUEST, the effect of the size distribution of aggregates and particles on sediment concentration is separated out through recognition in the theory of the role of the soil's settling velocity characteristics (or depositability). Although the erodibility factor β evaluated in the Type B version of GUEST is designed to represent properly the effect of soil strength in reducing sediment concentration, its magnitude also incorporates the role of any other factors or processes which contribute to the net erosion of soil. Because soil erodibility can be significantly affected by soil management (Rose, 1995), and even wetting and drying (Misra & Teixera, 2001), the ability to predict the magnitude of β is currently limited to general soil characteristics of the type illustrated in Fig. 11.8. However, the ACIAR project results (in particular) provide advice on the role of soil management in affecting β, and support the move towards minimizing tillage in agricultural land management (Goddard *et al.*, 2007).

The main purpose and requested aim of this chapter is to describe the GUEST program which (in its simpler Type-B form) enables efficient derivation of the physically-based erosion parameter β from experimental data on soil and water loss from an essentially plane bare soil plot in any single erosion event. Of course, data from repeated events can add information on time-related change in erodibility, but application of GUEST is not limited to longer-term datasets, as is recommended for the USLE methodology (Edwards, 1987). In comparison to WEPP, GUEST is also not limited to information only available in certain regions.

Information on the erodibility of soil in a bare condition is important in soil loss prediction and decision-making on the suitability of land for uses which expose it, and in the choice from a range of possible soil-conserving management systems. However, given the many factors involved in the design and adoption of land-using systems, it is recognised that information on soil erodibility is but *one* basic desirable input.

   Where direct experience is lacking, soil characteristics, soil type, expected soil management and cropping regimes, landscape factors, and the climatic/hydrological context can all be combined to provide broad guidelines on the danger of soil erosion, and thus the emphasis required in implementing soil-conserving management strategies. Of the wide range of such strategies, the agronomically-based experimentation reviewed in this chapter support the effectiveness of soil contact cover (Fig. 11.5) which can be implemented by all types of mulching strategies. The use of vegetative barriers has also been found to be most effective in reducing soil loss, and their implementation in steeper tropical lands, for example, can be encouraged by recognition of their potential role in adding to soil nutrition and productivity. More recent research, based on the same theory as employed in GUEST, has provided a physically-based understanding of the effectiveness of vegetative barriers in reducing soil loss (Rose *et al.*, 2002; Ghadiri *et al.*, 2002; Hussein *et al.*, 2007a,b).

   Ongoing work is designed to build on the GUEST approach in ways that address the spatial complexity of natural landscapes (e.g. Fentie *et al.*, 2004b).

Symbols List

Symbol	Description
A	Cross-sectional area of flow (m^2)
c	Sediment concentration ($kg\ m^{-3}$)
$\bar{c}$	Average sediment concentration for an erosion event ($kg\ m^{-3}$)
c_t	Sediment concentration at the transport limit ($kg\ m^{-3}$)
$\bar{c}_t$	Average transport limit sediment concentration for an erosion event ($kg\ m^{-3}$)
c_b	Bare soil sediment concentration ($kg\ m^{-3}$)
C	The fraction of bare soil surface in which soil aggregates are not fully immersed by overland flowing water
C_f	Surface fractional coverage by contact cover
d	Rate of sediment deposition ($kg\ m^{-2}\ s^{-1}$)
d_i	Sediment deposition rate of sedimentary particles of size range i ($kg\ m^{-2}\ s^{-1}$)
D	Depth of overland flow (m)
e_i	Rainfall detachment rate for particles of size range i ($kg\ m^{-2}\ s^{-1}$)
e_{ri}	Rate of rainfall re-detachment for particles of size range i ($kg\ m^{-2}\ s^{-1}$)
F	Fraction of streampower effective in erosive processes of erosion event
i	as a subscript refers to a particular sediment size range
$\bar{I}$	Spatial average value of infiltration rate ($m\ s^{-1}$)
I_m	Maximum possible value of $\bar{I}$ for a plot ($m\ s^{-1}$)
J	Specific energy of entrainment ($J\ kg^{-1}$)
k	Defined by Equation (11.14) ($kg\ m^{-3.4}\ s^{0.4}$)
K	Defined by Equation (11.20)
L	Length of plane
n	Manning's roughness coefficient
P	Rainfall rate ($m\ s^{-1}$)
q	Volumetric water flux per unit width of plane ($m^3\ m^{-1}\ s^{-1}$)
Q	Runoff rate per unit area of plane ($m\ s^{-1}$)
Q_e	Effective steady-state runoff rate for a rainfall event ($m\ s^{-1}$)
r_i	Rate of flow entrainment ($kg\ m^{-2}\ s^{-1}$)
r_{ri}	Rate of flow re-entrainment ($kg\ m^{-2}\ s^{-1}$)
R	Hydraulic radius of a rill (m)
S	Slope of the land surface (the sine of the slope angle)
t	Time (s)
V	Velocity of overland flow ($m\ s^{-1}$)

v_i	Settling velocity of sedimentary particle of size range i (m s^{-1})
W_r	Average rill spacing (m)
W_b	Bottom rill width (m)
β	Soil erodibility parameter
ϕ	Depositability (mean particle settling velocity) (m s^{-1})
ϕ_e	Effective depositability (m s^{-1})
ρ	Density of water (kg m^{-3})
ρ_e	Effective fluid density (kg m^{-3})
σ	Density of sedimentary material (kg m^{-3})
Σ	Summation sign
Ω	Stream power of flow (W m^{-2})

References

Asadi, H., Ghadiri, H., Rouhipour, H. & Rose, C.W. (2007) Interrill soil erosion processes and their interaction on low slopes. *Earth Surface Processes and Landforms* **32**: 711–24.

Cheng, N.S. (1997) Simplified settling velocity formula for sediment particle. *Journal of Hydraulic Engineering ASCE* **123**: 149–52.

Ciesiolka, C.A., Coughlan, K.J., Rose, C.W., *et al.* (1995) Methodology for a multi-country study of soil erosion management. *Soil Technology* **8**: 179–92.

Ciesiolka, C.A.A. & Rose, C.W. (1998) The measurement of soil erosion. In Penning de Vries, F.W.T., Agus, F. & Ker, J. (eds), *Soil Erosion at Multiple Scales – Principles and Methods for Assessing Causes and Impacts*. CABI Publishing in association with IBSRAM, Wallingford, UK: 287–301.

Ciesiolka, C.A.A., Yu, B., Rose, C.W., *et al.* (2006) Improvement in soil loss estimation in USLE type experiments. *Journal of Soil and Water Conservation* **61**: 223–9.

Coughlan, K.J. (1997) Description of sites, experimental treatments and methodology. In Coughlan, K.J. & Rose, C.W. (eds), *A New Soil Conservation Methodology and Application to Cropping Systems in Tropical Steeplands*. ACIAR Technical Report 40. Australian Centre for International Agricultural Research, Canberra: 3–8.

Coughlan, K.J. & Rose, C.W. (eds.) 1997a. *A New Soil Conservation Methodology and Application to Cropping Systems in Tropical Steeplands*. ACIAR Technical Report 40. Australian Centre for International Agricultural Research, Canberra.

Coughlan, K.J. & Rose, C.W. 1997b. Field experimental results- runoff, soil loss and crop yield. In Coughlan, K.J. & Rose, C.W. (eds), *A New Soil Conservation Methodology and Application to Cropping Systems in Tropical Steeplands*. ACIAR Technical Report 40. Australian Centre for International Agricultural Research, Canberra: 9–23.

Edwards, K. 1987. *Runoff and Soil Loss Studies in New South Wales*. Conservation Service of New South Wales and Macquarie University Technical Handbook No. 10, Sydney, NSW, Australia.

Fentie, B., Rose, C.W. & Coughlan, K.J. (1999) *GUEST 3.0: A program for calculating a soil erosion parameter in a physically-based erosion model*. ENS Working Paper 1/99, Faculty of Environmental Sciences, Griffith University, Brisbane, Australia.

Fentie, B., Yu, B. & Rose, C.W. (2004a) Comparison of seven particle settling velocity formula for erosion modelling. *ISCO 13th International Soil Conservation Organisation Conference, Brisbane, July 2004*, Paper No. 611.

Fentie, B., Ciesiolka, C.A.A., Silburn, D.M., *et al.* (2004b) Soil erosion and deposition modelling in a semi-arid grazing catchment in North Central Queensland. *Proceedings of the 13th International Soil Conservation Organisation Conference, Brisbane, July 2004*, Paper No. 610.

Foster, G.R. (1982) Modelling the erosion process. In Hann, C.T (ed.), *Hydrologic Modelling of Small Watersheds*. American Society of Agricultural Engineers Monograph No. 5, St. Joseph, MI: 297–379.

Ghadiri, H., Rose, C.W. & Misra, R.K. (2002) Buffer-strip induced flow retardation and sediment deposition. In *Proceedings of 12th International Soil Conservation Organisation Conference. May 26–31*. Volume III. Technology and Method of Soil and Water Conservation: 163–8.

Goddard, T., Zoebisch, M., Gan, Y., *et al.* (eds) (2007) *No-till Farming Systems*. World Association of Soil and Water Conservation Special Publication 3.

Hairsine, P.B. & Rose, C.W. (1991) Rainfall detachment and deposition: sediment transport in the absence of flow-driven processes. *Soil Science Society of America Journal* **55**: 320–24.

Hairsine, P.B. & Rose, C.W. (1992a) Modelling water erosion due to overland flow using physical principles: I. Uniform flow. *Water Resources Research* **28**: 237–43.

Hairsine, P.B. & Rose, C.W. (1992b) Modelling water erosion due to overland flow using physical principles: II. Rill flow. *Water Resources Research* **28**: 245–50.

Hashim, G.M., Ciesiolka, C.A., Yusoff, W.A., *et al.* (1995) Soil erosion processes in sloping land in the east coast of Peninsular Malaysia. *Soil Technology* **8**: 215–33.

Hashim, G.M., Ciesiolka, C.A.A., Rose, C.W., *et al.* (1997) Loss of chemical nutrients by soil erosion. In Coughlan, K.J. & Rose, C.W. (eds), *A New Soil Conservation Methodology and Application to Cropping Systems in Tropical Steeplands.* ACIAR Technical Report 40. Australian Centre for International Agricultural Research, Canberra: 79–100.

Heilig, A., De Bruyn, D., Walter, M.T., *et al.* (2001) Testing a mechanistic soil erosion model with a simple experiment. *Journal of Hydrology* **244**: 9–16.

Hussein, J., Yu, B., Ghadiri, H. & Rose, C. (2007a) Impact of vetiver grass strip on surface water hydrology and sediment deposition. *Journal of Hydrology* **338**: 261–72.

Hussein, J., Ghadiri,H., Yu, B. & Rose, C. (2007b) Sediment retention by a stiff grass hedge under sub-critical flow conditions. *Soil Science Society of America Journal* **71**: 1516–23.

Lisle, I., Coughlan, K.J. & Rose, C.W. (1995) *GUDPRO 3.1. A program for calculating particle size and settling characteristics. User guide and reference manual.* Technical publication. Faculty of Environmental Sciences, Griffith University, Nathan Campus, Queensland 4111, Australia.

Lovell, C.J. & Rose, C.W. (1988a) Measurement of soil aggregate settling velocities I. A modified bottom withdrawal tube. *Australian Journal of Soil Research* **26**: 55–71.

Lovell, C.J. & Rose, C.W. (1988b) Measurement of soil aggregate settling velocities II. Sensitivity to sample moisture content and implications for studies of structural stability. *Australian Journal of Soil Research* **26**: 73–85.

Misra, R.K. & Rose, C.W. (1989) *Manual for Use of Program GUEST.* Division of Australian Environmental Studies, Griffith University, Brisbane, Queensland, Australia 4111.

Misra, R.K. & Rose, C.W. (1992) *A Guide for the Use of Erosion-Deposition Programs.* Division of Environmental Sciences, Griffith University, Brisbane, Queensland, Australia 4111.

Misra, R.K. & Rose, C.W. (1995) An examination of the relationship between erodibility parameters and soil strength. *Australian Journal of Soil Research* **33**: 715–32.

Misra, R.K. & Rose, C.W. (1996) Application and sensitivity analysis of process-based erosion model GUEST. *European Journal of Soil Science* **47**: 593–604.

Misra, R.K. & Teixeira, P.C. (2001) The sensitivity of erosion and erodibility of forest soils to structure and strength. *Soil & Tillage Research* **59**: 81–93.

Nearing, M.A., Norton, L.D., Bulgakov, D.A., *et al.* (1997) Hydraulics and erosion in eroding rills. *Water Resources Research* **33**: 865–76.

Nelson, R.A., Dimes, J.D., Silburn, D.M., *et al.* (1997) Long-term effects of land management on soil erosion, crop yield and on-farm economics in the Philippines. In Coughlan, K.J. & Rose, C.W. (eds), *A New Soil Conservation Methodology and Application to Cropping Systems in Tropical Steeplands.* ACIAR Technical Report 40. Australian Centre for International Agricultural Research, Canberra: 111–40.

Paningbatan, E.P., Ciesiolka, C.A., Coughlan, K.J. & Rose, C.W. (1995) Alley cropping for managing soil erosion of hilly lands in the Philippines. *Soil Technology* **8**: 193–204.

Presbitero, A.L., Escalante, M.C., Rose, C.W., *et al.* (1995) Erodibility evaluation and the effect of land management practices on soil erosion from steep slopes in Leyte, the Philippines. *Soil Technology* **8**: 205–13.

Presbitero, A.L., Rose C.W., Yu, B, *et al.* (2005) Investigation of soil erosion from bare steep slopes of the humid tropic Philippines. *Earth Interactions,* Vol. 9, Paper No. 5.

Proffitt, A.P.B. & Rose, C.W. (1991a) Soil erosion processes I. The relative importance of rainfall detachment and runoff entrainment. *Australian Journal of Soil Research* **29**: 671–83.

Proffitt, A.P.B. & Rose, C.W. (1991b) Soil erosion processes II. Settling velocity characteristics of eroded sediment. *Australian Journal of Soil Research* **29**: 685–95.

Proffitt, A.P.B., Rose, C.W. & Hairsine, P.B. (1991) Rainfall detachment and deposition: Experiments with low slopes and significant water depths. *Soil Science Society of America Journal* **55**: 325–32.

Proffitt, A.P.B., Rose, C.W. & Lovell, C.J. (1993a) Settling velocity characteristics of sediment detached from a soil surface by raindrop impact. *Catena* **20**: 27–40.

Proffitt, A.P.B., Hairsine, P.B. & Rose, C.W. (1993b) Modelling soil erosion by overland flow: Application over a range of hydraulic conditions. *Transactions of the American Society of Agricultural Engineers* **36**: 1743–53.

Rose, C.W. (1993) Erosion and sedimentation. In Bonell, M., Hufschmidt, M.M. & Gladwell, J.S. (eds), *Hydrology and Water Management in the Humid Tropics – Hydrological Research Issues and Strategies for Water Management*. Cambridge University Press, Cambridge: 301–43.

Rose, C.W. (ed.) (1995) Soil erosion and conservation. *Soil Technology*. Special Issue, Vol. 8, No. 3. Elsevier Science, Amsterdam.

Rose, C.W. (2004) *An Introduction to the Environmental Physics of Soil, Water and Watersheds*. Cambridge University Press, Cambridge.

Rose, C.W. & Dalal, R.C. (1988) Erosion and runoff of nitrogen. In Wilson, J.R. (ed.), *Advances in Nitrogen Cycling in Agricultural Ecosystems*. CAB International: Wallingford, UK: 212–33.

Rose, C.W. & Hairsine, P.B. (1988) Processes of water erosion. In Steffen, W.L. & Denmead, O.T. (eds), *Flow and Transport in the Natural Environment*. Springer-Verlag, Berlin: 312–16.

Rose, C.W. & Yu, B. (1998) Dynamic process modelling of hydrology and soil erosion. In Penning de Vries, F.W.T., Agus, F. & Ker, J. (eds), *Soil Erosion at Multiple Scales – Principles and Methods for Assessing Causes and Impacts*. CABI Publishing in association with IBSRAM, Wallingford, UK: 269–86.

Rose, C.W., Williams, J.R., Sander, G.C. & Barry, D.A. (1983a) A mathematical model of soil erosion and deposition processes. I. Theory for a plane land element. *Soil Science Society of America Journal* **47**: 991–5.

Rose, C.W., Williams, J.R., Sander, G.C. & Barry, D.A. (1983b) A mathematical model of soil erosion and deposition processes. II. Application of data from an arid zone catchment. *Soil Science Society of America Journal* **47**: 996–1000.

Rose, C.W., Hairsine, P.B., Proffitt, A.P.B. & Misra, R.K. (1990) Interpreting the role of soil strength in erosion processes. *Catena Supplement* **17**: 153–65.

Rose, C.W., Coughlan, K.J., Ciesiolka, C.A.A. & Fentie, B. (1997) Program GUEST (Griffith University Erosion System Template). In Coughlan, K.J. & Rose, C.W. (eds), *A New Soil Conservation Methodology and Application to Cropping Systems in Tropical Steeplands*. ACIAR Technical Report 40. Australian Centre for International Agricultural Research, Canberra: 34–58.

Rose, C.W., Hogarth, W.L., Ghadiri, H., *et al.* (2002) Overland flow to and through a segment of uniform resistance. *Journal of Hydrology* **255**: 134–50.

Rose, C.W., Yu, B., Hogarth, W.L., *et al.* (2003) Sediment deposition from flow at low gradients into a buffer strip – a critical test of re-entrainment theory. *Journal of Hydrology* **280**: 33–51.

Sombatpanit, S., Rose, C.W., Ciesiolka, C.A. & Coughlan, K.J. (1995) Soil and nutrient loss under rozelle (*Hibiscus subdariffa* L. var. *altissima*) at Khon Kaen, Thailand. *Soil Technology* **8**: 235–41.

Yu, B. (1997) *A Program for Calculating Runoff Rates Given Rainfall Rates and Runoff Amount*. ENS Working Paper 2/97. Faculty of Environmental Sciences, Griffith University, Brisbane, Qld., Australia 4111.

Yu, B. (1999) A comparison of the Green-Ampt and a spatially variable infiltration model for natural storm events. *Transactions of the American Society of Agricultural Engineers* **42**: 89–97.

Yu, B. (2003) Unified framework for erosion and deposition equations. *Soil Science Society of America Journal* **67**: 251–7.

Yu, B. & Rose, C.W. (1997) *GUEPS: A Program for Calculating the Soil Erodibility Parameter β and Predicting the Amount of Soil Loss using GUEST Methodology*. ENS Working Paper 3/97. Faculty of Environmental Sciences, Griffith University, Nathan, Queensland, Australia 4111.

Yu, B. & Rose, C.W. (1999) Application of a physically based soil erosion model, GUEST, in the absence of data on runoff rate. I. Theory and methodology. *Australian Journal of Soil Research* **37**: 1–11.

Yu, B., Rose, C.W., Coughlan, K.J. & Fentie, B. (1997a) Plot-scale runoff modelling for soil loss prediction. In Coughlan, K.J. & Rose, C.W. (eds), *A New Soil Conservation Methodology and Application to Cropping Systems in Tropical Steeplands*. ACIAR Technical Report 40. Australian Centre for International Agricultural Research, Canberra: 24–33.

Yu, B., Rose, C.W., Ciesiolka, C.A.A., *et al.* (1997b) Towards a framework for runoff and soil loss using GUEST technology. *Australian Journal of Soil Research* **35**: 1191–1212.

Yu, B., Rose, C.W., Coughlan, K.J. & Fentie, B. (1997c) Plot-scale rainfall-runoff characteristics and modeling at six sites in Australia and Southeast Asia. *Transactions of the American Society of Agricultural Engineers* **40**: 1295–1303.

Yu, B., Cakurs, B. & Rose, C.W. (1998) An assessment of methods for estimating runoff rates at the plot scale. *Transactions of the American Society of Agricultural Engineers* **41**: 653–61.

Yu, B., Sajjapongse, A., Yin, D., *et al.* (1999) Application of a physically based soil erosion model, GUEST, in the absence of data on runoff rate. II. Four case studies from China, Malaysia and Thailand. *Australian Journal of Soil Research* **37**: 13–31.

Yu, B., Sombatpanit, S., Rose, C.W., *et al.* (2000a) Characteristics and modelling of runoff hydrographs for different tillage systems. *Soil Science Society of America Journal* **64**: 1763–70.

Yu, B., Rose, C.W., Ciesiolka, C.A.A. & Cakurs, U. (2000b) The relationship between runoff rate and lag time and the effects of surface treatments at the plot scale. *Hydrological Sciences Journal* **45**: 709–26.

12 Evaluating Effects of Soil and Water Management and Land Use Change on the Loess Plateau of China using LISEM

R. HESSEL[1], V.G. JETTEN[2], B. LIU[3] AND Y. QIU[3]

[1]*Soil Science Centre, Alterra, Wageningen University and Research Centre, Wageningen, The Netherlands*
[2]*Department of Earth Systems Analysis, International Institute of Geoinformation Science and Earth Observation, Enschede, The Netherlands*
[3]*School of Geography, Beijing Normal University, Beijing, China*

12.1 Introduction

The Yellow River is China's second largest river and derives its name from the sediment suspended in its waters. These sediment contents pose a major problem because sedimentation in the lower course of the river has raised the river bed to several metres above the surrounding landscape, and because reservoirs are being filled with sediment. The Chinese government is committed to combating these problems. Much attention is being directed at decreasing the erosion rates on the Loess Plateau, since this is the source of about 90% of all the sediment that enters the Yellow River (Douglas, 1989; Wan & Wang, 1994). Lowering the erosion rates should decrease downstream sedimentation problems, and at the same time reduce the loss of agricultural land on the Loess Plateau itself.

The Loess Plateau has some of the highest erosion rates on Earth, especially in the so-called hilly part, which is heavily dissected by gullies. Plate 8 shows a typical landscape for the hilly part of the Loess Plateau. Jiang *et al.* (1981) estimated that erosion rates may be as much as 18,000 t km^{-2} y^{-1} for the hilly loess region of the Wuding catchment. Sediment concentrations in runoff on the Loess Plateau of over 1000 g l^{-1} have been recorded regularly (Jiang *et al.*, 1981; Zhang *et al.*, 1990; Wan & Wang, 1994). There are several reasons for these very high erosion rates:

• The loess is very erodible, especially when wet.
• The area's rainfall is characterized by heavy storms in summer (mainly July and August). Single storms can produce 10% of yearly precipitation and 40% of erosion (Gong & Jiang, 1979; Zhang *et al.*, 1990). Although the saturated conductivity of the loess is generally higher than rainfall intensity, crusting prevents the water from infiltrating (Douglas, 1989).
• The area has considerable relief.
• Vegetation cover is generally sparse. This is partly caused by a semi-arid climate with cold winters, but also by removal of the natural vegetation and by grazing (Jiang *et al.*, 1981).

In 1999, the Chinese government formulated new ambitious policies about the Loess Plateau. These policies aim to decrease erosion rates through changes in land use (McVicar *et al.*, 2007). In particular, they aim at a large decrease

Handbook of Erosion Modelling, 1st edition. Edited by R.P.C. Morgan and M.A. Nearing. © 2011 Blackwell Publishing Ltd.

in cropland area so that all fields on slopes above a certain slope degree should be changed to other uses. The decrease in cropland should be accompanied by an intensification of the remaining cropland and by an increase in orchards (cash trees). The idea is that in the long term the income of the farmers should increase once they get better yields from the remaining cropland as well as income from fruit trees and other cash trees. Since it takes time before the new land use can start to benefit the farmers, the government pays compensation to make the change economically feasible for them.

The Loess Plateau is likely to remain an area of considerable erosion despite all efforts to reduce erosion rates. It will remain a high-relief, low vegetation-cover area with heavy storms on erodible soils. Since the gully erosion has very markedly increased local relief, it is unrealistic to think that proper conservation methods will reduce erosion rates to pre-cultivation levels. Nevertheless, such conservation methods could achieve large reductions in current erosion rates.

Soil erosion modelling is potentially a powerful tool for combating soil erosion. It helps us to understand erosion better, to locate erosion hotspots, to predict erosion and to evaluate the effects of different soil and water conservation measures before implementing them. Even though research on the Loess Plateau has been intense for the past 50 years, process-based erosion models have not often been applied. Instead, more attention has been given to monitoring.

For erosion modelling on the Loess Plateau, several characteristics of the Plateau need to be addressed:
- Slopes in the erodible loess can be very steep, which may have consequences for flow velocity and transport capacity of the flow.
- Sediment concentrations in runoff may be extremely high. At such concentrations the fluid properties might differ from those of clear water.
- The area is heavily dissected by gullies. Thus, erosion models should be able to cope with gully erosion, or at least with gullies as a source of sediment.

The LISEM (Limburg Soil Erosion Model) model is chosen for this exercise because it is, in principle, suitable for simulating erosion on the Loess Plateau for several reasons. Firstly, LISEM is storm-based, so that it should be able to handle the storm-dominated water erosion of the Plateau. Secondly, LISEM is a distributed model, so that spatial predictions of erosion inside a catchment are possible. Thirdly, LISEM is a process-based model. As the source code was available to us, this meant that process descriptions in the model could be adapted if the specific characteristics of the Loess Plateau required this. Fourthly, LISEM is integrated with the geographical information system (GIS) PCRaster (Wesseling *et al.*, 1996), and reads GIS maps as input and produces GIS maps as output. Finally, LISEM was developed for loess soils. This combination of characteristics made LISEM more suitable than other well-known erosion models, such as WEPP, EUROSEM and KINEROS. The principles of LISEM have been described in several papers (De Roo *et al.*, 1994, 1996a,b; Jetten & De Roo, 2001).

The aims of this study were:
(**1**) to evaluate the effects of the particular characteristics of the Loess Plateau on soil erosion processes;
(**2**) to adapt the LISEM model to Loess Plateau conditions if this proves necessary;
(**3**) to calibrate and validate the LISEM model for a small catchment on the Chinese Loess Plateau; and
(**4**) to simulate the effects that different soil and water conservation methods have on soil erosion.

12.2 Study Area

The study area was the Danangou catchment, a typical small (3.5 km²) Loess Plateau catchment in Shaanxi Province, northern China, with steep slopes and a loess thickness of more than 100 metres (Plate 9). The soils are mainly erodible silt loams that classify as Calcaric Regosols/Cambisols in the FAO-system (Messing *et al.*, 2003a). Median grain size of the loess is about 35 μm. The climate is semi-arid, with occasional heavy thunderstorms

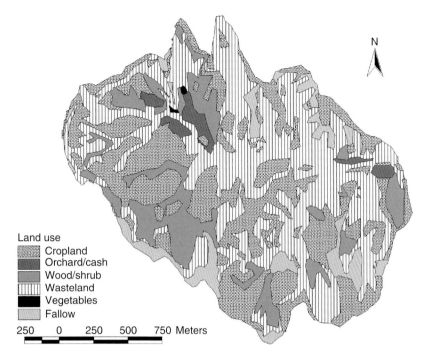

Fig. 12.1 Land use of the Danagou catchment in 1998.

Land use
- Cropland
- Orchard/cash
- Wood/shrub
- Wasteland
- Vegetables
- Fallow

250   0   250   500   750 Meters

in summer. On average three to four storms each year are large enough to cause runoff, but the actual number varies widely from year to year. In the Danangou catchment, discharge only occurs during these heavy storms, when peak discharges of over $10\,m^3\,s^{-1}$ can be reached within 15 minutes of the onset of channel runoff. The catchment is deeply dissected by gullies, which, according to the digital elevation model (DEM), have slope angles of up to 250% (68°). Gullies occupy about 25% of the catchment area.

The main land uses in the Danangou catchment are wild grassland (wasteland) and cropland. Wild grassland is mainly located on the steeper parts of the gully slopes as well as on the gully bottoms, and was until recently used for grazing goats. This practice has ceased since grazing was prohibited in September 1999. Cropland is located mainly on the hilltops, on slopes that are usually between 25 and 50% (14–27°), but sometimes even steeper, and on the relatively gentle slopes at lower elevation. The most common crops in the area are potato,

millet, soybean, buckwheat and maize. Fallow land is mostly situated along the hilltops and woodland in the upper parts of some of the valleys. Figure 12.1 shows land use in 1998.

12.3   LISEM Model

12.3.1   *Structure of LISEM*

Figure 12.2 shows a simplified flow chart of the LISEM model. LISEM can be divided into two parts: a water part and an erosion part. Rainfall is the basic input of the water part. Interception is subtracted from the rainfall. The remaining rainfall reaches the soil surface, where it can infiltrate or form a surface storage. Since LISEM is a storm-based model, the infiltrated water is essentially a loss of water in the sense that infiltrated water cannot resurface. Infiltration can be simulated using one of several available equations. Partly empirical equations such as the Green and Ampt and Holtan equations can be chosen, as

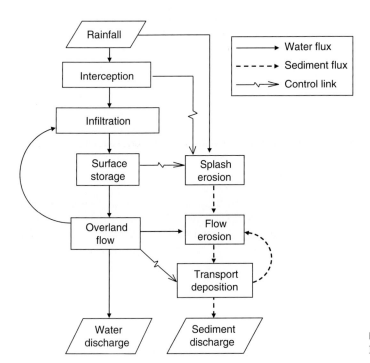

Fig. 12.2 Simplified flowchart of the LISEM soil erosion model.

well as the physically-based Richards equation (using the SWATRE submodel: Belmans *et al.*, 1983). Surface storage will result in surface runoff once a certain threshold is exceeded. Flow velocity is calculated with the Manning equation and surface runoff is routed over the landscape with the kinematic wave equation. The user can specify a separate channel network. Overland flow can flow into the channel and is then routed to the catchment outlet as channel flow.

Overland flow and channel flow are both routed with the kinematic wave, which is solved by a four-point finite difference solution using an implicit method (Chow *et al.*, 1988). LISEM routes the sediment explicitly using water fluxes at the beginning and end of each time step to determine the sediment concentration in each pixel.

LISEM simulates erosion by rainfall and erosion by overland flow and channel flow. Rainsplash erosion is calculated as a function of rainfall and throughfall kinetic energy and depth of the surface water layer. Sediment transport only occurs by overland flow and channel flow. For both overland flow and channel flow, LISEM uses the transport equation developed by Govers (1990) for slopes of up to 12°. The equation is based on a stream-power approach:

$$TC = c(SV - SV_{cr})^d \cdot \rho_s \qquad (12.1)$$

where TC is the transport capacity (g l^{-1}), S is the slope (m m^{-1}), V is the mean velocity (m s^{-1}), SV_{cr} is the critical unit stream power (m s^{-1}), ρ_s is the density of solids (kg m^{-3}), and c and d are coefficients.

According to Govers (1990) the critical unit stream power is 0.004 m s^{-1}. The coefficients c and d depend on the median of the grain-size distribution and can be calculated with equations given by Morgan *et al.* (1998). Net flow detachment and deposition are calculated with two equations based on the EUROSEM model (Morgan *et al.*, 1998), but reformulated for use with pixels:

$$D_f = y \cdot (TC - C) \cdot Q \qquad (12.2a)$$

$$D_p = (TC - C) \cdot \omega \cdot w \cdot DX \qquad (12.2b)$$

where $D_{f/p}$ is the net rate of sediment detachment/deposition by flow (kg s^{-1}), y is the efficiency coefficient (value between 0 and 1), TC is the sediment transport capacity (kg m^{-3}), C is the sediment concentration (kg m^{-3}), Q is the discharge (m^3 s^{-1}), ω is the particle settling velocity (m s^{-1}), w is the flow width (m) and DX is the pixel length (m).

To avoid mass balance problems, the detachment in a timestep cannot exceed the remaining transport capacity (TC–C), while the deposition cannot be more than there is sediment in the flow. Current versions of LISEM (2.5 and above) have a possibility of including a stationary baseflow in the simulations.

12.3.2 LISEM calibration and validation

Although, theoretically, fully physically-based models should not have to be calibrated, reality is different (see Chapter 3 for a more general discussion of this issue). Models are never fully physically-based, and many authors have demonstrated the need to calibrate process-based erosion models to obtain an acceptable predictive quality (e.g. Jetten *et al.*, 1999). In the case of hydrological/erosion models, calibration has mostly been done using measured data at the outlet of the plot or catchment.

LISEM was developed for the province of Limburg, The Netherlands, and has been previously calibrated and validated for several catchments in the Loess region of northwestern Europe. De Roo *et al.* (1996b) found that the most sensitive variable in the prediction of runoff is saturated hydraulic conductivity. For the prediction of soil erosion, LISEM was most sensitive to changes in Manning's n and transport capacity. De Roo *et al.* (1996b) found that LISEM gave reasonable results for 60% of the storms that were modelled. They attributed the discrepancy for the other 40% to spatial and temporal variability in saturated hydraulic conductivity and initial moisture content, and to differences between summer storms and winter storms. De Roo and Jetten (1999) identified saturated conductivity, suction at the wetting front, and initial moisture content as sensitive parameters.

More recently, several authors have pointed to the necessity of calibrating process-based, distributed models in a spatial way (e.g. Jetten *et al.*, 1996, 2003; Takken *et al.* 1999; Beven, 2002; see also Section 3.5). Such a calibration is a logical step since the main advantage of distributed models over lumped models should be their ability to predict spatial patterns. Also, there are circumstances where the location of erosion or deposition in a catchment is more important than the precise amount of water/sediment passing the outlet: for example, to design effective anti-erosion measures.

12.4 LISEM in relation to the Loess Plateau

Some of the main algorithms of LISEM are only valid for a given range of circumstances. Since the circumstances on the Loess Plateau are very different from the circumstances for which LISEM was developed, it is possible that application limits will be encountered when applying LISEM to the Loess Plateau. At present LISEM does not specifically take into account the effects of steep slopes, high sediment concentrations and the presence of gullies.

Soil erosion models have not been applied to steep slopes very often. Since their focus has been on predicting erosion from arable land in Europe and the US, which is generally on gentle to moderate slopes, not much attention has been paid to slope steepness. Slopes of 10% are usually considered 'steep', while in many other areas of the world, including China, cropland occurs on much steeper slopes. The velocity equations and the sediment transport equations that are commonly used in erosion models have not been developed or tested for such steep slopes as those of the Chinese Loess Plateau. LISEM calculates flow velocity with Manning's equation and subsequently calculates stream power and transport capacity. Since unit stream power is the product of the velocity and the (energy) slope, the slope angle influences the stream power considerably. According to Govers (1990), the transport capacity equation derived from stream power is valid for slopes up to 20%, and this might pose one of the largest potential problems in the application of LISEM to this area, since erosion and deposition are modelled as transport deficits and surpluses and are therefore strongly determined by the flow conditions.

Soil erosion models have so far not paid specific attention to high sediment concentrations. For most regions, concentrations in runoff will not be very high, so that no special attention is needed. For the Loess Plateau, however, very high concentrations have been reported regularly. These kinds of flow occupy intermediate positions between clear water flow and debris flow, and could have properties that differ significantly from clear water flow. More specifically, high sediment concentrations could change density, viscosity, resistance to flow, velocity profile and transport capacity (Hessel, 2006). Obviously, erosion models that deal with high concentrations should at least consider these effects, and some of the effects can be relatively easily incorporated into erosion models. Examples are corrections to settling velocity, fluid density and viscosity, as most existing equations that relate viscosity and settling velocity to sediment concentration give fairly similar results (Hessel, 2006). For other fluid properties, such as velocity, velocity profile, flow resistance and transport capacity, the evidence is more scant and even partly contradictory, so that corrections will be difficult. Care should also be taken to avoid comparing model results with measurements without taking into account that models use clear water values (e.g. g per litre of clear water), while the raw measurement data are usually fluid values (e.g. g per litre of fluid).

Gully erosion is a topic that has received more attention in erosion modelling. Some of the present-day erosion models do simulate some sort of gully erosion, but even models that have been specifically developed to model gully erosion require that some properties of the gullies be set in advance (see Chapter 19). Modelling changes in gully morphology is outside the scope of the cur-

rent research, but sediment yield coming from gullies needs to be taken into account.

12.5   Method

In 1998, the EROCHINA project started, with the aim of finding ways to decrease erosion rates in a small catchment on the Loess Plateau. The project consisted of two main parts: participatory research into farm economy, and soil erosion modelling. The results of the participatory work have been discussed elsewhere (Messing & Hoang Fagerström, 2001; Hoang Fagerström *et al.*, 2003), and this chapter focuses on soil erosion modelling. Soil erosion modelling consisted of data collection, adaptation of the LISEM model, calibration of the model, and simulation of the effect of land use scenarios.

12.5.1   *Data collection*

As a process-based distributed model, LISEM needs a large amount of input data. During the study period (1998–2000) most of the input parameters needed were measured repeatedly in the Danangou catchment. Plant and soil characteristics were measured on a fortnightly basis, except for Manning's *n*, which was measured in two separate campaigns using small runoff plots (Hessel *et al.*, 2003b). Soil physical characteristics such as saturated hydraulic conductivity, soil moisture retention curves and the water content–conductivity relationships were determined using samples taken in the catchment, for the land uses shown in Fig. 12.1. All these measurements are discussed elsewhere (Wu *et al.*, 2003; Liu *et al.*, 2003; Stolte *et al.*, 2003). The field data were converted to input maps for LISEM using the land-use map as a base, so that, for a given storm, these variables were constant within a land use, but differed between land uses. For variables that clearly also depend on soil type (e.g. cohesion), a combination of land use and soil type was used to extrapolate the measurements. Initial moisture content was predicted with multiple regression equations based on aspect and slope, and was therefore spatially variable. The resulting moisture contents were yearly averages, but these were corrected for particular events using data from a portable Delta-T theta probe. These data were collected close to the date of the event.

Rainfall was measured using six tipping-bucket rain gauges (1998–2000) and four simple rain gauges that measure total rainfall only (1999–2000). The rain gauges were distributed throughout the catchment. Thus, the number of rain gauges used in LISEM was between six and ten. Discharge and sediment concentration were measured at a V-shaped weir built in 1998 (position indicated in Plate 9). The area upstream of the weir is slightly over 2 km², but the total area of the catchment is 3.5 km². In the three-year study period only six events could be measured, one of which was not used for calibration because no sediment concentration data were available.

12.5.2 Adaptations to LISEM

Because of steep slopes, high sediment concentrations and permanent gullies in the area, the following changes were made to LISEM.

(i) Steep slopes
- Correction for overland flow distance. Previously, LISEM used the distance between pixel centres as flow distance. The grid is, however, essentially a horizontal grid. For steep slopes the overland flow distance is not equal to the distance between pixel centres. For example, if the slope is 45° and the distance according to the grid is 10 m, the actual distance over the surface is 14.4 m. To correct for this, a map showing the overland flow distance was calculated from the slope map.
- Use of sine instead of tangent, both in the Manning equation, and in equations for stream power. This is theoretically better since the slope in these equations is the energy slope. The sine of the slope angle gives the actual distance over which friction is exerted on the flow. For steep slopes, the tangent is much larger than the sine. Therefore, flow velocity and stream power will be smaller for steep slopes when sine is used instead of tangent.
- Use of a slope-dependent Manning's *n*. Hessel *et al.* (2003b) found that for the steep

erodible croplands of the Danangou catchment, flow velocity was independent of slope angle. The Manning equation, however, predicts an increase of velocity with slope angle. The most pragmatic solution to this problem was to allow Manning's *n* to vary for cropland as a function of slope, instead of taking a value that is constant for each particular land use. For other land uses, no slope dependency was found, and a constant value for Manning's *n* was used.
- Use of alternative transport equations. So far, the LISEM model has only used the stream-power based equation developed by Govers (1990) to predict transport capacity. A number of other equations were tested for the Danangou catchment (Hessel & Jetten, 2007). These equations were selected on their ability to deal with steep slopes and with high sediment concentrations. Concentration-dependent fluid density and viscosity were used in these equations.

(ii) High sediment concentrations
- Introduction of a sediment concentration-dependent particle fall velocity. At high sediment concentrations, settling velocity is significantly lower than settling velocity in clear water. The Chien and Wan (1983; as reported in Shen & Julien, 1993) equation was implemented because it was developed for Chinese conditions.

(iii) Gullies
- Introduction of a map with loose material present in gullies. Observations in the Danangou catchment showed that loose material accumulates on gully floors in between runoff events due to soil falls from the gully walls. A daily-based gully model was developed by Hessel and van Asch (2003) to model the amount of loose material available on gully floors. During the LISEM run the only factor determining whether or not the material is removed is the availability of transport capacity. The remaining available material is recalculated during each timestep and erosion will stop when there is no material remaining.

For some changes it is difficult to assess whether the improvement to the model also results in an improvement in simulation. If one starts with a calibrated model and then implements a theoretical improvement, it can be expected that the adapted model gives less good predictions. To evaluate whether the implemented change is an improvement in terms of simulation accuracy, one has to recalibrate the model. If this results in either a better fit with observations or the use of more realistic calibrated parameter values, the change can be considered an improvement. This method most easily applies to changing Manning's n and to slope correction because these changes affect the hydrograph. Therefore, the hydrographs can be compared with the predicted hydrographs of the original version. The other changes affect only sediment transport and are more difficult to test since the measurements of sediment concentration are less frequent and probably less reliable, so that there is also uncertainty about the accuracy of the measurements.

12.5.3   *Calibration and validation*

In this study, the prediction of both catchment soil loss and spatial erosion patterns was evaluated. To do this, a two-step approach was used. Firstly, the LISEM model was calibrated on runoff and sediment yield measured at the catchment outlet. Then, the simulation results of the calibrated model were evaluated in a spatial way using field observations on erosion patterns. Both the original version of LISEM (LISEM 163) and the adapted version (LISEM LP) were calibrated and the results of both versions were compared with each other.

Model calibration had several objectives: to simulate correctly the peak discharge, the total discharge and the total soil loss. The LISEM model was calibrated first on peak discharge (including time to peak and hydrograph shape) to obtain the correct shape of the hydrograph, and after that an adjustment was made to obtain the correct total discharge. Once the discharge prediction could not be improved any more the model was calibrated on sediment yield. Several parameters were used to calibrate on peak discharge:

(1) Saturated conductivity (K_{sat}). K_{sat} is the model parameter with the largest influence on discharge.
(2) Initial suction. Initial suction determines the unsaturated conductivity (and thus infiltration) during the start of a rainfall event. Initial suction was only used for calibration if calibrating on saturated conductivity proved insufficient.
(3) Manning's n. This influences the velocity of runoff and therefore affects the shape and timing of the hydrograph.
(4) Channel length. In LISEM, pixels can be defined that contain a channel characterized by a separate Manning's n. The width of these channels can be defined by the user, but must be smaller than the pixel size. Flow velocity in the channel will generally be higher as a result of different hydraulic radius. Changing the channel length therefore influences timing and shape of the simulated hydrograph.

All these parameters were changed within reasonable boundaries; that is, within boundaries that could be argued to be realistic given the available amount of data and its uncertainty. Because of the limited number of storms, a split in calibration and validation events was not possible. Instead, each event was calibrated separately. This resulted in five different calibration sets. Each calibration set was validated by applying it to the other four events.

Peak discharge calibrations are most suited to evaluate the performance of LISEM because they use time to peak, peak discharge and the shape of the hydrograph. The use of a goodness-of-fit coefficient (e.g. Nash & Sutcliffe, 1970) was less appropriate because these coefficients are very sensitive to a time shift in runoff. Therefore the fitting was done by eye, and the Nash-Sutcliffe coefficient was only calculated to compare the final calibrated versions of LISEM 163 and LISEM LP.

Total runoff volume calibrations were done because sediment loss is calculated as the product of runoff volume and concentration. The total runoff volume calibrations used the peak discharge calibration as a starting point. For the total runoff calibrations, only saturated conductivity was changed. Where the fit between predicted soil loss and measured soil loss was unsatisfactory, the LISEM model was calibrated on sediment yield (Hessel, 2002). To do this, cohesion, aggregate stability and median grain-size can be used because these parameters affect only sediment yield and have no influence on predicted discharge. In this chapter, results are only shown for calibration on peak discharge.

In accordance with Hessel (2005), pixel size and timestep length were chosen before calibration started. For all simulations, LISEM was used with a pixel size of 10 m and a timestep length of 15 s. Since the upper few decimetres of the soil are crucial for infiltration during a storm, ten calculation layers were used in the finite difference solution of the Richards equation, with node spacing increasing with depth. A single median grain-size (D_{50}) of 35 μm was used in all cases in the sediment transport equations of LISEM.

At the end of each rainy season (September) the occurrence and intensity of rilling was mapped throughout the 3.5 km² catchment. Rill intensity was classified in three classes: slight rill erosion, moderate rill erosion, and severe rill erosion. Quantification of the amount of erosion for each class was possible due to a number of measurements of rill frequency, width and depth that were conducted for each rill erosion class. For most years such mapping will give an aggregated result for all events, but in 1999 only a single rill-producing event occurred, so that the rill erosion map made in that year can be used directly to evaluate the performance of LISEM.

A single erosion plot was installed in 1999 to determine the amount of erosion occurring in arable fields. The plot was assumed to be representative of the cropland area in the Danangou catchment. Its dimensions were about 34 × 6.5 m, while slope steepness ranged from 15% at the top to 55% at the bottom. The total amount of water and sediment was measured on an event basis using a divisor and barrels. Since rill measurements were also conducted on the plot, it was possible to calculate the total sheet erosion by subtracting rill erosion from total plot erosion. This estimate is expressed as an amount of erosion per unit area of cropland, and was added to

the quantified rill erosion map to obtain a field erosion map. The resulting map was then compared with the erosion map produced by LISEM.

12.5.4 Land use scenarios

LISEM was used to evaluate the effects of different land use scenarios. The scenarios were developed based on a biophysical resource inventory (Messing *et al.*, 2003a,b), farmer's perceptions (Messing & Hoang Fagerström, 2001; Hoang Fagerström *et al.*, 2003) and the plans of the authorities to re-green the Loess Plateau. These plans include the gradual restriction of cropland to slopes of less than 15°, and the prohibition of grazing.

Four scenarios were developed, one using the 1998 land-use map (scenario 0), and three using respectively 25°, 20° and 15° as the upper slope limit for cropland (scenarios 1 to 3). The 15° limit scenario (already proposed by Fu & Gulinck, 1994) is considered a long-term scenario and the 25° and 20° limits are therefore short-term intermediate scenarios (Chen *et al.*, 2003). These scenarios had subscenarios that considered the effects of biological measures (such as cropland mulching and improved fallow) and mechanical measures (e.g. contour ridges on cropland). These measures are relatively simple and inexpensive, but labour-intensive.

Compared with the present land use (Fig. 12.1), the scenario land-use maps (15° map in Fig. 12.3) all have much more woodland/shrubland, while the cropland area is decreased according to the specified slope limits. Table 12.1 shows that for the 25°, 20° and 15° land-use maps there is a gradual decrease in cropland and fallow land and a gradual increase in orchard/cash trees. By definition, the change was limited to areas below 25°. The other land uses (including all slopes of more than 25°) remained unaffected for these scenario groups, so for these land uses and slopes, only the present land-use scenario differed. The economic consequences of the large changes in land use specified in Table 12.1 were discussed by Chen *et al.* (2003) and Hoang Fagerström *et al.* (2003).

To simulate the effect of land-use scenarios, a real storm event was used, because the use of a

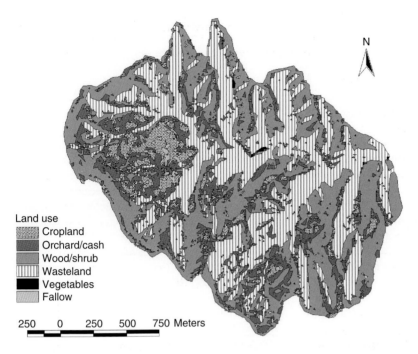

Land use
▦ Cropland
▨ Orchard/cash
▩ Wood/shrub
▥ Wasteland
■ Vegetables
▦ Fallow

250   0   250   500   750 Meters

Fig. 12.3   Scenario land-use (15° cropland limit) for the Danangou catchment.

Table 12.1   Areas (%) occupied by the different land uses for the different land-use maps. Catchment area is 3.52 km².

Land use	Present	25°	20°	15°
Cropland	35.4	21.0	13.0	6.7
Orchard/cash tree	2.4	0.0	9.5	17.8
Wood/shrubland	13.4	38.2	38.4	38.3
Wasteland	41.4	35.5	35.7	35.6
Vegetables	0.1	0.1	0.1	0.1
Fallow	7.3	5.2	3.2	1.6

so-called design storm would not allow calibration. An event that occurred on 1 August 1998 was chosen because it was large enough to cause erosion, and because it produced rainfall over all the catchment. The simulations were carried out with LISEM LP.

The total catchment area of 3.5 km² made simulation with pixels smaller than 10 m impractical. Many soil conservation measures are much smaller in scale than 10 × 10 m and therefore

cannot be implemented directly in LISEM (e.g. by changing the DEM). Instead, such measures were incorporated by changing other parameters that are influenced by them. For example, it can be expected that the use of contour ridges will increase infiltration because water storage on the slope is increased. One can then increase, for example, saturated conductivity to produce such an increase in infiltration. In this way all the proposed measures were translated into changes of input parameters for the LISEM model (Table 12.2). Table 12.2 gives assumed multiplication factors for the original LISEM input dataset: for example, a value of 1.25 for 'fallow biological' plant cover means that the data that apply to the 1 August storm event were multiplied by 1.25.

12.6   Results

12.6.1   Data collection

The measured discharge events occurred on: 1 August 1998, 23 August 1998, 20 July 1999, 21 July 1999, 11 August 2000 and 29 August 2000.

Table 12.2 Effects of conservation measures on LISEM input parameters.

	Cropland biological	Fallow biological	Cropland mechanical	Orchard ridges with grass strips
Plant height	1	0.5*	1	1
Plant cover	1.05	1.25	1	2
Leaf area index	1.05	1.25	1	2
Cohesion	1	1	1	1
Added cohesion	1	1.25	1	1.25
Random roughness	2	1	1.5	1.5
Aggregate stability	2	1	1	1.25
Manning's n	0.15*	1.1	1.25	1.25
K_{sat}	1.25	1.25	1.1	1.25

*These values are not multiplication factors but real values.

Table 12.3 Event characteristics.

	980801	980823	990720	990721	000811	000829
Average rainfall (mm)	15.1	13.0	14.1	3.5	11.6	17.8
Max 1-min intensity (mm h^{-1})[a]	69.9	47.2	66.2	35.8	49.5	84.9
Time to peak (minutes)[d]	15	34	19	32	31	15
Peak discharge (l s^{-1})	5125	701	3589	453	214	8757
Total discharge (m^3)	3982[b]	735	3282	488	199	5893
Total sediment yield (tonnes)[c]	1280	96	770	NA	16	2630

[a] The given value is a weighted average for the entire catchment. Intensities at individual rain gauges can be much higher (up to about 120 mm h^{-1} for 1-minute intervals in the case of both the 980801 and 990720 events, and up to 190 mm h^{-1} for the 000829 event).
[b] Hydrograph was incomplete, value is estimated by assuming a linear reservoir.
[c] Calculated from total discharge and measured sediment concentration.
[d] Time difference between maximum rainfall intensity and peak discharge.

Table 12.3 shows the importance of rainfall intensity; the lower intensities of the 980823 and 000811 events clearly resulted in much lower discharges, despite similar total rainfall amounts. The data suggest that runoff from the catchment only occurs if more than about 11.5 mm of high-intensity rain falls. Only the event of 990721 deviated from this trend. This is probably because it occurred shortly after the 990720 event, so the soil was still very wet. The 990720 event was special in that the rainfall amounts in the eastern and western part of the Danangou catchment were very different: at the eastern border about 30 mm of rain fell, while at the western border (about 2 km away) only 3 mm fell. Water samples taken during events revealed concentrations of up to 500 g l^{-1}, which resulted in large total sediment yields. Table 12.4 shows the LISEM input dataset as used for the 990720 event. Datasets for the other events were similar.

Table 12.4 Measured LISEM input dataset (plant and soil characteristics) for the 990720 event.

	Crop[a]	Fallow	Orchard	Shrub	Waste	Forest
Aggregate stability (median drop no)	6	5	6	6	8	7.25
Cohesion (kg cm^{-2})	0.08	0.10	0.10	0.09	0.11	0.11
Random roughness (cm)	1.75	1.11	1.28	1.03	1.66	0.88
Manning's n	SD[b]	0.079	0.092	0.153[c]	0.091	0.214
Leaf Area Index	0.06	0.12	1.46	1.25	0.54	1.63
Plant cover (fraction)	0.06	0.10	0.18	0.40	0.23	0.35
Plant height (m)	0.28	0.11	3.1	0.97	0.25	13.6
K_{sat}-meas (cm day^{-1})[d]	55.9	82.2	96.9	164	153	122
K_{sat}-fitted (cm day^{-1})[d]	1	1	25	10	5	13
Theta-init	Calculated with regression equation					

[a] Cropland was subdivided into five types. Here the values for foxtail millet are given. The other types were pearl millet, potato, tall crops (maize, sorghum) and beans.
[b] SD, slope dependent; see Hessel *et al.* (2003b).
[c] Average of wasteland and forest.
[d] K_{sat}-meas, K_{sat} measured with constant head method; K_{sat}-fitted, K_{sat} value resulting from wind evaporation method (Halbertsma & Veerman, 1997)

12.6.2 *Evaluation of adaptations*

The effects of the adaptations to LISEM were evaluated. This section discusses the effects without recalibration of the model, while Section 12.6.3 shows the results after recalibration.

(i) **Flow distance** The combined effect of the slope correction and the use of sine in the Manning equation proved to have a large effect on simulated discharge; peak discharge decreased by about 50% (Hessel, 2002). A decrease in discharge was to be expected since:

- On steep slopes the pixel areas have increased. The same amount of rainfall is therefore spread out over a larger area, so that the water layer will be thinner. The hydraulic radius will also be smaller, so that flow will be slower according to the Manning equation.
- The flow distance between pixels is larger.
- Since on steep slopes the pixel area is larger, infiltration will be greater as well.

Factors that affect flow distance or flow velocity will also affect discharge because longer flow distance and lower flow velocity allow more time for infiltration.

(ii) **Sine versus tangent** The effect of using sine instead of tangent is small for simulations for the entire catchment. Simulated total soil loss from the catchment decreased from about 1050 tonnes to 1000 tonnes. Because the slopes near the catchment outlet are fairly gentle, transport capacity close to the outlet will not have changed much by using sine instead of tangent. It seems likely that the effect would be larger when only steep areas are simulated.

(iii) **Manning's *n*** Using a slope-dependent Manning's *n* retarded the peak discharge slightly. The effect was much smaller than that of using a slope correction.

(iv) **Transport equations** Figure 12.4 shows simulated concentrations at the catchment outlet for different transport equations (Hessel, 2002). As can be seen from the figure, the differences are large. In the simulations the maximum

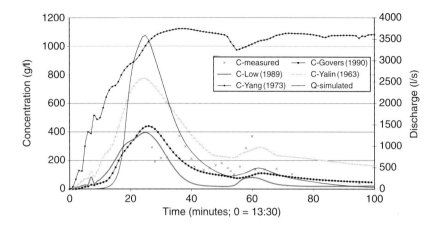

Fig. 12.4 Measured and predicted sediment concentrations using different transport equations.

concentration was restricted to 1060 g l⁻¹ because higher concentrations were assumed to be physically unrealistic. This restriction affected most results considerably (e.g. the Yang equation), but the effect of the restriction on the results obtained with the Govers equation was small. Thus, the equation that is at present being used by LISEM was found to perform best and was therefore used in this study.

(v) Particle fall velocity The settling velocity correction resulted in an increase of predicted sediment concentrations during the runoff peaks, but not at other times. The result was slightly unexpected since the settling correction could be expected to slow down settling after the sediment peak. This would result in higher concentrations after the sediment peak. The overall predicted increase in sediment yield seems reasonable.

(vi) Loose material map The results showed that in some places a considerable amount of material was removed. For the entire catchment the amount of loose material declined from 372 to 265 tonnes, so that 107 tonnes of loose sediment was taken up by the flow. Nevertheless, the total amount of sediment leaving the catchment only increased by 1–2 tonnes. The extra eroded sediment is deposited before reaching the catchment outlet, as shown by the deposition map produced by LISEM. Most pixels with loose

sediment showed only a small decline in loose material. This can be explained by the fact that many of these pixels did not experience much runoff because they had fairly small upstream areas. The results indicate that including a loose material map is likely to have more effect when the loose material is present close to the outlet of the catchment.

12.6.3 Calibration and validation

(i) Results for peak discharge calibration Table 12.5 shows the calibrated values of LISEM input parameters for all five events, both for LISEM 163 and for LISEM LP. It shows that the calibration gave different results for the different storms. Figure 12.5 shows the measured discharge as well as the calibrated discharge for each of the events, while Table 12.6 gives a summary of the simulation results. These calibration results show a number of features that are common to most events, as well as some features that were event-specific.

Saturated conductivity: Calibrated saturated conductivity was always much lower than the measured values. A possible explanation for this would be disturbance during sampling. Soil sealing/crusting could also be important. The effect of sealing/crusting is hard to measure on samples taken from the field. It is also possible that in the

Table 12.5 Peak discharge calibrated values for all events.

	980801	980823	990720	000811	000829
LISEM 163					
K_{sat} a LISEM[a]	0.90	0.96	0.85	0.85	0.81
Initial suction (˙ori)[b]	1	0.45	1	1	1
Manning's n[c]	m-0.2*s	m-1*s	m-1*s	m-1*s	m-1*s
Total channel length (m)	2576	2114	1319	1319	2114
Manning's n channel	0.04	0.06	0.05	0.05	0.05
LISEM LP					
K_{sat} a LISEM[a]	0.95	0.98	0.90	0.92	0.89
Initial suction (˙ori)[b]	1	0.45	1	1	1
Manning's n[c]	m-0.2*s	m-1*s	m-1*s	m-1*s	m-1*s
Total channel length (m)	2576	2114	1319	1319	2114
Manning's n channel	0.04	0.05	0.05	0.05	0.04

[a] The a indicates the extent to which K_{sat}-meas and K_{sat}-fitted were used, e.g. 0.90 means that 90% of K_{sat}-fitted was used, and 10% of K_{sat}-meas.
[b] The original value is multiplied by the value in the table; hence, for the 980823 event the suction becomes smaller and the soil is wetter.
[c] m, mean of measured data; s, standard deviation of measured data. Values of Manning's n below 0.03 were set to 0.03.

Table 12.6 Simulation summary (measured values are found in Table 12.3).

Simulated (LISEM LP)	980801	980823	990720	000811	000829
Peak discharge (l s^{-1})	4967	591	3330	189	8886
Total discharge (m^3)	5918	1054	3931	361	7191
Total sediment yield (tonnes)	1285	76	898	18	1784

field complete saturation was not reached. Since for very wet soils the difference in conductivity is very large for a very small change in water content, this could also be an important factor in explaining the much lower conductivities that need to be used during simulation. Another complicating factor is that saturated conductivity appears to increase with an increase in rainfall intensity (e.g. Morgan, 1996; Van Dijck, 2000), which might help to explain differences between the events.

Manning's n: The calibrated values of Manning's n were always lower than measured. The reason for this is not certain, but it seems possible that measured values were too high (Hessel, 2003b).

Peak time: The graphs (Fig. 12.5) of simulated discharge show that the discharge peak almost always occurred too early. Several possible explanations for this can be given. Firstly, it could be caused by large macropores (such as fissures and sinkholes). The effect of macropores is not simulated with LISEM, since the Richards equation is only valid for matrix flow. In reality the first runoff on hillslopes might well infiltrate by way of fissures or sinkholes. Another possible cause, for LISEM 163, would be overland flow distance (Section 12.5.2). LISEM LP, however, corrects for this effect. A third possibility would be storage in the channels or infiltration into the channel bed. Since the streams in the catchment are usually dry depressions in the channel bed have to be filled

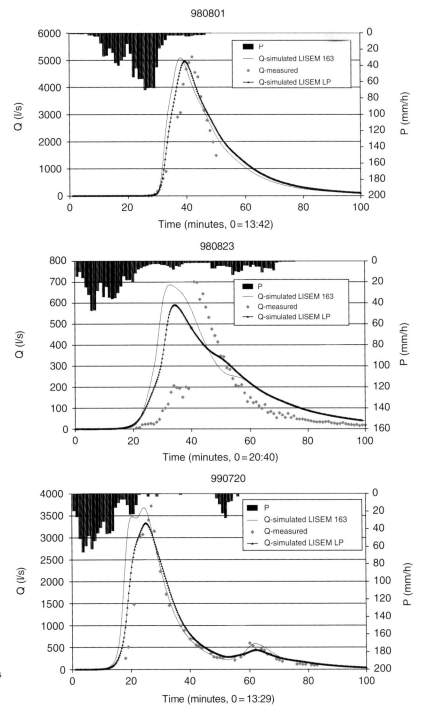

Fig. 12.5 Calibration results for the five measured events.

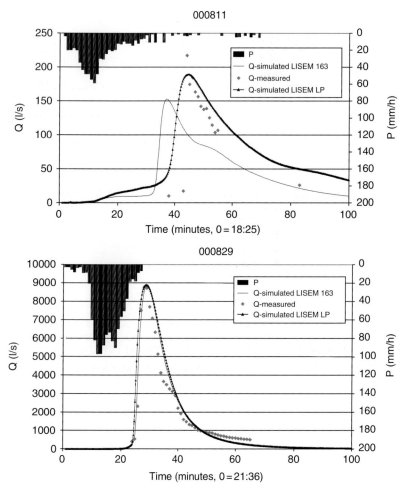

Fig. 12.5   cont'd.

before the water can advance further. Furthermore, it seems likely that in that case, infiltration of channel flow could be important. This process is not simulated in either of the LISEM versions used here, but has been added to later versions.

Simulated total discharge was always too high when LISEM was calibrated on peak discharge (Table 12.6). Figure 12.5 shows that LISEM was generally unable to predict accurately the very rapid rise and fall of water level that occurred in reality. This was especially the case for the smaller events.

(ii) LISEM version In LISEM LP, calibrated saturated conductivity was always lower than it was for LISEM 163. As shown in Section 12.6.2, this is mainly due to the slope correction that has been applied. Comparison of the calibrated hydrographs of LISEM 163 and LISEM LP shows that for most events the runoff peak arrived a little later, although still before the measured peak. The overall fit was usually similar, because the predicted water level also tended to decline a little less rapidly. This shows that using a model that is theoretically better does not guarantee that the prediction will also be better. Nevertheless, the Nash-Sutcliffe coefficient was higher for four out of five events (Table 12.7). This was mainly due to the fact that the time shift

Table 12.7 Nash-Sutcliffe coefficients for LISEM 163 and LISEM LP.

	980801	980823	990720	000811	000829	Average
LISEM 163	0.74	−0.28	0.64	−0.50	0.94	0.31
LISEM LP	0.86	0.24	0.91	0.33	0.85	0.64

Table 12.8 Simulated peak discharge (l s⁻¹) using the calibrated datasets for the different events.

	Calibration dataset					
Event	980801	980823	990720	000811	000829	Measured
980801	4967	12,004	3617	4346	3873	5125
980823	35	591	13	13	24	701
990720	4649	10,536	3330	4151	3301	3589
000811	612	4573	83	189	95	214
000829	9608	21,329	14,331	15,471	8886	8757

between the rise of the measured and simulated hydrographs decreased somewhat.

For the 990720 storm the 'calibration' channel length was much shorter than for most other events. This can be explained by the fact that this storm only produced high intensity rain in the areas close to the catchment outlet. The most striking difference between the simulation with LISEM 163 and LISEM LP is that LISEM LP no longer simulated a double peak for this event. The double peak was probably caused by water from different parts of the catchment arriving at the weir slightly out of phase. The hydrograph predicted with LISEM LP suggests that the first peak has been retarded by the adaptations to LISEM, so that both peaks are now in phase.

Calibration of the 000829 event did not show an early peak (Fig. 12.5). The rising limb of the hydrograph was reproduced almost perfectly, especially by LISEM 163, but the falling limb of the hydrograph went down a little too slowly. Three possible causes for the early peaks have been mentioned earlier. Because of the very abrupt nature of the 000829 storm, where very high-intensity rain suddenly occurred, it is possible that the aforementioned effects of infiltration in macropores (such as fissures and pipes) and the channel bed did not play

a large role in this case. The effect of overland flow distance should still occur. In contrast to the other events, the result for LISEM 163 was slightly better than for LISEM LP, as was also indicated by the Nash-Sutcliffe coefficient (Table 12.7).

(iii) Results for peak discharge validation To validate LISEM LP, the calibrated datasets for each storm were applied to the other four events. Using the calibration dataset of one event for the other four events usually gave worse fits than those presented in Fig. 12.5. The results of this validation are shown in Plate 10 for the 990720 event. Table 12.8 gives a summary of the results for peak discharge for all events. Plate 10 shows that applying calibration settings from another event almost always gave results that differed much from those obtained by calibrating on that particular event. Table 12.8 shows that this was the case for all events. It is therefore concluded that a separate calibration is necessary for events of different magnitudes, and probably even for each event separately.

(iv) Spatial patterns A minor event that occurred at the sediment plot on 21 July 1999 was used to derive a sheet erosion rate on an event basis. This minor storm did not result in any rill formation

Table 12.9   Average observed rill erosion rates compared with LISEM simulations.

Erosion class	Erosion rate (tonnes km^{-2})	Class boundaries (tonnes km^{-2})	Correctly predicted by LISEM (% of observed class area)
No rill erosion	251	0–800	51
Slight rill erosion	1450	800–3000	25
Moderate rill erosion	4822	3000–8000	21
Severe rill erosion	17,195	Over 8000	14

on the sediment plot. Nevertheless, the sediment concentration as determined from the barrels at the bottom of the plot was in excess of 700 g l^{-1}, resulting in a sheet erosion rate of 836 t km^{-2} (8.4 t ha^{-1}). The sediment delivery ratio of the fields was assumed to be 0.3. This gives erosion rates (on event basis) of 251 t km^{-2} for croplands that do not show evidence of rill erosion.

Table 12.9 shows the average erosion rates obtained from the erosion plot data as well as the rill measurements on the fields. The erosion rate for the 'no rill erosion' class is thus a single event estimate based on sediment plot data, while the other rates are based on rill mapping. Since only one event produced rills in 1999, all these rills must have formed on 20 July. The resulting 1999 rill erosion map is shown in Fig. 12.6a.

LISEM produces maps of erosion and deposition rates in t ha^{-1}. Since the field mapping only involved erosion and not deposition, it would appear logical to use the LISEM erosion map only. Erosion and deposition, however, cannot be treated as separate entities in LISEM simulations; deposition and re-entrainment can occur during the simulation. Thus, the same sediment can be eroded several times and be deposited several times. Therefore, the net erosion map should be used to assess the performance of the LISEM model in a spatial way. The result should be judged more on patterns than on amounts, because during mapping deposition was ignored. The net erosion map was also based on the classification given in Table 12.9 and is shown in Fig. 12.6b. Since the rill erosion map (Fig. 12.6a) only shows erosion on

fields, only the cropland areas were used for classification of the LISEM net erosion map.

Comparing the maps in Fig. 12.6, it is obvious that both maps have the highest erosion rates in the southeastern part of the catchment. This is also in agreement with the observed distribution of rainfall on 990720 (Section 12.6.1). The lack of heavy rain in the western part of the catchment is probably the reason why the LISEM prediction for the 'no rill erosion' class is much better than for the other classes (Table 12.9). Closer inspection of Fig. 12.6 shows that although the overall pattern was similar, the pattern in detail was very different. This is also reflected by the data given in Table 12.9, which show that for the classes with more severe rill erosion, only about 20% of the observed area for that particular class is predicted correctly.

(v) Land use scenarios   Figure 12.7 shows a classified version of the present land use erosion map produced by LISEM. Without classification the map appearance would be totally dominated by a few very high values. For a large area in the southern part of the catchment, no serious erosion was predicted. This was caused by the fact that, according to the measured rainfall data, less rain fell in this area during the 980801 event. Comparison of Figs 12.7 and 12.1 indicates that other areas with negligible erosion rates for the 980801 storm were mainly under woodland. A zone along both sides of the main valleys also had little erosion since this area is underlain by bedrock, which has a much higher cohesion, both in reality and in the model. The hilltop areas generally had slight or

(a) Rill erosion mapping

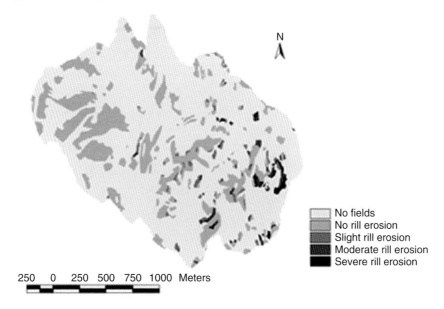

(b) Lisem simulation

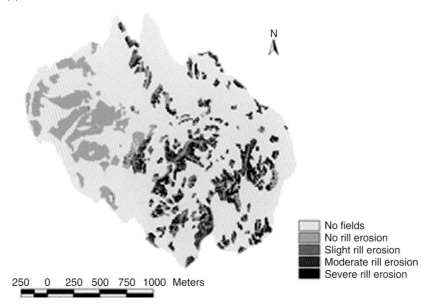

Fig. 12.6 (a) Mapped rill erosion of 1999. Pixel size is 5 metres. (b) LISEM simulation results for the 990720 event showing net erosion (erosion–deposition). Pixel size is 10 metres. For both maps the classification scheme given in Table 12.9 was used.

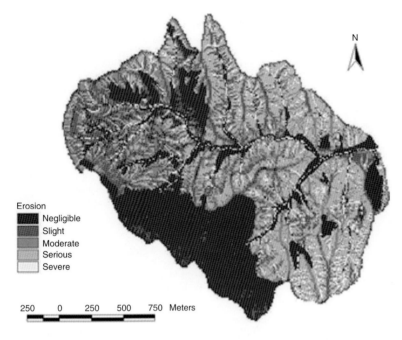

Fig. 12.7 Classified LISEM erosion map for the present land use and the 1 August 1998 storm.

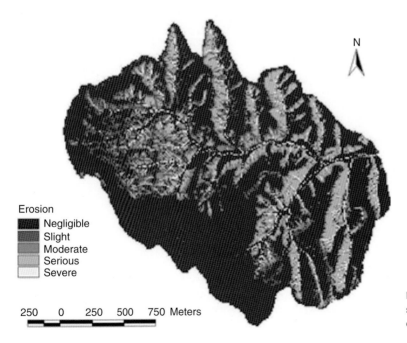

Fig. 12.8 Predicted erosion for scenario 3 (land use with 15° cropland slope limit).

Table 12.10 Erosion (tonnes) for different land-use scenarios.

Land use	Present	25°	20°	15°
Cropland	2949	1644	942	437
Orchard/cash tree	97	0	312	524
Wood/shrubland	173	198	173	140
Wasteland	6833	5080	4881	4727
Vegetables	31	62	63	62
Fallow	870	499	360	130
Total	10,953	7483	6732	6019

Table 12.11 Percentage decrease for different scenarios compared with scen0 results, catchment outlet, 980801 event.

	scen0	scen0a	scen0b	scen1	scen2	scen3
Peak discharge	0.0	17.5	10.6	42.3	51.8	56.4
Total discharge	0.0	12.8	6.5	33.4	42.2	50.1
Total soil loss	0.0	21.1	11.1	54.8	64.0	71.4

scen0a = use of biological measures (mulching, improved fallow).
scen0b = use of mechanical measures (contour ridges).

moderate erosion rates, while the steeper parts of the slopes had serious or severe erosion rates. Apart from woodland, not much difference in erosion is visible for the different land uses.

Figure 12.8 shows the results of scenario 3, using the same classification scheme as used for Fig. 12.7. It is evident that the erosion in the catchment has decreased. The area with negligible erosion rates has more than doubled in size, while the other erosion classes have all decreased in area. Comparison with Fig. 12.3 shows that the negligible erosion rates again mainly occurred under woodland/shrubland. The large decrease in predicted erosion was therefore probably a direct consequence of the increase in woodland/shrubland area.

Table 12.10 shows that erosion in woodland decreased from present land use to that of the 15° scenario, even though the woodland area increased significantly (Table 12.1). Comparison of Tables 12.1 and 12.10 shows that erosion for cropland and fallow land steadily decreased with decreasing area occupied by these land uses. Table 12.10 also shows a decrease in erosion for wasteland/wild grassland from the present land use to the 25° scenario. However, from the present land use to the 15° scenario, erosion in the catchment became increasingly dominated by erosion in wasteland/wild grassland.

Apart from the erosion/deposition maps, LISEM also generates time series of discharge and erosion for the outlet of the catchment. These time series showed that using conservation measures in the present land use (scenario 0) decreased the peak discharge by 10% to 18%,

while the reduction was much larger when applying scenarios 1, 2 and 3 (40% to 60%). The sedigraphs showed much the same trend as the hydrographs.

Table 12.11 shows that the decrease in total discharge was smaller than the decrease in peak discharge and total soil loss. Total soil loss showed the largest decrease. Relatively small decreases are shown for the conservation measures (ranging from 2% to 21%), whereas large decreases (ranging from 33% to 71%) are shown for the alternative land-use scenarios compared with scenario 0. The differences between scenarios 1, 2 and 3 were much smaller than those between scenario 1 and scenario 0. These differences between scenarios 1, 2 and 3 were caused by conversion of cropland to orchard/cash tree, while the larger difference between scenario 1 and scenario 0 was caused by a major redistribution of land uses.

12.7 Discussion

12.7.1 Adaptation of LISEM

The adaptations to LISEM that were made to make it able to deal with the Loess Plateau characteristics of steep slopes, high sediment concentrations and the presence of gullies, influenced the outcome of the model considerably. It was found that especially the choice of transport equation, and simple improvements like using the actual flow distance, had great effects. After recalibration, the adapted model performed slightly better than the original version.

12.7.2   *Calibration*

Calibration of LISEM showed that the model can simulate runoff and soil loss from the catchment for each event separately. However, it was not possible to find a general calibration set that could be used for all measured events. Validation indicated that it might be possible to find calibration sets for low-magnitude and high-magnitude events when data for more storms become available. LISEM gave the worst results for small events with low average rainfall intensity. Several authors (e.g. Nearing, 1998; Jetten *et al.*, 2003) also found that erosion models have difficulty in predicting small events. They ascribed this to spatial variability and uncertainty in the input data. The problem with small storms in the Danangou catchment is probably caused by several factors:
(1) For smaller storms it is much more important to get the initial conditions right. For large storms, a smaller or larger initial loss of water to infiltration probably does not matter much for the total amount of runoff, while for small storms it might be a large percentage of total runoff.
(2) Spatial distribution of rain in the catchment, such that rainfall measured at the gauges might not represent the actual spatial distribution of rainfall, and maximum rainfall intensities might not have been measured.
(3) If runoff was caused by heavy rainfall in part of the catchment, it is also crucial to get the other input data for this region right. For the events of 980823 and 000811, for example, heavy rain occurred in the gullied northern part of the catchment. Most LISEM parameters were not measured there, but were extrapolated from elsewhere.
(4) Finally, for smaller storms, storage in pools will have a larger effect.
These results indicate that the usefulness of LISEM for predictive purposes is limited to events that are large enough to cover the entire catchment. Our finding that a separate calibration is necessary for small and for large events means

that one should be cautious when applying the LISEM model to predict runoff for future events. Such predictions might be possible when these events are similar in size to the ones used here. Even then, initial conditions might well be different, so that it would probably be necessary to do simulations with different initial moisture contents. At present LISEM might be more suited to evaluate certain land use and management scenarios for their effects on erosion, since in that case all scenarios will use the same rainfall data and the same initial conditions.

Rill erosion intensity was mapped in the field and compared with LISEM simulations of erosion distribution. This comparison shows that the general appearance of simulated and mapped erosion patterns is similar, but also that the patterns are very different in detail. Many explanations are possible for the discrepancies between observed and simulated erosion rates and patterns inside the catchment (see e.g. Takken *et al.* 1999). For the topographically complex Danangou catchment, the following factors are likely to be important:
• DEM inaccuracy. The DEM determines the flow direction in the model, which, in turn, determines where erosion will occur according to the model. On relatively flat areas, the tillage direction can determine the direction of water flow (Ludwig *et al.*, 1996; Takken *et al.*, 1999; Van Dijck, 2000), but also in steep terrain, the flow direction is influenced by small topographic features (like pathways, gullies, cut-off drains, local escarpments) which are smaller than the pixel size, and therefore cannot be captured by the DEM. For topographically complex areas such as the Danangou catchment, it seems questionable whether it would be possible to obtain a DEM with sufficient detail to extract flow directions accurately.
• Limitations of LISEM. The pixel-based approach used in LISEM has some consequences for the calculation process. The most important is probably related to inertia. All processes are calculated per pixel, and only the kinematic wave distributes water and sediment. There is no momentum transfer between cells. Thus a change in slope may cause abrupt changes in streampower and a large amount

of deposition in a pixel can occur. The downstream pixel then has a lot of detachment because $TC-C$ is large, resulting in alternating detachment–deposition patterns. This problem is generally more pronounced where there is more water, hence in the channels, and has larger influence on the distribution of erosion than on total sediment yield. It is also more pronounced for topographically complex areas with strong alternations in slope angle, and can therefore be only partly solved by changing computational procedures.

• Incomplete or incorrect process descriptions. The issue of steep slopes, for example, was discussed above. All currently available erosion models use sediment transport equations that have been developed for slopes of no more than 20%. Much steeper slopes are, however, very common in the Danangou catchment. Similarly, other characteristics of the catchment might also be outside the range of conditions for which equations that are being used were developed. Besides, process knowledge is usually incomplete. When simulation errors appear to be systematic, there might be incomplete process descriptions in LISEM. For example, the fact that the discharge peak always seems to arrive too early might indicate this. Incomplete and incorrect process descriptions might affect both distribution and amount of erosion. As Beven (2001) points out, such errors in theory might be masked by calibration and can therefore be hard to find.

• Data inaccuracy. There can be inaccuracies in the data used to evaluate model performance as well as in the input data for the model. Morgan and Quinton (2001) suggested that such inaccuracies are a more important cause of incorrect model predictions than model flaws. Inaccuracies in input data can be caused by several factors. The first is incorrect measurements. Inaccuracies can also be caused by non-representative measurements. Input data for the LISEM model were collected on fields that were supposed to be representative of their respective land uses. Finally, some parameters (e.g. soil moisture content) are liable to rapid fluctuations, while others (e.g. saturated conductivity) are notoriously heterogeneous in space. Simulation results also indicate that the rainfall distribution has a large influence on simulated runoff and erosion. Data inaccuracy is a fundamental problem with distributed modelling. Nowadays, distributed models can contain several tens of thousands of pixels for which the different calculations are performed. Data on saturated conductivity, soil roughness, plant characteristics, and so on, are needed for all these pixels. Furthermore, even if such data were available, one can for reasons of spatial variability and upscaling seriously question whether the used values are indeed representative for the given pixel (see Chapter 6 for a general discussion related to upscaling). It seems likely that the actual amount of runoff and erosion occurring is controlled by many variations in parameters operating on subgrid scale. One has to face reality: there will always be a lack of accurate input data.

Often, a combination of factors could be operating, so that it will be difficult to find out what exactly causes an observed discrepancy between simulation and measurement. To evaluate the LISEM model (or any other process-based, distributed erosion model) in a spatial way, very detailed data both on model input and on erosion and deposition patterns distribution are needed. Such datasets are very hard to obtain for catchment-size areas, especially when topography is complex. Data for the catchment outlet are easier to obtain, so that calibration on the outlet alone will often be the only possibility. Our data, however, confirm the findings of Takken *et al.* (1999) that an erosion model calibrated on outlet data might well predict spatial patterns incorrectly.

12.7.3 Scenarios

The present study is one of the first attempts to use process-based soil erosion modelling as a tool for optimizing land use and management strategies to reduce runoff and erosion rates on the Chinese Loess Plateau. To perform the simulations, several assumptions were made.

The first is that LISEM can be used for scenario simulations. Such applicability might be threatened by the need to use different calibrations for different storms, and by the apparent inability of LISEM to simulate erosion patterns correctly (Takken *et al.*, 1999; Hessel *et al.*, 2003a; Jetten *et al.*, 2003). The first problem might be less significant if the same calibrated storm is used for all scenarios. The implications of the second problem should be investigated further, and are likely to differ depending on the aim of the scenario simulation.

The second assumption is our use of multiplication factors. It is important to realize that the effect that is predicted for soil and water conservation (SWC) measures is determined by the values of the multiplication factors that were assumed (Table 12.2). More quantitative data on the effects of SWC measures should be gathered to be able to select values with a higher degree of certainty, especially for sensitive parameters such as saturated conductivity and Manning's n.

A third assumption is that the selected storm is representative. In reality, it seems likely that the effect of soil and water conservation measures will depend on the intensity and size of a storm. Gong and Jiang (1979), for example, reported that reforestation and planting grasses had similar effects for low-intensity rain, but that planting grasses was less effective for heavy rain. Farmers in our study area also indicated that for really large storms it does not matter what conservation measures you have, because there will be severe erosion anyway. In LISEM, the decreasing effectiveness of conservation measures in large storms is only because rainfall intensity is higher, but saturated conductivity remains the same. LISEM cannot, however, simulate other reasons for their decreasing effectiveness in larger storms, such as exceedance of storage capacities (e.g. of ditches), and destruction of measures. Despite the fact that LISEM only simulates part of the effect of changed storm size, Hessel (2002) showed that if a storm of half the size of the original storm was used, only 1% of the runoff was produced, while a storm of double the size showed a 6.6-fold increase in predicted soil loss.

Thus, more research is needed before we can say to what degree the simulation results of LISEM reflect reality. LISEM is a state-of-the-art

erosion model, and other erosion models will suffer from similar limitations. Therefore, care must be taken not to read too much into scenario simulation results. Scenario simulations with erosion models give us useful insights into what might happen, but they do not tell us what will happen.

It should also be realised that this chapter only discussed the physical effectiveness of measures, while for adoption, the socio-economy should also be considered. Measures can be very effective, but if they are not acceptable to stakeholders, they cannot be implemented.

12.8   Conclusions

LISEM was adapted to conditions on the Chinese Loess Plateau, such as steep slopes, high sediment concentrations and the occurrence of gullies. These adaptations improved the performance of LISEM slightly after calibration.

Calibration was done both for the catchment outlet, and for spatial patterns inside the catchment. It was found that LISEM can be calibrated well for large events at the outlet, but that small events cannot be simulated properly. The main reason for this seems to be that small discharge events in the catchment are caused by localized heavy rainfall. Observed and simulated erosion patterns showed general similarities, but were quite different in detail. These results indicate that LISEM is more reliable for larger events, and that simulated erosion patterns should be regarded with caution.

Therefore, a calibrated large event was used to evaluate the effects of different land use scenarios, and observed patterns were analysed in relative terms only. Although such simulations cannot tell us what will happen after land use change, they do indicate that a major change in land use would be far more effective than a change in management practices.

Acknowledgements

The research described in this chapter was part of the INCO-DEV Erochina project (Contract IC18-CT97-0158). We would like to thank all students and colleagues from Utrecht University, Alterra, Swedish University of Agricultural Sciences, Institute for Soil and Water Conservation, Beijing Normal University and Research Center for Eco-Environmental Sciences, who assisted in collecting the data used in this chapter. We also thank the farmers of Danangou catchment for their hospitality.

References

Belmans, C., Wesseling, J.G. & Feddes, R.A. (1983) Simulation model of the water balance of a cropped soil: SWATRE. *Journal of Hydrology* **63**: 271–86.

Beven, K. (2001) On modelling as collective intelligence. *Hydrological Processes* **15**: 2205–7.

Beven, K. (2002) Towards an alternative blueprint for a physically based digitally simulated hydrologic response modeling system. *Hydrological Processes* **16**: 189–206.

Chen, L., Messing, I., Zhang, S., *et al.* (2003) Land use evaluation and scenario analysis towards sustainable planning on the Loess Plateau in China – case study in a small catchment. *Catena* **54**: 303–16.

Chow, V.T., Maidment, D.R. & Mays, L.W. (1988) *Applied Hydrology*. McGraw-Hill, New York.

De Roo, A.P.J. & Jetten, V.G. (1999) Calibrating and validating the LISEM model for two data sets from The Netherlands and South Africa. *Catena* **37**: 477–93.

De Roo, A.P.J., Wesseling, C.G., Cremers, N.H.D.T., *et al.* (1994) LISEM: a new physically-based hydrological and soil erosion model in a GIS-environment, theory and implementation. *IAHS Publication* **224**: 439–48.

De Roo, A.P.J., Wesseling, C.G. & Ritsema, C.J. (1996a) LISEM: a single-event physically based hydrological and soil erosion model for drainage basins: I: Theory, input and output. *Hydrological Processes* **10**: 1107–17.

De Roo, A.P.J., Offermans, R.J.E. & Cremers, N.H.D.T. (1996b) LISEM: a single-event, physically based hydrological and soil erosion model for drainage basins. II: Sensitivity analysis, validation and application. *Hydrological Processes* **10**: 1119–26.

Douglas, I. (1989) Land degradation, soil conservation and the sediment load of the Yellow River, China: review and assessment. *Land Degradation and Rehabilitation* **1**: 141–51.

Fu, B. & Gulinck, H. (1994) Land evaluation in an area of severe erosion: the Loess Plateau of China. *Land Degradation & Rehabilitation* **5**: 33–40.

Gong, S. & Jiang, D. (1979) Soil erosion and its control in small watersheds of the loess plateau. *Scientia Sinica* **22**: 1302–13.

Govers, G. (1990) Empirical relationships for the transport capacity of overland flow. *IAHS Publication* **189**: 45–63.

Halbertsma, J.M. & Veerman, G.J. (1997) Determination of the unsaturated conductivity and water retention characteristic using the Wind's evaporation method. In Stolte, J. (ed.), *Manual for Soil Physical Measurements*. Technical Document 37, DLO Winand Staring Centre, Wageningen: 47–55.

Hessel, R. (2002) *Modelling soil erosion in a small catchment on the Chinese Loess Plateau*. PhD thesis, Utrecht University. *Netherlands Geographical Studies* **307**.

Hessel, R. (2005) Effects of grid cell size and time step length on simulation results of the Limburg soil erosion model (LISEM). *Hydrological Processes* **19**: 3037–49.

Hessel, R. (2006) Consequences of hyperconcentrated flow for process based erosion modelling on the Chinese Loess Plateau. *Earth Surface Processes and Landforms* **31**: 1100–14.

Hessel, R. & van Asch, T. (2003) Modelling gully erosion for a small catchment on the Chinese Loess Plateau. *Catena* **54**: 131–46.

Hessel, R. & Jetten, V. (2007) Suitability of transport equations in modelling soil erosion for a small Loess Plateau catchment. *Engineering Geology* **91**: 56–71.

Hessel, R., Jetten, V., Liu, B., *et al.* (2003a) Calibration of the LISEM model for a small Loess Plateau catchment. *Catena* **54**: 235–54.

Hessel, R., Jetten, V. & Zhang, G. (2003b) Estimating Manning's *n* for steep slopes. *Catena* **54**: 77–91.

Hoang Fagerström, M.H., Messing, I., Wen, Z.M., *et al.* (2003) A participatory approach for integrated conservation planning in a small catchment in Loess Plateau, China. Part II: Analysis and findings. *Catena* **54**: 271–88.

Jetten, V. & de Roo, A.P.J. (2001) Spatial analysis of erosion conservation measures with LISEM. In Harmon, R. & Doe, W.W. (eds), *Landscape Erosion and Evolution Modeling*. Kluwer Academic/Plenum, New York: 429–45.

Jetten, V., Boiffin, J. & de Roo, A. (1996) Defining monitoring strategies for runoff and erosion studies in agricultural catchments: a simulation approach. *European Journal of Soil Science* **47**: 579–92.

Jetten, V., de Roo, A. & Favis-Mortlock, D. (1999) Evaluation of field-scale and catchment-scale soil erosion models. *Catena* **37**: 521–41.

Jetten, V., Govers, G. & Hessel, R. (2003) Erosion models: quality of spatial predictions. *Hydrological Processes* **17**: 887–900.

Jiang, D., Qi, L. & Tan, J. (1981) Soil erosion and conservation in the Wuding River Valley, China. In Morgan, R.P.C. (ed.), *Soil Conservation: Problems and Prospects*. Wiley, Chichester: 461–79.

Liu, G., Xu, M.X. & Wen, Z.M. (2003) A study of soil surface characteristics in a small watershed in the hilly, gullied area on the Chinese Loess Plateau. *Catena* **54**: 31–44.

Low, H.S. (1989) Effect of sediment density on bed-load transport. *Journal of Hydraulic Engineering* **115**: 124–38.

Ludwig, B., Daroussin, J., King, D. & Souchère, V. (1996) Using GIS to predict concentrated flow erosion in cultivated catchments. *IAHS Publication* **235**: 429–36.

McVicar, T.R., Li, L., van Niel, T.G., et al. (2007) Developing a decision support tool for China's re-vegetation program: Simulating regional impacts of afforestation on average annual streamflow in the Loess Plateau. *Forest Ecology and Management* **251**: 65–81.

Messing, I. & Hoang Fagerström, M.H. (2001) Using farmers' knowledge for defining criteria for land qualities in biophysical land evaluation. *Land Degradation and Development* **12**: 541–53.

Messing, I., Chen, L. & Hessel, R. (2003a) Soil conditions in a small catchment on the Loess Plateau in China. *Catena* **54**: 45–58.

Messing, I., Hoang Fagerström, M.H., Chen, L. & Fu, B. (2003b) Criteria for land suitability evaluation in a small catchment on the Loess Plateau in China. *Catena* **54**: 215–34.

Morgan, R.P.C. (1996) *Soil Erosion and Conservation* (2nd edn). Longman, Harlow, UK.

Morgan, R.P.C. & Quinton, J.N. (2001) Erosion modelling. In Harmon, R.S. & Doe, W.W. (eds), *Landscape Erosion and Evolution Modeling*. Kluwer Academic/Plenum, New York: 117–43.

Morgan, R.P.C., Quinton, J.N., Smith, R.E., et al. (1998) The European Soil Erosion Model (EUROSEM): a dynamic approach for predicting sediment transport from fields and small catchments. *Earth Surface Processes and Landforms* **23**: 527–44.

Nash, J.E. & Sutcliffe, J.V. (1970) River flow forecasting through conceptual models, I: A discussion of principles. *Journal of Hydrology* **10**: 282–90.

Nearing, M.A. (1998) Why soil erosion models overpredict small soil losses and under-predict large soil losses. *Catena* **32**: 15–22.

ReVegIH (2005) Re-Vegetation Impacts on Hydrology (of the Loess Plateau, China). ACIAR, CAS and CSIRO. Available at http://www.clw.csiro.au/ReVegIH/ [accessed 18 September 2008].

Shen H.W. & Julien P.Y. (1993) Erosion and sediment transport. In Maidment, D.R. (ed.), *Handbook of Hydrology*. McGraw-Hill, New York.

Stolte, J., van Venrooij, B., Zhang, G., et al. (2003) Land-use induced spatial heterogeneity of soil hydraulic properties on the Loess Plateau of China. *Catena* **54**: 59–76.

Takken, I., Beuselinck, L., Nachtergaele, J., et al. (1999) Spatial evaluation of a physically-based distributed erosion model (LISEM). *Catena* **37**: 431–47.

Van Dijck, S. (2000) Effects of agricultural land use on surface runoff and erosion in a Mediterranean area. *Netherlands Geographical Studies* 263.

Wan, Z. & Wang, Z. (1994) *Hyperconcentrated Flow*. IAHR Monograph, Balkema, Rotterdam.

Wesseling, C.G., Karssenberg, D., Burrough, P.A. & van Deursen, W.P.A. (1996) Integrating dynamic environmental models in GIS: The development of a Dynamic Modelling language. *Transactions in GIS* **1**: 40–48.

Wu Y., Xie, K., Zhang, Q., et al. (2003) Crop characteristics and their temporal change on the Loess Plateau of China. *Catena* **54**: 7–16.

Yalin, M.S. (1963) An expression for bedload transportation. *Journal of the Hydraulics Division, Proceedings of the American Society of Civil Engineers* **89**: 221–50.

Yang, C.T. (1973) Incipient motion and sediment transport. *Journal of the Hydraulics Division, Proceedings of the American Society of Civil Engineers* **99**: 1679–1703.

Zhang, J., Huang, W.W. & Shi, M.C. (1990) Huanghe (Yellow River) and its estuary: sediment origin, transport and deposition. *Journal of Hydrology* **120**: 203–23.

13 Modelling the Role of Vegetated Buffer Strips in Reducing Transfer of Sediment from Land to Watercourses

J.H. DUZANT[1], R.P.C. MORGAN[1],
G.A. WOOD[2] AND L.K. DEEKS[1]

[1]National Soil Resources Institute, Cranfield University, Cranfield, Bedfordshire, UK
[2]Integrated Environmental Systems Institute, Cranfield University, Cranfield, Bedfordshire, UK

13.1 Introduction

Phosphorus is normally the primary limiting factor in the process of freshwater eutrophication. In the United Kingdom more than 50% of the phosphorus entering surface waters is derived from agriculture and, on arable land, as much as 80% of this is in particulate form (Defra, 2002). This means that controlling the transfer of sediment from agricultural land to water bodies is an important component of the protection of the aquatic environment. Sediment control is therefore a policy issue for the UK government if it is to meet the requirements set out in the European Water Framework Directive, the European Freshwater Fish Directive, the European Habitats Directive and the Convention for the Protection of the Marine Environment of the North-East Atlantic (OSPAR) (Owens & Collins, 2008). In response, various schemes with financial support have been made available to farmers to manage land in environmentally responsible ways, including the Agri-Environmental Schemes and Single Farm Payment Schemes supported by the Department for Environment, Food and Rural Affairs (Defra),

for example the Environmental Stewardship scheme (Defra, 2005). Among the options available is support for planting vegetated buffer strips or barriers.

The term buffer usually describes a vegetated strip placed between a river, stream or creek and an adjacent upland land-use activity (Hickey & Doran, 2004) to protect water quality and act as a nutrient filter (Nieswand et al., 1990). The main functions of the buffer are to reduce surface runoff, filter sediment, promote infiltration of water into the soil and groundwater, and protect the river bank against erosion (Muscutt et al., 1993). In addition, the buffer provides a habitat and a wildlife corridor for fauna and flora (Parkyn, 2004). Buffers can also be placed on hillslopes to control erosion in upstream areas closer to the sediment source. They often comprise vegetated barriers, usually on or close to the contour, where they perform similar functions to the riparian barriers but they are usually much narrower, often only 2 to 6 m wide and rarely more than 25 m wide (van Dijk et al., 1996). The other major difference is that runoff leaving the barrier passes on to land downslope instead of discharging into the watercourse.

Vegetated buffer strips at the edge of arable fields have been shown to trap as much as 70–90% of the inflowing sediment (Dillaha et al., 1987; Magette

Handbook of Erosion Modelling, 1st edition. Edited by R.P.C. Morgan and M.A. Nearing. © 2011 Blackwell Publishing Ltd.

et al., 1987; Haan *et al.*, 1994), but these results have generally been obtained on planar slopes. Their applicability to general field situations is questionable, particularly where convergent slope shapes cause runoff to concentrate and enter rivers at a limited number of points in the landscape, rather than flowing uniformly into the buffer along the whole of its length. In certain situations, the greater depth of flow arising from the concentration of the runoff can lead to submergence of the grass barrier, rendering it less effective.

Numerous studies have shown that the length of the buffer is important from upslope to downslope edge. Lalonde (1998) found that trapping efficiency varied between 68% and 98% as the length increased from 2 to 10m; Abu-Zreig (2001) observed that a 15m long buffer was three times more efficient than a 1m long barrier; while Dillaha *et al.* (1989) showed that trapping efficiency ranged from 53% to 86% with 4.6m long buffers and from 70% to 98% with 9.1m long buffers. Other studies, however, show that since most sediment is deposited within the first few metres of the strip, increasing the length beyond 3–5m has little effect on trapping efficiency (Line, 1991; van Dijk *et al.*, 1996).

The effectiveness of different lengths depends on the nature of the sediment being carried in the flow. Schwer and Clausen (1989) found that a 26m wide grass strip trapped 92% of the total phosphorus (particulate and dissolved) associated with the runoff on a sandy soil, but only 33% on a silty clay. Barriers of 5–10m length may trap nearly all the sediment on sandy soils, whereas clay particles are transported greater distances and may form 80–90% of the material leaving the buffer. However, even if all the clay particles and associated phosphorus are not trapped, a sufficient quantity may be which will enable a vegetated barrier to serve as a pollution control measure. Kronvang *et al.* (2000) found that a 29m long buffer was able to trap 100% of the sediment and particulate phosphorus on soils in Denmark, and that a 12m long strip was enough to reduce the delivery of particulate phosphorus to water courses to an acceptable level.

Recent research has shown that the effectiveness of the buffer is greatly influenced by the architecture of its vegetation, and therefore care has to

be taken in selecting appropriate plant species. For grasses, sturdy, tall, perennial species are considered the most suitable, and short, flexible grasses less so (Grismer *et al.*, 2006). Taller grasses that grow vertically often produce a sward with a high density of interwoven stems, forming a porous filter, whereas shorter grasses generally bend from the vertical and consist of individual stems with a more porous structure. Fescue (*Festuca ovina*) was found in laboratory experiments to trap 90% of the sediment moving on slopes of 7–9°, compared with only 70% for meadow grass (*Poa pratensis*) (Lakew & Morgan, 1996). Adding switch grass (*Panicum virgatum*) to barriers of fescue makes them even more effective (Blanco-Canqui *et al.*, 2004, 2006). An open structure to the barrier can sometimes enhance rather than protect against erosion. Concentrations of runoff through gaps in the barrier can result in localized increases in flow velocity. Runoff can therefore leave the barrier at a high velocity but relatively free of sediment. Under these conditions the presence of the barrier can create more erosion downslope that would have occurred without the barrier (Emama Ligdi & Morgan, 1995; Ghadiri *et al.*, 2001; Spaan *et al.*, 2005).

A further uncertainty in the performance of buffers is that, whilst they can trap sediment during storms, the deposition is not permanent and the sediment can be re-suspended during subsequent events (McKergow *et al.*, 2006). Thus the barriers can be both a sediment sink and a sediment source, depending on the magnitude of the storm (Daniels & Gilliam, 1996; Verchot *et al.*, 1997).

Before planning the installation of new buffer features in the landscape, it is important to note that in many agricultural areas of the UK, vegetated areas already exist. These include hedgerows along field boundaries, and areas of trees and woodland either at the edge of or between fields. Riparian buffers are often the remnants of former river plain forests with willows, alder and some hardwood trees. Although not designed as soil-protecting buffers, they may perform that function to varying extents depending on their structure and density. Where a buffer already exists, it makes sense to evaluate its present performance and make recommendations to enhance its function rather than replace it with a purpose-designed buffer, particularly if the buffer is broadly in an appropriate place in the landscape. Although field observations and measurements could be used to evaluate an individual buffer, a more generic approach is required for developing farm management plans. This can best be achieved by using erosion models.

This chapter describes the application of an erosion model for the assessment of buffer performance at two field sites, where measurements were made of sedimentation within the buffer and the characteristics of the buffer architecture over a 14-month period. The first step is to decide which of the many erosion models found in the literature is the most appropriate to use. A selection procedure is developed based on establishing criteria that the models should satisfy. The chosen model is then applied to the field conditions. Attention is paid to the way the model is set up to give the best representation possible of the pattern of runoff movement over the land surface and the resultant spatial distribution of erosion and sedimentation. The model output is compared with the measured data and a judgment is made on whether the model is suitable for evaluating existing buffers and for making recommendations on how their performance might be improved.

13.2   The Study Area

The study area comprises two arable fields within the basin of the River Tone, Somerset, England, where the hilly land is known for its high rate of erosion and high delivery of sediment to watercourses (McHugh *et al.*, 2002; Murdoch & Culling, 2003). The mean annual rainfall of the basin ranges from 800 to 1000 mm, with highest monthly totals in December and January and the driest months from April to July. July and August are the warmest months when the mean daily maximum temperature is 21°C; January is the coldest month with mean daily minimum temperatures of 1–2°C (Meteorological Office, 2005).

The first field, on a sandy clay loam soil of the Crediton Series, has an area of 39,000 m², a slope length of 195 m at an angle of 2°, and a grass buffer at the downslope end established in 2001/2002 under the Countryside Stewardship scheme. The grass varies between 50 and 200 mm in height in the winter and 300 and 500 mm in the summer with a ground and canopy cover of some 80–100%. The buffer seems to be well-maintained. During the study period, the field above the buffer was under maize during the summer months and bare over the winter.

The second field of 97,000 m² on a sandy clay loam soil of the Hodnet Series has a 6.9 m long grass strip at its downslope end, below which is a fence, followed by an area of shrubs, small trees and an irrigation pond. The field is subject to erosion when bare of crop, and in 2004 a gully with an average width of 1.07 m and an average depth of 0.45 m had developed perpendicular to the buffer. At the base of the gully, covering the buffer strip from front to back, was a 33 m wide area of sedimentation (Plate 11). The field slopes down to the buffer at 7°, but the slopes also converge across the field towards the gully forming contributing side slopes at an angle of 3–4°. Sediment also covered the buffer at the corner of the field below a set of farm tracks where the land was used for turning vehicles. Along the rest of the buffer, the grass covered 60–100% of the ground surface, although the grass within the back 3 m was rougher and more tussocky than that at the front, and there was a 50 mm step from the field surface up to the buffer edge, most probably brought about by ploughing. The field was subsequently planted to wheat, after which the gully became less deep and the grass started to re-emerge through the sediment in the buffer. In the spring of 2006, a grass waterway was planted along the line of the gully. This appeared to prevent any further gully development and reduce the transport of sediment (Plate 12).

At both field sites the buffer areas were instrumented to assess their ability to trap and retain sediment eroded from the slopes above (Owens *et al.* 2006). Sediment was collected on astroturf mats installed within the buffer at the front (upslope boundary), mid-buffer and back (downslope edge) (Plate 13). The mats were installed in December 2004 and were inspected for sediment

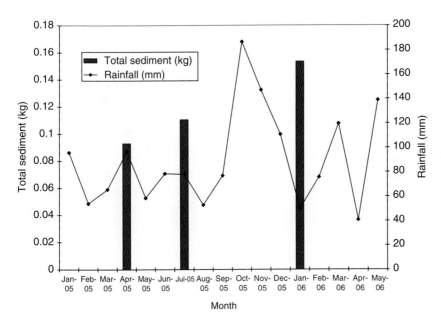

Fig. 13.1 Relationship between sedimentation at Site 2 and rainfall.

Table 13.1 Sediment collected on the astroturf mats.

Site	Number of mats installed	Sediment deposition in each observation period (g cm⁻²)					
		Jan–Apr 2005	Apr–Jul 2005	Jul–Oct 2005	Oct–Jan 2005	Jan–May 2005	Total
1	10	0	0	0	0.62	0	0.62
2	12	5.79	1.51	0	0.01	0	7.31

in April, July and October 2005, and January and May 2006. Between 1 February 2005 and 30 April 2006 there were 112 rain days and 3718 mm of rain recorded in the Somerset area; this included a number of heavy rainfall events, for example, over 25 mm of rainfall on 24 and 28 June 2005. Overall there is a clear relationship between rainfall and sediment deposition, with most of the deposition occurring following the highest rainfall amounts (Fig. 13.1). At the first site, the land dipped slightly towards the centre of the field, indicating an area where flow was likely to concentrate before entering the buffer. Ten mats were placed in the buffer at this point of which five collected sediment, two at the front, one in the middle and two at the back of the buffer. Deposition was measured only

in January 2006 when the field was bare. The land was under maize in the summer of 2005 and field beans in the spring of 2006.

In the second field, 12 mats were located at the base of the gully, six in the corner of the field below the farm tracks, and six in a sedimentation area at the field edge. Sediment was collected from mats at the front, middle and back of the buffer in the first and second locations in the first, second and fourth measurement periods. The step at the field edge prevented sediment from reaching the mats at the third location.

At the end of each observation period, the mats containing sediment were removed and replaced. The mats taken from the field were placed in plastic bags and transferred to the laboratory where they were air-dried at room temperature for a minimum of 48 hours. The sediment was then carefully removed and the dry weight obtained. The particle size distribution (<1 mm) was determined by mechanical sieving down to 63 μm and then by sedigraph analysis. The cumulative particle size distribution was determined, from which the percentages of sand (>63 μm), silt (2–63 μm) and clay (<2 μm) were obtained. Table 13.1 shows the amount of sediment collected in the buffers in the two fields. This information provided the measured data for comparison with model predictions.

13.3   Model Selection

The usefulness of a model for evaluating vegetated buffers as a method of erosion control depends on its ability to simulate the main functions performed by the buffer, namely enhancing infiltration, increasing sedimentation and increasing the filtration of material suspended in the runoff. Particular attention is therefore given to determining which processes a model simulates and how the role of vegetation is described. The following criteria were established as design requirements that a model should satisfy for use in the field study area:

(1) it must be physically-based in order to offer greater potential than a conceptual or empirical model in predicting the spatial distribution of runoff and sediment;

(2) it must be simple and easily understood without requiring a trained user;

(3) it must estimate infiltration, runoff, erosion and sediment deposition in order to allow evaluation of sediment and water retained by, and passing through, the buffer; and

(4) it must take vegetative cover into account explicitly so that guidance can be given on the design of buffers to improve their performance.

Table 13.2 lists a number of models in which an attempt is made to represent the relevant filtration and sedimentation processes. Many of these have not been fully tested using independently collected field data, and none has

Table 13.2   List of models considered.

Model	Reference
CREAMS	Flanagan *et al.* (1989)
WEPP	Nearing *et al.* (1989)
RUSLE	Renard *et al.* (1997)
VFSMOD	Muñoz-Carpena & Parsons (2004)
MMF	Morgan & Duzant (2008)
EUROSEM	Morgan *et al.* (1998)
TRAVA	Deletic (2001)
SEDIMOT	Wilson *et al.* (1981)
GRASSF	Barfield *et al.* (1979)
AGNPS	Tim & Jolly (1994)
REMM	Lowrance *et al.* (2000)

been tested using standard datasets, so comparison between them is difficult. Based on an understanding of the bioengineering role of vegetation (Coppin & Richards, 1990), it is clearly important that at least the effects of ground cover and the density of the plant leaves and stems are modelled explicitly. Of the models listed, only EUROSEM and the modified MMF model use these parameters directly. The majority of the models express plant cover effects through some form of coefficient, such as the *C* factor of the Universal Soil Loss Equation, which, as indicated by Styczen and Morgan (1995), is not a process-based approach. While these models may represent the conditions for which the *C* factor values have been experimentally derived, they cannot be used to predict the effect of the same or different vegetation in other climatic and soil conditions, and *C* factors are not designed to account for deposition processes.

Flanagan *et al.* (1989) demonstrated that, as long as a number of assumptions are met, a simplified version of the CREAMS model could effectively simulate the processes of sediment deposition within a vegetated buffer. These assumptions are that the flow is shallow and uniformly distributed along the upslope edge of the buffer, concentrated flow effects are minimal, the grass is not submerged or flattened by the

Table 13.3 Additional criteria used for model selection.

Criterion	Model evaluation		
	WEPP	EUROSEM	MMF
Physically-based	***	***	***
Easy to use	*	*	**
Describes erosion, runoff and deposition	*	*	***
Accounts for effects of vegetation explicitly	**	***	***
Proven applicability to UK conditions	***	***	***
Describes soil particle sizes	***	*	***
Annual/seasonal model	***	*	***
Available or readily attainable data requirements	**	**	***
Enables routing of water and sediment over multiple elements from field to farm and small catchment scales	***	***	***
Can be combined with land cover and land management options	***	***	***
Reasonable simulation time requirements	***	***	***
Has been applied to vegetated buffer design	**	**	*
Has been applied to vegetated buffer placement	**	**	*
Low cost to obtain and use	***	***	***

Suitability scoring: *, low; **, moderate; ***, high.

flow, and previously trapped sediment does not affect future depositional capacity (Wenger, 1999). These assumptions are likely to be particularly limiting where, as in the second field site, concentrated flow dominates.

Deletic (2000) showed that the methods currently used in models for simulating the transport of sediment through a grass buffer are very poor. This is attributed to previous work focusing on the overall performance of the buffers rather than the processes involved. For example, it is questionable whether any of the models adequately define the erosive potential of the runoff once it leaves a grass strip. EUROSEM is based on the assumption that, if transport capacity has not been reached, then the flow is still erosive. If all or much of the sediment is deposited within or upslope of a buffer, then a flow's erosion potential should be increased, as demonstrated in laboratory experiments by Emama Ligdi and Morgan (1995). Styzcen and Morgan (1995) suggested that as sediment builds up within the buffer, the foreslope of the resultant sediment wedge becomes steeper than

the ground slope, as simulated in the GRASSF model. Therefore, once the foreslope has migrated to the downslope edge of the strip, the velocity of the flow leaving the buffer will be increased.

Despite the need to select a model that simulates correctly the processes operating within the buffer, there is also a need to choose a model that matches the availability of the input data. Generally, model users require simple models that are easy to operate and understand rather than parameter-intensive models with large data requirements. Merritt *et al.* (2003) also noted that model accuracy does not always increase with model complexity because there can be an accumulation of error through inaccuracies in the large number of parameters required. Thus models like REMM are restricted to trained users as well as requiring large amounts of data.

With the exception of TRAVA and EUROSEM, most of the models listed in Table 13.2 were developed in the US and, of these, only WEPP has been widely used in Europe. Combining the requirements of the need to account for vegetation explicitly and to have been applied in Europe with the criteria listed above, only WEPP, EUROSEM and MFF were selected for more detailed consideration. A further ten criteria were now added to the original criteria in order to aid selection between these. The extra criteria were selected based on what was perceived to be useful in providing guidance and supporting decisions on the design and placement of vegetated buffers in the field. It can be seen from Table 13.3 that the MMF model (Morgan & Duzant (2008) version) scores highly against these new criteria. It is therefore selected as the preferred model.

13.4   Model Application

The MMF model simulates the movement of water and sediment over the landscape from source to delivery to the river system. The original version of the model was developed to predict mean annual soil loss from field-sized areas on hillslopes (Morgan *et al.*, 1984), but with the addition of simple routing procedures, it could also be applied to small catchments (Morgan, 2001). The model operates by separating the erosion into two phases: a water phase in which the energy of the rainfall and the volume of runoff are estimated, and a sediment phase, which considers the detachment of soil particles by raindrop impact and runoff and their transport over the land surface by overland flow. The predictions of total particle detachment and transport capacity of the runoff are compared and the lower of the two values represents the annual soil loss, thereby indicating whether detachment or transport processes are the limiting factor. In the Morgan and Duzant (2008) version, the effects of vegetation are described explicitly by the parameters of percentage canopy cover, percentage ground cover, plant height, effective hydrological depth, density of the plant stems and stem diameter. In addition to detachment and transport, sediment deposition is modelled through a particle fall number which considers particle settling veloc-

ity, flow velocity, flow depth and slope length. The detachment, transport and deposition of the soil particles are also calculated separately for the clay, silt and sand fractions of the soil.

When operated over small catchments, the MMF model requires the landscape to be divided into elements of reasonably uniform soil type, slope and land cover. The user needs to determine the pathways that the runoff follows in moving from one element to another. Each element is therefore assigned a number (e.g. element 1, element 2, etc.) and then, for each element, determining the number of the element from which it receives water and sediment and the number of the element to which it discharges water and sediment. For example, Fig. 13.2 shows the processes represented by the model as operating on a single element. If this was element number 2, it would receive runoff and sediment from an element upslope (e.g. element 1) and contribute runoff and sediment to an element downslope (e.g. element 3). The model calculates the runoff and sediment budgets for each element, taking account of the runoff and soil particle detachment from raindrop impact generated on the element itself, and the runoff and sediment inputs from upslope. Together these determine the total runoff on the element, soil particle detachment by runoff, and the sediment transport capacity.

Previous experience with the model has shown that better results are generally achieved when measured data rather than guide values are used for the input parameters (Morgan *et al.*, 1984) and that care needs to be taken when determining the element structure. A particular problem is that very different results can be obtained according to how the landscape is described. For example, for a 100 m long uniform slope, the model produces different results if this is treated as a single element compared with describing it by two 50-m long elements, or routing the water and sediment over five 20-m long elements. In order to keep these differences within acceptable limits, it is recommended that element lengths should normally be about 10 m and no shorter than 5 m or longer than 50 m.

Table 13.4 lists the input parameters required to operate the model and the sources of informa-

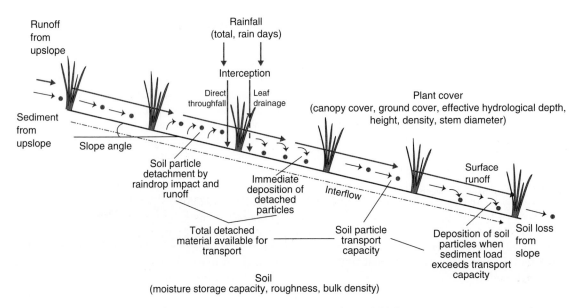

Runoff from upslope

Rainfall (total, rain days)

Interception

Direct throughfall

Leaf drainage

Plant cover (canopy cover, ground cover, effective hydrological depth, height, density, stem diameter)

Sediment from upslope

Slope angle

Soil particle detachment by raindrop impact and runoff

Immediate deposition of detached particles

Interflow

Surface runoff

Total detached material available for transport

Soil particle transport capacity

Deposition of soil particles when sediment load exceeds transport capacity

Soil loss from slope

Soil (moisture storage capacity, roughness, bulk density)

Fig. 13.2 Erosion and sedimentation processes simulated by the MMF model (after Morgan & Duzant, 2008).

Table 13.4 Model input parameters.

Parameter	Source of information
Annual or seasonal rainfall (mm)	Meteorological records for Yeovilton
Mean annual or seasonal temperature (°C)	Meteorological records for Yeovilton
Rainfall intensity (mm h^{-1})	Guide value (10 mm h^{-1})
Annual or seasonal number of rain days	Meteorological records for Yeovilton
Soil moisture at field capacity (% w/w)	Field measurement
Soil bulk density (Mg m^{-3})	Field measurement
Effective hydrological depth (m)	Guide value according to vegetation cover
Slope angle (°)	Field measurement
Slope length (m)	Field measurement
Slope width (m)	Field measurement
Soil surface roughness (cm m^{-1})	Field measurement
Permanent interception (proportion)	Guide value according to vegetation cover
E_t/E_o ratio	Guide value according to vegetation cover
Canopy cover (proportion)	Field measurement
Ground cover (proportion)	Field measurement
Plant height (m)	Field measurement
Plant stem diameter (m)	Field measurement
Plant stem density (number m^{-2})	Field measurement
Detachability of clay particles by rain (g J^{-1})	Guide value
Detachability of silt particles by rain (g J^{-1})	Guide value
Detachability of sand particles by rain (g J^{-1})	Guide value
Detachability of clay particles by runoff (g mm^{-1})	Guide value
Detachability of silt particles by runoff (g mm^{-1})	Guide value
Detachability of sand particles by runoff (g mm^{-1})	Guide value
Manning's n	Guide value according to vegetation
Flow depth (m)	Guide value (0.005 m) for unconcentrated flow; measured channel depth for concentrated flow.

Guide values are taken from the model description in Morgan and Duzant (2008).

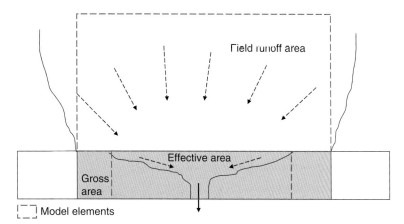

Fig. 13.3 Relationship between gross buffer area (element configuration 1 in text) and effective buffer area (element configuration 2 in text) (after Dosskey *et al.*, 2002).

Table 13.5 Model results for element configurations 1 and 2.

Site	Measured deposition in buffer (kg m^{-2})	Predicted deposition in buffer (kg m^{-2})	
		Element configuration 1	Element configuration 2
1	3.1	1.83	3.65
2	10.25	0.79	7.78

Note: The measured values of deposition differ from those given in Table 13.1 because they have been corrected for the effective buffer width.

tion used for the study area. For each site the model was set up to run with only two elements, comprising the field above the buffer and the buffer strip respectively. This approach, however, assumes that all the runoff from the field enters the buffer uniformly across the slope. Since field observations showed that this was not the case and that runoff concentrated in local depressions and approached the buffer in a single or a limited number of flow paths, an alternative approach was also tried, taking account of the effective area of the buffer strip. This was determined by examining the presence of sediment deposition within the buffer. These observations indicated that in the first field, only 50% of the buffer was likely to receive sediment and for the second field, only 10%. The difference between the two approaches is illustrated in Fig. 13.3 (Dosskey *et al.*, 2002).

Although the model predicts the runoff and soil loss leaving each field below the buffer, these cannot be used to indicate whether the model gives reasonable results since neither was measured in the field. The field data are only for deposition of sediment within the buffer. The model output can, however, be used to calculate this by comparing the value of the soil loss from the slope element above the buffer with that below the buffer. The difference between the two represents either net erosion or deposition within the buffer. The predicted and measured values of deposition for the two element configurations are shown in Table 13.5, from which it is clear that using the effective buffer strip area improves the predictions. However, even with the second approach, the model still underpredicts the amount of sediment deposited within the buffer.

A third approach was therefore tried whereby, in addition to determining the effective buffer area, field evidence was used to define the actual contributing area more precisely. At the first field site, where no concentrated flow channels were observed, the contributing area was defined by the extent of the localized depression upslope of the barrier, giving an area of 20 m by 20 m. Since there were no channels, the flow depth used was that recommended for unchannelled flow (0.005 m). A further refinement with this approach was to reduce the effective buffer area further to consider only the area where deposition was recorded. At this site, the effective buffer area was considered equal to the 20 m of the contributing field area and a 6 m buffer length, reflecting the recording of deposition in the front, middle and back of the buffer.

At the second field site, it is clear that runoff and associated sediment transport are concentrated in the gully. The contributing element is therefore modelled based on the length (60 m) of the gully and the width of the contributing gully side slopes (10 m in total) instead of the area of the whole field, that is, an area of only 600 m² instead of 97,000 m². With sedimentation occurring on mats at the front, middle and back of the buffer, a buffer length of 8 m is used. Effective buffer width is taken as equal to the width of the contributing area (i.e. 10 m).

Table 13.6 gives the dimensions of the upslope (element 1) and buffer (element 2) used in these simulations. It can be seen from Table 13.7 that a detailed understanding of the effective area of the processes operating in the field, followed by setting up the model properly to reflect these, gives substantially better predictions.

In addition to the total amount of deposition, the model predicts the particle size distribution of the deposited material. Table 13.8 shows the predicted and measured values. It can be seen that the predictions are poor for these two sites. The main reason for this is that the model substantially underpredicts the sand fraction deposited within the buffers. Whereas the field data demonstrate that it is mostly the sand fraction which is trapped by the buffers, the model

Table 13.6 Element dimensions (m) used with configuration 3 based on effective contributing area and effective buffer area.

Site	1	2
Length of contributing element	20	60
Width of contributing element	20	10
Length of buffer element	6	8
Width of buffer element	20	10

Table 13.7 Model results for element configuration 3.

Site	Measured deposition in buffer (kg m⁻²)	Predicted deposition in buffer (kg m⁻²)
1	6.21	6.35
2	4.85	4.29

Note: The measured values differ from those in Table 13.5 because they have been corrected for the effective buffer widths used in configuration 3.

Table 13.8 Model results for the particle size distribution of sediment deposited in the buffer.

Site	Particle size	Measured	Predicted
1	% clay	3	24
	% silt	7	76
	% sand	90	0
2	% clay	4	17
	% silt	16	83
	% sand	80	0

output shows that no sand particles are delivered to the buffer. Instead, most are deposited close to the point of detachment and only small amounts are picked up and transported by the runoff. All of the latter are then deposited in the upslope element and do not reach the buffer. To some extent this reflects reality in that sandy material is deposited on the lower slope as evidenced by the sediment fan at the base of the gully at site 2. Within the buffer, the model shows that the transport capacity is greater than detachment for sand, but not for clay or silt. This explains the deposition of some clay and silt particles within the buffer as the sediment load reduces to equal the transport capacity. As expected, the model indicates that sediment movement through the buffer is limited by the transport capacity.

The failure of the model to predict the delivery of sand to the buffer was somewhat surprising because when applied to field sites in Bedfordshire and Cambridgeshire, the eroded material leaving the base of the fields simulated on sandy loam soils was predicted to contain as much as 79–83% sand (Morgan & Duzant, 2008). The difference may well reflect the greater slope of the latter sites which were 4–9° compared with only 2–3° at the sites in this study.

Having shown that the chosen model can be set up to give acceptable predictions of sediment deposition within a vegetated buffer, as compared to field measurement, the model was then run to determine whether the predicted mean annual soil loss from the field at Site 2 fell within acceptable limits. Using meteorological data from Taunton, the following input data were used: mean annual rainfall, 1075.5 mm; mean annual temperature, 9.86 °C; and the mean annual number of rain days, 84. The mean annual soil loss from the field's contributing area (60 m long and 33 m wide) was predicted as 18.2 t ha^{-1}, which is far higher than the 1 t ha^{-1} proposed as the upper limit to control downstream pollution arising from runoff from agricultural land (Moldenhauer & Onstad, 1975; Morgan, 2005). As already shown, much of this soil is trapped within the downslope buffer. Below the buffer, the mean annual soil loss from the field was predicted at 0.8 t ha^{-1}, indicating that the buffer is very effective. However, this calculation assumes that the sediment passes uniformly through the whole of the buffer width. If, given the field observations above, the model is set up to simulate the concentration of the runoff and sediment into only 8 m of the buffer width, the effective area becomes 2044 m^2 (1980 m^2 for the field and 64 m^2 for the buffer) and the predicted mean annual soil loss

below the buffer then increases slightly to 0.9 t ha^{-1}, which is still acceptable. This result indicates that the present buffer feature should be capable of reducing sediment delivery to the adjacent watercourse to a level which will also control pollution from phosphorus transported by surface runoff in particulate form. In reality, however, the recent installation of the grass waterway in the contributing area upslope of the buffer (Plate 12) has probably made the buffer redundant. The grass waterway controls the detachment and transport of soil particles by runoff closer to source rather than relying on the vegetated buffer to create a sediment sink.

13.5   Conclusions

The best results with the model are obtained when both the effective contributing area and the effective buffer area are defined from field observations and measurements. Only where the slope is planar and there are no concentrations of flow is the model likely to give reasonable predictions using gross areas of the individual slope elements. When operated with field observations for the input data, and set up to reflect the patterns of the flow paths and the spatial distribution of erosion and sedimentation observed in the field, the model gives good predictions of the amount of sedimentation within the vegetated buffers. The model thus deals reasonably well with concentrated flow in the contributing upslope area and the delivery of runoff and sediment to the buffer at a limited number of points rather than uniformly across the buffer width. At present the model does not predict the particle size of the deposited material particularly well because it simulates the deposition of most of the sand particles close to their point of detachment. Further development of the model will be needed to represent better the detachment, transport and deposition of sand. One approach would be to use lower fall velocities than those conventionally recognized for sand.

Despite its limitations, the model does simulate the effectiveness of vegetated buffer strips using explicit input parameters to describe the vegetation. It is therefore appropriate to use in the design of vegetated buffers. By setting a target for the mean annual soil loss discharging downslope of the buffer, it is possible to evaluate whether an existing buffer is sufficient to meet that target. If not, scenarios can be run with the model to determine what stem diameters, plant densities and percentage covers are required to meet the target. The outcome can then be matched to plants that meet those requirements and which will grow in the study area. Such scenarios can be adopted to determine how to improve existing buffers and to design new buffers where none previously existed.

Acknowledgements

The research described in this chapter was carried out as part of a research project on the strategic placement and design of buffering features for mitigating the effects of sediment and phosphorus in the landscape. The authors are grateful to Defra for funding under Research Project PE0205, and to Dr P.N. Owens for his advice and assistance with the field measurement programme.

References

Abu-Zreig, M. (2001) Factors affecting sediment trapping in vegetated filter strips: simulation study using VFSMOD. *Hydrological Processes* **15**: 1477–88.

Barfield, B.J., Tollner, E.W. & Hayes, J.C. (1979) Filtration of sediment by simulated vegetation. Part I: steady-state flow with homogeneous sediment. *Transactions of the American Society of Agricultural Engineers* **22**: 540–45.

Blanco-Canqui, H., Gantzer, C.J., Anderson, S.H., et al. (2004) Grass barriers and vegetative filter strip effectiveness in reducing runoff, sediment, nitrogen and phosphorus losses. *Soil Science of America Journal* **68**: 1670–78.

Blanco-Canqui, H., Gantzer, C.J. & Anderson, S.H. (2006) Performance of grass barriers and filter strips under interrill and concentrated flow. *Journal of Environmental Quality* **35**: 1969–74.

Coppin, N.J. & Richards, I.G. (1990) *Use of Vegetation in Civil Engineering*. CIRIA/Butterworths, London.

Daniels, R.B. & Gilliam, J.W. (1996) Sediment and chemical load reduction by grass and riparian filters. *Soil Science Society of America Journal* **60**: 246–51.

Defra (2002) *The Government's strategic review of diffuse pollution from agriculture in England: agriculture and water. A diffuse pollution review.* Defra, London.

Defra (2005) Water friendly farming: target areas announced to curb agricultural pollution. England Catchment Sensitive Farming Delivery Initiative announced. News Release, 19 December 2005. Defra, London. Available at http://www.defra.gov.uk/news/2005/051219a.htm

Deletic, A. (2000) *Sediment behaviour in overland flow over grassed areas.* PhD Thesis, University of Aberdeen.

Deletic, A. (2001) Modelling of water and sediment transport over grassed areas. *Journal of Hydrology* **248**: 168–82.

Dillaha, T.A., Reneau, R.B., Mostaghimi, S., et al. (1987) *Evaluating nutrient and sediment losses from agricultural lands: vegetative filter strips.* Report No. CBP/TRS 2/87, Washington DC: US Environment Protection Agency.

Dillaha, T.A., Reneau, R.B., Mostaghimi, S. & Lee, D. (1989) Vegetative filter strips for agricultural nonpoint source pollution control. *Transactions of the American Society of Agricultural Engineers* **32**: 513–19.

Dosskey, M.G., Helmers, M.J., Eisenhauer, D.E., et al. 2002. Assessment of concentrated flow through riparian buffers. *Journal of Soil and Water Conservation* **57**: 336–43.

Emama Ligdi, E. & Morgan, R.P.C. (1995) Contour grass strips: a laboratory simulation of their role in soil erosion control. *Soil Technology* **8**: 109–17.

Flanagan, D.C., Foster, G.R., Neibling, W.H. & Burt, J.P. (1989) Simplified equations for filter strip design. *Transactions of the American Society of Agricultural Engineers* **32**: 2001–7.

Ghadiri, H., Rose, C.W. & Hogarth, W.L. (2001) The influence of grass and porous barrier strips on runoff hydrology and sediment transport. *Transactions of the American Society of Agricultural Engineers* **44**: 259–68.

Grismer, M.E., O'Geen, A.T. & Lewis, D. (2006) Vegetative filter strips for nonpoint source pollution control in agriculture. *University of California, Division of Agriculture and Natural Resources Publication* No. 8195.

Haan, C.T., Barfield, B.J. & Hayes, J.C. (1994) *Design Hydrology and Sedimentology for Small Catchments.* Academic Press, San Diego.

Hickey, M.B.C. & Doran, B. (2004) A review of the efficiency of buffer strips for the maintenance and enhancement of riparian ecosystems. *Water Quality Research Journal of Canada* **39**: 311–17.

Kronvang, B., Laubel, A.R., Larsen, S.E. & Iversen, S.L. (2000) Soil erosion and sediment delivery through buffer zones in Danish slope units. *International Association of Hydrological Sciences Publication* **263**: 67–73.

Lakew, D.T. & Morgan, R.P.C. (1996) Contour grass strips: a laboratory simulation of their role in erosion control using live grasses. *Soil Technology* **9**: 83–9.

Lalonde, M. (1998) *Filter strips: impact of design parameters on removal of non-point source pollutants from cropland runoff.* PhD thesis, University of Guelph.

Line, D.E. (1991) Sediment trapping effectiveness of grass strips. *Proceedings of the Fifth Federal Interagency Sedimentation Conference*, March 18–21, Las Vegas NV: PS56–PS63.

Lowrance, R., Altier, L.S., Williams, R.G., et al. (2000) REMM: The riparian ecosystem management model. *Journal of Soil and Water Conservation* **55**: 27–34.

Magette, W.L., Brinsfield, R.B., Palmer, R.E., et al. (1987) *Vegetated filter strips for agriculture and runoff treatment.* Report No. CBP/TRS 2/87-003314-01. US Environment Protection Agency, Washington, DC.

McHugh, M., Morgan, R.P.C., Walling, D.E., et al. (2002) *Prediction of sediment delivery to watercourses from land. Phase II.* Environment Agency R&D Report P2-209. Environment Agency, Bristol, UK.

McKergow, L.A., Prosser, I.P., Weaver, D.M., et al. (2006) Performance of grass and eucalyptus riparian buffers in a pasture catchment, Western Australia, Part 2: water quality. *Hydrological Processes* **20**: 2327–46.

Merritt, W.S., Letcher, R.A. & Jakeman, A.J. (2003) A review of erosion and sediment transport models. *Environmental Modelling and Software* **18**: 761–99.

Meteorological Office (2005) http://www.metoffice.com/climate/uk/locations/southwestengland.html. Accessed October 2005.

Moldenhauer, W.C. & Onstad, C.A. (1975) Achieving specified soil loss levels. *Journal of Soil and Water Conservation* **30**: 166–8.

Morgan, R.P.C. (2001) A simple approach to soil loss prediction: a revised Morgan-Morgan-Finney model. *Catena* **44**: 305–22.

Morgan, R.P.C. (2005) *Soil Erosion and Conservation* (3rd edn). Blackwell, Oxford.

Morgan, R.P.C. & Duzant, J.H. (2008) Modified MMF (Morgan-Morgan-Finney) model for evaluating effects of crops and vegetation cover on soil erosion. *Earth Surface Processes and Landforms* **33**: 90–106.

Morgan, R.P.C., Morgan, D.D.V. & Finney, H.J. 1984. A predictive model for the assessment of soil erosion risk. *Journal of Agricultural Engineering Research* **30**: 245–53.

Morgan, R.P.C., Quinton, J.N., Smith, R.E., et al. (1998) The European Soil Erosion Model (EUROSEM): a dynamic approach for predicting sediment transport from fields and small catchments. *Earth Surface Processes and Landforms* **23**: 527–44.

Muñoz-Carpena, R. & Parsons, J.E. (2004) A design procedure for vegetative filter strips using VFSMOD-W. *Transactions of the American Society of Agricultural Engineers* **47**: 1933–41.

Murdoch, N. & Culling, S. (2003) *Nutrient budgets in the Parrett catchment.* Environment Agency, Bristol, UK.

Muscutt, A.D., Harris, G.L., Bailey, S.W. & Davies, D.B. (1993) Buffer zones to improve water quality: a review of their potential use in UK agriculture. *Agriculture, Ecosystems and Environment* **45**: 59–77.

Nearing, M.A., Foster, G.R., Lane, L.J. & Finckner, S.C. (1989) A process-based soil erosion model for USDA-Water Erosion Prediction Project technology. *Transactions of the American Society of Agricultural Engineers* **32**: 1587–93.

Nieswand, G.H., Hordon, R.M., Shelton, T.B., et al. (1990) Bufferstrips to protect water supply reservoirs: a model and recommendations. *Water Resources Bulletin* **26**: 959–66.

Owens, P.N. & Collins, A.J. (2008) Soil erosion and sediment redistribution in river catchments: summary, outlook and future requirements. In Owens, P.N. & Collins, A.J. (eds), *Soil Erosion and Sediment Redistribution in River Catchments.* CAB International, Wallingford: 297–317.

Owens, P.N., Duzant, J.H., Deeks, L.K., et al. (2006) The use of buffer features for sediment and phosphorus retention in the landscape: implications for sediment delivery and water quality in river basins. *International Association of Hydrological Sciences Publication* No. **306**: 243–8.

Parkyn, S. (2004) Review of riparian buffer zone effectiveness. *MAF Technical Paper* No. 2004/05.

Renard, K.G., Foster, G.R., Weesies, G.A., et al. (1997) Predicting soil erosion by water: a guide to conservation planning with the Revised Universal Soil Loss Equation (RUSLE). *USDA Agricultural Handbook* No. 703.

Schwer, C.B. & Clausen, J.C. (1989) Vegetative filter treatment of dairy milkhouse wastewater. *Journal of Environmental Quality* **18**: 446–51.

Spaan, W.P., Sikking, A.F.S. & Hoogmoed, W.B. (2005) Vegetation barrier and tillage effects on runoff and sediment in an alley crop system on a Luvisol in Burkina Faso. *Soil & Tillage Research* **83**: 194–203.

Styczen, M.E. & Morgan, R.P.C. (1995) Engineering properties of vegetation. In Morgan, R.P.C. & Rickson, R.J. (eds), *Slope Stabilization and Erosion Control: a Bioengineering Approach*. E & FN Spon, London: 5–58.

Tim, U.S. & Jolly, R.W. (1994) Evaluating agricultural nonpoint source pollution using integrated GIS and hydrological water quality models. *Journal of Environmental Quality* **23**: 25–35.

Van Dijk, P.M., Kwaad, F.J.P.M. & Klapwijk, M. (1996) Retention of water and sediment by grass strips. *Hydrological Processes* **10**: 1069–80.

Verchot, L.V., Franklin, E.C. & Gilliam, J.W. (1997) Nitrogen cycling in Piedmont vegetated filter zones. I. Surface soil processes. *Journal of Environmental Quality* **26**: 327–36.

Wenger, S. (1999) *A review of the scientific literature of riparian buffer width, extent and vegetation.* Institute of Ecology, University of Georgia, Athens, GA.

Wilson, B.N., Barfield, B.J., Ward, A.D. & Moore, I.D. (1981) University of Kentucky surface mine model: hydrologic component. American Society of Agricultural Engineers Winter Meeting, Chicago ILL, Paper No. 81-2505.

14 Predicting Impacts of Land Use and Climate Change on Erosion and Sediment Yield in River Basins using SHETRAN

J.C. BATHURST

School of Civil Engineering and Geosciences, Newcastle University, Newcastle upon Tyne, UK

14.1   Introduction

Erosion is a natural process that causes soil loss and generates sediment yields from river basins even in the absence of humans. Yields vary from the very low in well-vegetated lowland catchments such as in the UK (e.g. Walling & Webb, 1981), to the very high in areas of mountain building, glacial activity and volcanism (e.g. Griffiths, 1981; Lavigne, 2004). However, ever since humans have modified the Earth's environment for their own gain, erosion rates have risen above natural levels, a phenomenon known as accelerated erosion (e.g. Dunne & Leopold, 1978: 510; Vanacker et al., 2007; Wilkinson & McElroy, 2007). Some well-documented examples include the impacts of Viking farming activity in Sweden (e.g. Gaillard et al., 1991), widespread tree-felling to make way for agricultural development by European settlers in the US (e.g. Trimble, 1976) and New Zealand (e.g. Glade, 2003), and hydraulic gold mining in California (Gilbert, 1917, cited by, among others, Trimble, 1995). As a result, human activity has long been considered to be probably the greatest single influence on the impact and balance of the various erosion processes in a river basin (e.g. Dunne & Leopold, 1978: 510; Simons et al., 1979; Sundborg, 1983). The consequences of accelerated erosion are generally undesirable and are evident both on-site (in the area of the erosion) and off-site (distant from the erosion). On-site effects include loss of soil fertility, loss of land itself (through river bank erosion), and destruction of infrastructure and loss of life caused by landslides and debris flows. Off-site impacts include reservoir sedimentation, deterioration of aquatic habitat, and increased frequency of flooding caused by river sedimentation and clogging of water abstraction plants. There is strong interest, therefore, in minimizing the amount of accelerated erosion. This is likely only to increase in the face of climate change, with predictions variously of more intense rainfall (increasing erosion potential) and longer periods of drought (decreasing vegetation cover and thence protection against erosion).

An important requirement in the development of strategies for minimizing erosion is a framework which integrates the many processes determining sediment yield in a river basin and which can be used (a) to test our understanding of those processes and their linkages, and (b) to predict the impacts of proposed courses of action. Such a framework can be provided by mathematical modelling, the essence of which, in this context, is the linking of on-site rates of erosion and soil loss in a basin to the sediment yield at the basin outlet. Field and laboratory studies are essential for improving our understanding of the means by which sediment yield is generated, for example by identifying and quantifying individual

Handbook of Erosion Modelling, 1st edition. Edited by R.P.C. Morgan and M.A. Nearing. © 2011 Blackwell Publishing Ltd.

processes and by indicating cumulative, basin-scale effects. They also provide the data with which to create and test models. However, the sheer complexity and natural variation of sediment yield controls make it difficult to transfer experimental results between even similar basins with more than order-of-magnitude accuracy. Furthermore, field experiments cannot quantify the impacts of a land-use change on basin-scale sediment yield in a period of less than several years. Modelling, by contrast, is able to investigate the integrated effects of the various processes, including changes in the relative importance of the processes. Modelling can also deliver rapid, systematic investigation of a range of land management strategies, allowing selection of the strategy that minimizes the accelerated erosion before there is any intervention on the ground. A difficulty, though, with models is that there are a number of sources of uncertainty which must be considered carefully when interpreting simulation results.

This chapter considers the use of physically-based models for predicting the impacts of land use and climate change on sediment yield at spatial scales ranging from the hillslope to river basins of up to 1000 km^2. It reviews the requirements of such models, discusses their capabilities and limitations (including uncertainty) and demonstrates their application, focusing on the SHETRAN basin modelling system in particular. The overall aim is to provide a sufficient background for a user to understand the issues behind model selection, parameterization and application and the interpretation of model results.

14.2   Model Requirements

The overall requirement is to be able to predict the impacts of possible future changes in land use, vegetation cover and climate on the erosion, sediment transport and sediment yield of a basin. Typical sediment-related management problems which need to be addressed include:
• the effect of forest logging in upland areas on the incidence of shallow landslides and thus the supply of sediment into the river system, with its potentially detrimental consequences for salmon-spawning and sport fishing;
• the extent to which soil loss can be reduced through the introduction of sympathetic agricultural techniques such as seed ploughing and buffer strips;
• the effect of agricultural fertilizer application on non-point source pollution, involving the transport of phosphorus adsorbed to sediment particles;
• the impact on reservoir sedimentation of the replacement of native forest by cropland in the upstream basin; and
• the impact of future changes in rainfall regime on erosion and sediment yield.

Tackling such problems involves repeat modelling of the focus area with different management and climate scenarios and the comparison of output data in such forms as annual sediment yields, peak sediment concentrations in rivers and spatial variation in soil erosion. A number of specific modelling requirements then need to be satisfied.

(1) Capability to predict the impacts of future change   Of the available modelling approaches (e.g. empirical, conceptual, stochastic, physically-based), only the physically-based can truly be said to have this capability (e.g. Abbott et al., 1986a). These models are based on the fundamental equations of physics, such as the equations of mass, momentum and energy conservation, and theoretically and experimentally derived relationships (such as the factor-of-safety equation for slope stability). Their parameters have a physical meaning (such as soil permeability, vegetation root cohesion and percentage of bare soil) and can therefore be changed to represent specified future conditions on the basis of physical reasoning, available databases and expert judgment. It is not possible so to change the parameters of, for example, an empirically-derived regression equation as they are a product of the regression procedure and have no physical meaning. This chapter is therefore concerned solely with physically-based modelling.

(2) Capability to represent spatial variability
Erosion is a spatially non-uniform process and it is likely to be the case that a large part of the sediment yield comes from only a small part of the basin. As an example, it has been suggested that targeted reforestation of a small part of a basin (the critical areas for shallow landslide occurrence) could have a disproportionately large effect in reducing sediment yield (Reid & Page, 2002). Similarly, a sediment problem in one part of the basin may be caused by activity in another (e.g. erosion triggered by headwater deforestation might cause siltation of the downstream river channel) (e.g. Glade, 2003). Models intended for land management applications or for investigating sediment yield processes therefore need to account for spatial variability in erosion and sediment transport. Typically this is achieved with some form of grid system.

(3) Inclusion of the relevant processes The main sediment-generating processes are erosion by raindrop impact, overland flow, rilling, gullying, landsliding and bank erosion. These, and the sediment transporting processes, are hydrologically driven, and important hydrological processes therefore include the generation of overland flow by rainfall exceeding infiltration (Hortonian overland flow) and by upward saturation of the soil, generation of river flow from surface and subsurface inputs, and variation of the soil porewater pressures which affect slope stability. Different processes are likely to dominate in different parts of a basin and at different basin scales (e.g. de Vente & Poesen, 2005). Model relevance should therefore be clearly defined, in terms of the erosion and transport processes that are represented, and erosion models are likely to retain the greatest flexibility if based on a general hydrological model. At the same time, attempts to represent erosion processes in ever finer detail must be carefully balanced against the ability to distinguish the results in the face of uncertainties associated with model parameterization, data collection and the natural variability in erosion and sediment yield (Nearing, 2004; Quinton, 2004; see Chapter 4 for a more general discussion of this issue).

(4) Simulation period Generation of long-term sediment yields requires a capability for simulating periods of several years. Generation of extreme sediment loads requires a capability for simulating individual storm events. Both requirements can be met by continuous simulation models. These models also have an advantage over single-event models in being able to generate the appropriate antecedent soil moisture conditions for individual events.

(5) Data availability Physically-based models require meteorological input data to drive the simulation, basin property data (such as elevations, soil characteristics and vegetation characteristics), and calibration or validation data consisting of hydrological, erosion, sediment transport, landslide and other response data. The input data and model parameters have to be evaluated at every grid element, so the greater the spatial distribution in the available data, the more accurate or more representative that evaluation will be. However, it is currently impractical to take measurements for every model grid element, and some form of estimate is therefore required for the majority of elements. On the other hand there is a wide range of data sources which can be used, including soil property measurements, soil hydrology maps, land-use maps, runoff and erosion plot experiments, gully experiments, gauging station and rain gauge records, well logs, river channel surveys and landslide inventories. Remote-sensing techniques can provide increasingly relevant data on a spatially distributed basis and in the future should be able to reduce the extent of parameter estimation. The input data and many of the calibration data consist of time series. In principle, physically-based models do not require lengthy data records for calibration. However, there are sufficient approximations in their design (for example, the use of one-dimensional instead of three-dimensional formulations) that a short period of record (especially if containing an extreme response) is helpful in adjusting the final values of the model parameters (see Chapter 3).

14.3   SHETRAN

SHETRAN is a descendant of the original SHE (Système Hydrologique Européen) hydrological modelling system, which was built in a collaborative programme in the late 1970s and early 1980s by the then UK Institute of Hydrology (now the Centre for Ecology and Hydrology, Wallingford), the then Danish Hydraulic Institute (now DHI) and the French consultancy SOGREAH (Abbott *et al.*, 1986a,b). SHETRAN was created at Newcastle University after the transfer there of the UK SHE programme in the late 1980s, and its development paralleled that of MIKE SHE (Refsgaard & Storm, 1995), the Danish SHE descendant. SHETRAN is now a general, physically-based, spatially-distributed modelling system that can be used to construct and run models of all or any part of the land phase of the hydrological cycle, including sediment and contaminant transport (Ewen *et al.*, 2000). It is physically based in the sense that the various flow and transport processes are modelled either by finite difference representations of the partial differential equations of mass, momentum and energy conservation, or by empirical equations derived from experimental research. The model parameters have a physical meaning and can be evaluated by measurement. Spatial distributions of basin prop-erties, inputs and responses are represented on a three-dimensional, finite-difference mesh, and the channel system is represented along the boundaries of the mesh grid squares as viewed in plan (Fig. 14.1). Table 14.1 summarizes the main processes modelled and the equations used to describe them.

Detailed descriptions of the SHE/SHETRAN system can be found in Abbott *et al.* (1986b), Bathurst *et al.* (1995), Ewen (1995), Birkinshaw and Ewen (2000) and Ewen *et al.* (2000). In addition, a website (http://www.ceg.ncl.ac.uk/shetran/) provides details of the system, access to the software, and a user guide with test datasets. As computational power has increased, the hardware required for running the system has evolved from mainframe computer to Unix-based workstation, to desktop computer and laptops.

14.3.1   SHETRAN hydrological component

Subcomponents account for evapotranspiration and interception, overland and channel flow, subsurface flow, snowmelt and channel/surface aquifer exchange (Fig. 14.1 and Table 14.1). SHETRAN is continually evolving as new process descriptions and solution schemes are introduced. The current most advanced version is Version 5, although this does not yet support some of the

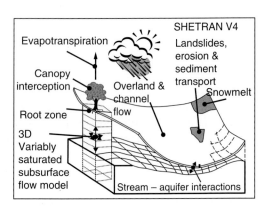
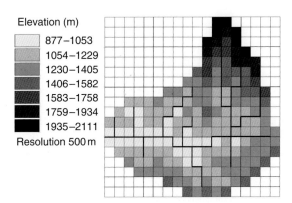

Fig. 14.1   SHETRAN schematic and example of a model grid network, channel system (in solid black lines) and elevation distribution. The schematic has evolved through time; this particular version was created by Dr Isabella Bovolo.

Table 14.1 Hydrological, erosion and sediment transport processes modelled by SHETRAN Version 4.

(1)   Interception of rainfall on vegetation canopy (Rutter storage model)
(2)   Evaporation of intercepted rainfall, ground surface water and channel water; transpiration of water drawn from the root zone
      (Penman-Monteith equation, or the ratio of actual to potential evapotranspiration as a function of soil moisture tension)
(3)   Snowpack development and snowmelt (temperature-based or energy budget methods)
(4)   Three-dimensional variably saturated subsurface flow (Darcy-Buckingham approach)
(5)   Two-dimensional overland flow; one-dimensional channel flow (Saint Venant equations)
(6)   Subsurface zone/surface water interaction and subsurface zone/channel interaction
(7)   Soil erosion by raindrop impact, leaf drip impact, overland flow and bank erosion (see text for equations)
(8)   Soil erosion by shallow landslide erosion (see text for equations)
(9)   Two-dimensional total load convection in overland flow by size fraction, including input to the channels; deposition and
      resuspension of sediments in overland flow (mass conservation equation incorporating Engelund-Hansen total load and Yalin
      bedload transport capacity equations)
(10) One-dimensional convection of cohesive and non-cohesive sediments in channel flow by size fraction; deposition and resuspension
      of non-cohesive sediments in channel flow; channel bed erosion by channel flow (mass conservation equation incorporating
      Ackers-White and Engelund-Hansen transport capacity equations)
(11) Transport of contaminants adsorbed to sediment particles (convection-dispersion equation with partition distribution coefficient)

processes, such as landslide erosion, available in Version 4. The chief distinguishing feature of this version, in terms of physical processes, is the representation of the subsurface by a fully three-dimensional variably saturated subsurface scheme which enables such features as perched water tables and hypodermic flow (i.e. just below the ground surface) to be modelled. Importantly for erosion modelling, SHETRAN generates overland flow both by an excess of rainfall over infiltration, and by upward saturation of the soil column.

14.3.2   SHETRAN erosion and sediment yield component

Subcomponents account for soil erosion by raindrop impact, leaf drip impact and overland flow, channel bed and bank erosion by channel flow, and sediment transport by overland and channel flow (Table 14.1). The component is driven by inputs from the hydrological (water flow) simulations, but feedback to the flow simulations is not modelled as the effects are unlikely to be significant at the sediment concentrations and relative scales of erosion and deposition typically considered. Detailed descriptions of the subcomponents are provided in the above references and Wicks

and Bathurst (1996). However, it will be helpful to know that the equations for determining soil erosion are:

for raindrop and leaf drip impact:

$$D_r = k_r F_w \left(1 - C_g - C_r\right)\left(M_r + M_d\right) \quad (14.1)$$

and for overland flow:

$$D_f = k_f \left(1 - C_r\right)\left(\frac{\tau}{\tau_c} - 1\right) \quad \text{for } \tau > \tau_c \quad (14.2a)$$

$$D_f = 0 \quad \text{for } \tau \le \tau_c \quad (14.2b)$$

where D_r and D_f are the respective rates of detachment of material per unit area (kg m^{-2} s^{-1}); k_r is the raindrop impact soil erodibility coefficient (J^{-1}); k_f is the overland flow soil erodibility coefficient (kg m^{-2} s^{-1}); C_g is the proportion of ground protected from drop/drip erosion by near-ground cover such as low vegetation (range 0–1); C_r is the proportion of ground protected against drop/drip erosion and overland flow erosion by, for example, a cover of loose rocks (range 0–1); M_r is the momentum squared for raindrops falling directly on the ground ((kg m s^{-1}) m^{-2} s^{-1}); M_d is the momentum squared for leaf drip ((kg m s^{-1}) m^{-2} s^{-1});

F_w accounts for the effect of a surface water layer in protecting the soil from raindrop impact (dimensionless); τ is the overland flow shear stress (N m^{-2}); and τ_c is the critical shear stress for initiation of soil particle motion (N m^{-2}). The soil erodibility coefficients k_r and k_f increase in value as the soil becomes easier to erode (i.e. sandy soils have larger values than clayey soils). However, they cannot yet be determined from a directly measurable soil property and therefore require calibration (e.g. Wicks *et al.*, 1992; Adams & Elliott, 2006).

A simple estimate of bank erosion by channel flow is made as follows:

$$E_b = k_b \left(\frac{\tau_b}{\tau_{bc}} - 1 \right) \quad \text{for } \tau_b > \tau_{bc} \quad (14.3a)$$

$$E_b = 0 \quad \text{for} \quad \tau_b \le \tau_{bc} \quad (14.3b)$$

where E_b is the rate of detachment of material per unit area of bank (kg m^{-2} s^{-1}); k_b is the bank erodibility coefficient (kg m^{-2} s^{-1}); τ_b is the shear stress acting on the bank (N m^{-2}); and τ_{bc} is the critical shear stress for initiation of motion of bank material (N m^{-2}).

Although the SHETRAN contaminant transport component is not presented here, it may be noted that it allows for the transport of contaminants adsorbed to sediment particles (Ewen, 1995).

14.3.3 SHETRAN landslide erosion and sediment yield component

Through its integrated surface and subsurface representation of river basins, SHETRAN provides not only the overland and channel flows needed for modelling the transport of eroded soil, but also soil moisture conditions and hence a basis for simulating rain- and snowmelt-triggered landsliding. The SHETRAN landslide component was thus developed to simulate the erosion and sediment yield associated with shallow landslides at the basin scale (Burton & Bathurst, 1998). The occurrence of shallow landslides is determined as a function of the time- and space-varying soil sat-

uration conditions simulated by SHETRAN, using factor of safety analysis. This includes an allowance for the effect of vegetation root cohesion. For each landslide the volume of eroded material is determined and, depending on conditions, is routed down the hillslope as a debris flow. The proportion of the material reaching the channel network is then calculated and fed to the SHETRAN sediment transport component for routing to the basin outlet.

The central feature of the landslide model is the use of derived relationships (based on a topographic index) to link the SHETRAN grid resolution (which may be as large as 1 or 2 km), at which the basin hydrology and sediment yield are modelled, to a subgrid resolution (typically around 10–100 m) at which landslide occurrence and erosion is modelled. Through this dual resolution design, the model is able to represent landsliding at a physically realistic scale while remaining applicable at the basin scales (up to 500 km^2) likely to be of interest, for example feeding a reservoir.

Full detail of the landslide component is provided in Burton and Bathurst (1998). However, it will be helpful to know that the critical soil saturation conditions for landslide occurrence are determined using the one-dimensional, infinite-slope factor of safety equation:-

$$FS = \frac{\left[\dfrac{2[C_s + C_r]}{\gamma_w \, d \sin(2\beta)} + \dfrac{(L - m)\tan\phi}{\tan\beta} \right]}{L} \quad (14.4)$$

where

$$L = \frac{q_o}{\gamma_w d} + m \frac{\gamma_{sat}}{\gamma_w} + (1 - m) \frac{\gamma_m}{\gamma_w} \quad (14.5)$$

and FS is the factor of safety ($FS < 1$, unsafe; $FS \ge 1$, safe); C_s is the effective soil cohesion; C_r is the root cohesion; ϕ is the effective angle of internal friction of soil on an impermeable layer; d is the soil depth above the failure plane; β is the slope angle; q_o is the vegetative surcharge per unit plan area; γ_{sat} is the weight density of the saturated soil; γ_m is the weight density of soil at field moisture content; γ_w is the weight density of water; and m is the relative saturated depth (thickness of the saturated zone divided by soil depth above the failure plane) (Ward *et al.*, 1981).

14.3.4 Simulation procedure

A mesh is set up which defines the spatial extent of the basin model and which is used for representing spatial variability in the basin properties (Fig. 14.1). The number of mesh grid squares and the mesh resolution are limited primarily by computational power but may also depend on the length of the simulation and the size of the basin. With current capabilities, a mesh of around 400 squares will comfortably allow simulations of decades to a century with run times of a few hours. The size of the square then determines the basin scale which can be simulated. However, as the fundamental model equations typically represent small-scale physics, there is a potential conflict with the use of large grid squares. The maximum size considered physically reasonable has not been defined, but mesh resolutions of 2 km have been used in a number of applications, allowing basins of up to around 1500 km² to be simulated. A trade-off would enable a larger number of squares to be used for a shorter simulation, allowing either a finer grid resolution for a given basin size or a larger basin for a given grid resolution. Smaller basins similarly enable finer grid resolutions to be imposed. Adams and Elliot (2006), for example, used 3755 squares of resolution 0.5 m to model an area of 939 m².

The appropriate meteorological data, spatially and temporally distributed, are fed into the model. Each SHETRAN component is applied at each grid square to generate a response (e.g. phreatic surface rise, overland flow, soil erosion and landslide occurrence). These responses interact and both surface and subsurface waters, and surface sediments, are routed from square to square as a function of gradient. Eventually these products reach the river system and are routed towards the basin outlet. Model outputs may be obtained for any part of this procedure on a spatially and temporally distributed basis. They may include time-varying data for phreatic surface level, snowpack depth, overland flow depth, soil erosion or any other variable at any grid square or channel link within the basin. Alternatively, synoptic or time-integrated views of the spatial distribution of any variable across the basin can be produced (Fig. 14.2).

14.3.5 Data provision

In order to represent a basin realistically, physically-based models require a wide range of data. While on the one hand this requirement is demanding and not always achievable through direct measurement, on the other it is an advantage of physically-based models that they can make use of all available data, from digital terrain models to soil property measurements to historical information (e.g. newspaper articles and photographs) on basin responses. The data required by SHETRAN are:

(i) precipitation and potential evaporation input data to drive the simulation, preferably at hourly intervals;

(ii) topographic, soil, vegetation, sediment and geotechnical properties to characterize the basin on a spatially distributed basis; and

(iii) basin response data, such as discharge records, sediment yield, landslide inventories and photographic evidence, for testing the model output.

It is usually possible to obtain information on the basin property data through field measurements, laboratory analysis, national agencies and literature sources. For example, soil hydraulic and geotechnical properties can be obtained directly from field measurements and laboratory analysis of field samples, or indirectly using pedotransfer relationships published in the literature (e.g. Saxton *et al.*, 1986). Vegetation distributions are increasingly available from remote sensing surveys, and digital terrain data are provided by national agencies (or increasingly from websites, e.g. the HydroSHEDS data at http://hydrosheds. cr.usgs.gov (Lehner *et al.*, 2008)). Runoff and erosion plot studies are helpful in evaluating the

270   J.C. BATHURST

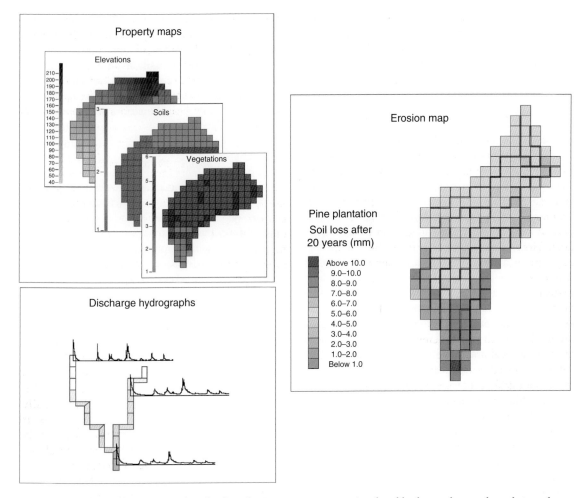

Fig. 14.2 Examples of SHETRAN data displays: basin property maps; simulated hydrographs at selected sites along a channel network (represented in plan); and an output map of simulated soil loss across a basin. Reproduced from Bathurst (2002).

overland flow resistance coefficient and soil erodibility coefficients (e.g. Engman, 1986; Wicks *et al.*, 1992). However, it is rare outside small research basins to be able to obtain all the required time-varying data. Rainfall data are often at the daily, rather than hourly, scale, evaporation may have to be determined from temperature data rather than direct measurement, and a basin with a landslide inventory map may not also have, for example, sediment yield data. Various techniques

such as rainfall disaggregation and regional scaling of water and sediment discharges then need to be employed to fill the gaps (e.g. Bathurst *et al.*, 2005). Nevertheless, physically-based models can potentially use a wide range of data for calibration and validation, including discharge and sediment transport gaugings, reservoir outflow and water-level records, reservoir sedimentation surveys, well levels, soil moisture measurements, snowline data, landslide inventories, gully growth rates and, more indirectly, crop yields (as an indicator of soil moisture levels), volumes of sediment removed by local authorities in clean-up operations after a flood event, and information on expected basin responses obtained from research basins.

14.4　Model Calibration and Uncertainty

In principle, the parameters of a physically-based, spatially-distributed model should not require calibration. They are supposedly based on measurements and already truly representative of that part of the catchment for which they are evaluated. However, within a model there may be approximations in the representation of the physical processes (e.g. the use of one-dimensional instead of three-dimensional formulations) and potential inconsistencies between the model grid scale, the scale at which property measurements are made, and the scale relevant to each particular hydrological process (e.g. saturated conductivity is measured at the scale of an auger hole but the model grid scale may be tens to hundreds of metres). Also, the current impossibility of measuring all the model parameters at every grid square across the entire basin means that many parameter values must be estimated or assumed from data obtained at a coarser resolution or at a few sampling points or from the literature. A degree of calibration or adjustment of parameter values is therefore likely to be needed to minimize the differences between observed and simulated responses. Such calibration, though, should be constrained by physical plausibility, so that the parameter values either lie in a physically realistic range or can otherwise be explained by physical reasoning. Thus, the term 'calibration' refers to a much more restricted and physically informed procedure than that associated with other types of model.

In certain cases, it has been found that a satisfactory simulation requires a model parameter to lie outside the physically realistic range. This is most likely where the model grid scale significantly exceeds the scale at which the parameter measurement is made. The solution is then to use a so-called 'effective' parameter value, representative at the grid scale. This is a pragmatic approach and it is recognized that the concept may not allow an accurate reproduction of the observed response in all circumstances (as shown, for example, by Binley *et al.*, 1989). However, experience to date suggests that the SHETRAN simulations are robust in this regard and that scale dependency needs to be considered only for certain parameters. Most notably, model saturated soil conductivity may increase beyond measured values, probably to compensate for the reduction in simulated groundwater gradients caused by the use of large grid squares in catchments with hilly terrain (see Section 6.4 for a more general treatment of this issue).

As a result of such problems as scaling and use of estimates instead of measurements, it is generally acknowledged that the parameterization (i.e. parameter evaluation) of physically-based, spatially-distributed models involves uncertainty (e.g. Beven & Binley, 1992; Beven, 2001: 19–23; Guimarães *et al.*, 2003). This uncertainty creates the potential for multiple parameterizations with possibly quite different but apparently equally acceptable combinations of parameter values. It is therefore also accepted that the uncertainty and its implications for model output should be explicitly recognized in the modelling procedure (e.g. Beven & Binley, 1992; Ewen & Parkin, 1996; Quinton, 1997, 2004; Christiaens & Feyen, 2002). Typically this is achieved by quantifying some form of uncertainty envelope. A popular approach is the Generalized Likelihood Uncertainty Estimation (GLUE) methodology (Beven & Binley, 1992; see also Chapter 4 for a general discussion and Chapter 5 for a worked example), in which model runs with several thousand parameter sets drawn from a defined range are tested as the basis for generating uncertainty bounds on the output. For physically-based, spatially-distributed models, though, the computing requirements associated with many thousands of runs are prohibitive. The particular approach for SHETRAN, therefore, is to set bound values on the more important model parameters and, through a limited set of

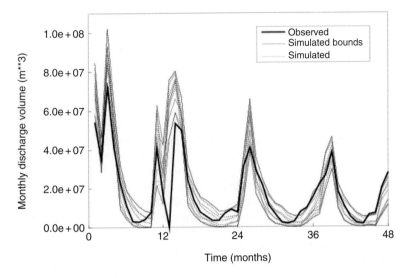

Fig. 14.3 Example comparison of observed discharge record with output uncertainty bounds created from a series of individual simulations: the Agri basin in southern Italy. Reproduced from Bathurst *et al.* (2002) by permission of John Wiley and Sons.

simulations, create corresponding bounds on the model output (Ewen & Parkin, 1996; Lukey *et al.*, 2000; Bathurst *et al.*, 2004) (Fig. 14.3). The model parameters or functions to which the simulation results are typically most sensitive in full basin simulations are: saturated zone hydraulic conductivity, overland flow resistance coefficient, soil hydraulic property curves (the variations of moisture tension and hydraulic conductivity with moisture content), the relationship between the ratio of actual to potential evapotranspiration and soil moisture content, root cohesion and the soil erodibility coefficients.

With the introduction of uncertainty envelopes, the aim of the calibration is then not to reproduce the observed hydrological and sediment yield responses as accurately as possible with one simulation, but to bracket the observed responses with several simulations. Between them, these simulations should represent the uncertainty in the key inputs.

Finally, the traditional calibration of models against outlet water and sediment discharges concerns only the integrated basin response and does not test for internal basin response. A good outlet simulation could be obtained on the basis of compensating internal errors (e.g. Anderton *et al.*, 2002). Spatially-distributed models cannot

therefore be considered to be fully calibrated unless they are shown to be able to reproduce internal as well as outlet responses. Incorporation of internal data can also help to constrain parameter uncertainty (e.g. Franks *et al.*, 1998; Christiaens & Feyen, 2002; Blazkova *et al.*, 2002; Ebel & Loague, 2006).

14.5   Model Fitness for Purpose

Validation tests the applicability of the calibrated model. However, Ewen and Parkin (1996) have criticized standard techniques of model validation for not clearly testing a model's fitness for purpose. The particular purpose of interest here is predicting the impacts of future changes in land use and climate. In this case, the prediction essentially concerns a hypothetical basin which will exist (if at all) only after the specified changes take place (Ewen & Parkin, 1996). Model validation should therefore involve the test basin as if it were hypothetical. The modeller should not be allowed sight of measured output data for the basin and thus should not be able to calibrate the model for that basin (hence the term 'blind' testing). Ewen and Parkin (1996) consequently proposed a blind testing procedure in which output uncertainty bounds are generated on the basis of a sequence of blind simulations. The measured data are then made available and conclusions on model performance are drawn according to the width of the uncertainty envelope and the degree to which it contains the measured data. Applications of the procedure to the SHETRAN system are described by Parkin *et al.* (1996), Thorne *et al.* (2000) and Bathurst *et al.* (2004) but so far the procedure has not been fully tested for sediment transport simulations.

14.6 SHETRAN in the Context of Physically-based, Basin-scale Sediment Yield Models

The advent of powerful computational facilities in the 1960s and 1970s made possible the numerical solution of the fundamental partial differential equations which describe water and sediment movement in river basins. On this basis, a number of physically-based models have been developed, with new models still appearing as a function of advances in conceptual understanding, database provision and solution techniques. A number of recent reviews have compared their capabilities (e.g. Merritt *et al.*, 2003; Aksoy & Kavvas, 2005; de Vente & Poesen, 2005), that of Merritt *et al.* (2003) being particularly detailed. Merritt *et al.* (2003) noted the difficulty of developing strict criteria for assessing different models without concentrating on a particular issue, such as future land management options. Even then, the model choice may not be clear as, for example, a model which applies well at small scales (e.g. of the order of 1 km²) may not be suitable for a larger scale. A successful model is therefore likely to be one which has inherent flexibility and adaptability. Nevertheless, some broad comparisons are possible, which highlight both the variations in individual capability and the evolution of overall capability over the last 20 years or so.

Table 14.2 extends Bathurst's (2002) comparison of the SHETRAN system with a number of physically-based, basin-scale models which have established a presence in the literature. These are the Areal Non-point Source Watershed Environment Response Simulation (ANSWERS), developed in the US contemporaneously with SHETRAN and subsequently used extensively in a range of countries (Park *et al.*, 1982; Silburn & Connolly, 1995); the more recent US Department of Agriculture's Water Erosion Prediction Project (WEPP) (considering the basin version) (Lane *et al.* 1992; see also Chapters 9 and 10); two further European developments, the European Soil Erosion Model (EUROSEM) (Morgan *et al.*, 1998; see Chapter 5) and the Limburg Soil Erosion Model (LISEM) (De Roo *et al.*, 1996; see Chapter 12); and a recent Californian development, the Integrated Hydrology Model (InHM) (Heppner *et al.*, 2006). The table suggests that, in many ways, the conceptual design of physically-based models has not changed greatly over the last two decades. All the models are applicable to single events but only SHETRAN, InHM and WEPP can simulate continuous periods containing sequences of runoff events separated by dry periods, although increasing computer power may enable previously event-based models to become continuous. All the models can generate overland flow as a result of rainfall exceeding infiltration rate but only SHETRAN, LISEM and InHM have the coupled surface/subsurface design which allows overland flow to be generated by upward saturation of the soil. The sediment routing scheme is not shown in the table since all the models use a similar approach involving a balance between sediment availability and the sediment transport capacity of the flow. The upper limit on the basin area that can be simulated is not well defined (and in any case is partly a function of available computer power). However, many of the models are designed for small basins; ANSWERS has been applied at scales of tens to perhaps low hundreds of square kilometres, InHM has so far been applied up to low hundreds of square kilometres, and SHETRAN has been applied to 1500 km². Most of the models can provide time-varying and spatially distributed output. ANSWERS, EUROSEM and LISEM were designed especially to simulate erosion on lands growing crops; however, they could probably be adapted to represent the wider range of covers, including rangeland

Table 14.2 Comparison of SHETRAN with five physically-based erosion and sediment yield models.

Model feature	SHETRAN	ANSWERS	WEPP	EUROSEM	LISEM	InHM
Simulation type:						
Continuous	Y	N	Y	N	N	Y
Single event	Y	Y	Y	Y	Y	Y
Maximum basin size	2000 km^2	Order 100 km^2	2.6 km^2	Small basin	Small basin	Order 100 km^2
Spatial distribution	Grid	Grid or GIS raster	Grid	Uniform slope planes	GIS raster	Finite element mesh
Overland flow:						
Rainfall excess	Y	Y	Y	Y	Y	Y
Upward saturation	Y	N	N	N	Y	Y
Erosion process:						
Raindrop impact/ overland flow	Y	Y	Y	Y	Y	Y
Rilling	N	N	Y	Y	Y	N
Crusting	N	Y	N	Y	Y	N
Channel banks	Y	N	N	Y	N	N
Gullying	N	N	N	N	N	N
Landsliding	Y	N	N	N	N	N
Roads	N	N	N	N	N	Y
Output:						
Time-varying sedigraph	Y	Y	N	Y	Y	Y
Time-integrated yield	Y	Y	Y	Y	Y	Y
Erosion map	Y	Y	Y	N	Y	Y
Land use	Most vegetation covers	Mainly agricultural	Wide range of land uses	Mainly agricultural	Mainly agricultural	Most vegetation covers

Y, yes; N, no.

and forests, which can be simulated by SHETRAN, WEPP and potentially InHM.

As noted by Merritt *et al.* (2003), most models tend to represent only one of the broad erosional processes. All those in Table 14.2 consider soil erosion by raindrop impact and overland flow in some form. WEPP, EUROSEM and LISEM represent the process in considerable detail, accounting for rilling, crusting and other effects. However, these effects are most significant at small scales and may merge with or become dominated by a wider range of erosion controls at larger scales (de Vente & Poesen, 2005). Other specialist classes of models consider erosion by gullying (e.g. Bull &

Kirkby, 1997; Sidorchuk, 1999; see Chapter 19) and by landsliding (e.g. Montgomery & Dietrich, 1994; Wu & Sidle, 1995; Dietrich *et al.*, 2001; Dhakal & Sidle, 2003; Sidle & Ochiai, 2006). However, of the physically-based, basin-scale models, only SHETRAN currently provides a framework within which components have been developed for raindrop impact/overland flow erosion, landslide erosion, channel bank erosion (albeit at a simple level) and transport of contaminants adsorbed to sediment particles, and within which a preliminary design has been developed for a gully erosion component (Bathurst *et al.*, 1998a). A class of semi-quantitative models that consider the integrated effect of a range of erosion processes at the catchment scale is reviewed by de Vente & Poesen (2005). These models apply factor scores to each process to quantify an index which is related to an erosion risk class or sediment yield and, therefore, unlike SHETRAN, are not physically-based.

A full comparison of SHETRAN with its sister model, MIKE SHE, is not presented in Table 14.2 as the models have a common heritage. Both models simulate basin hydrology and sediment yield, although some of the components have a different basis. The principal difference between them is that SHETRAN has perhaps seen greater process development (e.g. landslide erosion, bank erosion) and has more generally been used in research studies (but also for applied work), while MIKE SHE has been developed for industry users and is widely used for applied work (but also for research studies).

Advances in model design over the last two decades have tended to be incremental rather than fundamental. For example, the more recent models represent the same broad erosional processes but tend to include greater process detail, such as macropore flow, depression storage and representation of microtopography in InHM, and soil crusting in EUROSEM and LISEM. More fundamentally, InHM's use of a finite element scheme for spatial discretization enables linear features such as roads to be represented (Heppner *et al.*, 2007). Such features are often important sediment sources (e.g. Reid & Dunne, 1984; Grayson *et al.*, 1993) but are not easily incorporated in the finite difference or grid square schemes used by earlier models. Another advance is that, compared with the early developments, most models now make use of Geographical Information Systems (GIS), either directly in their design or in the process of parameterization.

Despite model developments, many of the difficulties recognized in the application of the earlier physically-based models remain unsolved or only partially solved:
• the data assembly and detailed considerations required for model parameterization are, as presented by Heppner *et al.* (2007) for InHM, reminiscent of the early SHE applications (e.g. Bathurst, 1986; Jain *et al.*, 1992). Remote sensing data hold considerable promise for parameterization in the future (e.g. King *et al.*, 2005) but are not yet able to characterize crucial soil properties and conditions (e.g. the hydraulic property curves or moisture content);
• most of the models represent the ease with which the soil can be eroded using an erodibility coefficient but at present these coefficients cannot be evaluated from a directly measurable soil property and must therefore be calibrated;
• the uncertainties associated with parameterizing physically-based, spatially-distributed models (such as the potential for multiple acceptable parameterizations) are only partially addressed by the use of uncertainty envelopes and effective parameter values, as discussed above;
• spatially-distributed models are still only rarely tested for their ability to represent basin internal as well as outlet responses, leaving open the possibility that an apparently good outlet simulation may be based on compensating internal errors and limiting confidence in the ability of the model to be extrapolated to wider ranges of conditions; and
• in any model where the erosion and sediment transport simulation is driven by the output from a hydrology or flow hydraulics simulation, the accuracy of the erosion simulation is heavily dependent on the accuracy of the hydrology simulation.

In conclusion, SHETRAN still compares favourably with other popular models, especially in its ability to simulate the impacts of land use and climate change at a range of basin scales and as a function of several broad erosion processes. Of the more recently developed models, InHM is probably the closest to SHETRAN in its design (e.g. for integrated surface/subsurface response) but also incorporates important new developments (e.g. a finite element discretization scheme).

14.7   Experience of Model Application

The practical problem under consideration is the use of the model to support the development of strategies for minimizing the adverse effects of land use and climate change on soil erosion and basin sediment yield. Experience in the use of SHETRAN to represent different land uses and climates, including scenario simulation of possible future basin conditions, is therefore summarized here. Full details of the applications are in the cited references, and only the most pertinent points are presented below.

It should be noted that model simulations are only one contribution to the process of decision-making, which is also likely to include political, social and economic dimensions. The particular contribution from the modelling is the prediction of basin response (e.g. runoff and sediment yield) for different scenarios of land use and climate change, so enabling optimum solutions from the point of view of the basin environment to be identified. It is important therefore to establish the extent to which the model can correctly represent basin responses to different land use and climate conditions.

14.7.1 Use of erosion plots for parameter evaluation

A requirement common to all the physically-based erosion models is for the erodibility coefficients to be calibrated, as currently they cannot be determined directly from measurable soil properties. Erosion plots are an excellent basis for such calibration as they are relatively simple to model, and measurements often represent a range of rainfall and land use conditions. Bathurst et al. (1996) noted that plots allow the calibration of not only the soil erodibility coefficients but also the overland flow resistance coefficient and the soil saturated conductivity, provided that, for the latter, overland flow is generated by rainfall excess over infiltration rate rather than upward saturation and that runoff is a significant percentage of the rainfall. Furthermore, plot calibrated values already represent a spatially integrated response, albeit at a relatively small scale of tens to hundreds of square metres, and may form a sounder basis for extrapolation to larger scales than values determined at a point scale (e.g. conductivity measured using an auger hole). Studies

with SHE, SHETRAN and ANSWERS suggest that the same model parameters can be applied at both plot (1–100 m²) and microbasin (order of 1 ha) scales, using small grid spacings (20 m or less) and with a good availability of field data (Wicks et al., 1988; Connolly & Silburn, 1995; Figueiredo, 1998; Figueiredo & Bathurst, 2002). For larger basins, scale effects in evaluating soil saturated conductivity appear not to be significant, or at least to be masked by uncertainty in parameter evaluation, as long as basin topography is subdued and there is a general homogeneity of land use, soil characteristics and hydrological response within the basin. Thus Bathurst et al. (1996) found that conductivities evaluated at the plot scale could be successfully applied with a model grid spacing of 2 km for the 701 km² Cobres basin in Portugal. Even in dissected badlands terrain, Bathurst et al. (1998b) found that any scale effects which may distinguish simulations at the scales of 0.133 and 86 ha were small enough to be masked by uncertainty in parameter evaluation. However, for basins with hilly terrain, experience is that model conductivity values are relatively high compared with point measurements (e.g. Bathurst et al., 2002, for the 1532 km² Agri basin in Italy; Bathurst et al., 2005, for the 160 km² Valsassina basin in Italy; Bathurst et al., 2007, for the 45 km² Ijuez basin in Spain).

Table 14.3 shows the SHE/SHETRAN erodibility coefficients calibrated for a number of plots. The results are not fully comparable as the basic rainfall erosion equation was changed slightly between the SHE and SHETRAN versions. Also, there is a mixture of natural and artificial rainfall application. However, the results indicate an excellent ability to represent the effects of significantly different soil conditions. The coefficients calibrated by Wicks et al. (1992) clearly distinguish between tilled ground and natural brush and grass cover grazed by cattle. Bathurst et al. (1996) found a need to vary the coefficients between moderate rainfall events and extreme events, with the suggestion that rilling occurred under the extreme conditions. Likewise Adams and Elliott (2006) calculated different coefficient values for the more and less erodible conditions

Table 14.3 Plot calibrated values of the model erodibility coefficients.

Location	Plot area (m²)	Rainfall	Raindrop impact erodibility coefficient (J⁻¹)	Overland flow erodibility coefficient (mg m⁻² s⁻¹)	Notes
Reynolds Creek, Idaho[a]	32.54	Artificial (single pattern)	1.3/11.8	0.65/5.9	Function of land use
Vale Formoso, Portugal[b]	167	Natural	0.13/2	1.3/20	Function of rainfall event
North Island, New Zealand[c]	970	Artificial	2.3/12.2	0.00045–0.00188	Function of season

References: [a]Wicks *et al.* (1992); [b]Bathurst *et al.* (1996); [c]Adams & Elliott (2006).

of winter and summer respectively. The raindrop impact coefficient varies in the two orders of magnitude range 0.1–$10\,J^{-1}$ but with the lower end of the range perhaps more relevant to natural rather than artificial rainfall. The calibrations of Wicks *et al.* (1992) and Bathurst *et al.* (1996) for the overland flow erodibility coefficient are also in reasonable agreement. However, considerably lower values are obtained by Adams and Elliott (2006), possibly because their plot had a dense cover of grass. In general the results of the plot calibrations are consistent with physical reasoning (e.g. that loose tilled soil should have a higher erodibility coefficient than compacted grazed ground). They therefore provide confidence in the ability of the model to represent the effect of differences in soil conditions and land use on soil erosion and sediment yield. On the other hand, they also show that there can be significant variations in erodibility even at a single site, for example as a function of season or rainfall/runoff characteristics, and these variations need to be accounted for.

Table 14.4 shows the erodibility coefficients applied in a series of basin simulations. Mostly these represent the upper and lower uncertainty bounds, although there is some calibration or adjustment in the light of measurements for the Iowa and Rimbaud simulations and the Cobres values are the same as the Vale Formoso plot calibrations. The high raindrop impact coefficient for the Iowa sites may reflect the easily eroded soil of an arable field area. Otherwise,

there is a tendency for the coefficient to adopt values from the lower end of the plot calibrated range, again perhaps a function of natural rainfall characteristics.

14.7.2 *Representation of spatial variability*

Tests of the ability of SHETRAN (or other distributed models) to represent spatial variability in soil erosion and sediment yield are very limited, largely because of a lack of the necessary response data. It is hard enough to find a suitable basin with sediment transport measurements at its outlet, let alone on a nested basis! Similarly, the measurement of spatial variability of erosion over more than a limited area is currently too time-consuming to be practical, although remote-sensing techniques hold considerable hope for the future. Data currently depend on measurements of caesium-137 activity in the soil for mapping patterns of erosion and soil redistribution (e.g. Ritchie & McHenry, 1990; Walling & Quine, 1992; Higgitt, 1995). A number of studies have explored the use of such maps for model validation (e.g. De Roo & Walling, 1994; Sidorchuk & Golosov, 1996; Ferro *et al.*, 1998). Data provided by the technique were therefore used to validate the capability of SHETRAN for predicting long-term (30-year) erosion rates and their spatial variability (Norouzi Banis *et al.*, 2004). Simulations on a 20-m grid were carried out for two arable farm sites (area 3–5 ha) in central England for which average annual erosion rates had already

Table 14.4   Values of model erodibility coefficients applied in field and basin simulations.

Location	Area (km²)	Raindrop impact erodibility coefficient (J⁻¹)	Overland flow erodibility coefficient (mg m⁻² s⁻¹)	Reference
UK fields	0.034/0.047	1.3–11.8	0.65–5.9	Norouzi Banis *et al.* (2004)
Iowa fields	0.051/0.064	28–82	0.14–0.33	Wicks and Bathurst (1996)
Laval, France	0.86	0.1–10	1–20	Lukey *et al.* (2000)
Rimbaud, France	1.46	0.01–5	0.1–10	Lukey *et al.* (1995)
Ijuez, Spain	45	0.05–0.2	0.5–2	Bathurst *et al.* (2007)
Valsassina, Italy	180	0.05–0.2	0.5–2	Bathurst *et al.* (2005)
Llobregat, Spain	505	0.05–0.2	0.5–2	Bathurst *et al.* (2006)
Cobres, Portugal	701	0.13/2	1.3/20	Bathurst *et al.* (1996)
Agri, Italy	1532	0.1–10	1–20	Bathurst *et al.* (2002)

been determined using caesium-137 measurements. These rates were successfully contained within the range of simulated values that represented the uncertainty in model output derived from uncertainty in the model parameterization. In addition, the spatial variability in the long-term erosion and deposition rates was reproduced excellently at one site and partially at the other. This suggests that the model does have the potential to represent spatial variability correctly, given the appropriate data. It also shows how such field data can improve the severity of the validation procedure, accounting for internal as well as outlet conditions.

14.7.3   Land use and climate change studies

The SHETRAN erosion model has been used to simulate the impacts of river basin change in two main ways. The first is testing the ability to represent change using measured data, either from paired basins with different land covers or from a single basin that has undergone a change; generally these concern the effect of forest logging or planting. The second is calibration for the current basin conditions and then application to scenarios for possible future land uses or climates. A third approach has also been proposed in which the model is used in a completely hypothetical manner to allow the initial exploration of a potential problem.

The tests using measured data are, almost by necessity, limited to small basins. Lukey *et al.*

(2000) simulated two of the Draix research basins administered by the French agency CEMAGREF in southeast France. A model was first constructed for the 86 hectare Laval basin which is severely affected by badlands erosion. The model was then altered to represent the Laval basin as if it were equivalent to the neighbouring 108 ha Brusquet basin, which was successfully rescued from badlands erosion by reforestation in the early 1900s. Changes in the model parameters included specification of forest vegetation (e.g. with appropriate leaf drip parameters), an increase in overland flow resistance and elimination of erosion by overland flow. Uncertainty bounds were developed both for the simulation results and for the measured data. The results showed a good ability to simulate the observed two orders of magnitude reduction in sediment yield from the Laval (127 t ha⁻¹ y⁻¹) to the Brusquet (0.03–2.75 t ha⁻¹ y⁻¹) basin, with the simulated changes exceeding the output uncertainty. Lukey *et al.* (1995) tested the ability of SHETRAN to simulate sediment yield in the 1.46 km² Rimbaud basin (administered by CEMAGREF in southeast France) following a fire in August 1990 that destroyed approximately 85% of the original cover of maquis and chestnut plantation. Output uncertainty bounds were derived as a function of specified uncertainty in the model erodibility coefficients. The results showed a good ability to contain measured sediment discharge values within the simulation bounds for individual

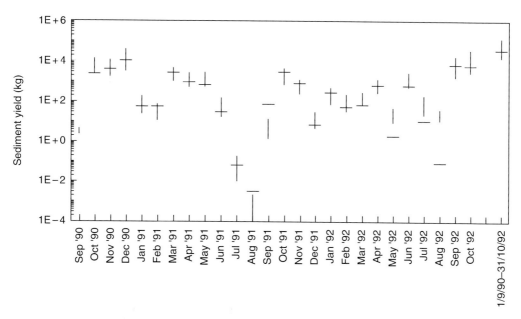

Fig. 14.4 Comparison of the simulated bulk monthly sediment yields with sediment yields generated directly from measurements for the Rimbaud basin in southeast France. The vertical lines represent the simulated bounds; the horizontal lines represent the measured yields. Reproduced from Lukey *et al.* (1995).

events, and an excellent ability to represent inter-month variability and long-term sediment yield over a two-year period (Fig. 14.4).

Scenario simulations are reported for a number of basins in the Mediterranean area. Typically the model is first calibrated for the current state of the basin. Land use scenarios are then specified as a function of possible changes in vegetation cover, such as partial or complete deforestation or a change of agricultural basis. The model is then rerun with the new parameter values and the outputs are compared. Climate change scenarios are developed on the basis of predictions for the rest of this century by general circulation models. This is a complex procedure, requiring for example the linking of the general circulation model data, at spatial and temporal resolutions of typically tens to hundreds of kilometres and months respectively, to the model grid (hundreds to thousands of metres) and time (hours) resolutions (e.g. Kilsby *et al.*, 2007). The model is then rerun with the altered rainfall and evaporation data and the

outputs are compared. Bathurst *et al.* (2002) calibrated SHETRAN for the upper 1532 km² of the Agri basin in southern Italy for the four-year period 1985–88, and then carried out simulations (unpublished) for land use and climate scenarios. For the land use scenario, all croplands (including pasture lands) were changed to native vegetation, the cover provided by the latter thereby increasing from 28% to 71% of the basin. The model was then run for the same four-year period. As the native vegetation has a lower actual evapotranspiration rate than crops, the change caused an increase in annual runoff (17%) and a corresponding increase in sediment yield (8%) (Fig. 14.5). For the climate change simulations, three scenarios were generated from the UK Hadley Centre Climate Model 2 (HADCM2) for 1970–79 (representing the current conditions), 2030–39 (a relevant planning horizon) and 2090–99 (for a full global warming effect). Relative to the first period, annual rainfall decreases by 5% and 16% for the second and third periods (most notably

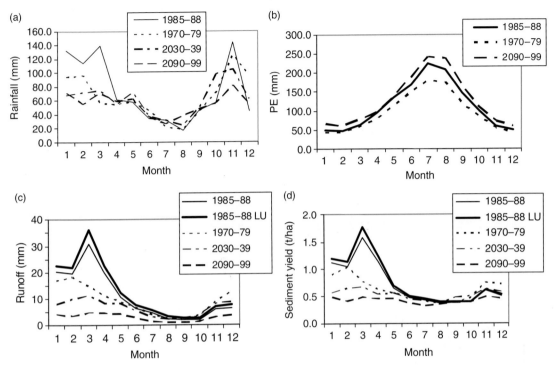

Fig. 14.5 Scenario simulations for the Agri basin in southern Italy. (a) Mean monthly rainfall for 1985–88 (measured) and three generated scenarios; (b) mean monthly potential evapotranspiration for 1985–88 (measured) and the bounding generated scenarios; (c) mean monthly runoff and (d) mean monthly sediment yield simulated for 1985–88 with existing land use, 1985–88 with land use scenario (LU) and the three climate scenarios.

in the winter) and annual evapotranspiration increases by 8% and 33% (most notably in the summer). Consequently the simulated annual runoff decreases by about 26% and 66% for the two periods, and annual sediment yield decreases by 14% and 32% (Fig. 14.5). The biggest changes occur in the winter; the summer runoff is already low and therefore unable to fall much further. Tests showed some differences between the generated and measured rainfalls and potential evapotranspiration for the recent historical period, and therefore the relative values arising from the simulations are likely to be more reliable than the absolute values. Nevertheless the results indicate a capability for providing quantified data of direct use in developing strategies to cope with possible future change.

Bathurst et al. (2005) applied the full SHETRAN sediment model, including landslide erosion, to the 180 km² Valsassina-Esino basins in the Italian Southern Alps. The model reproduced the observed spatial distribution of landslides from a 50-year record very well but overestimated the annual rate of landsliding. Uncertainty bounds were derived for the number of landslides: the upper bound considerably overestimated the observed number but was useful in visualizing spatial distribution, while the lower bound was generally more representative of reality (Fig. 14.6). Simulated sediment yields could not be checked directly against measured data but were within the range observed in a wider region of northern Italy. The results suggested, though, that the supply of shallow landslide material to the channel network contributes

Fig. 14.6 Example of landslide simulation: comparison of uncertainty bounds for the SHETRAN simulation (upper diagrams) with the 50-year map of observed landslides (lower diagram) in the Valsassina/Esino basin. Landslide locations are shown as dots. Reproduced from Bathurst *et al.* (2005).

relatively little to the overall long-term sediment yield compared with other sources. The model was then applied for scenarios of possible future climate change based on the UK Hadley Centre HadRM3 GCM for the period 2070–99 (drier and warmer than the present) and possible land use change (hillslope meadows abandoned to forest). The land use change produced a modest reduction in runoff while the climate change cut runoff by almost 50%. In both cases there was a modest reduction in simulated shallow landslide occurrence and the overall sediment yield. The scenario results can be explained in terms of model design and capability. In other words they are physically realistic, within the limitations of the model design and scenario characteristics. However, given the uncertainties in the scenario formulation and parameter evaluation, the relative variations between the simulations are likely to be more reliable than the simulated output magnitudes. Comparison of the scenario results with the simulation for the current period therefore provides an indication of the potential future changes in catchment response and thus provides a context within which guidelines for land management can be developed to minimize debris flow impacts.

14.7.4   Use of model for hypothetical exploration in the absence of data

A criticism frequently levelled at physically-based, spatially-distributed models is that data provision does not match the models' requirements for parameterization and calibration (e.g. Merritt *et al.*, 2003). However, this criticism misses two important points. The first is that decisions have to be made, whatever the state of the data. In such cases physically-based models make full use of whatever data are available and can be used to quantify the uncertainty associated with making a decision on the basis of those data. The second is that, even if there are no data for the basin, a physically-based model can still be parameterized using estimates and data from the literature, an option not available to other types of model. Of course, the uncertainty associated with such a parameterization is large but the model can provide an important first step in exploring a potential problem area. The extent to which the uncertainty is acceptable has to be weighed against the cost of its reduction through the collection of more detailed spatially and temporally varying data. If the exploration suggests a potential for an undesirable or unacceptable basin response, then it is clear that the more detailed application is required. In that case the model can be used to indicate which data are most needed to reduce the uncertainty and can thus guide the field measurements. An example is presented by Bathurst *et al.* (1998c) to demonstrate the potential impact of forest plantation on runoff and soil erosion in southern Chile. A SHETRAN model of a hypothetical basin was created using basin data and meteorological data from other applications in similar climatic environments. Simulations were carried out for a period of 20 years, a typical plantation cycle. First a simulation was run for native forest, calibrated to produce equilibrium conditions. Then a simulation was carried out for a plantation forest of exotic species, growing from a bare soil condition through to full cover and mature height. Comparison of the simulations showed the plantation conditions to produce a potential for enhanced flood flows in the early years of forest growth (because of the reduced interception and transpiration). More importantly, the increased interception and transpiration of the later years reduced streamflows to levels that could be detrimental to aquatic habitats and threatened drinking water abstraction. Soil erosion was correspondingly enhanced in the early years but reduced in the later years of the plantation cycle (Fig. 14.7).

Another form of exploration was carried out by Birkinshaw and Bathurst (2006) to investigate the scaling relationship between specific sediment yield and basin area, as a function of sediment source, land use and rainfall distribution. Two basins which had already been modelled by SHETRAN were used in the study. However, the simulations were not representations of the real-world basins. They simply used the topographies and river networks as a framework within which to investigate systematically the scaling relationship for a number of specified conditions. The study showed that the relationship could be inverse or direct, depending on the conditions.

14.8   Future Research Needs

Challenging but exciting research needs concern model parameterization, design and use. Some examples of parameterization developments relevant to erosion modelling include:

• use of airborne LIDAR (LIght Detection And Ranging), which has the capability of providing vertical resolutions of a few centimetres and horizontal resolutions of a few decimetres to metres (e.g. McKean & Roering, 2004), to produce accurate microtopography maps and long-term erosion and deposition measurements;

• use of additional remote-sensing techniques to characterize soil surface conditions and spatial variability in the factors controlling erosion and transport (e.g. King *et al.*, 2005);

• a means of quantifying the soil erodibility coefficients from measurable soil properties, to reduce the current reliance on calibration;

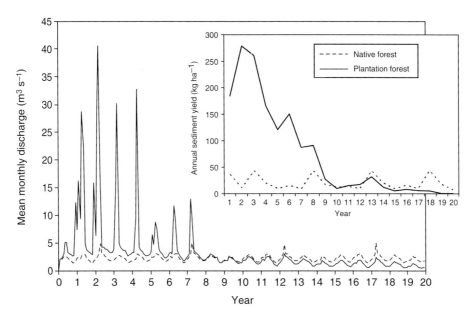

Fig. 14.7 Example of application to a hypothetical basin in Chile: comparison of mean monthly discharges and (inset) annual sediment yields for covers of native forest and plantation forest over the 20-year plantation cycle. Adapted from Bathurst *et al.* (1998c).

• a protocol for using plot studies to quantify model parameters and for applying the values at different spatial scales; and

• an emphasis on obtaining internal basin response data, including sediment discharges, gully growth and landslide inventories on both event and long-term bases.

There are many areas in which model design can be improved, from ever-more-detailed representation of the process of raindrop impact erosion to more efficient solution techniques. However, there is still considerable scope for integrating a wider range of the basic erosion processes in a single model than is currently the case. Some possible areas of advance are:

• the incorporation of a gully erosion and sediment yield component;

• the representation of linear features such as roads and tracks;

• the development of a sediment tracking capability, enabling sediment particles arriving at the basin outlet to be related to their source areas within the basin. This capability could be built upon recent developments in tracking water movement from source to outlet (O'Donnell, 2008) and could be tested using data on sediment sources obtained by chemical 'fingerprinting' techniques (e.g. Collins *et al.*, 1997);

• the coupling of a forest architecture and growth model with a hydrological-geotechnical landslide model, allowing feedback between forest growth (including root systems and water demand) and soil moisture content, with implications for overland flow transport of sediment and landslide occurrence; and

• the incorporation of the two-dimensional rotational slip model for landslide simulation (Collison & Griffiths, 2004). Current basin-scale models typically employ the one-dimensional infinite slope stability model because of its practicality but this restricts accurate representation of slope conditions where groundwater flow or topography produce forces that are significant in directions other than slope-normal.

Considering model use, a major challenge is to educate potential users in how to apply physically-based, spatially-distributed basin models most advantageously. Some important points include:

• the problems of data provision and parameter uncertainty associated with such models must be acknowledged but should not be taken to prevent model use. Instead, users must learn how to interpret outputs in the form of uncertainty envelopes. This allows in any case a more realistic appreciation of the limitations of all types of sediment model (not just physically-based) and provides a means of quantifying the implications of making decisions on the basis of imperfect data;

• even in the absence of data it is possible to apply physically-based models to explore problem areas and to carry out 'what if?' studies of different potential courses of action;

• satisfactory calibration and validation require the incorporation of internal basin response data and appropriate tests for a model's fitness for purpose;

• most reported model applications have been in temperate and semi-arid regions. However, tropical regions are currently suffering from severe problems of accelerated erosion related to land use changes and demonstrations of model applicability to these regions are therefore needed.

14.9   Conclusion

Physically-based, spatially-distributed models provide the strongest basis for predicting the impacts of future changes in land use and climate on soil erosion and sediment yield in river basins. However, current computational constraints limit their applicability to basins of a few thousands of square kilometres. New models continue to be developed but the basic process representation has shown only modest improvements over the past two decades. SHETRAN remains a viable and relevant model with particular advantages in integrating a range of broad erosional processes. Considerable experience has been accumulated in its application at scales ranging from small erosion plots to basins of 1500 km^2 and it has been used in scenario studies and the exploration of problem areas. As with other physically-based models, uncertainties are associated with data provision and parameterization; this uncertainty should be accounted for in the modelling procedure, for example with uncertainty envelopes (see Chapter 5 for a case study), and users must learn to interpret model output accordingly. As yet, a new generation of models involving fundamental developments in process understanding and solution technique has still to emerge. However, there are a number of exciting opportunities for enhancing existing capabilities, especially in the areas of data provision, model parameterization and process representation.

Acknowledgements

The continued evolution and application of SHETRAN over the last two decades has depended on the efforts of a team of researchers at Newcastle University and other centres. These researchers have been acknowledged in, or have been co-authors of, previous publications. However, there have been a number of important developments over the last few years (not all associated with the erosion model) and this author would particularly like to acknowledge the work of Steve Birkinshaw, Isabella Bovolo, Ahmed El-Hames, John Ewen and Greta Moretti. The SHETRAN website is maintained by Steve Birkinshaw.

References

Abbott, M.B., Bathurst, J.C., Cunge, J.A., *et al.* (1986a) An introduction to the European Hydrological System – Système Hydrologique Européen, "SHE", 1: History and philosophy of a physically-based, distributed modelling system. *Journal of Hydrology* **87**: 45–59.

Abbott, M.B., Bathurst, J.C., Cunge, J.A., *et al.* (1986b) An introduction to the European Hydrological System – Système Hydrologique Européen, "SHE", 2: Structure of a physically-based, distributed modelling system. *Journal of Hydrology* **87**: 61–77.

Adams, R. & Elliott, S. (2006) Physically based modelling of sediment generation and transport under a large rainfall simulator. *Hydrological Processes* **20**: 2253–70.

Aksoy, H. & Kavvas, M.L. (2005) A review of hillslope and watershed scale erosion and sediment transport models. *Catena* **64**: 247–71.

Anderton, S., Latron, J. & Gallart, F. (2002) Sensitivity analysis and multi-response, multi-criteria evaluation of a physically based distributed model. *Hydrological Processes* **16**: 333–53.

Bathurst, J.C. (1986) Physically-based distributed modelling of an upland catchment using the Système Hydrologique Européen. *Journal of Hydrology* **87**: 79–102.

Bathurst, J.C. (2002) Physically-based erosion and sediment yield modelling: the SHETRAN concept. In Summer, W. & Walling, D.E. (eds), *Modelling Erosion, Sediment Transport and Sediment Yield*. International Hydrological Programme, IHP-VI, Technical Documents in Hydrology, No. 60. UNESCO, Paris: 47–67.

Bathurst, J.C., Wicks, J.M. & O'Connell, P.E. (1995) The SHE/SHESED basin scale water flow and sediment transport modelling system. In Singh, V.P. (ed.), *Computer Models of Watershed Hydrology*, Water Resources Publications, Highlands Ranch, Colorado, US: 563–94.

Bathurst, J.C., Kilsby, C. & White, S. (1996) Modelling the impacts of climate and land-use change on basin hydrology and soil erosion in Mediterranean Europe. In Brandt, C.J. & Thornes, J.B. (eds), *Mediterranean Desertification and Land Use*, John Wiley & Sons, Chichester: 355–87.

Bathurst, J.C., Gonzalez, E. & Salgado, R. (1998a) Physically based modelling of gully erosion and sediment yield at the basin scale. In Wheater, H. & Kirby, C. (eds), *Hydrology in a Changing Environment*, Vol III, John Wiley & Sons, Chichester: 253–9.

Bathurst, J.C., Lukey, B., Sheffield, J., *et al.* (1998b) Modelling badlands erosion with SHETRAN at Draix, southeast France. In *Modelling Soil Erosion, Sediment Transport and Closely Related Hydrological Processes*, International Association of Hydrological Sciences Publ. No. 249, Wallingford, Oxon, UK: 129–36.

Bathurst, J.C., Birkinshaw, S.J., Evans, J. & Francke, S. (1998c) Modelo de bases físicas para la predicción de los impactos hidrológicos y la erosión de los suelos provocados por opciones de la gestión forestal en Chile. (Physically based modelling for predicting hydrological and soil erosion impacts of forest management options in Chile.) *Chile Forestal* 265, Documento Técnico 120, CONAF, Chile.

Bathurst, J.C., Sheffield, J., Vicente, C., *et al.* (2002) Modelling large basin hydrology and sediment yield with sparse data: the Agri basin, southern Italy. In Geeson, N.A., Brandt, C.J. & Thornes, J.B. (eds), *Mediterranean Desertification: a Mosaic of Processes and Responses*, John Wiley & Sons, Chichester: 397–415.

Bathurst, J.C., Ewen, J., Parkin, G., *et al.* (2004) Validation of catchment models for predicting land-use and climate change impacts. 3. Blind validation for internal and outlet responses. *Journal of Hydrology* **287**: 74–94.

Bathurst, J.C., Moretti, G., El-Hames, A., *et al.* (2005) Scenario modelling of basin-scale, shallow landslide sediment yield, Valsassina, Italian Southern Alps. *Natural Hazards & Earth System Sciences* **5**: 189–202.

Bathurst, J.C., Burton, A., Clarke, B.G. & Gallart, F. (2006) Application of the SHETRAN basin-scale, landslide sediment yield model to the Llobregat basin, Spanish Pyrenees. *Hydrological Processes* **20**: 3119–38.

Bathurst, J.C., Moretti, G., El-Hames, A., *et al.* (2007) Modelling the impact of forest loss on shallow landslide sediment yield, Ijuez river catchment, Spanish Pyrenees. *Hydrology & Earth System Sciences* **11**: 569–83.

Beven, K.J. 2001. *Rainfall-Runoff Modelling: The Primer*. John Wiley & Sons, Chichester.

Beven, K. & Binley, A. (1992) The future of distributed models: model calibration and uncertainty prediction. *Hydrological Processes* **6**: 279–98.

Binley, A., Beven, K. & Elgy, J. (1989) A physically based model of heterogeneous hillslopes 2. Effective hydraulic conductivities. *Water Resources Research* **25**: 1227–33.

Birkinshaw, S.J. & Bathurst, J.C. (2006) Model study of the relationship between sediment yield and river basin area. *Earth Surface Processes and Landforms* **31**: 750–61.

Birkinshaw, S.J. & Ewen, J. (2000) Nitrogen transformation component for SHETRAN catchment nitrate transport modelling. *Journal of Hydrology* **230**: 1–17.

Blazkova, S., Beven, K.J. & Kulasova, A. (2002) On constraining TOPMODEL hydrograph simulations using partial saturated area information. *Hydrological Processes* **16**: 441–58.

Bull, L.J. & Kirkby, M.J. (1997) Gully processes and modelling. *Progress in Physical Geography* **21**: 354–74.

Burton, A. & Bathurst, J.C. (1998) Physically based modelling of shallow landslide sediment yield at a catchment scale. *Environmental Geology* **35**: 89–99.

Christiaens, K. & Feyen, J. (2002) Constraining soil hydraulic parameter and output uncertainty of the distributed hydrological MIKE SHE model using the GLUE framework. *Hydrological Processes* **16**: 373–91.

Collins, A.L., Walling, D.E. & Leeks, G.J.L. (1997) Sediment sources in the Upper Severn catchment : a fingerprinting approach. *Hydrology and Earth System Sciences* **1**: 509–21.

Collison, A. & Griffiths, J. (2004) Modelling slope instability. In Wainwright, J. & Mulligan, M. (eds), *Environmental Modelling: Finding Simplicity in Complexity*, John Wiley & Sons, Chichester: 197–209.

Connolly, R.D. & Silburn, D.M. (1995) Distributed parameter hydrology model (ANSWERS) applied to a range of catchment scales using rainfall simulator data II: Application to spatially uniform catchments. *Journal of Hydrology* **172**: 105–25.

De Roo, A.P.J. & Walling, D.E. (1994) Validating the ANSWERS soil erosion model using [137]Cs. In Rickson, R.J. (ed.), *Conserving Soil Resources – European Perspectives*, CAB International, Wallingford, Oxon, UK: 246–63.

De Roo, A.P.J., Wesseling, C.G. & Ritsema, C.J. (1996) LISEM: A single-event physically based hydrological and soil erosion model for drainage basins. I: Theory, input and output. *Hydrological Processes* **10**: 1107–17.

De Vente, J. & Poesen, J. (2005) Predicting soil erosion and sediment yield at the basin scale: scale issues and semi-quantitative models. *Earth-Science Reviews* **71**: 95–125.

Dhakal, A.S. & Sidle, R.C. (2003) Long-term modelling of landslides for different forest management practices. *Earth Surface Processes and Landforms* **28**: 853–68.

Dietrich, W.E., Bellugi, D. & Real de Asua, R. (2001) Validation of the shallow landslide model, SHALSTAB, for forest management. In Wigmosta, M.S. & Burgess, S.J. (eds), *Land Use and Watersheds: Human Influence on Hydrology and Geomorphology in Urban and Forest Areas*, Water Science and Application, Vol. 2. American Geophysical Union, Washington, DC: 195–227.

Dunne, T. & Leopold, L.B. (1978) *Water in Environmental Planning*. Freeman, San Francisco.

Ebel, B.A. & Loague, K. (2006) Physics-based hydrologic-response simulation: Seeing through the fog of equifinality. *Hydrological Processes* **20**: 2887–2900.

Engman, E.T. (1986) Roughness coefficients for routing surface runoff. *Proceedings of the American Society of Civil Engineers, Journal of Irrigation & Drainage Engineering* **112**: 39–53.

Ewen, J. (1995) Contaminant transport component of the catchment modelling system SHETRAN. In Trudgill, S.T. (ed.), *Solute Modelling in Catchment Systems*, John Wiley & Sons, Chichester: 417–41.

Ewen, J. & Parkin, G. (1996) Validation of catchment models for predicting land-use and climate change impacts. 1. Method. *Journal of Hydrology* **175**: 583–94.

Ewen, J., Parkin, G. & O'Connell, P.E. (2000) SHETRAN: distributed river basin flow and transport modeling system. *Proceedings of the American Society of Civil Engineers, Journal of Hydrologic Engineering* **5**: 250–58.

Ferro, V., Di Stefano, C., Giordano, G. & Rizzo, S. (1998) Sediment delivery processes and the spatial distribution of caesium-137 in a small Sicilian basin. *Hydrological Processes* **12**: 701–11.

Figueiredo, E. de (1998) *Scale effects and land use change impacts in sediment yield modelling in a semi-arid region of Brazil.* PhD thesis, University of Newcastle upon Tyne, UK.

Figueiredo, E.E. de & Bathurst, J.C. (2002) Sediment yield modelling at various basin scales in a semiarid region of Brazil using SHETRAN. In *Hydroinformatics 2002: Proceedings of the 5th International Conference on Hydroinformatics*: 316–21.

Franks, S.W., Gineste, P., Beven, K.J. & Merot, P. (1998) On constraining the predictions of a distributed model: the incorporation of fuzzy estimates of saturated areas into the calibration process. *Water Resources Research* **34**: 787–97.

Gaillard, M.-J., Dearing, J.A., El-Daoushy, F., *et al.* (1991) A multidisciplinary study of the lake Bjäresjösjön (S Sweden): land-use history, soil erosion, lake trophy and lake-level fluctuations during the last 3000 years. *Hydrobiologia* **214**: 107–14.

Gilbert, G.K. (1917) Hydraulic mining debris in the Sierra Nevada. *USGS Professional Paper* 105, US Geological Survey, Washington, DC.

Glade, T. (2003) Landslide occurrence as a response to land use change: a review of evidence from New Zealand. *Catena* **51**: 297–314.

Grayson, R.B., Haydon, S.R., Jayasuriya, M.D.A. & Finlayson, B.L. (1993) Water quality in mountain ash forests – separating the impacts of roads from those of logging operations. *Journal of Hydrology* **150**: 459–80.

Griffiths, G.A. (1981) Some suspended sediment yields from South Island catchments, New Zealand. *Water Resources Bulletin* **17**: 662–71.

Guimarães, R.F., Montgomery, D.R., Greenberg, H.M., *et al.* (2003) Parameterization of soil properties for a model of topographic controls on shallow landsliding: application to Rio de Janeiro. *Engineering Geology* **69**: 99–108.

Heppner, C.S., Ran, Q., VanderKwaak, J.E. & Loague, K. (2006) Adding sediment transport to the integrated hydrology model (InHM): Development and testing. *Advances in Water Resources* **29**: 930–43.

Heppner, C.S., Loague, K. & VanderKwaak, J.E. (2007) Long-term InHM simulations of hydrologic response and sediment transport for the R-5 catchment. *Earth Surface Processes and Landforms* **32**: 1273–92.

Higgitt, D.L. (1995) The development and application of caesium-137 measurements in erosion investigations. In Foster, I.D.L., Gurnell, A.M. & Webb, B.W. (eds), *Sediment and Water Quality in River Catchments*, John Wiley & Sons, Chichester: 287–305.

Jain, S.K., Storm, B., Bathurst, J.C., *et al.* (1992) Application of the SHE to catchments in India Part 2. Field experiments and simulation studies with the SHE on the Kolar subcatchment of the Narmada River. *Journal of Hydrology* **140**: 25–47.

Kilsby, C.G., Jones, P.D., Burton, A., *et al.* (2007) A daily weather generator for use in climate change studies. *Environmental Modelling and Software* **22**: 1705–19.

King, C., Baghdadi, N., Lecomte, V. & Cerdan, O. (2005) The application of remote-sensing data to monitoring and modelling of soil erosion. *Catena* **62**: 79–93.

Lane, L.J., Nearing, M.A., Laflen, J.M., *et al.* (1992) Description of the US Department of Agriculture water erosion prediction project (WEPP) model. In Parsons, A.J. & Abrahams, A.D. (eds), *Overland Flow: Hydraulics and Erosion Mechanics*, UCL Press, University College, London: 377–91.

Lavigne, F. (2004) Rate of sediment yield following small-scale volcanic eruptions: a quantitative assessment at the Merapi and Semeru stratovolcanoes, Java, Indonesia. *Earth Surface Processes and Landforms* **29**: 1045–58.

Lehner, B., Verdin, K. & Jarvis, A. (2008) New global hydrography derived from spaceborne elevation data. *Eos, Transactions of the American Geophysical Union* **89**(10): 93–4.

Lukey, B.T., Sheffield, J., Bathurst, J.C., *et al.* (1995) Simulating the effect of vegetation cover on the sediment yield of Mediterranean catchments using SHETRAN. *Physics & Chemistry of the Earth* **20**: 427–32.

Lukey, B.T., Sheffield, J., Bathurst, J.C., *et al.* (2000) Test of the SHETRAN technology for modelling the impact of reforestation on badlands runoff and sediment yield at Draix, France. *Journal of Hydrology* **235**: 44–62.

McKean, J. & Roering, J. (2004) Objective landslide detection and surface morphology mapping using high-resolution airborne laser altimetry. *Geomorphology* **57**: 331–51.

Merritt, W.S., Letcher, R.A. & Jakeman, A.J. (2003) A review of erosion and sediment transport models. *Environmental Modelling & Software* **18**: 761–99.

Montgomery, D.R. & Dietrich, W.E. (1994) A physically based model for the topographic control on shallow landsliding. *Water Resources Research* **30**: 1153–71.

Morgan, R.P.C., Quinton, J.N., Smith, R.E., *et al.* (1998) The European Soil Erosion Model (EUROSEM): a dynamic approach for predicting sediment transport from fields and small catchments. *Earth Surface Processes and Landforms* **23**: 527–44.

Nearing, M.A. (2004) Soil erosion and conservation. In Wainwright, J. & Mulligan, M. (eds), *Environmental Modelling: Finding Simplicity in Complexity*, John Wiley & Sons, Chichester: 277–90.

Norouzi Banis, Y., Bathurst, J.C. & Walling, D.E. (2004) Use of caesium-137 data to evaluate SHETRAN simulated long-term erosion patterns in arable lands. *Hydrological Processes* **18**: 1795–1809.

O'Donnell, G.M. (2008) *Information tracking for flood impact of land use and management change*. PhD thesis, Newcastle University, UK.

Park, S.W., Mitchell, J.K. & and Scarborough, J.N. (1982) Soil erosion simulation on small watersheds: a modified ANSWERS model. *Transactions of the American Society of Agricultural Engineers* **25**: 1581–8.

Parkin, G., O'Donnell, G., Ewen, J., *et al.* (1996) Validation of catchment models for predicting land use and climate change impacts: 2. Case study for a Mediterranean catchment. *Journal of Hydrology* **175**: 595–613.

Quinton, J.N. (1997) Reducing predictive uncertainty in model simulations: a comparison of two methods using the European Soil Erosion Model (EUROSEM). *Catena* **30**: 101–17.

Quinton, J.N. (2004) Erosion and sediment transport. In Wainwright, J. & Mulligan, M. (eds), *Environmental Modelling: Finding Simplicity in Complexity*, John Wiley & Sons, Chichester: 187–96.

Refsgaard, J.C. & Storm, B. (1995) MIKE SHE. In Singh, V.P. (ed.), *Computer Models of Watershed Hydrology*. Water Resources Publications, Highlands Ranch, Colorado, US: 809–46.

Reid, L.M. & Dunne, T. (1984) Sediment production from forest road surfaces. *Water Resources Research* **20**: 1753–61.

Reid, L.M. & Page, M.J. (2002) Magnitude and frequency of landsliding in a large New Zealand catchment. *Geomorphology* **49**: 71–88.

Ritchie, J.C. & McHenry, J.R. (1990) Application of radioactive fallout cesium-137 for measuring soil erosion and sediment accumulation rates and patterns: a review. *Journal of Environmental Quality* **19**: 215–33.

Saxton, K.E., Rawls, W.J., Romberger, J.S. & Papendick, R.I. (1986) Estimating generalized soil-water characteristics from texture. *Soil Science Society of America Journal* **50**: 1031–6.

Sidle, R.C. & Ochiai, H. (2006) *Landslides: Processes, Prediction, and Land Use*. Water Resources Monograph 18, American Geophysical Union, Washington, DC.

Sidorchuk, A. (1999) Dynamic and static models of gully erosion. *Catena* **37**: 401–14.

Sidorchuk, A.Y. & Golosov, V.N. (1996) Calibration of soil-erosion models based on study of radioactive fallout. *Eurasian Soil Science* **28**: 383–95.

Silburn, D.M. & Connolly, R.D. (1995) Distributed parameter hydrology model (ANSWERS) applied to a range of catchment scales using rainfall simulator data I: Infiltration modelling and parameter measurement. *Journal of Hydrology* **172**: 87–104.

Simons, D.B., Ward, T.J. & Li, R.-M. (1979) Sediment sources and impacts in the fluvial system. In Shen, H.W. (ed.), *Modeling of Rivers*, John Wiley & Sons, New York: 7.1–7.27.

Sundborg, Å. (1983) Sedimentation problems in river basins. *Nature & Resources* **19**: 10–21.

Thorne, M.C., Degnan, P., Ewen, J. & Parkin, G. (2000) Validation of a physically based catchment model for application in post-closure radiological safety assessments of deep geological repositories for solid radioactive wastes. *Journal of Radiological Protection* **20**: 403–21.

Trimble, S.W. (1976) Sedimentation in Coon Creek Valley, Wisconsin. *Proceedings of the 3rd Federal Inter-Agency Sedimentation Conference*, Denver, CO, Symposium 5, Sedimentation Committee, Water Resources Council: 5.100–5.112.

Trimble, S.W. (1995) Catchment sediment budgets and change. In Gurnell, A. & Petts, G. (eds), *Changing River Channels*, John Wiley & Sons, Chichester: 201–15.

Vanacker, V., von Blanckenburg, F., Govers, G., *et al.* (2007) Restoring dense vegetation can slow mountain erosion to near natural benchmark levels. *Geology* **35**: 303–6.

Walling, D.E. & Quine, T.A. (1992) The use of caesium-137 measurements in soil erosion surveys. In *Erosion and Sediment Transport Monitoring Programmes in River Basins*, International Association of Hydrological Sciences Publ. No. 210. Wallingford, Oxon: 143–52.

Walling, D.E. & Webb, B.W. (1981) Water quality. In Lewin, J. (ed.), *British Rivers*, Allen & Unwin, London: 126–69.

Ward, T.J., Li, R.-M. & Simons, D.B. (1981) Use of a mathematical model for estimating potential landslide sites in steep forested basins. In Davies, T.R.H. & Pearce, A.J. (eds), *Erosion and Sediment Transport in Pacific Rim Steeplands*. International Association of Hydrological Sciences Publ. No. 132, Wallingford, Oxon: 21–41.

Wicks, J.M. & Bathurst, J.C. (1996) SHESED: a physically based, distributed erosion and sediment yield component for the SHE hydrological modelling system. *Journal of Hydrology* **175**: 213–38.

Wicks, J.M., Bathurst, J.C., Johnson, C.W. & Ward, T.J. (1988) Application of two physically-based sediment yield models at plot and field scales. In Bordas, M.P. & Walling, D.E. (eds), *Sediment Budgets*. International Association of Hydrological Sciences Publ. No. 174, Wallingford, Oxon: 583–91.

Wicks, J.M., Bathurst, J.C. & and Johnson, C.W. (1992) Calibrating SHE soil-erosion model for different land covers. *Proceedings of the American Society of Civil Engineers, Journal of Irrigation & Drainage Engineering* **118**: 708–23.

Wilkinson, B.H. & McElroy, B.J. (2007) The impact of humans on continental erosion and sedimentation. *Geological Society of America Bulletin* **119**: 140–56.

Wu, W. & Sidle, R.C. (1995) A distributed slope stability model for steep forested basins. *Water Resources Research* **31**: 2097–2110.

15 Modelling Impacts of Climatic Change: Case Studies using the New Generation of Erosion Models

J.P. NUNES[1] AND M.A. NEARING[2]

[1]*Centre for Environmental and Marine Studies (CESAM), Department of Environment and Planning, University of Aveiro, Aveiro, Portugal*
[2]*USDA-ARS, Southwest Watershed Research Center, Tucson, AZ, USA*

15.1 Introduction

There is a growing consensus in Earth systems sciences that global temperatures are increasing and will continue to do so during the next century, leading to changes in global climate patterns (IPCC, 2007). Although different regions of the globe could respond differently to global warming, most are expected to suffer significant changes to the amount and variability of rainfall and temperature (Giorgi, 2006), associated with an increase in the frequency of extreme episodes such as heat waves and high-intensity storms (Tebaldi *et al.*, 2006).

These changes have the potential to alter significantly the driving forces and parameters behind soil erosion; examples include changes to rainfall intensity, vegetation cover and surface runoff generation (Kundzewicz *et al.*, 2007). There is therefore a need to quantify the impacts of climate change on soil erosion and on the most important erosion drivers, to estimate on- and off-site consequences, and support the development of adequate adaptation measures. This can be a challenging problem due to the non-linear relationships between soil erosion drivers and processes and the complex interactions between climate change impacts. Erosion models can be useful assessment tools to support climate change studies, since they codify the existing knowledge on soil erosion processes and their response to climate forcing, allowing the quantification of the impacts of changed climate patterns in a feasible and, hopefully, robust way.

The purpose of this chapter is to explore these challenges. It begins by discussing potential impacts of climate change on soil erosion drivers and processes, and potential interactions between them. It then proceeds with a systematization of soil erosion modelling applied to climate change studies, discussing issues such as climate change scenario and model selection, or calibration and validation issues. This is followed by a number of case studies from around the globe which exemplify climate change impact assessment supported by a modelling framework. The chapter concludes with a discussion of research results in this area, current limitations and potential avenues of future research.

15.2 Potential Impacts of Climatic Change on Erosion Processes

Climate change is expected to impact upon a number of soil erosion drivers and processes, which should be taken into account when

Handbook of Erosion Modelling, 1st edition. Edited by R.P.C. Morgan and M.A. Nearing. This chapter © 2011 M.A. Nearing. Published 2011 by Blackwell Publishing Ltd.

designing a modelling strategy. The fourth assessment report of the Intergovernmental Panel for Climate Change (IPCC) (Parry *et al.*, 2007; Solomon *et al.*, 2007) reviews a number of potential changes to soil erosion drivers and processes. This chapter summarizes these impacts as changes to rainfall erosivity, water runoff, vegetation cover and soil erodibility, with a focus on the combined changes caused by desertification.

15.2.1   Rainfall erosivity

One of the most direct impacts of climate change could be an increase in the erosive power of rainfall. In the IPCC's fourth assessment report, Meehl *et al.* (2007) reported that global general circulation models (GCMs) point to an average increase in rainfall of 5% over land masses by 2100, but unevenly distributed, with the high latitudes, the tropics and the monsoon region of southeast Asia experiencing the highest increases (up to +20%) and with the largest decreases over the Caribbean and Mediterranean seas and in the western subtropical coasts of each continent (down to –20%). Rainfall increases are expected to reflect disproportionately in heavy precipitation events, with average rainfall intensity increasing, which is also a trend that has been observed in the global climate record (Groisman *et al.*, 2005; Trenberth *et al.*, 2007). Even in regions where rainfall decreases, this is expected to lead to an increase in the length of dry periods, with rainfall intensity in wet periods increasing; in some of these regions this fact could be particularly significant due to the contribution of rare extreme events for overall soil erosion (e.g. González-Hidalgo *et al.*, 2007). Finally, rainfall increases could accumulate with a shift from snowfall to rainfall due to the warmer climate (Kundzewicz *et al.*, 2007).

As an example of current climate change scenarios for extreme events, Tebaldi *et al.* (2006) analysed historical and future simulations of precipitation extremes indicators by nine GCMs, under a range of emission scenarios. The authors reported a significant global trend of greater pre-

cipitation intensity emerging from model results, although with a high regional, interannual and inter-model variability. The largest and most significant changes were found in days with rainfall over 10 mm, and 5-day maximum rainfall, which indicate more precipitation for a given event, resulting from the greater moisture-holding capacity of a warmer atmosphere and a polewards shift of storm tracks. Precipitation intensity is expected to increase over all land masses, with significant increases in the mid to high latitudes of the northern hemisphere and the tropical regions of Africa and South America. In the subtropical regions, an increase in the number of dry days is coupled with no significant changes to rainfall intensity.

Climate change projections from GCMs should be treated with care, since the non-linear nature of the climate system and natural forcings, compounded with differences in the formulation of different GCMs, causes an intrinsic level of uncertainty in GCM-based climate change predictions (Stott & Kettleborough, 2002; Giorgi, 2005). Nevertheless, the consistency of the predictions among the GCMs as well as with the historical climate record indicates that rainfall erosivity will increase in many regions throughout the globe. Ongoing research focusing on regional climate change predictions and climate extremes (e.g. Hanson *et al.*, 2007) should provide better estimates of the impacts of climate change on rainfall erosivity in the near future.

15.2.2   Water runoff

The estimated impacts of climate change on runoff are more complex than on rainfall. The IPCC's fourth assessment report (Kundzewicz *et al.*, 2007) points to significant changes in river runoff, due to changes in rainfall coupled with an increase in potential evapotranspiration, as higher temperatures increase the atmospheric vapour pressure deficit. Changes to runoff are generally expected to follow changes in rainfall, increasing in high latitudes, southeast Asia and the tropics (where rainfall is expected to increase more than evapotranspiration), and decreasing in the Mediterranean coastline and in the western subtropical regions of each continent. However, the IPCC report and other global studies on climate change impacts (e.g. Wetherald & Manabe, 2002; Nohara *et al.*, 2006) do not separate surface runoff from total runoff; changes to surface runoff have consequences for soil erosion processes at the field and hillslope scales, including gully erosion, while changes to total runoff can impact upon channel erosion and deposition processes and watershed sediment yield.

One of the expected impacts of an increase in rainfall intensity is greater surface runoff generation through infiltration-excess processes, especially when coupled with soil surface crusting (Bronstert *et al.*, 2002). However, soil moisture rates are also an important factor for runoff generation in humid and semi-arid catchments, particularly for low- and medium-intensity storms (e.g. Cammeraat, 2002; Boix-Fayos *et al.*, 2006). The IPCC report (Meehl *et al.*, 2007) points to a global decrease in soil moisture; this is more marked in regions where rainfall decreases, but also predicted to occur in high latitudes despite an increase in rainfall due to the earlier start of snowmelt. Global modelling results obtained by Wetherald and Manabe (2002) and Manabe *et al.* (2004) point to a high seasonal variability of soil moisture changes, which are expected to occur mostly in the spring-to-autumn period. For example, soil moisture in the Mediterranean is not expected to decrease significantly in the winter despite a large annual decrease, while in the mid latitudes, soil moisture is expected to increase in winter and decrease in summer leading to small changes in annual averages.

These changes are likely to have an impact on surface runoff generation and therefore on soil erosion, especially during low- to medium-intensity storms; in general terms, surface runoff generation can be expected to follow overall runoff trends. However, changes to runoff generation processes are expected to have a high degree of spatial variability, with the spatial distribution of soil hydrological properties playing a significant role (e.g. Bronstert *et al.*, 2002; van den Hurk *et al.*, 2005; Nunes *et al.*, 2008). Another important difference could be a shift in the most important runoff generation processes for regions where climatic aridity surpasses desertification thresholds (Cammeraat, 2002); this issue is discussed further below. In summary, the processes linking rainfall, soil moisture and surface runoff generation are non-linear and often particular to a small catchment or region, and therefore it is difficult to generalize impacts at the continental or global scale (Kleinen & Petschel-Held, 2007). It should be noted that most soil erosion modelling studies presented in this chapter also focus on surface runoff generation.

Furthermore, Kundzewicz *et al.* (2007) pointed to a greater irregularity of streamflow throughout the globe, both in seasonal and daily terms, coupled with an increase in flash flood frequency, especially in mid to high latitudes. Higher flow seasonality is expected due to changes in evapotranspiration and seasonal soil moisture patterns, coupled with shifts in rainfall seasonality and, in the mid to high latitudes, by a shorter snow accumulation season and earlier onset of snowmelt (Meehl *et al.*, 2007). These changes, compounded with the increases in rainfall intensity described earlier, can combine to increase significantly the probability of occurrence of large floods (Kundzewicz *et al.*, 2007); in a global study, Kleinen and Petschel-Held (2007) found that up to 20% of the world population could be affected by a significant increase in the occurrence of large-scale inundations of flood plains. Meehl *et al.* (2007) also pointed out the impact of an increase in snowfall extremes on the occurrence of large spring floods. In short, climate change could lead to an increase of high peak flow events in many river basins, with potential impacts on channel erosion processes.

15.2.3 *Vegetation cover*

Climate change is also expected to have complex impacts on both natural and agricultural vegetation, affecting the protection given by canopy cover from the erosive power of rainfall. A review in the IPCC's fourth assessment report (Fischlin et al., 2007) points to changes in vegetation productivity patterns, resulting from the interaction between increased atmospheric CO_2 concentrations, generally leading to increased vegetation productivity, and rising temperatures, whose effect on productivity depends on vegetation species and current adaptability to the local climate. Rainfall decreases and changed soil moisture patterns could also lead to a decrease in vegetation productivity in water-stressed regions. At the global scale, the report points to a global increase in productivity for mild climate change conditions, with vegetation benefiting from CO_2 fertilization and longer growing seasons, followed by negative impacts of more severe climate change scenarios where these benefits are counterbalanced by higher evapotranspirative demands and temperature inhibition to growth.

Ecosystems are expected to dampen the impacts of modest amounts of climate change, but changes above a certain threshold can lead to major transitions or productivity collapses. While shifts in biogeographical regions could be expected, landscape fragmentation associated with evolving human land uses is likely to impede migrations and therefore reduce the potential for natural adaptation to climate change. Furthermore, the increase in climate instability, especially drought frequency and intensity (Meehl et al., 2007), could lead to a greater frequency of vegetation disturbances such as droughts, wildfires or pest outbreaks (Martínez-Vilalta et al., 2002; Mouillot et al., 2002), leading to vegetation mortality and exposing the soil to erosive forces. The IPCC report (Fischlin et al., 2007) points to a potential for extensive forest and woodland decline in mid to high latitudes and the tropics associated with an increase in disturbance regimes.

These changes will also have impacts on agricultural systems, which are particularly vulnerable to soil erosion and land degradation. The IPCC review by Easterling et al. (2007) pointed to a negative impact of even a moderate increase in temperature upon crop yields in low latitudes. In the mid to high latitudes, an increase in crop yields

can be expected with moderate warming, but more severe climate change can have increasingly negative effects. An increase in the frequency of disturbances can also lead to lower overall agricultural productivity. However, it is likely that cropping systems can be adapted to some climate changes as long as new options for cultivation are available, although smallholders and poor farmers in low-latitude regions have significantly less adaptive capacity than farmers in developed countries (Berry et al., 2006; Easterling et al., 2007). Potential adaptations include changes to cropping practices, such as different planting and harvesting dates to adapt to the new growing season, or irrigation rescheduling to adapt to new rainfall patterns; these changes could alter the seasonal patterns of vegetation cover.

Geographical changes in cropping systems could also follow climate shifts. It is difficult to predict the extent of these changes due to the complexity of the processes involved, crossing between biophysical and socio-economic systems; however, combined models exploring interactions between these systems are currently being developed and applied to climate change impact assessment. For example, the modelling approach developed by Wu et al. (2007), combining climate impacts with crop prices, points to an increase in the cropping areas of rice (in Asia and South America) and wheat (in all regions except Oceania) by 2035, while maize shows an uneven trend. This would lead to a combined spatial and temporal shift in vegetation cover patterns, with complex consequences for soil erosion. These changes would be superimposed over other processes with socio-economic motivations, for example the increase in cereal demand for food and biofuel production, leading to the cultivation of more marginal lands (Garbrecht et al., 2007).

15.2.4 Soil erodibility

Climate change can have multiple impacts on the ability of soils to resist erosion. One impact that can be predicted with some certainty is the deeper permafrost thawing in the high latitudes (Meehl et al., 2007), which will expose soil that was previously protected from erosion processes. The changes in vegetation productivity, described in the previous paragraphs, would also lead to changes in ground cover by vegetation residue, with additional impacts from changes in microbial activity driven by temperature and soil moisture availability (Kundzewicz *et al.*, 2007).

The impacts of climate change upon soil organic matter and structural properties are more difficult to estimate, since the processes involved are complex and not completely understood (e.g. Dawson & Smith, 2007). A number of studies along climatic gradients (e.g. Lavee *et al.*, 1998; Boix-Fayos *et al.*, 2001; Sarah, 2006) have demonstrated how soil structural stability is related to soil organic matter content, which varies with temperature and soil moisture in a non-linear way, although with high spatial heterogeneity due to relationships with vegetation cover patterns. The IPCC report points to climate change impacts on soil carbon dynamics, however, due to a combination of changes to vegetation productivity and an increase in soil respiration; this could lead to a global decrease in soil organic carbon content, with soil becoming a net source of CO_2 during the early stages of climate change until a carbon equilibrium is eventually re-established (Fischlin *et al.*, 2007). While there is significant uncertainty in the magnitude of this process, it could lead to a decrease in soil structural stability in many regions.

15.2.5 *Crossing desertification thresholds*

The potential of climate change to aggravate desertification processes in drylands illustrates how the different impacts of climate change on soil erosion drivers can combine and reinforce each other and lead to severe problems of land degradation. Desertification can be defined as the degradation of biophysical and socio-economic conditions in dry regions, leading to land degradation, reduced vegetation productivity and human abandonment (Thornes, 1998; Fernández, 2002). This process usually occurs when biophysical conditions (e.g. soil and climate) are insufficient to support existing natural and socio-economic

systems, due to either overexploitation or a reduction in the carrying capacity for these systems (Puigdefábregas, 1998; Fernández, 2002; Herrmann & Hutchinson, 2005).

As described above, climate change could lead to decreased water availability and increased physical constraints on ecosystem productivity in drylands, especially in subtropical regions and Mediterranean-type ecosystems in the mid latitudes, increasing their vulnerability to desertification (Fischlin *et al.*, 2007). An increase in climatic aridity can surpass a threshold where available water cannot support full vegetation canopy cover; ecosystems adopt strategies to harvest water and nutrients by adopting a pattern of vegetated and bare patches, with the latter acting as runoff and sediment sources for the former, leading to an increase in runoff generation and soil erosion when compared with drylands above the threshold (Bergkamp *et al.*, 1999; Imeson & Prinsen, 2004; Ludwig *et al.*, 2005). This process can be self-reinforcing, as reduced biological activity in the bare patches promotes an increase in runoff generation and a decrease in soil structural stability, resulting in greater erosion and poorer conditions for vegetation support in these patches (Imeson & Lavee, 1998; Yair & Kossovsky, 2002).

This process can be exacerbated by soil erosion, due to a reduction in the soil's capacity to support vegetation when compared with non-eroded soils in a similar climate as a result of nutrient losses and, in severely eroded soils, to a decrease in the soil water-holding capacity (Arora, 2002; Bakker *et al.*, 2004; Boer & Puigdefábregas, 2005). Intensive agricultural, forestry and grazing practices are common in many dryland regions, usually leading to increased soil erosion and land degradation; high market prices, government subsidies and other socio-economic factors can extend these practices to unsuitable regions, such as marginal areas with steep slopes and low water availability, and maintain them even after the onset of land degradation and consequential decreases in crop yield (Martínez-Fernández & Esteve, 2005; Audsley *et al.*, 2006; Vogiatzakis *et al.*, 2006). Furthermore, disturbances common in drylands such as severe droughts and wildfires can increase soil erosion during the subsequent window of vulnerability, while vegetation recovers (e.g. Imeson & Lavee, 1998; Shakesby & Doerr, 2006).

Eventually, degraded regions subjected to these pressures can be pushed beyond their resilience threshold by extreme disturbances, leading to land abandonment (Puigdefábregas, 1998; Martínez-Fernández & Esteve, 2005). This can result in the recovery of natural vegetation and gradual increase in soil quality (e.g. Vicente-Serrano et al., 2004; Martínez-Fernández & Esteve, 2005), but severe land degradation and continuing climatic instability can reinforce desertification processes in abandoned lands, and as described above, the process itself is self-reinforcing (Puigdefábregas, 1998; Bakker et al., 2005). This conclusion is sustained by several observations of increased soil erosion processes, particularly gully erosion and vegetation patchiness in degraded landscapes (e.g. Cammeraat & Imeson, 1999; Oostwoud Wijdenes et al., 1999; Seixas, 2000; Ries & Hirt, 2008). In these cases, land degradation and desertification can be irreversible without extensive human intervention, and misapplied intervention practices (e.g. intensive irrigation, afforestation) can even increase the problem (Thornes, 1998; Puigdefábregas & Mendizabal, 1998; Martínez-Fernández & Esteve, 2005).

In the past, drylands have experienced periods of alternating human expansion and contraction due to changes in climate; humid periods, leading to increased pressure on natural resources, alternated with dry periods, leading to irreversible degradation if pressure is not released before resilience thresholds are exceeded (Puigdefábregas, 1998). As described above, climate change can be expected to bring about a transition into a drier period for these regions, leading to lower support for vegetation productivity and soil structure stability, combined with constant or increasing rainfall intensity and increased frequency of disturbances. The interaction between these three factors is likely to determine the impacts of climate change on soil erosion in drylands (Imeson & Lavee, 1998).

Desertification is an example of how the combined impacts of climate change on vegetation cover and soil properties can interact and reinforce each other, increasing soil erosion in a region where lower rainfall rates might indicate otherwise. This complexity should be taken into account in modelling studies where the aim is to assess accurately the multiple impacts of climate change on soil erosion processes.

15.3 Erosion Modelling Approaches and Climatic Change

The previous section described a number of potential impacts of climate change on soil erosion drivers; this section presents modelling strategies used to assess these impacts for particular problems or locations. Model-based impact assessment studies can evaluate the potential and magnitude of these impacts. In the context of climate change vulnerability assessment frameworks, such as the one proposed by Adger (2006), modelling studies can be useful to:
• assess the sensitivity of soil erosion processes to climate shifts;
• evaluate whether climate change will move soil erosion beyond existing thresholds for desertification, land degradation, or loss of ecosystem services;
• test eventual adaptation measures that can mitigate the impacts of climate change.
Most modelling studies so far have focused on sensitivity assessment, with a small number (e.g. O'Neal et al., 2005) testing adaptation measures. Threshold evaluation studies have been hampered by the difficulty in their delineation, due to the current lack of knowledge on the interactions between soil erosion, soil quality and vegetation support (Herrmann & Hutchinson, 2005; Boardman, 2006). Sensitivity assessment has usually focused on coupled hydrological and erosion prediction, attesting to the fact that an accurate estimate of runoff depth and velocity is at least as important as the correct estimation of other soil erosion parameters (Aksoy & Kavvas, 2005). Modelling exercises have differed in terms of objectives, model used, spatial and temporal extent of the study, and climate change scenario strategy. The processes used to represent

Table 15.1 Example of modelling studies on the impacts of climate change on soil erosion.

Temporal scale	Spatial scale	Climate change scenario	Geographical region	Reference
Continuous	Slope	Downscaling	South America (Amazon river basin)	Favis-Mortlock & Guerra (1999)
			Europe	Mantel *et al.* (2003)
			US	Pruski & Nearing (2002b)
			US (Midwest)	O'Neal *et al.* (2005)
			US (Oklahoma)	Zhang & Nearing (2005)
			China (Yellow river basin)	Zhang & Liu (2005)
		Hypothetical	US	Pruski & Nearing (2002a)
	Catchment	Downscaling	Europe (Finland)	Bouraoui *et al.* (2004)*
			Europe (Britain)	Lane *et al.* (2007)
			Europe (Denmark)	Thodsen *et al.* (2008)
		Hypothetical	US	Istanbulluoglu & Bras (2006)
			US (Iowa and Texas)	Chaplot (2007)
			Europe (Portugal)	Nunes *et al.* (2008)
			China (Yangtze river basin)	Zhu *et al.* (2008)**
Event-based	Slope	Downscaling	Europe (Germany)	Michael *et al.* (2005)
	Catchment	Downscaling	Europe (Portugal)	Nunes (2007)
		Hypothetical	US (Arizona) and Europe (Belgium)	Nearing *et al.* (2005)
			Europe (Portugal)	Nunes (2007)

*Climate in the late 20th century, with variability removed.
**Empirical model using Artificial Neural Networks (ANN).

hydrology and soil erosion in most models (Favis-Mortlock *et al.*, 2001) are similar enough to allow for a comparative analysis, grouping similar models into more general categories. These categories can be matched to a particular approach to climate change impact assessment, and can be a useful guide when devising modelling strategies for other studies.

Table 15.1 summarizes a number of soil erosion studies reviewed in this chapter, grouped according to temporal scale, spatial scale, and climate change scenarios. Almost all studies used models based on a conceptual description of water and sediment sources and sinks, also called process-based models. The terms used for each group follow the model reviews by Aksoy and Kavvas (2005) for the first two categories, and Xu and Singh (2004) for the third. Temporal and spatial scales in the Table refer to the model extent (the period or area of simulation encompassed by the model) rather than the model resolution (the level of discretization with which time or space are

represented). The categories used in the table can be described as follows:
• *Temporal scale*: Modelling studies can be divided into (i) continuous, if the model is applied to consecutive rainfall events occurring during a season or longer period; and (ii) event-based, if the model is applied to a single rainfall event. The processes governing the long-term temporal variability of hydrological and erosion processes are quite different from those operating within an extreme event (Imeson & Lavee, 1998; Favis-Mortlock *et al.*, 2001), which has led to a structural distinction in soil erosion models. Models which operate continuously usually incorporate some sort of vegetation modelling component as well as long-term hydrological processes such as evapotranspiration and subsurface runoff, while in models operating for single events, these parameters are considered to be constant or negligible (Morgan & Quinton, 2001) This scale difference means that continuous models usually do not simulate at the within-event scale, as the

inclusion of these processes would make a model too cumbersome.

• *Spatial scale*: Modelling approaches exist at the (i) slope-scale, representing processes occurring mostly at the field and hillslope scale, such as splash and rill erosion; and (ii) catchment-scale, also representing processes operating in regions of accumulated runoff, such as gully erosion and channel sediment processes (e.g. Lane et al., 1997; de Vente & Poesen, 2005); there is often some juxtaposition between scales. It should be noted that this distinction is between the simulated processes rather than the extent of the area of application; slope-scale models can be applied to catchments without representing other processes. As an example, Mantel et al. (2003) applied the PESERA model to large areas in northern France and Belgium and southern Iberia, but the model did not represent gully erosion and channel processes within these regions and therefore the study was classified as slope-scale.

• *Climate change scenario*: A common problem in climate change studies is the mismatch between GCM results, with higher quality at coarse spatial scales and for annual and seasonal values, and those required for model application, at fine spatial scales and for daily averages (Xu & Singh, 2004). Therefore, GCM results are usually taken as a starting point to generate climate change scenarios using two processes: (i) downscaling of GCM results to the desired spatial and temporal scale; and (ii) hypothetical, where GCM results are used to provide a range of possible changes to climate variables and scenarios are artificially built within these ranges. The most common downscaling methods referred to by Xu and Singh (2004) are (a) dynamic downscaling, where GCM results are used to force regional simulations of climate change at finer spatial and temporal scales using regional climate models (RCM); and (b) statistical downscaling, which uses a statistical relationship between GCM 'control' runs (for current conditions) and the observed climate patterns in a given location to provide future climate scenarios at the desired spatial and temporal scale. The choice of downscaling method can have significant impacts on the

results given by erosion models (Zhang, 2007). In contrast, hypothetical scenarios are usually perturbations of current climate conditions with several degrees of change, aiming to obtain a response function of soil erosion to changes in climate parameters, in effect studying the sensitivity of soil erosion to changes in climate given a reasonable interval (Xu & Singh, 2004).

Selecting between the modelling approaches summarized in Table 15.1 depends upon the overall objectives of the study. It should be taken into account that increasing the complexity of a modelling study – in terms of both process description and spatial and temporal discretization – does not lead to improved results, due, to a great extent, to the uncertainty associated with the input parameters required by complex models which often lead to a greater uncertainty in the results without providing additional predictive power (Jetten et al., 1999, 2003; see also Section 6.4). Therefore, the complexity of the selected approach should match the questions which the modelling study wishes to answer. Two examples of this selection process can be taken from the studies in Table 15.1:

• studies with a continuous modelling approach focus on interactions between climate, vegetation growth and soil erosion at longer temporal scales, while those with an event-based approach focus on non-linear processes such as gully erosion and peak discharge/sediment yield relationships;

• studies at the catchment scale usually focus on within-watershed erosion patterns (e.g. gully erosion and sediment deposition) and channel processes, while studies focusing on soil erosion in agricultural fields constrained simulations to the slope scale.

Other factors should also influence the selection of a modelling approach, such as the dominant erosion processes in the study area or the availability of data for model parameterization and validation (Jetten et al., 1999, 2003; Section 6.4). In some cases, a multiscale modelling framework could be selected, using different models to study different problems with the required degree of complexity; this approach can be exemplified by the work of Nunes (2007), who studied the interaction between vegetation growth (using a continuous model) and soil erosion processes at different spatial scales (using an event-scale model).

As for the selection of a specific model to implement the study, some recommendations by Jetten *et al.* (1999) can be taken into account:

• input data quality (both quantitative measurements and qualitative knowledge of the study area), calibration procedures, and the knowledge of modellers can be more important than model structure for successful erosion simulations (see Chapter 3 for further discussion);

• models usually perform better for the processes and at the spatial and temporal scale they were designed to operate in, so evaluating different model structures and applications can provide an insight on the most appropriate model for a given case study;

• when modelling for changing conditions, process-based models can accommodate processes that do not currently occur, while empirical approaches are constrained by currently operating processes.

Finally, when designing a modelling study, some attention should also be given to calibration and validation strategies for climate change scenarios. Calibration and validation is a complex process, especially for models requiring large amounts of input data. Typical problems include the lack of measured data at the appropriate scale used by the model; parameter equifinality, where different sets of calibrated parameters provide equally good results; and over-calibration, where model parameterization is optimized using an excessively small sample of observations (e.g. Quinton, 1997; Beven, 2000; Boardman, 2006). Since models usually perform best for the range of conditions for which they were calibrated (Favis-Mortlock *et al.*, 2001), calibrating and validating a model for future conditions presents a number of additional problems. For example, calibrated model parameters can have limited transferability in time, particularly in the face of significant changes to climate parameters or watershed conditions (Apaydin *et al.*, 2006). Parameter equifinality can present a similar

challenge, since parameter sets performing equally well for current conditions can lead to significant differences in climate change predictions (Wilby 2005). Toy *et al.* (2002) defined a robust model as one able to perform reasonably well with similar parameter values, including highly dynamic ones, for the widest possible range of conditions; calibration and validation problems for uncertain future conditions call into question the robustness of runoff and erosion models for climate change analysis (Beven, 2000; Morgan & Quinton, 2001).

To address this problem, models used for climate change studies should demonstrate an increased degree of robustness considering both current conditions and those as close to possible changes as achievable. One important strategy to increase robustness is to demonstrate that the model can simulate alterations to hydrological and erosion processes caused by changes in climate; this can be achieved using a 'space-for-time' approach, where the consequences of future climate change are studied using a comparative analysis between one study area and another with climatic characteristics resembling GCM predictions (Imeson & Lavee, 1998). In practice, this strategy can be implemented by calibrating and validating a model for different study areas with different climates and hydrological and erosion processes operating, or by using periods with different climate conditions in the calibration and validation process, especially if these conditions represent in some way the expected climate change scenarios (e.g. Bronstert, 2004; Xu & Singh, 2004). This approach can be further detailed by reproducing climate change in the calibration and validation process; for example, Xu and Singh (2004) proposed that, if the goal is to simulate a drier climate scenario, a model should be calibrated for a wet year and validated for a dry year, thus demonstrating its ability to simulate a wet/dry transition.

Another approach to increase model robustness is multi-process validation, i.e. to calibrate and validate a model for the highest possible number of variables representing different processes occurring at different scales, such as splash,

rill and gully erosion, sediment yield, soil moisture and runoff at different spatial and temporal
scales; this strategy can also address the problem
of parameter equifinality (Ebel & Loague, 2006).
However, this approach requires an increase in
the data used for the calibration and validation
process, which contrasts with the generally poor
availability of data; hydrological data are often
only available for catchment outlets, while erosion data are often not available at all, preventing
a calibration and validation study of this kind
in most catchments (Beven, 2000; Morgan &
Quinton, 2001). To overcome this problem a
qualitative evaluation of model performance can
be used when quantitative data are not available;
this approach consists of comparing model outputs with expected results in terms of process
knowledge to assert the rationality of model
behaviour, and therefore the model's capacity to
simulate responses to changes in environmental
conditions (Favis-Mortlock et al., 2001; Ebel &
Loague, 2006). In this approach, soft knowledge
of the impacts of climate change – from observations in different sites, laboratory experimentation or extrapolation from observations in
different climatic regions – can be useful to judge
model performance under changed climates.

15.4   Case Studies

This section presents a number of case studies to
demonstrate the application of different modelling strategies to different problems. Each case
study includes a number of different studies and
publications (listed in Table 15.1) grouped the-
matically, in order to illustrate how different
modelling approaches were used to answer different questions.

15.4.1   Continuous modelling at the slope scale

This case study reviews work using the WEPP
model – Water Erosion Prediction Project
(Flanagan & Nearing, 1995). WEPP simulates
hillslope processes such as inter-rill and rill erosion, sediment transport and deposition, as well
long-term processes such as vegetation growth,

plant residue generation and decomposition, or
soil consolidation. An additional model feature is
the capacity to simulate agricultural operations
and their impact on soil properties, making this a
good tool to simulate agricultural hillslopes
(although the model also simulates catchment-
scale processes operating in small watersheds
driven by infiltration-excess surface runoff). The
following applications of WEPP focused on the
model's strengths, namely the continuous simulation of cultivated hillslopes.

Several publications have reported on climate
change impact assessment in the US using WEPP
(Table 15.1; for a partial review see Nearing et al.,
2004). A first approach was reported by Pruski and
Nearing (2002a); this comprised applying hypothetical changes to annual rainfall (from –20% to
+20%) for three soils, three slopes, and four crops
in three locations in the US with different climate
characteristics. The results include a ratio of sensitivity to climate change; WEPP indicates a 2.0
ratio of surface runoff increase to rainfall increase,
and a 1.7 ratio of erosion increase to rainfall
increase, showing the enhanced sensitivity of
these parameters to changes in climate.
Furthermore, this ratio also depends on the mode
in which rainfall changes; the ratios reported
above assume that half of rainfall changes are due
to changes in intensity, with the remainder due to
changes in the number of rain days. However,
when the model is applied with rainfall intensity
changes only, the sensitivity ratios increase to 2.5
and 2.4 for runoff and erosion respectively. This
points to the importance of knowing how climate
change may impact upon individual rainfall
events before estimating impacts on soil erosion.

Pruski and Nearing (2002b) also applied the
WEPP model with a climate change scenario
downscaled from the HadCM3 GCM. WEPP was
modified to take into account plant fertilization
by CO_2, and applied to eight locations in the US
with the climate change scenario. The results
point to the complex interactions between different erosion drivers, particularly rainfall and vegetation biomass production; rainfall changes were
often not the dominant impact on soil erosion.
The importance of different drivers changed with

location and was complicated by changes to sea-sonal climate patterns. However, one trend emerged from this work: in the US, erosion can be expected to increase where rainfall increases significantly, but where rainfall decreases the impacts are more complex and erosion can either increase or decrease, depending upon the interac-tions between the impacts of plant biomass and rainfall on erosion.

Finally, this work also focused on the impacts of adaptations to climate change on soil erosion. O'Neal *et al.* (2005) studied the combined impact of changes to climate and crop management on soil erosion in the Midwestern US using WEPP, with a similar climate change scenario to the one described above, and downscaling results using a stochastic weather generator, CLIGEN (using a method similar to the one described by Zhang *et al.*, 2004). Management practices were adapted to fit the climate scenarios by adjusting planting, tillage and harvesting dates, and changing crop rotations; the scenario used a future shift from maize and wheat to soybeans. Model results point to an increase in soil erosion between 33% and 274% by the 2050s in most of the study areas; the increase in erosion can be attributed to higher rain-fall, later planting dates leaving the soil exposed for longer, and shifts towards greater cultivation of soybeans. Vegetation changes led to more erosion even in regions with lower rainfall. Zhang and Nearing (2005) used WEPP to study the impacts of three climate change scenarios (A2a, B2a and GGa1) on soil erosion in central Oklahoma. The climate scenarios were downscaled from HadCM3 predictions for the 2070s, also with CLIGEN, and predicted less rainfall and higher temperatures. However, WEPP predicted an increase in soil ero-sion of between 18% and 82% due to the combined impacts of higher rainfall variability (resulting in increased frequency of large storms) and, in some scenarios, a decrease in wheat yield. The authors also studied the impacts of adopting conservation tillage and no tillage to counteract soil erosion increase, with model results indicating their effec-tiveness as adaptation measures.

A similar approach was subsequently applied to the Yellow River basin in China, focusing on the Loess plateau drylands, a region which already experiences high levels of soil erosion (Zhang & Liu, 2005) and where climate change is expected to increase rainfall and, in particular, rainfall ero-sivity (Zhang *et al.*, 2005). High soil erodibility and different climate and cropping systems pre-sented different challenges to this work. Zhang and Liu (2005) applied the WEPP model to two slopes in this region, using a stochastic weather generator (CLIGEN) to downscale three climate change scenarios (A2a, B2a, and GGa1) from the HadCM3 GCM for the 2080s. The results point to an increase in soil erosion of between 2% and 81% despite a significant increase in crop yield; in this region the rise in rainfall was the domi-nant driving force for soil erosion changes. The authors also concluded that the adoption of con-servation tillage could be sufficient to adapt to climate change and reduce the negative impacts on soil erosion. In a subsequent work, Zhang (2007) tested the impacts of different downscal-ing methods on WEPP predictions. The downs-caling approach described above was refined by introducing an intermediate step, where GCM results were first downscaled spatially using cur-rent climate data for local stations, using a trans-fer function; the spatially downscaled results were then used to drive the stochastic weather generator. Using this approach, the author reached soil erosion predictions of 4 to 10 times higher than previously. These results point to the impor-tance of correctly downscaling GCM predictions when studying the impacts of climate change on soil erosion.

Finally, the WEPP model was applied to hillslopes cultivated with soya on a tropical hillslope in Brazil (Favis-Mortlock and Guerra, 1999). Future climate scenarios were taken from three GCMs (HADCM2, CSIRO9 Mk2 and ECHAM3TR) for 2050; two of the models predict a large increase in summer rainfall, while the third points to a slight decrease. These scenarios were also downscaled using a statistical approach based on CLIGEN. WEPP predicted changes to soil erosion from –9% to +55%, following rela-tively modest changes in rainfall (–2 to +10%), a result also of increased water stress during the growing period for soya leading to less vegetation cover. This was accompanied by an increase in the spatial and temporal variability of erosion.

Overall, these studies provide a good example of a comprehensive modelling approach to evaluate the impacts of climate change on soil erosion. They are also representative of a type of early modelling study (e.g. Favis-Mortlock & Boardman, 1995) not discussed in this section. The results for the US cover the range of possible impacts using hypothetical climate change scenarios (Pruski & Nearing, 2002a) and evaluate interactions between changes in rainfall, runoff and vegetation productivity for different climatic conditions (Pruski & Nearing, 2002b), as well as the impacts of agricultural land use changes (O'Neal et al., 2005) and the potential to implement adaptation measures (Zhang & Nearing, 2005). In China, the uncertainty inherent in climate change predictions was also explored (Zhang, 2007). The main results point to the complex interactions between different impacts of climate change, which can lead to increases in soil erosion even where rainfall is expected to decrease; the importance of vegetation biomass productivity in these regions is highlighted. However, these studies were constrained to the slope scale and agricultural fields; possible impacts on rangelands, gully erosion or catchment sediment yield were not studied, and should not be inferred from the results due to the complex nature of the processes involved.

15.4.2 Continuous modelling at the catchment scale

Another model used to assess the impacts of climate change on soil erosion has been SWAT – Soil and Water Assessment Tool (Neitsch et al., 2002). Like WEPP, SWAT simulates hillslope erosion processes, sediment transport and deposition, vegetation growth and residue processes, and agricultural operations. Unlike WEPP, however, SWAT was designed to simulate mesoscale catchments, trading detail at the slope scale for the ability to represent more complex catchment and river network structures, including large

reservoirs and irrigation schemes. The following SWAT applications were focused on processes linking hillslopes to the river network to take advantage of these features. In contrast with the single-slope WEPP applications detailed above, SWAT was applied to complex catchments.

The first example is an application of SWAT to a watershed in Finland by Bouraoui et al. (2004). In contrast with the usual approach for climate change impact assessment, the authors looked at the impacts of recent climate change (1965 to 1998) on river flow and sediment yield, by comparing model results with observed climate and a synthetic climate series where rainfall and temperature increases were removed using nonparametric methods. The model results pointed to an increase in winter runoff and suspended sediment caused by a combination of increasing rainfall and decreasing snow cover. These results indicate the likely trend for future climate change impacts in this region. Another example is given by Chaplot (2007), who applied SWAT to two watersheds in the US, one with a humid climate and agricultural land (in Iowa), and the other with a semi-arid climate and a significant proportion of pasture (in Texas). The author simulated the impacts on runoff and soil erosion of two CO_2 and temperature change scenarios, combined with rainfall changes from –40% to +40%; the scenarios were stochastically generated using CLIGEN. Model results point to a dominant impact of rainfall in soil erosion rates at the humid watershed, with the wettest scenario leading to an increase of 157%, and similar decreases for the drier scenarios. Soil erosion in the semi-arid watershed, however, did not show great sensitivity to changes in climate, except by decreasing for the –40% rainfall scenario; surprisingly, soil erosion decreased in the wetter scenarios, possibly due to improved conditions for vegetation cover in the winter.

A final example is given by Nunes et al. (2008), who applied SWAT to two groups of watersheds in Portugal, also with humid and semi-arid climates. Climate change scenarios were generated using CLIGEN with the intention of simulating the range of climate change predictions by GCMs for this area, with temperature increases from 2°C to 6°C and rainfall decreases from 2.5% to 40%. The authors simulated changes to each parameter separatcly; they also simulated two sets of scenarios of combined rainfall and temperature change with wetter and drier conditions (rainfall decreasing by 1.6% and 6.2% respectively per 1°C increase in temperature). The results point to rainfall changes as the main driving forces for soil erosion in all landcovers except in wheat croplands, where temperature increases were more important due to the negative impact on biomass production and soil cover. For the combined changes, model results varied significantly between vegetation cover types; rangelands and managed forests showed a decrease in soil erosion in all scenarios, while agricultural lands (wheat croplands and vineyards) responded differently according to the combination of rainfall and temperature changes. For the drier scenarios, soil erosion decreased in both agricultural landcovers; for the wetter scenarios, soil erosion decreased slightly in vineyards (–25%) and increased in wheat croplands (up to 149%). These results are important since these landcover types represent the most important sediment sources in the study area. The authors also found greater responses in the humid watershed, as in the previous study, and note that in one of the test sites the shallow soils (c. 10 mm) were responsible for a relatively low sensitivity of surface runoff to rainfall decreases.

In short, these studies are not as comprehensive as those presented in the previous section, but they do provide additional information at the large catchment scale. The simulated areas are quite large (up to 1000 km² in all studies), and include different vegetation cover and soil types, which impact differently upon similar climate change scenarios; these different impacts combine to determine changes to watershed sediment yield. In particular, these results confirm one of the conclusions of the previous case studies, namely that the impact of decreasing rainfall rates on soil erosion is complex and depends upon the impact on vegetation biomass growth; however, these results also indicate that croplands in

drying climates are particularly vulnerable to increases in soil erosion rates.

A final note should be made on the robustness of both this and the previous modelling approaches. While the SWAT approach has a larger spatial domain, there was in all cases a lack of data for validating the erosion simulations; the model was assessed using sediment yield measurements in the channel network, which is not sufficient to ensure that sediment sources and sinks are being correctly simulated (Jetten *et al.*, 2003). The WEPP approach was more robust, since it was applied to heavily monitored slopes with data available for calibration and validation. This difference illustrates how the lack of measured erosion data may hamper climate change impact studies using watershed-scale models.

15.4.3 Grid-based continuous modelling

A different example of erosion modelling at the slope scale was performed using the PESERA – Pan-European Soil Erosion Risk Assessment model (Kirkby *et al.*, 2004). PESERA simulates erosion at the slope scale in a similar way to SWAT and WEPP, including a vegetation growth component; however, the model has been applied using a 1×1 km grid for western and central Europe, taking into account the spatial variability of rainfall, relief, soil and vegetation properties. This grid-based approach allows an estimation of soil erosion for large areas while taking into account some degree of spatial variability, although the processes represented are still at the slope scale (e.g. gully erosion and channel processes are not taken into account by PESERA). This approach could also be used to generate global predictions for the impacts of climate change on soil erosion, similar to the work done for surface runoff in recent years (e.g. Manabe *et al.*, 2004).

This approach was performed by Mantel *et al.* (2003) using the A2b climate scenario based on the HadRM3 RCM. The study was performed for two windows in northwestern and southwestern Europe, with contrasting climates (humid and dry, respectively) and land uses. Land-use change scenarios (switching other arable crops to maize) were also assessed. The results for northwestern Europe indicate an increase in soil erosion ranging from 1 to 15 t ha^{-1} y^{-1}, due in a large part to an increase in winter and spring rainfall, but with very significant spatial variability which can be partially explained by different topography and land-use patterns. In southwestern Europe, the results point to a change in the spatial patterns of soil erosion, leading to an increase in the area for which significant erosion risk is expected (especially sparsely vegetated areas), coupled with a great decrease in soil erosion rates for the rest of the study area. Nevertheless, the overall erosion rates remain low, of the order of 0.5 to 1 t ha^{-1} y^{-1}. No significant differences were found for the land-use change scenarios. These results illustrate how a grid-based approach can analyse the superimposition of different spatial patterns – climate, topography, land use and soil – which when combined lead to complex patterns of soil erosion change that do not directly correlate with the spatial patterns of climate change.

This study also illustrates an important limitation on soil erosion model application: the lack of long-term soil erosion measurements at the slope scale (Boardman, 2006). Van Rompaey et al. (2003) evaluated the PESERA model for several areas in Europe; while the model gives acceptable results for agricultural areas in northern and central Europe, results for Mediterranean Europe have a poor correlation with estimates based on sediment yield. Part of this uncertainty is due to the difficulty in accurately determining the relationship between hillslope erosion and sediment yield due to the importance of gully and channel erosion processes in Mediterranean regions. Again, the lack of erosion data at the field scale makes a grid-based approach less robust than the modelling studies on highly monitored slopes described earlier; in many regions there are insufficient data to assess the robustness of upscaling model results from the slope to the regional scale.

15.4.4 Modelling channel processes

A number of studies have also focused specifically on suspended sediment transport and deposition in channels, which requires different models from the ones used in the previously described applications. Thodsen et al. (2008) looked at the impacts of combined land-use and climate change scenarios (A2 scenario from the HIRHAM RCM) on suspended sediment transport in two Danish rivers, using sediment rating curves adjusted for rainfall, runoff and season (to take into account vegetation cover inside the catchment). They found that a warmer and rainier scenario led to increases of 9% to 27% for the 2080s, mostly due to increased river flow in winter with greater sediment transport capacity; the longer growing season for annual crops had a minor impact on these predictions. Finally, the non-alluvial river was more sensitive to changes in climate than the alluvial river, possibly due to the greater irregularity of the flow regime.

A different approach was tested by Zhu et al. (2008), who used artificial neural networks (ANNs) instead of the process-based modelling approaches that dominate climate change impact studies; ANNs, while constrained to observations under current climate patterns, may take into account interactions between climate, hydrology and soil erosion not represented in models. The authors tested hypothetical scenarios of rainfall changes, from –20% to +20%, combined with temperature changes from –1 to +3°C, evaluating the impacts on sediment flux in the Yangtze river basin, China. The ANN predicts higher sediment flux with higher rainfall, while for higher temperatures the response is more complex, with lower runoff and therefore lower sediment flux, but higher sediment concentration in rivers thanks to increased soil erosion and sediment delivery. A combination of warmer and wetter climates is expected to lead to greater sediment flux due to higher soil erosion and sediment transport in the study area.

Another modelling approach was applied to US catchments by Istanbulluoglu and Bras (2006). They applied a river sediment dynamics model with a stochastic model linking rainfall amount and frequency, soil moisture and within-catchment vegetation dynamics to study the relationship between climate, vegetation cover and soil loss potential. The results indicate that soil erosion is not only dependent upon changes to mean annual rainfall, but also that an increase in soil erosion can be expected under lower storm frequency, especially in humid catchments, when considering lower rainfall rates. The relationship between climate and soil loss appears to be controlled by soil texture characteristics in shape and magnitude. This is one of the few studies focusing specifically on the impacts of drought frequency and length on soil erosion processes within the context of climate change.

Finally, Lane *et al.* (2007) used a more detailed modelling approach to study the feedback between impacts of climate change on flood frequency and sedimentation in Britain. The modelling framework included a detailed inundation model coupled with an estimate of channel aggradation from suspended sediment deposition; the model was applied using the A2 climate change scenario for the 2050s and 2080s using the HADRM3 RCM. Results point to an increase in inundated area, due to rainfall changes alone, of 12.2% to 14.7% during relatively frequent floods (1-in-0.5 and 1-in-2 year events). When considering also the impact of sedimentation in the river bed, the inundated area increases by 38.2% to 52.1%. The results indicate that in-channel sedimentation increases the sensitivity of flood inundation to climate change, and measures to prevent streambank erosion might aggravate this problem as the river would require enlargement to compensate for the rising channel bed. This study highlights one possible off-site effect of increased soil erosion rates, a subject which has not received much attention in recent research.

15.4.5   Modelling extreme events

The previous examples focused on long-term continuous modelling; there are fewer case studies specifically focusing on individual extreme events. This can be attributed in part to the complexity of the processes involved, particularly when compared with the low spatial and temporal resolution of current climate prediction approaches. Furthermore, studies at the extreme event scale are dependent on longer-term predictions for vegetation cover and land use.

Nevertheless, there are a number of examples in the literature. One case study is the Soil Erosion Network's model intercomparison exercise (Nearing *et al.*, 2005), aimed at investigating the response of different soil erosion models, with different methods and levels of process representation, to key variables expected to be impacted by climate change: precipitation and vegetation. Seven different models were applied to two watersheds, one humid (in Belgium) and one semi-arid (in Arizona, US); they utilize different approaches to erosion process description, and temporal and spatial discretization, and include continuous as well as event-based models. This modelling approach was used to study the response of three storms per catchment to hypothetical changes in storm rainfall, vegetation cover and ground cover, from –20% to 20%.

The model response to these changes was coherent, with all models responding more strongly to changes in rainfall. The median ratio of sensitivity of sediment yield to rainfall changes was around 8 in the humid catchment, and around 5 in the semi-arid catchment, with models responding more strongly to changes in rainfall amount and intensity than to changes in rainfall amount alone. For vegetation and ground cover, the sensitivity was around –2 in both catchments. The coefficients of variation between models are significant, but most models responded within a similar range, especially for stronger storms; these results indicate that the tested models give, in relative terms, similar responses to climate forcings, which increases the credibility of the different modelling approaches.

Another extreme event study was made by Nunes (2007), who applied the MEFIDIS model (Nunes *et al.*, 2005) to one humid and one semi-arid catchment in Portugal. MEFIDIS is a model optimized for extreme events, with a high process discretization in space and time. The author performed a first approach using hypothetical scenarios of rainfall and vegetation cover change, similar to the one presented above. The results for sediment yield sensitivity were similar, but the author also analysed differences in erosion response with spatial scale. Catchment-scale sediment yield and gully erosion were found to be more sensitive to changes in rainfall than erosion at the slope scale, due to non-linear relationships with changes to surface runoff rates and channel peak flow rates; however, changes to vegetation cover had similar impacts on soil erosion at all scales. The author also analysed the impact of changes to soil water deficit (–20 to +20%), which can have important consequences for water and sediment connectivity between hillslopes and the river network. While erosion at the slope scale showed a relatively low sensitivity to this parameter, gully erosion and sediment yield were significantly more sensitive. Finally, the author noted the relationship between low soil depth and increased sensitivity of soil erosion to climate change, as a low water-holding capacity of soils increases the response of water runoff generation to storm rainfall characteristics.

Nunes (2007) also applied the MEFIDIS model for the A2 and B2 climate change scenarios, downscaled with the PROMES RCM; scenarios for changes to vegetation cover and average soil water deficit were created by applying the SWAT model (described above). For the study areas, the scenarios combine higher temperatures (+2 to +4°C) with lower rainfall (–30 to –40%) and higher rainfall intensity in extreme events (10 to 20% in the A2 scenario, for winter and spring). The SWAT model indicates that this would lead to slightly higher vegetation cover (from c. 5% for wheat and vineyards to c. 30% for Mediterranean oaks and shrubs) and lower soil water content (–40% in winter and spring, –80% in autumn). These scenarios combine to cause different impacts on soil erosion according to climate scenario and season. Soil erosion is expected to decrease in both catchments (–20 to –60%), due to the combined impacts of higher vegetation cover, lower soil moisture and modest increases in rainfall intensity. However, this decrease is more marked for gully erosion and sediment yield due to the lower water and sediment connectivity; at the slope scale, erosion decreases are expected to be halved (–10 to –30%). Also, this decrease is mostly under forests and Mediterranean vegetation cover types; wheat croplands experience little or no reduction in soil

erosion rates for the A2 scenario. These changes are coupled with increased seasonal differences in soil erosion; in the humid catchment, erosion is expected to increase by 20% in winter and spring. These results highlight how different soil erosion processes have different responses to climate change, and the role of soil moisture in determining changes to sediment connectivity as well as to soil erosion, an issue which is not explored in most studies. However, an analysis of these results must take into account the lack of data to validate gully erosion simulations in the study watersheds, a problem which is widespread in erosion studies (Boardman, 2006).

A final example of extreme event modelling approaches was given by Michael et al. (2005), who applied the EROSION 2D model (Schmidt, 1990) to two agricultural slopes in Germany. The authors used GCM scenarios for the B2 emission scenario, 2030–2050, downscaled to 5 min rainfall data using a statistical approach driven by prevailing weather types; the scenario points to an increase of 23% in the intensity of the most extreme events, but a 38% decrease in frequency. The results point to an increase in soil erosion between 22% and 66%, but it should be noted that changes to vegetation cover were not taken into account.

Overall, these studies highlight the complexity and potential of an event-based modelling approach for climate change impact assessment. The models are often difficult to parameterize and evaluate, and require both high-resolution climate change scenarios and predictions for longer-term impacts (in these examples, vegetation cover and soil moisture). However, predictions can be made for within-storm processes dependent upon surface runoff concentrations (gully erosion) or peak flow rates (sediment transport) with more detail than that achieved by continuous models.

15.5 Conclusions, Limitations and Research Needs

The case studies presented in the previous section are representative of the typical modelling approaches used to study the impacts of climate change on soil erosion. It should be noted that the most complete studies were applied using continuous models at the slope scale, possibly due to the availability of data to evaluate these models; studies at the watershed scale and using extreme event models are rarer and appear, in comparison, to be less developed. Despite the higher number of slope applications, current modelling approaches are still limited in space and scope, and therefore it is difficult to extrapolate the results to more general conclusions. It can perhaps be said that this branch of climate change impact science is not yet fully developed, and that modelling approaches still need further testing, refinement and discussion before robust results can be presented.

However, one overall conclusion indicated by the results of the different studies is that the relationship between soil erosion and climate change is complex and depends upon a number of impacts highlighted at the start of this chapter. Furthermore, soil erosion processes are themselves highly sensitive to changes in driving forces, making it difficult to exclude complexity from the analysis without invalidating the main conclusions. These issues should be taken into account when designing a modelling approach to be applied in a particular case study. Two broad conclusions emerging from this work relate the patterns of changes in climate with erosion response, at least at the slope scale:

• where rainfall is expected to increase significantly, this dominates erosion response; and
• where rainfall is not expected to change or is expected to decrease, more complex processes take hold, with the dominant processes involving a relationship between changes to rainfall and vegetation biomass.

The results are not sufficient for a quantitative estimate of these impacts, and there are still many knowledge gaps surrounding these estimates, especially when transferring results from slope-scale studies to larger scales. Some of these gaps are related to more general issues in soil erosion science, such as the lack of data and the uncertainties surrounding estimates in erosion magnitude, location of hotspots, on- and off-site impacts and conservation measures (Boardman,

2006). The knowledge gaps on the impacts of climate change on soil erosion can be, in broad terms, systematized in a few questions:

• Can we upscale model results at the individual hillslope and watershed scales to regional and global scales?
• What is the uncertainty surrounding the estimates?
• Which are the links and feedbacks between soil erosion and land use/land cover that can be affected by climate change, and which adaptation measures can be taken?

15.5.1 Upscaling results to the regional and global scales

Most of the studies presented earlier in the chapter focus on single hillslopes or, at most, watersheds. While these applications have been extremely useful to increase our understanding of the processes behind the impacts of climate change, one can question whether the results can be upscaled. Soil erosion is a phenomenon with high variability in space and time, and different processes intervening at different scales. The PESERA study (Mantel *et al.*, 2003) highlighted the high spatial variability of climate change impacts on soil erosion, even at the hillslope scale; the within-watershed results by Nunes (2007) showed how impacts may vary with spatial scale. Both studies have been hampered by the lack of data to evaluate the modelling approach used.

These issues, especially the lack of erosion data, appear to limit the feasibility of grid-based impact assessment studies at the regional or global scale, such as those currently done for surface runoff (e.g. Nohara *et al.*, 2006). The lack of regional or global-scale erosion estimates for current conditions (Boardman, 2006) should be overcome before attempting to upscale climate change impacts. Furthermore, the studies presented in this chapter are not evenly distributed throughout the globe. Most studies have focused on the mid-latitudes, with temperate humid and semi-arid climates; subtropical regions and the high latitudes are poorly represented by comparison, which could limit the understanding of particular interactions between climate change, runoff, vegetation cover, and so on, which are required to upscale model predictions in these regions. It can therefore be argued that more slope-scale studies are still needed in order to increase our understanding of the processes linking climate change and soil erosion, thereby increasing our confidence in subsequent upscaling exercises.

15.5.2   Uncertainty in climate change impact estimates

Knowing the uncertainty of a climate change impact assessment is necessary to provide robust adaptation measures, i.e. measures that can be expected to provide acceptable results under a large range of conditions. While sources of uncertainty in soil erosion studies have been discussed since the first modelling experiments (e.g. Favis-Mortlock & Guerra, 1999), published studies usually only quantify uncertainty due to model errors; even the assessment of this kind of uncertainty may be hampered by the lack of soil erosion data, an issue which is a broader problem in soil erosion modelling (Boardman, 2006).

An example of other sources of uncertainty which can significantly hinder modelling results can be taken from the water resources sciences (e.g. Dessai & Hulme, 2007). These include uncertainties in: greenhouse gas emission scenarios, different GCM results for a similar emission scenario, downscaling (particularly important for extreme event predictions), model results for surface runoff generation and vegetation response, and future land-use changes; these uncertainties may propagate through the modelling approach. Some of these sources of uncertainty have been taken into account in the studies presented in this chapter, for example by driving erosion models with outputs from different GCMs or different climate change scenarios; the remaining sources of uncertainty are rarely taken into account. While progress has been made on quantifying and reducing these uncertainties, such as the PRUDENCE project (Déqué et al., 2005) which provided downscaled climate change estimates for Europe, soil erosion is downstream from a number of climate change

studies (e.g. changes to the hydrological cycle and vegetation patterns) and therefore this issue is likely to remain a problem in the near future.

Furthermore, it is doubtful if some sources of uncertainty can be quantified since they can be defined as 'deep uncertainties': processes which are not yet fully understood (Dessai & Hulme, 2007) and therefore not well simulated. The studies shown in this chapter focus mainly on long-term hillslope erosion rates and, to a lesser degree, channel processes. Gully erosion and sediment connectivity issues are poorly represented; however, the current lack of data and knowledge on these processes has limited their integration in soil erosion models (Boardman, 2006), which may severely hamper progress in this area. Disturbances are also poorly represented in these studies. While continuous models can represent the impacts of drought on vegetation cover, other important impacts of severe droughts – such as woody plant mortality, changes in vegetation patterns, desertification – are not well described; this issue might potentially be addressed by associating erosion models with more detailed vegetation models. The impacts on soil erosion of increased wildfire frequency are also poorly represented; in this case, while links between climate change and wildfires have been studied, the impacts upon soil erosion are still not well understood (Shakesby & Doerr, 2006) and require additional research before being included in climate change impact studies. Finally, the relationships between climate change, soil carbon processes and erodibility have been studied mainly through space-for-time approaches (e.g. Lavee et al., 1998), as described earlier in this chapter. Soil structure processes are poorly represented in erosion models, and some attention to this issue is required in order to obtain improved predictions, including more data on climate–soil structure relationships.

15.5.3   Links and feedbacks between erosion and land use/land cover

Links between soil erosion, land use/land cover and socio-economic issues are also barely touched by the studies presented in this chapter. Future land-use changes are not usually considered in climate change studies, but existing scenarios can be taken into account by models, as shown in a few examples in this chapter. Feedbacks between soil erosion, agricultural productivity and land-use changes (e.g. Bakker *et al.*, 2004, 2005; Avni, 2005) are less well known and more difficult to take into account; for example, Nunes (2007) suggested that in regions of degraded soils, long-term soil erosion might have as much impact on vegetation productivity as changes in climate, with a possible feedback relation with soil erosion. In particular, one poorly understood subject is the relationship between desertification and climatic or land degradation thresholds; more data and research on this issue would allow its inclusion in soil erosion studies. This will be particularly important when the focus is on semi-arid 'threshold regions', with dry climates tending to become drier. One possible approach to consider these interactions is to link physical erosion models with socio-economic models; however, interdisciplinary models of this nature are still rare, and existing approaches (e.g. Wu *et al.*, 2007) are still not developed enough to be combined with relatively complex erosion models.

Furthermore, most of the studies do not consider adaptation to climate change. The studies that did tested different adaptation methods (e.g. conservation and no tillage) to assess whether they are efficient tools to counterbalance the negative impacts of climate change. A further issue that needs to be addressed is the interaction between measures to adapt to different climate change impacts. For example, the study by O'Neal *et al.* (2005), reported above, shows that agricultural adaptations to future climate aimed at increasing productivity might also lead to increased erosion. On the other hand, Lacombe *et al.* (2008) showed how the adoption of extensive soil and water conservation measures in the past has led to a decrease in available water downstream. These issues need to be addressed, particularly at the watershed scale and in light of proposing integrated watershed management methods to adapt to multiple impacts of climate change on water resources, floods, agricultural productivity, soil erosion, nutrient exports, and so on.

These and other limitations must be addressed before the impacts of climate change can be evaluated with some measure of confidence. Some ongoing research projects are proposing modelling strategies that take these issues into account; one such example is the ongoing MESOEROS21 project, aiming to study the impact of global climate change on soil erosion in the Mediterranean (MESOEROS21, 2006). The modelling approach takes into account soil erosion drivers such as land-use changes, intensification of irrigation and desertification, as impacted by climate change, as well as direct impacts on soil erosion; in some cases, more complex vegetation models are used to generate vegetation cover scenarios. Changes to erosion processes are explored with complex models in intensively monitored small catchments, but results are also upscaled to the regional scale using simpler models validated with long-term soil erosion databases for the region. Finally, this project also studies the vulnerability to soil erosion – the expected impact of changes to soil erosion rates on soil water storage, crop productivity, and so on. MESOEROS21 shows an example of how more complex, and hopefully more robust, approaches can tackle a number of the research gaps outlined above.

In summary, this chapter has hopefully shown how climate change can impact soil erosion through a number of processes, many of which are not linearly dependent upon changes to rainfall patterns. Current soil erosion modelling approaches have been developed and applied to test the impacts of climate change for different case studies involving different climate scenarios and locations, providing an insight into the processes linking climate and soil erosion. However, a significant number of research gaps are still present, including the upscaling of results for larger spatial scales; uncertainties in climate change scenarios and their impacts, particularly on soil erosion drivers not fully taken into account by current models; and the links between climate change, soil erosion and land-use changes involving socio-economic as well as biophysical processes. More complex modelling approaches can be developed to address these limitations; however, some effort in other areas such as data collection and process understanding are still needed to improve our knowledge of how soil erosion can be affected by changes in climate. This issue is critical to provide robust adaptation measures, especially when considering that global change is currently one of the greatest challenges for soil and water conservation (Garbrecht *et al.*, 2007).

References

Adger, W.N. (2006) Vulnerability. *Global Environmental Change* 16: 268–81.

Aksoy, H. & Kavvas, M.L. (2005) A review of hillslope and watershed scale erosion and sediment transport models. *Catena* 64: 247–71.

Apaydin, H., Anli, A.S. & Ozturk, A. (2006) The temporal transferability of calibrated parameters of a hydrological model. *Ecological Modelling* 195: 307–17.

Arora, V.K. (2002) The use of the aridity index to assess climate change effect on annual runoff. *Journal of Hydrology* 265: 164–77.

Audsley, E., Pearn, K.R., Simota, C., *et al.* (2006) What can scenario modelling tell us about future European scale agricultural land use, and what not? *Environmental Science & Policy* 9: 148–62.

Avni, Y. (2005) Gully incision as a key factor in desertification in an arid environment, the Negev highlands, Israel. *Catena* 63: 185–220.

Bakker, M.M., Govers, G. & Rounsevell, M.D.A. (2004) The crop productivity–erosion relationship: an analysis based on experimental work. *Catena* 57: 55–76.

Bakker, M.M., Govers, G., Kosmas, C., *et al.* (2005) Soil erosion as a driver of land-use change. *Agriculture, Ecosystems and Environment* 105: 467–81.

Bergkamp, G., Cerda, A. & Imeson, A.C. (1999) Magnitude-frequency analysis of water redistribution along a climate gradient in Spain. *Catena* 37: 129–46.

Berry, P.M., Rounsevell, M.D.A., Harrison, P.A. & Audsley, E. (2006) Assessing the vulnerability of agricultural land use and species to climate change and the role of policy in facilitating adaptation. *Environmental Science and Policy* 9: 189–204.

Beven, K. (2000) *Rainfall-Runoff Modelling – The Primer*. John Wiley & Sons, Chichester.

Boardman, J. (2006) Soil erosion science: reflections on the limitations of current approaches. *Catena* 68: 73–86.

Boer, M.M. & Puigdefábregas, J. (2005) Assessment of dryland condition using spatial anomalies of vegetation index values. *International Journal of Remote Sensing* 26: 4045–65.

Boix-Fayos, C., Calvo-Cases, A., Imeson, A.C. & Soriano-Soto, M.D. (2001) Influence of soil properties on the aggregation of some Mediterranean soils and the use of aggregate size and stability as land degradation indicators. *Catena* 44: 47–67.

Boix-Fayos, C., Martínez-Mena, M., Arnau-Rosalén, E., *et al.* (2006) Measuring soil erosion by field plots: Understanding the sources of variation. *Earth-Science Reviews* 78: 267–85.

Bouraoui F., Grizzetti, B., Granlund, K., *et al.* (2004) Impact of climate change on the water cycle and nutrient losses in a Finnish catchment. *Climatic Change* 66: 109–26.

Bronstert, A. (2004) Rainfall-runoff modelling for assessing impacts of climate and land-use change. *Hydrological Processes* 18: 567–70.

Bronstert, A., Niehoff, D. & Bürger, G. (2002) Effects of climate and land-use change on storm runoff generation: present knowledge and modelling capabilities. *Hydrological Processes* 16: 509–29.

Cammeraat, L.H. (2002) A review of two strongly contrasting geomorphological systems within the context of scale. *Earth Surface Processes and Landforms* 27: 1201–22.

Cammeraat, L.H. & Imeson, A.C. (1999) The evolution and significance of soil-vegetation patterns following land abandonment and fire in Spain. *Catena* 37: 107–27.

Chaplot, V. (2007) Water and soil resources response to rising levels of atmospheric CO_2 concentration and to changes in precipitation and air temperature. *Journal of Hydrology* 337: 159–71.

Dawson, J.J.C. & Smith, P. (2007) Carbon losses from soil and its consequences for land-use management. *Science of the Total Environment* 382: 165–90.

de Vente, J. & Poesen, J. (2005) Predicting soil erosion and sediment yield at the basin scale: scale issues and semi-quantitative models. *Earth-Science Reviews* 71: 95–125.

Déqué, M., Jones, R.G., Wild, M., *et al.* (2005) Global high resolution versus Limited Area Model climate change projections over Europe: quantifying confidence level from PRUDENCE results. *Climate Dynamics* 25: 653–70.

Dessai, S. & Hulme, M. (2007) Assessing the robustness of adaptation decisions to climate change uncertainties: A case study on water resources management in the East of England. *Global Environmental Change* 17: 59–72.

Easterling, W.E., Aggarwal, P.K., Batima, P., *et al.* (2007) Food, fibre and forest products. In Parry, M.L.,

Canziani, O.F., Palutikof, J.P., *et al.* (eds), *Climate Change 2007: Impacts, Adaptation and Vulnerability*. Contribution of Working Group II to the Fourth Assessment Report of the Intergovernmental Panel on Climate Change. Cambridge University Press, Cambridge: 273–313.

Ebel, B.A. & Loague K. (2006) Physics-based hydrologic-response simulation: seeing through the fog of equifinality. *Hydrological Processes* **20**: 2887–2900.

Favis-Mortlock, D. & Boardman, J. (1995) Nonlinear responses of soil erosion to climate change: a modelling study on the UK South Downs. *Catena* **25**: 365–87.

Favis-Mortlock, D.T. & Guerra, A.J.T. (1999) The implications of general circulation model estimates of rainfall for future erosion: a case study from Brazil. *Catena* **37**: 329–54.

Favis-Mortlock, D., Boardman, J. & MacMillan, V. (2001) The limits of erosion modelling: why we should proceed with care. In Harmon, R.S. & Doe, W.W. (eds), *Landscape Erosion and Evolution Modeling*. Kluwer Academic/Plenum Publishers, New York: 477–516.

Fernández, R.J. (2002) Do humans create deserts? *TRENDS in Ecology & Evolution* **17**: 6–7.

Fischlin, A., Midgley, G.F., Price, J.T., *et al.* (2007) Ecosystems, their properties, goods, and services. In Parry, M.L., Canziani, O.F., Palutikof, J.P., *et al.* (eds), *Climate Change 2007: Impacts, Adaptation and Vulnerability*. Contribution of Working Group II to the Fourth Assessment Report of the Intergovernmental Panel on Climate Change. Cambridge University Press, Cambridge: 211–72.

Flanagan, D.C. & Nearing, M.A. (1995) USDA-Water Erosion Prediction project: Hillslope profile and watershed model documentation. *NSERL Report No. 10*. USDA-ARS National Soil Erosion Research Laboratory, West Lafayette.

Garbrecht, J.D., Steiner, J.L. & Cox, C.A. (2007) The times they are changing: soil and water conservation in the 21st century. *Hydrological Processes* **21**: 2677–9.

Giorgi, F. (2005) Climate change prediction. *Climatic Change* **73**: 239–65.

Giorgi, F. (2006) Climate change hot-spots. *Geophysical Research Letters* **33**: L08707.

González-Hidalgo, J.C., Peña-Monné, J.L & de Luis, M. (2007) A review of daily soil erosion in Western Mediterranean areas. *Catena* **71**: 193–9.

Groisman, Y.P., Knight, R.W., Easterling, D.R., *et al.* (2005) Trends in precipitation intensity in the climate record. *Journal of Climate* **18**: 1326–50.

Hanson, C.E., Palutikof, J.P., Livermore, M.T.J., *et al.* (2007) Modelling the impact of climate extremes: an overview of the MICE project. *Climatic Change* **81**: 163–77.

Herrmann, S.M. & Hutchinson, C.F. (2005) The changing contexts of the desertification debate. *Journal of Arid Environments* **63**: 538–55.

Imeson, A.C. & Lavee, H. (1998) Soil erosion and climate change: the transect approach and the influence of scale. *Geomorphology* **23**: 219–27.

Imeson, A.C. & Prinsen, H.A.M. (2004) Vegetation patterns as biological indicators for identifying runoff and sediment source and sink areas for semi-arid landscapes in Spain. *Agriculture Ecosystems & Environment* **104**: 333–42.

IPCC: Intergovernmental Panel on Climate Change. 2007. *Climate Change 2007: The Physical Science Basis*. Contribution of Working Group I to the Fourth Assessment Report of the Intergovernmental Panel on Climate Change. Cambridge University Press, Cambridge.

Istanbulluoglu, E. & Bras, R.L. (2006) On the dynamics of soil moisture, vegetation, and erosion: Implications of climate variability and change. *Water Resources Research* **42**: W06418.

Jetten, V., de Roo, A. & Favis-Mortlock, D. (1999) Evaluation of field-scale and catchment-scale soil erosion models. *Catena* **37**: 521–41.

Jetten, V., Govers, G. & Hessel, R. (2003) Erosion models: quality of spatial predictions. *Hydrological Processes* **17**: 887–900.

Kirkby, M.J., Jones, R.J.A., Irvine, B., *et al.* (2004) *Pan-European Soil Erosion Risk Assessment: The PESERA Map*, Version 1 October 2003. Explanation of Special Publication Ispra 2004 No. 73 (S.P.I.04.73). European Soil Bureau Research Report No. 16, EUR 21176~, Office for Official Publications of the European Communities, Luxembourg.

Kleinen, T. & Petschel-Held, G. (2007) Integrated assessment of changes in flooding probabilities due to climate change. *Climatic Change* **81**: 283–312.

Kundzewicz, Z.W., Mata, L.J., Arnell, N.W., *et al.* (2007) Freshwater resources and their management. In Parry, M.L., Canziani, O.F., Palutikof, J.P., *et al.* (eds), *Climate Change 2007: Impacts, Adaptation and Vulnerability*. Contribution of Working Group II to the Fourth Assessment Report of the Intergovernmental Panel on Climate Change. Cambridge University Press, Cambridge: 173–210.

Lacombe, G., Cappelaere, B. & Leduc, C. (2008) Hydrological impact of water and soil conservation works in the Merguellil catchment of central Tunisia. *Journal of Hydrology* **359**: 210–24.

Lane, L.J., Hernandez, M. & Nichols, M. (1997) Processes controlling sediment yield from watersheds as functions of spatial scale. *Environmental Modelling & Software* **12**: 355–69.

Lane, S.N., Tayefi, V., Reid, S.C., *et al.* (2007) Interactions between sediment delivery, channel change, climate change and flood risk in a temperate upland environment. *Earth Surface Processes and Landforms* **32**: 429–46.

Lavee, H., Imeson, A.C. & Sarah, P. (1998) The impact of climate change on geomorphology and desertification along a Mediterranean-arid transect. *Land Degradation and Development* **9**: 407–22.

Ludwig, J.A., Wilcox, B.P., Breshears, D.D., *et al.* (2005) Vegetation patches and runoff-erosion as interacting ecohydrological processes in semiarid landscapes. *Ecology* **86**: 288–97.

Manabe, S., Milly, P.C.D. & Wetherald, R. (2004) Simulated long-term changes in river discharge and soil moisture due to global warming. *Hydrological Sciences Journal* **49**: 625–42.

Mantel, S., Van Lynden, G.J. & Huting, J. (2003) *PESERA, Deliverable 10: Scenario analysis*. ISRIC – World Soil Information, Wageningen.

Martínez-Fernández, J.M. & Esteve, M.A. (2005) A critical view of the desertification debate in southeastern Spain. *Land Degradation and Development* **16**: 529–39.

Martínez-Vilalta, J., Piñol, J. & Beven, K. (2002) A hydraulic model to predict drought-induced mortality in woody plants: an application to climate change in the Mediterranean. *Ecological Modelling* **155**: 127–47.

Meehl, G.A., Stocker, T.F., Collins, W.D., *et al.* (2007) Global climate projections. In Solomon, S., Qin, D., Manning, M., *et al.* (eds), *Climate Change 2007: The Physical Science Basis*. Contribution of Working Group I to the Fourth Assessment Report of the Intergovernmental Panel on Climate Change. Cambridge University Press, Cambridge: 747–845.

MESOEROS21. 2006. *MEditerranean SOils EROSion and vulnerability to global change during the 21st century*. Available at http://mesoeros21.brgm.fr/Files/dossier-ANR-VMC-erosion-2006.pdf [accessed September 2008].

Michael, A., Schmidt, J., Enke, W., *et al.* (2005) Impact of expected increase in precipitation intensities on soil loss – results of comparative model simulations. *Catena* **61**: 155–64.

Morgan, R.P.C. & Quinton, J.N. (2001) Erosion modeling. In Harmon, R.S. & Doe, W.W. (eds), *Landscape Erosion and Evolution Modeling*. Kluwer Academic/Plenum Publishers, New York: 117–44.

Mouillot, F., Rambal, S. & Joffre, R. (2002) Simulating climate change impacts on fire frequency and vegetation dynamics in a Mediterranean-type ecosystem. *Global Change Biology* **8**: 423–37.

Nearing, M.A., Pruski, F.F. & O'Neal, M.R. (2004) Expected climate change impacts on soil erosion rates: a review. *Journal of Soil and Water Conservation* **59**: 43–50.

Nearing, M.A., Jetten, V., Baffaut, C., *et al.* (2005) Modelling response of soil erosion and runoff to changes in precipitation and cover. *Catena* **61**: 131–54.

Neitsch, S.L., Arnold, J.G., Kiniry, J.R., *et al.* (2002) *Soil and Water Assessment Tool theoretical documentation*. TWRI Report TR-191, Texas Water Resources Institute, College Station, TX.

Nohara, D., Kitoh, A., Hosaka, M. & Oki, T. (2006) Impact of climate change on river discharge projected by Multimodel Ensemble. *Journal of Hydrometeorology* **7**: 1076–89.

Nunes, J.P. (2007) *Vulnerability of Mediterranean Watersheds to Climate Change: the Desertification Context*. PhD thesis, New University of Lisbon, Lisbon.

Nunes, J.P., Vieira, G., Seixas, J., *et al.* (2005) Evaluating the MEFIDIS model for runoff and soil erosion prediction during rainfall events. *Catena* **61**: 210–28.

Nunes, J.P., Seixas, J. & Pacheco, N.R. (2008) Vulnerability of water resources, vegetation productivity and soil erosion to climate change in Mediterranean watersheds. *Hydrological Processes* **22**: 3115–34.

O'Neal, M.R., Nearing, M.A., Vining, R.C., *et al.* (2005) Climate change impacts on soil erosion in Midwest United States with changes in crop management. *Catena* **61**: 165–84.

Oostwoud Wijdenes, D.J., Poesen, J., Vandekerckhove, L., *et al.* (1999) Gully-head morphology and implications for gully development on abandoned fields in a semi-arid environment, Sierra de Gata, southeast Spain. *Earth Surface Processes and Landforms* **24**: 585–603.

Parry, M.L., Canziani, O.F., Palutikof, J.P., *et al.* (eds) (2007) *Climate Change 2007: Impacts, Adaptation and Vulnerability*. Contribution of Working Group II to the Fourth Assessment Report of the Intergovernmental Panel on Climate Change. Cambridge University Press, Cambridge.

Pruski, F.F. & Nearing, M.A. (2002a) Runoff and soil-loss responses to changes in precipitation: A computer simulation study. *Journal of Soil and Water Conservation* 57: 7–16.

Pruski, F.F. & Nearing, M.A. (2002b) Climate-induced changes in erosion during the 21st century for eight U.S. locations. *Water Resources Research* 38: 1298.

Puigdefábregas, J. (1998) Ecological impacts of global change on drylands and their implications for desertification. *Land Degradation and Development* 9: 393–406.

Puigdefábregas, J. & Mendizabal, T. (1998) Perspectives on desertification: western Mediterranean. *Journal of Arid Environments* 39: 209–24.

Quinton, J.N. (1997) Reducing predictive uncertainty in model simulations: a comparison of two methods using the European Soil Erosion Model (EUROSEM). *Catena* 30: 101–17.

Ries, J.B. & Hirt, U. (2008) Permanence of soil surface crusts on abandoned farmland in the Central Ebro Basin/Spain. *Catena* 72: 282–96.

Sarah, P. (2006) Soil organic matter and land degradation in semi-arid area, Israel. *Catena* 67: 50–55.

Schmidt, J. (1990) A mathematical model to simulate rainfall erosion. *Catena Suppl.* 19: 101–9.

Seixas, J. (2000) Assessing heterogeneity from remote sensing images: the case of desertification in southern Portugal. *International Journal of Remote Sensing* 21: 2645–63.

Shakesby, R.A. & Doerr, S.H. (2006) Wildfire as a hydrological and geomorphological agent. *Earth-Science Reviews* 74: 269–307.

Solomon, S., Qin, D., Manning, M., *et al.* (eds) (2007) *Climate Change 2007: The Physical Science Basis.* Contribution of Working Group I to the Fourth Assessment Report of the Intergovernmental Panel on Climate Change. Cambridge University Press, Cambridge.

Stott, P.A. & Kettleborough, J.A. (2002) Origins and estimates of uncertainty in predictions of twenty-first century temperature rise. *Nature* 416: 723–6.

Tebaldi, C., Hayhoe, K., Arblaster, J.M. & Meehl, G.A. (2006) Going to the extremes: an intercomparison of model-simulated historical and future changes in extreme events. *Climatic Change* 79: 185–211.

Thodsen, H., Hasholt, B. & Kjærsgaard, J.H. (2008) The influence of climate change on suspended sediment transport in Danish rivers. *Hydrological Processes* 22: 764–74.

Thornes, J.B. (1998) Mediterranean desertification. In Mairota, P., Thornes, J.B. & Geeson, N. (eds), *Atlas of Mediterranean Environments in Europe.* John Wiley and Sons, Chichester: 2–5.

Toy, T.J., Foster, G.R. & Renard, K.G. (2002) *Soil Erosion: Processes, Prediction, Measurement, and Control.* John Wiley and Sons, New York.

Trenberth, K.E., Jones, P.D., Ambenje, P., *et al.* (2007) Observations: surface and atmospheric climate change. In Solomon, S., Qin, D., Manning, M., *et al.* (eds), *Climate Change 2007: The Physical Science Basis.* Contribution of Working Group I to the Fourth Assessment Report of the Intergovernmental Panel on Climate Change. Cambridge University Press, Cambridge: 235–336.

van den Hurk, B., Hirscho, M., Schär, C., *et al.* (2005) Soil control on runoff response to climate change in regional climate model simulations. *Journal of Climate* 18: 3536–51.

van Rompaey, A.J.J., Vieillefont, V., Jones, R.J.A., *et al.* (2003) *Validation of soil erosion estimates at European scale.* European Soil Bureau Research Report No. 13, EUR 20827 EN, Office for Official Publications of the European Communities, Luxembourg.

Vicente-Serrano, S.M., Lasanta, T. & Romo, A. (2004) Analysis of spatial and temporal evolution of vegetation cover in the Spanish central Pyrenees: Role of human management. *Environmental Management* 34: 802–18.

Vogiatzakis, I.N., Mannion, A.M. & Griffiths, G.H. (2006) Mediterranean ecosystems: problems and tools for conservation. *Progress in Physical Geography* 30: 175–200.

Wetherald, R.T. & Manabe, S. (2002) Simulation of hydrologic changes associated with global warming. *Journal of Geophysical Research* 107: 4379–93.

Wilby, R.L. (2005) Uncertainty in water resource model parameters used for climate change impact assessment. *Hydrological Processes* 19: 3201–19.

Wu, W., Shibasaki, R., Yang, P., *et al.* (2007) Global-scale modelling of future changes in sown areas of major crops. *Ecological Modelling* 208: 378–90.

Xu, C.-Y. & Singh, V.P. (2004) Review on regional water resources assessment models under stationary and changing climate. *Water Resources Management* 18: 591–612.

Yair, A. & Kossovsky, A. (2002) Climate and surface properties: hydrological response of small arid and semi-arid watersheds. *Geomorphology* 42: 43–57.

Zhang, G.-H., Nearing, M.A. & Liu, B.-Y. (2005) Potential effects of climate change on rainfall erosivity in the Yellow River basin of China. *Transactions of the American Society of Agricultural Engineers* 48: 511–17.

Zhang, X.-C. (2007) A comparison of explicit and implicit spatial downscaling of GCM output for soil erosion and crop production assessments. *Climatic Change* **84**: 337–63.

Zhang, X.C. & Liu, W.Z. (2005) Simulating potential response of hydrology, soil erosion, and crop productivity to climate change in Changwu tableland region on the Loess Plateau of China. *Agricultural and Forest Meteorology* **131**: 127–42.

Zhang, X.C. & Nearing, M.A. (2005) Impact of climate change on soil erosion, runoff, and wheat productivity in central Oklahoma. *Catena* **61**: 185–95.

Zhang, X.C., Nearing, M.A., Garbrecht, J.D. & Steiner, J.L. (2004) Downscaling monthly forecasts to simulate impacts of climate change on soil erosion and wheat production. *Soil Science Society of America Journal* **68**: 1376–85.

Zhu, Y.-M., Lu, X.X. & Zhou, Y. (2008) Sediment flux sensitivity to climate change: A case study in the Longchuanjiang catchment of the upper Yangtze River, China. *Global and Planetary Change* **60**: 429–42.

16 Risk-Based Erosion Assessment: Application to Forest Watershed Management and Planning

W.J. ELLIOT[1] AND P.R. ROBICHAUD[1]

[1]*USDA Forest Service, Rocky Mountain Research Station, Moscow, ID, USA*

This chapter discusses conditions where risk-based erosion modelling may be appropriate in forested watersheds. It then describes four modelling approaches for risk-based erosion modelling using WEPP-based erosion technology.

16.1   Background

In many applications of erosion modelling, the vegetation condition to be modelled is relatively similar year after year, as in continuous agriculture or grazing lands. In other cases, a known sequence of surface conditions occurs over a period of several years, like agricultural systems with fixed crop rotations or short rotation forestry. In such cases, describing erosion with an average annual value is usually an adequate approach to conservation planning.

In some conditions, however, erosion processes are dominated by extreme disturbance events followed by a prolonged period of minimal disturbance, like unmanaged forests or rangelands in fire-driven ecosystems (Fig. 16.1), or managed forests that experience a major harvesting or thinning operation only once every few decades. In these cases, erosion is minimal prior to the disturbance, potentially high immediately following the disturbance, and then returns to a

relatively low erosion risk within a few years. Wildfires can cause soils to become water repellent for a few years, increasing the risk of runoff and erosion immediately after fire. The repellency, however, dissipates in subsequent years on many soils (Doerr *et al.*, 2000; Robichaud, 2000). For example, Table 16.1 shows three different studies in which erosion rapidly declined during the three years following wildfires in all but one year in one study.

One characteristic of these highly disturbed conditions is a high spatial variability of the disturbance. The disturbance rather than soil properties dominates the erodibility of the soils (Robichaud *et al.*, 1993). The distribution of the disturbance following wildfires is seldom uniform or predictable (Robichaud *et al.*, 2007) as is the case with agricultural conditions. There will be sites following wildfire where the fire burned at a higher severity, leading to a complete loss of surface cover and most likely the generation or augmentation of a water repellent soil condition. There will be other sites where the fire burned very little, or not at all, resulting in an area of minimal erosion risk. There is often considerable spatial variability in erodibility on a hillslope. For example, in a study of hillslope erosion after a wildfire in the Bitterroot Valley, Montana, US, the four plots were within a 100-m wide hillside, yet sediment delivered from the 15 July 2001 storm ranged from 0.13 to 18 Mg ha^{-1} (Table 16.2).

The weather in the years following the disturbance is crucial in determining the erosion

Handbook of Erosion Modelling, 1st edition. Edited by R.P.C. Morgan and M.A. Nearing. © 2011 Blackwell Publishing Ltd.

Fig. 16.1 Scientist inspects the erosion occurring after a wildfire in northern California, US 2(photography courtesy of Natalie Copeland 2008).

Table 16.1 Observed annual erosion rates for three years following wildfires on three sites in the western US.

Site	Year of fire	Type of plot	Erosion (Mg ha^{-1})		
			Year 1	Year 2	Year 3
Wallowa-Whitman National Forest, Oregon[a]	1994	Silt fence	1.9	0.1	0.03
Bitterroot National Forest, Montana	2000	Silt fence[b]	29	0.8	0.07
		Small watershed[c]	0.64	0.93	0.09

[a]Robichaud & Brown (1999); [b]Robichaud *et al.* (2008a); [c]Robichaud *et al.* (2008b).

Table 16.2 Observed erosion rates (Mg ha^{-1}) from 20 m long plots in 2001 following the 2000 Bitterroot Valley Fire in Montana (Robichaud *et al.*, 2008a).

Plot	15-Jul	21-Jul	30-Jul	14-Sep	Total
A	2.9	20.9	0.09	0.06	23.9
F	17.9	–	0.08	0.03	18.0
I	0.33	8.4	0.23	0.08	9.0
N	0.13	15.8	0.25	0.02	16.2
Average	5.30	15.01	0.16	0.05	16.8
Pcp, mm[a]	6.6	15.7	22.1	3.8	
I-10, mm h^{-1} [b]	19.8	39.6	7.6	13.7	

[a]Total storm precipitation; [b]maximum 10-minute precipitation intensity of storm.

rate. If the precipitation is moderate, then erosion will probably be minimal. If the weather has storms or snowmelt rates that are above normal, then erosion can be severe. In the study summarized in Table 16.2, maximum precipitation intensity was more important than total precipitation amount at causing erosion. In other studies (e.g. Robichaud *et al.*, 2008b), total precipitation amount was more important than intensity in causing erosion.

In order to address erosion for disturbance-dominated conditions, an average annual value is of limited utility since the disturbed conditions are not average, and erosion following the disturbance is dependent upon the degree of the disturbance, the distribution of the disturbance, the weather immediately following the disturbance and the rate of vegetation and soil recovery. For these disturbed conditions, a risk-based approach is more appropriate. For example, one interpretation of the data in Table 16.2 is that the average erosion rate in the year following a wildfire is 16.8 Mg ha^{-1}. A more meaningful interpretation of these data might be that on one of these plots there is a 1 in 4 chance that the total erosion will exceed 23.9 Mg ha^{-1} from the four large storms in the year following the wildfire (Plot A total), and a 1 in 4 chance that erosion will exceed 20.9 Mg ha^{-1} from the largest single erosion event in the year following the wildfire (21 July on Plot A).

With risk-based erosion modelling, the modeller must estimate the probability distribution for a given set of conditions, and from that distribution, determine the probability of a given erosion rate occurring. Probability distributions should account for climate, soil properties and distribution of disturbance.

16.2   Risk-based Approach

This chapter will consider four different tools to use for risk-based erosion modelling, using interfaces developed for the Water Erosion Prediction Project (WEPP) model (Flanagan & Livingston, 1995). The interfaces are the Windows Interface (Flanagan *et al.*, 1998), the online Disturbed WEPP Interface (Elliot, 2004), the online Erosion Risk Management Tool (ERMiT; Robichaud *et al.*, 2007), and the GeoWEPP GIS wizard (Renschler, 2003).

An example application for each of these interfaces will be given to assist in understanding the technology. All examples will apply to an analysis of erosion following a high severity wildfire that occurred in forested mountains of the Bitterroot Valley in Western Montana, US, in July 2000. The soils in this area are gravelly sandy loam over granitic colluvium, with slopes typically from 20% to 50% (Robichaud *et al.*, 2008a). For the hillslope examples, a horizontal slope length of 20 m with a maximum steepness of 61% will be used, similar to the silt fence plots installed by Robichaud *et al.* (2008a). The ground cover is assumed to be 5%, as was observed on

these plots. In order to generate a climate for this remote area, the climate statistics of a nearby low-elevation weather station were modified with an online interface (Scheele *et al.*, 2001; Elliot, 2004). Monthly precipitation amounts were modified with data from a nearby high-elevation snow monitoring station, and the number of wet days was increased by half the proportionate increase in precipitation. The monthly maximum and minimum temperatures were decreased from the valley station by the adiabatic lapse rate (Scheele *et al.*, 2001). Observed erosion rates from Robichaud *et al.* (2008a) are presented for comparison with values predicted by the examples (Tables 16.1 and 16.2).

16.3   WEPP Windows

The weather file that drives the WEPP model contains daily data, and so WEPP predicts runoff and erosion on a storm-by-storm basis. All runoff events predicted by WEPP are stored in a single file. WEPP Windows accesses this file and determines the probability of exceeding a given amount of daily precipitation, daily runoff, peak runoff rate, or daily sediment delivery using a Weibull plotting formula (WEPP Help screen).

Example 16.1   To model the Bitterroot Valley site in the WEPP Windows interface, the following were selected from the downloaded databases and menus: a 50-yr stochastic weather file; a sandy loam, high-severity fire soil; the described topography; and the Return Period Analysis option. The management file was calibrated to ensure approximately 5% ground cover for every year of simulation.

The 'Return Period Analysis' output screen from this model run (Fig. 16.2) shows an estimate that for any given storm there is a 10% probability that erosion will exceed 2.2 Mg ha^{-1}. As a comparison, the 'Average' value predicted by this WEPP run was 0.55 Mg ha^{-1}.

In this example, the 10-year sediment delivery may or may not have been associated with the 10-year rainfall or 10-year peak runoff rate or

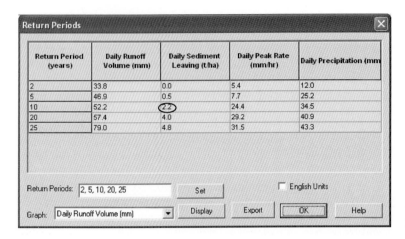

Fig. 16.2 Return period analysis output screen from WEPP Windows interface for Example 16.1.

volume, as WEPP considers daily conditions of the vegetation and soil water content before making runoff and erosion predictions. A detailed examination of the WEPP output for this sediment delivery event showed that it occurred on 2 August, year 32, when 43.3 mm of precipitation resulted in 4.5 mm of runoff. This storm was the 25-yr return period precipitation event, but the 24-h runoff depth was less than that for a 2-yr event (Fig. 16.2). The peak runoff rate predicted for the storm was 29.2 mm h^{-1}, the value for the 20-y return period peak runoff rate. Since the WEPP simulations show that there were no major runoff events in the four months prior to this storm, the hydrology was driven by the large precipitation event only.

If model users wish to consider the risk of exceeding a given level of erosion for an entire year, as a function of the variability in weather, then the user can request the detailed annual output from the WEPP model, and note the annual erosion rates for 50 or more years. These can either be analysed using a return period analysis technique, or simply ranked, with the year with the highest value serving as an estimate for the erosion with a probability of occurrence of one in the length of run, and the second largest value having a probability of occurrence of two in the length of run, and so on. This process has been programmed into the online Disturbed WEPP Interface (Elliot, 2004).

16.4   Online Interfaces

Two online interfaces have been developed that incorporate risk-based erosion prediction for forest conditions (Elliot, 2004), and they can be accessed at http://forest.moscowfsl.wsu.edu/fswepp. One interface, Disturbed WEPP, provides both the average annual runoff and erosion estimates, and the annual return period values for precipitation, runoff, upland erosion, and sediment delivery.

Example 16.2   The climate, soil, topography and vegetation cover conditions described for Example 16.1 were entered into the online input screen for Disturbed WEPP. A 50-year run was selected and the recommended cover calibration was carried out (Elliot, 2004). The model was then run and the output screen presented (Fig. 16.3).

The return period analysis (top part of Fig. 16.3) shows that following a wildfire, there is a 1 in 10 chance that annual sediment delivery from this hillslope will exceed 13.0 Mg ha^{-1}. The second output from Disturbed WEPP (bottom of Fig. 16.3) is that there is an 82% probability that there will be sediment delivered in the year following a wildfire. The average predicted erosion rate is 4.3 Mg ha^{-1}.

The predicted erosion rate is greater than the value predicted in Example 16.1 because it is for a full year and not a single storm, and maybe because there are different versions of WEPP

Return period analysis based on 50 years of climate

Return Period	Precipitation (mm)	Runoff (mm)	Erosion (t ha⁻¹)	Sediment (t ha⁻¹)
50 year	1212.40	79.98	39.32	39.3155
25 year	1068.60	34.90	23.80	23.8010
10 year	979.30	20.26	13.01	13.0085
5 year	947.20	14.77	6.29	6.1110
2.5 year	899.90	7.31	3.46	3.0960
Average	874.39	9.33	4.31	4.3100

Probabilities of occurrence first year following disturbance based on 50 years of climate

Probability there is runoff	84 %
Probability there is erosion	82 %
Probability there is sediment delivery	82 %

Fig. 16.3 Return period analysis from the Disturbed WEPP online interface output screen for Example 16.2.

associated with these interfaces (WEPP Windows was version 2008.907 and Disturbed WEPP version 2001.100) and differences in input files describing vegetation. Figure 16.3 shows that the average predicted erosion rate for these conditions is 4.3 Mg ha⁻¹, compared with 0.55 Mg ha⁻¹ from the WEPP Windows interface. Because of the skewed distribution of erosion events, observations of erosion rates need to be interpreted with care as erosion rates well below an average value are likely to be observed, while it is still possible to observe a greater than average erosion rate (Robichaud, 2005). As burned sites can quickly recover, only data collected the first year following a wildfire should be used to compare with these predicted values (Table 16.1).

The second online interface that incorporated probability into erosion prediction is the Erosion Risk Management Tool (ERMiT) (Robichaud et al., 2007). ERMiT predicts the probability of exceeding a given sediment delivery amount following a wildfire for a single event, and also estimates the benefits of several practices to reduce erosion risk. ERMiT considers not only variability in climate, but also variability in

severity of fire and the distribution of that severity on the hillslope (Robichaud et al., 2007).

Example 16.3 What is the erosion rate from a single event that would probably be exceeded once in ten years for the conditions described in Examples 16.1 and 16.2? The data were entered into the ERMiT input screen, and a high-severity wildfire specified.

The ERMiT output screen presents two tables and a figure. The first table (Fig. 16.4) shows the individual runoff events selected for the analysis. The 10-year runoff depth is 18 mm, from 40.9 mm of precipitation, occurring on 15 May. The second output from ERMiT is an erosion exceedance graph showing the probability associated with a given erosion rate for each of five years following the wildfire (Fig. 16.5). The figure shows that there is a 10% probability that sediment delivery from a single event will exceed approximately 2.5 Mg ha⁻¹ in the year following a wildfire. The final table on the ERMiT output screen is interactive, allowing the user to enter the desired exceedance probability. Once entered, the table displays the associated sediment delivery rate for each year following the wildfire and how that rate is impacted by common erosion mitigation treatments (Fig. 16.6). This table confirms the observation in Fig. 16.5, that there is a 10% probability that erosion will exceed 2.5 Mg ha⁻¹ on the example hillslope. It also shows that erosion risk drops quickly in the following years, similar to observed data presented in Table 16.1, and that mulching can be effective at reducing sediment delivery rates.

The ERMiT model predicted erosion rates similar to those observed on a small watershed (Table 16.1) and lower than those observed from silt fence plots (Table 16.1 silt fence plots, and Table 16.2) and predicted by the Disturbed WEPP interface. This is probably because the silt fence plots were located on a site where there were only high-severity fire conditions, and examples 16.1 and 16.2 modelled those conditions. ERMiT, however, considers a hillslope as a mosaic of fire severity, and internally is designed to consider a range of hillslope severity conditions in its

Rainfall Event Rankings and Characteristics from the Selected Storms

Storm Rank based on runoff (return interval)	Storm Runoff (mm)	Storm Precipitation (mm)	Storm Duration (h)	10-min Peak Rainfall Intensity (mm h⁻¹)	30-min Peak Rainfall Intensity (mm h⁻¹)	Storm Date
1	34.1	3.4	2.05	N/A	N/A	January 31 year 91
5 (20-year)	23.2	2.1	3.96	N/A	N/A	January 28 year 100
10 (10-year)	18	40.9	7.42	71.10	52.49	May 15 year 49
20 (5-year)	14.6	34.5	1.85	66.28	47.57	May 1 year 39
50 (2-year)	6.5	26.2	3.40	39.30	30.32	July 13 year 46
75 (1⅓-year)	4	20.4	8.50	50.37	32.49	July 15 year 77

Fig. 16.4 Runoff events selected for analysis with the ERMiT interface for Example 16.3.

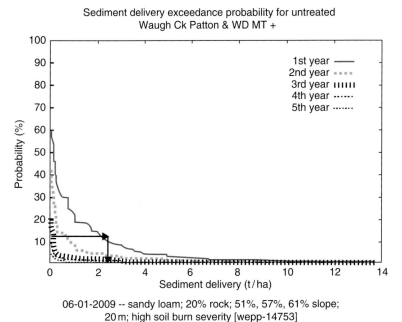

Sediment delivery exceedance probability for untreated
Waugh Ck Patton & WD MT +

06-01-2009 -- sandy loam; 20% rock; 51%, 57%, 61% slope;
20 m; high soil burn severity [wepp-14753]

Fig. 16.5 Probability versus sediment delivery for the five years following a wildfire predicted by the ERMiT interface for Example 16.3.

Mitigation Treatment Comparisons					
Probability that sediment yield will be exceeded	**Event sediment delivery (t ha⁻¹)**				
	Year following fire				
10 % 🔵	1st year	2nd year	3rd year	4th year	5th year
Untreated ⊟	2.5	0.86	0.15	0.05	0
Seeding ⊟	2.5	0.25	0.15	0	0
Mulch (1 t ha⁻¹) ⊟	0.22	0.2	0.15	0.05	0
Mulch (2 t ha⁻¹) ⊟	0.09	0.2	0.15	0.05	0
Mulch (3.5 t ha⁻¹) ⊟	0.07	0.2	0.15	0.05	0
Mulch (4.5 t ha⁻¹) ⊟	0.06	0.15	0.15	0.05	0
Erosion Barriers: Diameter 0.05 m Spacing 25 m 🔵 ?					
Logs & Wattles ⊟	2.5	0.86	0.15	0.05	0

Fig. 16.6 Erosion exceedance value and effectiveness of mitigation treatment comparison for the first five years following a wildfire predicted by the ERMiT interface for a 10% exceedance probability for Example 16.3.

sediment delivery prediction (Robichaud *et al.*, 2007). Hence the ERMiT-predicted erosion rates are lower than those observed on silt fence plots (2.5 Mg ha⁻¹ vs. 17 Mg ha⁻¹) and more typical of the values observed on the larger areas (Table 16.1, small watershed, <1 Mg ha⁻¹). For some of the selected events, ERMiT predicts more runoff than precipitation (20-year event, Fig. 16.4). This is due to melting snow contributing to runoff for these events. Erosion events in this area are frequently associated with large runoff events from rain falling on a snow pack (Tonina *et al.*, 2008), and such events are included in all the WEPP predictions.

16.5   GIS Interface

The GIS interface for WEPP technology is GeoWEPP (Renschler, 2003). GeoWEPP builds the stream network from a digital elevation model (DEM). The user selects the outlet for each sub-watershed of interest, generally limiting watershed areas to under 500 ha for a 30-m DEM. As with the WEPP technology, the user can either

use GeoWEPP to predict average annual erosion values, or can use it for risk analysis. GeoWEPP can be run in two modes, 'Watershed' or 'Flowpath' (Cochrane & Flanagan, 1999). The watershed mode is useful in determining sediment delivery to points of interest downstream from a major watershed disturbance. In Watershed mode, GeoWEPP predicts sediment delivery, surface runoff, and lateral flow from hillslope polygons, and routes the delivered sediment through the stream network (Dun *et al.*, 2009). In Flowpath mode, GeoWEPP determines distinct flow paths throughout the watershed, and determines the distribution of erosion along each flow path, estimating the erosion rate for each pixel in the analysis. The flowpath mode is useful for determining the location of the greatest risks of erosion within a watershed, so that erosion mitigation treatments can be targeted to those areas. With a 30-m DEM, there are usually two or three flow paths generated per hectare. For the example GeoWEPP flowpath run (Fig. 16.7), there were 286 flow paths identified on a 140-ha watershed.

Elliot *et al.* (2006) presented methods for applying risk-based erosion modelling to post-wildfire

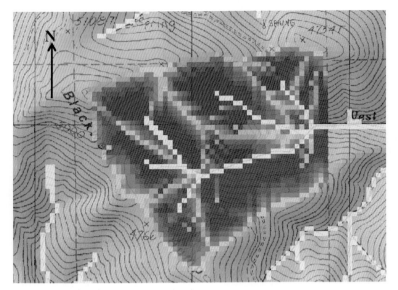

Fig. 16.7 Distribution of on-site erosion as predicted by GeoWEPP for Example 16.4, for a year with the 10-year sediment yield event. The stream network is white, and the darker the area, the greater the predicted erosion. The darkest areas have a predicted hillslope erosion rate exceeding $200\,Mg$ ha^{-1}, and the lightest erosion less than $12.5\,Mg\,ha^{-1}$. Pixel size is $30\,m$ and the watershed area is $140\,ha$.

Table 16.3 Return period analysis from GeoWEPP for Example 16.4. The watershed area was 140 ha.

Return period (years)	Sediment leaving (Mg)	Peak runoff rate ($m^3\,s^{-1}$)	Daily precipitation (mm)
1	2339	14.8	20
2	3164	18.7	24
5	4618	25.7	27
10	7505	36.2	33

conditions in forested watersheds. For the watershed analysis, one of the GeoWEPP output options is a return period analysis (e.g. Table 16.3). Carrying out a return period analysis with the flow path method is more complicated because the output is limited to hillslope polygon summaries of average annual erosion rates. The approach by Elliot *et al.* (2006) to apply probabilities to the flow path method was to determine from a hillslope (or watershed) return period analysis (Fig. 16.2 or Table 16.3) either the precipitation, runoff or sediment delivery amount for the desired return period, and then inspect the WEPP

or GeoWEPP output event files to determine what year the event occurred. Once the year was known, then a custom input climate could be developed containing only the year of interest. Because of carry-over of snow pack from November or December in one year to the next, if the event of interest is during spring snowmelt, then it may be necessary to include both the year with the event and the previous year in the climate file, and run the flowpath method for those two years only (Example 16.4). The ERMiT tool uses this approach, running both the year of interest (Fig. 16.4) preceded by the year before.

Example 16.4 The DEM for a steep forested watershed near the Bitterroot Valley, Montana, was obtained, and a small upland watershed was identified within that forest for post-wildfire erosion risk analysis. The watershed vegetative cover was assumed to average 30% following the wildfire for this example. Two runs were carried out for the same climate and soil texture as in the previous examples. The first analysis was a 50-year run for the entire watershed, using the 'Watershed' option in GeoWEPP.

The return period analysis from the watershed run (Table 16.3) predicted that the 10-year return period sediment yield was 7505 Mg. The area was determined by GeoWEPP to be 140 ha, leading to an erosion rate for the 10-year return period of 54 Mg ha^{-1} for the 10-yr return period sediment yield event. A review of the GeoWEPP 'Events' file showed that this event occurred on 9 April in year 15, a day when there was no precipitation, so it was a runoff event from snowmelt only. The stochastic climate file was then truncated to contain only years 14 and 15, and GeoWEPP was run for the same watershed with the 'Flowpath' option for those two years. The results of the flowpath run with the average annual erosion rate for two years, one of which contained the 10-yr event, are shown in Fig. 16.7. The erosion rate for the darkest pixels exceeded 200 Mg ha^{-1}, and on the lightest pixels it was less than 12.5 Mg ha^{-1}.

The GeoWEPP predictions were greater than the observed values or other predictions because the slope lengths were greater, averaging 200 m. The flowpath method showed that the areas with greater predicted sediment yields were the areas immediately adjacent to the streams, while the ridge tops had lower predicted erosion rates. It also showed that the more westerly-facing slopes were at a higher risk of erosion than the east-facing slopes (Fig. 16.7). If additional information about the spatial distribution of the severity of wildfire were known, this could also be incorporated into GeoWEPP by altering the soil properties and/or ground cover on each hillslope polygon to match the conditions determined by remote sensing or ground survey as described by Elliot *et al.* (2006).

16.6 Discussion

Four different predictive tools, all based on WEPP technology, have been presented. The results of each method are summarized in Table 16.4, for a ten-year return period erosion event. The ERMiT tool estimated a higher precipitation value, probably because it uses 100 years of stochastic

Table 16.4 Summary of risk-based predictions for a 10-year event. All data are daily values except Disturbed WEPP.

Interface	Precipitation (mm)	Runoff (mm)	Sediment yield (Mg ha^{-1})
WEPP Windows	34.5	52.2	2.2
Disturbed WEPP (annual)			13.01
ERMiT	40.9	19.0	2.5
GeoWEPP Watershed	33.4	99.1	53.6

weather whereas only 50 years were used for the other examples. The ERMiT tool and WEPP Windows predicted the lowest sediment delivery rates. This is probably due to the fact that ERMiT considers a number of different surface and soil conditions even for high severity, whereas all of the other tools considered a single high-severity condition. The WEPP Windows prediction may be lower than Disturbed WEPP because WEPP Windows modelled cover as perennial, whereas the Disturbed WEPP interface was developed when this feature was not available, and thus may have limited vegetation cover early in the spring when significant rain-on-snow events occur. Also, the estimate for Disturbed WEPP was for an entire year that included several events, whereas ERMiT and WEPP Windows predictions were for single events. The GeoWEPP flowpath method predicted a much higher erosion rate, probably due to the much longer slope lengths.

The output files portray another modelling challenge: many of the large runoff events were a combination of rainfall and snowmelt. This is especially evident when runoff exceeded precipitation for the 20-year runoff event for ERMiT (Fig. 16.4), and the 10-year event for the watershed example when the entire event was snowmelt and there was no precipitation. Because of the importance of snowmelt processes in this climate, traditional precipitation-based risk tools may not work as well as models that account for snowmelt processes.

16.7 Applicability to Climate Change

The WEPP model is the physically-based engine behind the interfaces that have been described. The climate input into the WEPP model includes daily precipitation amounts and maximum and minimum temperatures. These files are generally generated with a stochastic climate generator CLIGEN (Nicks et al., 1995) that is accessed by all of these interfaces. This interface allows incorporation of future climate scenarios into any WEPP technology.

The general approach to incorporate future climate scenarios for all of these applications is through the online 'RockClime' interfaces (Scheele et al., 2001). This interface allows users to access current climate station data from the CLIGEN database containing about 2600 stations, modify that climate for remote areas within the US using the PRISM monthly precipitation database (Daly et al., 1994), and further adjust the maximum and minimum temperatures, monthly precipitation amount, and number of wet days in a month to match future climate scenarios. Future temperatures and precipitation values are readily available from numerous sources (e.g. http://forest.moscowfsl.wsu.edu/climate/). Research is ongoing to determine the distribution of wet days in future climates.

It is generally predicted that future climates will be warmer, and in many areas, wetter in the winter months. This means that snowpack in the northern hemisphere will be less developed, and snowmelt or rain-on-snow events less severe, whereas runoff associated with large precipitation events may increase. Warmer summers will also likely lead to increased evapotranspiration and lower soil water contents, resulting in lower runoff from summer storms, unless those storms are more severe. Whatever the effect, the altered climate coupled with the WEPP technology will be able to predict the risk of a given amount of erosion from a single event, or from a year for any current or future climate. The biggest limitation is the ability to describe the future climate. The WEPP technologies are already providing average annual predictions and single storm predictions for these future scenarios (e.g. Nearing et al., 2005; Elliot, 2006; see also Chapter 15).

16.8 Summary

This chapter described the need for risk-based modelling to predict soil erosion associated with forest management disturbances and wildfires. It presented four different WEPP interfaces and demonstrated how they could be used for erosion risk analysis following a wildfire. These included a Windows interface, two online interfaces, and a GIS interface. In the examples provided, the online ERMiT tool, which considers a range of fire severities, and the WEPP Windows interface, estimated lower erosion rates than the Disturbed WEPP interface, which could only provide an annual estimate and not a single storm prediction. Each of the models predicted erosion rates within the wide range of those measured in field experiments in the modelled area. The WEPP model is well suited for making such risk-based predictions for current and future climate scenarios.

Acknowledgement

This chapter was written and prepared by US Government employees on official time, and therefore it is in the public domain and not subject to copyright.

References

Cochrane, T.A. & Flanagan, D.C. (1999) Assessing water erosion in small watersheds using WEPP with GIS and digital elevation models. *Journal of Soil and Water Conservation* **54**: 678–85.

Daly, C., Neilson, R.P. & Phillips, D.L. (1994) A statistical-topographic model for mapping climatological precipitation over mountainous terrain. *Journal of Applied Meteorology* **33**: 140–48.

Doerr, S.H., Shakesby, R.A. & Walsh, R.P.D. (2000) Soil water repellency: its causes, characteristics and hydro-geomorphological significance. *Earth-Science Review* **15**: 33–65.

Dun, S., Wu, J.Q., Elliot, W.J., *et al.* (2009) Adapting the Water Erosion Prediction Project (WEPP) model for forest applications. *Journal of Hydrology* **366**: 46–54.

Elliot, W.J. (2004) WEPP Internet interfaces for forest erosion prediction. *Journal of the American Water Resources Assoc.* **40**: 299–309.

Elliot, W.J. (2006) *Single storm analysis for conservation planning.* Presented at the Planning for Extremes Workshop, sponsored by the Soil and Water Conservation Society of America. 1–3 November 2006, Milwaukee, Wisconsin.

Elliot, W.J., Miller, I.S. & Glaza, B.D. (2006) *Using WEPP technology to predict erosion and runoff following wildfire.* Paper No. 068011, presented at the ASAE Annual International Meeting, 9–12 July, Portland, Oregon. American Society of Agricultural Engineers, St. Joseph, Michigan.

Flanagan, D.C. & Livingston, S.J. (1995) *WEPP User Summary, USDA-Water Erosion Prediction Project (WEPP).* US Department of Agriculture – Agricultural Research Service National Soil Erosion Research Laboratory, W. Lafayette, Indiana:

Flanagan, D.C., Fu, H., Frankenberger, J.R., *et al.* (1998) *A Windows interface for the WEPP erosion model.* Paper No. 98-2135. Presented at the Annual International Meeting of the American Society of Agricultural Engineers, St. Joseph, Michigan.

Nearing, M.A., Jetten, V., Baffaut, C., *et al.* (2005) Modeling response of soil erosion and runoff to changes in precipitation and cover. *Catena* **61**: 131–54.

Nicks, A.D., Lane, L.J. & Gander, G.A. (1995) Chapter 2. Weather generator. In Flanagan, D.C. & Nearing, M.A. (eds), *USDA Water Erosion Prediction Project Hillslope Profile and Watershed Model Documentation.* US Department of Agriculture Agricultural Research Service, W. Lafayette, Indiana: 2.1–2.22

Renschler, C.S. (2003) Designing geo-spatial interfaces to scale process models: The GeoWEPP approach. *Hydrological Processes* **17**: 1005–17.

Robichaud, P.R. (2000) Fire effects on infiltration rates after prescribed fire in Northern Rocky Mountain forests, USA. *Journal of Hydrology* **231-232**: 220–29.

Robichaud, P.R. (2005) Measurement of post-fire hillslope erosion to evaluate and model rehabilitation treatment effectiveness and recovery. *International Journal of Wildland Fire* **14**: 475–85.

Robichaud, P.R. & Brown, R.E. (1999) What happened after the smoke cleared: Onsite erosion rates after a wildfire in Eastern Oregon. In Olson, D.S. & Potyondy, J.P. (eds), *Proceedings AWRA Specialty Conference on Wildland Hydrology, 30 June–2 July 1999, Bozeman, Montana.* American Water Resources Association, Herndon, Virginia: 419–26.

Robichaud, P.R., Luce, C.H. & Brown, R.E. (1993) Variation among different surface conditions in timber harvest sites in the Southern Appalachians. In Larionov, J.A. & Nearing, M.A. (eds), *Proceedings from the Russia, U.S. and Ukraine International Workshop on Quantitative Assessment of Soil Erosion. Moscow, Russia. 20–24 Sept.* Center of Technology Transfer and Pollution Prevention, Purdue University, West Lafayette, Indiana: 231–41.

Robichaud, P.R., Elliot, W.J., Pierson, F.B., *et al.* (2007) Predicting postfire erosion and mitigation effectiveness with a web-based probabilistic erosion model. *Catena* **71**: 229–41.

Robichaud, P.R., Pierson, F.B., Brown, R.E. & Wagenbrenner, J.W. (2008a) Measuring effectiveness of three postfire hillslope erosion barrier treatments, western Montana, USA. *Hydrologic Processes* **22**: 159–70.

Robichaud, P.R., Wagenbrenner, J.W., Brown, R.E., *et al.* (2008b) Evaluating the effectiveness of contour-felled log erosion barriers as a post-fire runoff and erosion mitigation treatment in the western United States. *International Journal of Wildland Fire* **17**: 255–73.

Scheele, D.L., Elliot, W.J. & Hall, D.E. (2001) Enhancements to the CLIGEN weather generator for mountainous or custom applications. In Ascough II, J.C. & Flanagan, D.C. (eds), *Proceedings of the International Symposium of Soil Erosion Research for the 21st Century.* American Society of Agricultural Engineers, St. Joseph, Michigan: 392–5.

Tonina, D., Luce, C.H., Reiman, B., *et al.* (2008) Hydrological response to timber harvest in northern Idaho: implications for channel scour and persistence of salmonids. *Hydrological Processes* **22**: 3223–35.

17 The Future Role of Information Technology in Erosion Modelling

D.P. GUERTIN[1] AND D.C. GOODRICH[2]

[1]*Landscape Studies Program, School of Natural Resources, University of Arizona, Tucson, AZ, USA*
[2]*USDA-ARS, Southwest Watershed Research Center, Tucson, AZ, USA*

17.1 Introduction

Natural resource decision-making is a complex process requiring cooperation and communication between federal, state and local stakeholders, balancing biophysical and socio-economic concerns. Predicting soil erosion is common practice in natural resource management for assessing the effects of management practices and control techniques on soil productivity, sediment delivery and offsite water quality. Effective decision-making requires the integration of knowledge, data, simulation models and expert judgment to solve practical problems, and to provide a scientific basis for decision-making at the hillslope or watershed scale (National Research Council, 1999).

A user-friendly decision support system (DSS) would assist different professional or stakeholder groups to develop, understand and evaluate alternative soil conservation strategies. The DSS could integrate a suite of components consisting of database management systems (DBMS), geographic information systems (GIS), simulation models, decision models, and easy-to-understand user interfaces. The difficulty in developing a DSS is not a lack of available data or simulation models for erosion prediction, but rather making these models available to decision-makers, a key

Handbook of Erosion Modelling, 1st edition. Edited by R.P.C. Morgan and M.A. Nearing. © 2011 Blackwell Publishing Ltd.

observation made by the National Research Council's Committee on Watershed Management (National Research Council, 1999). Over the last 50 years the federal government has spent millions of dollars on the creation of spatial datasets and model development. While these simulation models are used extensively in research settings, they are infrequently incorporated into the decision-making process. Another aspect of erosion modelling is the continued use of simpler, empirically-based erosion models (e.g. USLE, RUSLE) instead of more complex, physically-based models (e.g. WEPP, EUROSEM). Reasons for this exclusion include: data requirements are usually only attained in a research setting; models are complex and underlying assumptions are poorly understood by resource managers; deriving model input parameters is extremely time-consuming and difficult; and the models are difficult to use with the current interfaces.

For example, Elliot (2004) reported that between 1993 and 1998 over 200 Forest Service specialists were trained to use the USDA Water Erosion Prediction Project model (WEPP) (Flanagan & Livingston, 1995; Elliot & Hall, 1997). Of those specialists, only three or four (or 2%!) subsequently applied the model because the interface was too difficult to operate and too much time was required to assemble the data and interpret the results. Occasional users found it difficult to keep track of which combinations of input files should be used for typical forest and range conditions. Some users were observed to specify unlikely combinations of soil and management files on these highly flexible interfaces, such as specifying a high severity fire soil in combination with a forest road management file.

Part of this problem can be addressed with improvements to model interfaces, lookup tables for model parameters, and internal file management (Hall & Elliot, 2001; Flanagan *et al.*, 2001). However, as erosion models continue to become more complex and integrate with other technologies, users will be required to have experience in DBMS, GIS, computer operating systems, remote sensing, Internet search engines for data gathering, and graphics, as well as a good foundation in erosion process knowledge. Few professionals have all of the above skills.

The solution to this problem is the development of Internet-based applications (Kingston *et al.*, 2000; Elliot, 2004; Flanagan *et al.*, 2004; Kirkby *et al.*, 2004; Miller *et al.*, 2004). Internet-based tools that support erosion modelling and conservation planning include applications to facilitate sharing datasets and software, direct data visualization, and online simulation. Applications exist for erosion modelling at the hillslope, watershed and regional scales. Many of the applications include a spatial component by either using spatial data or displaying the results in the form of maps representing the spatial distribution of erosion. Geospatial technology tools, such as Internet map services, are increasingly being used. Results from erosion models are also components in resource or environmental planning efforts, which are increasingly using Internet-based applications to support the planning process (Kingston *et al.*, 2000).

17.2   Characterization of an Internet Application

Traditional erosion models and applications are closed centralized systems that incorporate models, interfaces and data (Miller *et al.*, 2004). These desktop systems are platform- and application-dependent, and migrating into different operating systems (e.g. Apple vs. Microsoft) or platforms

(e.g. server, desktop, laptop, PDA, or mobile phone) can be time-consuming. Application of new erosion models often requires the use of other applications, such as DBMS, GIS, image processing systems (IPS), and graphic software, all of which must operate on the same OS and platform. The need for different software systems increases the computing requirements (i.e. RAM, disk storage) of the platform, rendering some applications inoperable on some platforms. Users also need expertise in the different applications, and all application data must be stored locally, increasing storage requirements.

Client/server systems are based on generic client/server architecture in network design (Tsou & Buttenfield, 1998; Peng & Tsou, 2003), referred to as 2- or 3-tier systems. A 2-tier system is where the data and applications are located on the same server. A 3-tier system is where the data and applications are located on separate servers. Client/server architecture allows distributed clients (i.e. users) to access a server remotely by using distributed computing techniques such as Remote Procedure Calls (RPCs) or database connectivity techniques such as Open Database Connectivity (ODBC). All necessary applications/ models and data are hosted on the server(s). Applications can be developed that integrate functionality from several different software products and models transparent to the client. The computer resource requirements for the client are considerably less since the client does not need to handle data storage and management, or to install the applications on the local platform. In many cases the client only needs an Internet browser and connection. However, the client-side components are usually platform-dependent, and each client component can access only one server at a time (Peng & Tsou, 2003).

Distributed systems (i.e. distributed computing, distributed services) can connect to, and interact with, multiple and heterogeneous systems and servers at the same time (n-tier systems) and without the constraints of traditional client/ server relationships (Montgomery, 1997; Peng & Tsou, 2003). Under a distributed architecture there is no difference between a client- or a server-based network. A client is defined as the requestor of a service in a network where a server then provides the service. The advantage is that a distributed architecture permits dynamic combinations and linkages between data services and application services via networking. Consequently, data can be stored on a suite of heterogeneous servers and dynamically accessed at an application's request. Potentially, distributed systems will promote the development of accessible institutional data nodes that provide information to Internet-based applications.

The Internet-GIS architecture determines the complexity and efficiency provided by the application. Currently, there are two types of Internet-GIS applications: client (user)-side and server-side. Client-side strategies require the majority of the processing to be conducted by the client on their own computer platforms. This typically requires the web browser to load a program (such as an applet or plug-in) the first time that users request a service. This 'thicker client' architecture provides the advantage of more functionality for users and requires fewer interactions with the server, potentially saving time and using less bandwidth. With this approach there are usually fewer security risks. However, applets are not persistent and must be downloaded at the inception of the application, and plug-ins are required to be downloaded and installed like traditional applications. This type of architecture is typically best for applications with dedicated application users (Plew, 1997) because users are required to have knowledge of handling and manipulating data. Server-side strategies perform all processing on the server, relying on the spatial server to conduct the analysis and generate output (Peng, 1997). These 'thin-client' applications require a high-performance server due to the computation intensity, and have higher network congestion since each operation performed by users must communicate to the server, increasing the need for bandwidth. However, users have transparent access to large and complex datasets, so they do not need either the software or the skills to manipulate data. Users are not required to have sophisticated computers since client machines

perform little processing (Foote & Kirvan, 1997). Since tradeoffs exist between functionality, efficiency and required knowledge, integrated decision support systems should support multiple weight clients, providing access to users with different backgrounds, experiences, and network connection speeds.

Most Internet-based applications that support erosion modelling and control are thin-clients, being distributed from a client/server system (2- or 3-tier server architecture). For most erosion model applications the user is requested to enter the input information, so the data service requirements are relatively small.

17.3   Advantages of Internet-based Applications

The goal of an Internet-based application is to provide information and tools to a user group in a cost-effective manner. Internet-based applications primarily save time and money by centralizing activities. Databases and models can be maintained and located in the same place, with a single update distributed to all potential client users. A few database specialists can also maintain the system for all client users. Advanced software, such as GISs or statistical programs, can also be centrally located on host computers. Consequently, client users do not need to purchase or maintain the software, assuming the licensing of proprietary software for Internet use is available. Client users can also rely on less powerful hardware systems in terms of processing and storage, since most activities are accomplished on the host computer. The system can also be made more secure.

The ultimate advantage of Internet-based applications is that they promote data sharing and equity between stakeholder groups. Internet-based data services have become the primary mechanism for the distribution of data and information. Internet-based decision support tools, such as erosion modelling applications, can provide advanced analysis capabilities to a wide and untraditional audience. The increase in access to information and analysis tools for all citizens encourages equity and shared governance, advancing transparency between different stakeholder groups when addressing potential conflicts. Internet-based tools provide opportunities for shared learning experiences between stakeholder groups, where modelling results based on different proposed alternatives can be quickly viewed by participants. In the near future most formal and informal learning activities will likely be based on Internet applications (Pickles, 1995; Bruckman, 2002; Aggett & McColl, 2006).

17.4   Issues Related to Internet-based Applications

As applications are developed, integrating models, GIS, decision support systems and the Internet, new issues are introduced that should be recognized. These problems range from incompatibilities of technologies used for integrating disparate applications, to security in Internet environments, and are discussed in more detail below.

Since natural resource decision-making requires a coordinated effort between stakeholders representing different groups and levels of government, integrated decision support systems should facilitate interaction and communication among agencies' information systems to make the decision-making process more efficient (Miller *et al.*, 2004). However, different competing application programming platforms (i.e. .NET, Java, C++, FORTRAN, PHP, etc.), operating systems (i.e. Windows, Unix, Linux, etc.), database management systems (i.e. ESRI ArcGIS Server, Oracle, MSSQL Server, MySQL, etc.) make communication difficult or impossible. Standardizing programming languages, operating systems and database management systems for soil conservation stakeholders is impractical since different groups have different budgets, legacy systems, and requirements of their IT infrastructure. Creating a centralized database repository containing environmental data for decision-making is a possibility, but leads to logistical issues such as what data are contained in the database, who

administers the database, how often the database is updated, and who pays for infrastructure. Component-based frameworks have been adopted such as Microsoft's .NET, but lack the inclusion of all programming languages and all operating systems. A standardized protocol that is programming language and platform independent should be utilized when developing integrated decision support systems.

Years of research and development have been spent on developing simulation models that encapsulate our understanding of environmental and erosion processes. These applications represent the current state of knowledge and should be leveraged in the decision-making process. However, these models are often developed using technologies which make interaction with today's object-oriented, web-based technologies cumbersome. Since different programming languages are developed for different purposes, languages that are computationally efficient are often not compatible with languages that have extensive libraries for Internet development, and no single language is ideal for all applications. Therefore, an integrated DSS must be capable of incorporating legacy applications that are built with technology that natively does not communicate with Internet-capable programming languages.

While deploying applications via the Internet drastically increases availability to users, there are still 37% of adult Americans without home broadband access (Horrigan, 2009). Moreover, Internet access is unequally distributed across the US, with only 54% of adult Americans in rural areas having broadband. Low-income Americans also have limited broadband access, with 65% of the households with annual incomes less then $20,000 having no access (Horrigan, 2009). Therefore, rural and low-income Americans are forced to find other alternatives, such as public libraries, to get access to Internet applications. However, the digital divide between the 'haves' and the 'have nots' is narrowing, with a 9% increase in home broadband access between May 2008 and April 2009 (Horrigan, 2009). Importantly, most businesses and government offices have broadband Internet access today, and in the US the Internet is the primary source of information from federal and state agencies.

A limitation in creating richer applications is the lack of bandwidth for Internet access. Bandwidth is the rate at which information can be transferred on a given transmission path (Miller *et al.*, 2004). As internet-based applications become larger and provide more features, the need for high-speed Internet access will increase, especially for server 'thick client' applications. While high-speed access is increasing, applications should target users with broadband. Thus, challenges exist for increasing application functionality while keeping applications available to the majority of Internet users.

Security is always a concern in Internet environments, and reports of security breaches are frequently documented (Grandison & Sloman, 2000; Palmer & Helen, 2001). If Internet-based applications are going to be integrated into the decision-making process, precautions need to taken to assure application security. Secure applications can lead to users trusting the design and architecture of the application; conversely, users are unwilling to expose themselves to unnecessary risks. With land managers storing data in central data warehouses used in Internet applications, data ownership questions arise. For example, does the data placed in a government data warehouse by a watershed group composed of private citizens belong to the private citizen, or become public property? These issues can be argued and must be recognized when using information technology in soil conservation.

17.5 Examples of Internet Applications

17.5.1 Data, information and model sharing

The common Internet-based applications being used today support the distribution of data, information and software. Websites (see Table 17.1) have been developed for professional societies (e.g. European Society for Soil Conservation; International Erosion Control Association; Soil and Water Conservation Society), individual sites for erosion control equipment, installation and

training companies (see the Erosion Control Magazine, Erosion Control Technology Council for potential vendors), and there is even an erosion control information clearinghouse site with an Erosion BLOG (Erosion Control Forum). There are numerous websites supported by federal and local government agencies or non-profit organizations that provide information on erosion control practices (e.g. US EPA – Polluted Runoff, Natural Resource Conservation Service; California Department of Transportation; Tennessee Department of Environmental & Conservation; Center for Watershed Protection).

Most of the traditional erosion models (standalone versions), and supporting information (e.g. documentation), are available for download from websites. Examples of models or modelling support tools used in erosion and water quality assessment currently available from websites include the Universal Soil Loss Equation (USLE), Revised USLE (RUSLE; Ouyang & Bartholic, 2001), Water Erosion Prediction Project model (WEPP; Flanagan *et al.*, 2001), Geo-spatial interface for WEPP (GeoWEPP; Renschler, 2003; Renschler *et al.*, 2002), Wind Erosion Prediction System (WEPS; Hagen, 1991; Wagner, 2001), European Soil Erosion Model (EUROSEM; Morgan *et al.*, 1998), Soil and Water Assessment Tool (SWAT; Arnold & Fohrer, 2005), Automated Geospatial Watershed Assessment tool (AGWA; Miller *et al.*, 2007), and US EPA's Better Assessment Science Integrating Point and Non-point Sources (BASINS) water quality tools and models portal (EPA, 1998; Di Luzio *et al.*, 2009). Several of the websites also provide support data for the different models or links to access data from other websites. Most websites also provide tutorials on using the tools and lists of available publications.

Most of our commonly needed datasets can now be found from source agency websites, including terrain (US Geological Survey), soils (Natural Resource Conservation Service) and land use/land cover (US Geological Survey). There are also data portals available where users can find different sources of geospatial data (e.g. NRCS Geospatial Data Gateway; GIS Data Depot).

Table 17.1 WEB Resources (October, 2009).

Descriptions	URL
California Department of Transportation Erosion Control Toolbox	http://www.dot.ca.gov/hq/LandArch/ec/
Center for Watershed Protection	http://www.cwp.org/
Erosion Control Forum	http://erosioncontrolforum.com/
Erosion Control Magazine International Erosion Control Association Free Trade Journal	http://www.erosioncontrol.com/
Erosion Control Technology Council	http://www.ectc.org/links.asp
European Commission European Soil Data Center	http://eusoils.jrc.ec.europa.eu/
European Society for Soil Conservation	http://www.essc.sk/
European Soil Bureau Pan European Soil Erosion Estimates (PESERA Map Server)	http://eusoils.jrc.ec.europa.eu/website/Pesera/viewer.htm
Kangwon National University, South Korea Sediment Assessment Tool for Effective Erosion Control	http://www.envsys.co.kr/~sateec/)
Lancaster University, UK European Soil Erosion Model, EUROSEM	http://www.es.lancs.ac.uk/people/johnq/EUROSEM.html
Michigan State University Revised USLE (RUSLE) Online Soil Erosion Assessment Tool	http://www.iwr.msu.edu/rusle/
MindSites Group, LLC GIS Data Depot	http://data.geocomm.com/
Minnesota Department of Transportation Approved/Qualified Product Lists	http://www.dot.state.mn.us/products/

Purdue University
  Sediment and Erosion Control Planning, Design and SPECification
    Information and Guidance Tool
	http://cobweb.ecn.purdue.edu/runoff/sedspec/

International Erosion Control Association	http://www.ieca.org/
Soil and Water Conservation Society	http://www.swcs.org/
Tennessee Department of Environmental & Conservation Erosion and Sediment Control Handbook	http://www.state.tn.us/environment/wpc/sed_ero_control handbook/
US Department of Agriculture Agricultural Research Service Universal Soil Loss Equation (USLE)	http://topsoil.nserl.purdue.edu/usle/index.html
US Department of Agriculture Agricultural Research Service Soil and Water Assessment Tool (SWAT)	http://www.brc.tamus.edu/swat/
US Department of Agriculture Agricultural Research Service Automated Geospatial Watershed Assessment Tool (AGWA)	http://www.tucson.ars.ag.gov/agwa/
US Department of Agriculture Agricultural Research Service Water Erosion Prediction Project (WEPP) Official Website	http://www.ars.usda.gov/Research/docs.htm?docid=10621
US Department of Agriculture Agricultural Research Service Water Erosion Prediction Project Climate Assessment Tool (WEPPCAT)	http://typhoon.tucson.ars.ag.gov/weppcat/

US Department of Agriculture Agricultural Research Service
  Water Erosion Prediction Project (WEPP) Web browser interface
    Online
	http://www.geog.buffalo.edu/~rensch/geowepp

US Department of Agriculture Agricultural Research Service
  Wind Erosion Research Unit
  Wind Erosion Prediction System (WEPS)
	http://www.weru.ksu.edu/new_weru/

Table 17.1   (cont'd).

Descriptions	URL

US Department of Agriculture
  US Forest Service
  Forest Service WEPP Interfaces
	http://forest.moscowfsl.wsu.edu/fswepp/

US Department of Agriculture Natural Resource Conservation Service Agronomy and Erosion	http://www.nrcs.usda.gov/technical/agronomy.html
US Department of Agriculture Natural Resource Conservation Service Geospatial Data Gateway	http://datagateway.nrcs.usda.gov/NextPage. aspx?HitTab=1&Progress=0
US Department of Agriculture Natural Resource Conservation Service Soil Data Access	http://soils.usda.gov/
US Department of Agriculture Natural Resource Conservation Service Soil Use	http://soils.usda.gov/use/

US Environmental Protection Agency
  Better Assessment Science Integrating Point and Non-point
  Sources (BASINS)
	http://www.epa.gov/waterscience/basins/b3webdwn.htm

US Environmental Protection Agency
  Multi-Resolution Land Characteristics Consortium (MRLC)
  2006 National land cover Data (NLCD 2006)
	http://www.epa.gov/mrlc/nlcd-2006.html

US Environmental Protection Agency
  Office of Wetlands, Oceans, and Watersheds
  Polluted Runoff (Nonpoint Source Pollution)
	http://www.epa.gov/owow/nps/

US Department of the Interior, US Geological Survey National Geospatial Program	http://www.usgs.gov/ngpo/

US Department of the Interior, US Geological Survey
  National Elevation Dataset
  National Map Seamless Server
	http://seamless.usgs.gov/

University of Buffalo – SUNY GeoWEPP	http://www.geog.buffalo.edu/~rensch/geowepp/

17.5.2 Direct Access

Direct Access (also called Direct Read) refers to applications that allow users to view erosion modelling results directly using a web browser. Non-profit organizations provide erosion potential or vulnerability maps, using a map server, as part of watershed information portals. For example, the Arizona Nonpoint Education for Municipal Officials program (AZNEMO) provides erosion potential in their watershed-based plans that cover Arizona. The erosion potential maps are used to identify watersheds at risk for water quality impairment (http://arizonanemo.org). The Watershed Center in the Grand Travers Bay Watershed on the northwest of Michigan's lower peninsula maintains an Interactive Maps website that includes maps on public lands, wetlands, watershed boundaries, water quality monitoring locations and erosion potential (http://www.gtbay.org/maps.asp). The US Department of Agriculture Natural Resources Conservation Service (NRCS) has a website with maps (in Adobe Portable Document Format) and charts developed using National Resource Inventory (NRI) data (http://soils.usda.gov/use/). NRCS also has information on world soil resources including maps on erosion.

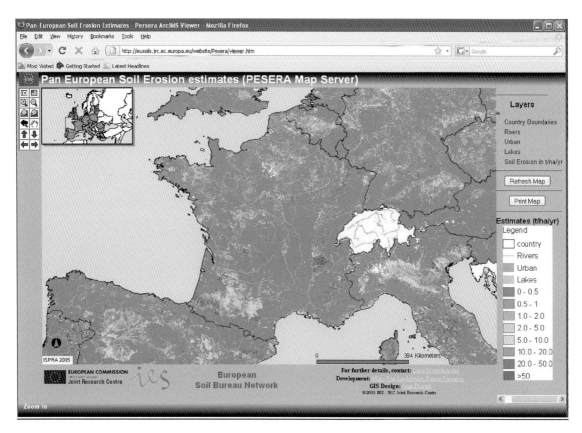

Fig. 17.1 The PESERA Map Server supported by the European Commission Joint Research Centre. The map server is based on ESRI's ArcIMS Internet technology.

Pan European Soil Erosion Estimates Map Server (Fig. 17.1) is an interactive application that allows the user to navigate in the Pan European Soil Erosion Estimates (PESERA) map and data. The PESERA Map Server (http://eusoils.jrc. ec.europa.eu/website/Pesera/viewer.htm) may represent the largest internal map on erosion. Soil erosion estimates (t ha^{-1} y^{-1}) are made by applying the PESERA GRID model at 1 km, using the European Soil Database, CORINE land cover, climate data and a digital elevation model. The resulting estimates of sediment loss are from erosion by water. The PESERA model produces results that depend crucially on land cover as identified by CORINE and the accuracy of the interpolated meteorological data.

17.5.3 Erosion model applications

One area where there has been considerable effort is the creation of hillslope erosion model applications. In most cases the user is expected to provide input information on the hillslope characteristics (e.g. soil type, cover, slope gradient) and there is little to no linkage to external databases, making the applications self-contained. Examples of web-based hydrological applications have been developed by scientists at the USDA Agricultural Research Service (USDA-ARS) and by scientists at the USDA-FS Rocky Mountain Research Station, in Moscow, ID.

In the later 1990s the USDA-ARS Southwest Watershed Research Center developed an

Internet-based hillslope erosion and sediment yield model (HEM: http://eisnr.tucson.ars.ag.gov/hillslopeerosionmodel). The model predicts runoff volume, sediment yield, inter-rill and rill detachment, rill deposition, and mean concentration of sediment for each hillslope segment, provided that the lengths, slopes, percentage canopy and surface ground cover for each hillslope segment, along with runoff volume and a soil erodibility value for the entire hillslope, are known. The HEM model produces graphs depicting the input hillslope profile and distribution of cover on the hillslope, and output for sediment discharge, detachment and deposition, and mean sediment concentration along the hillslope profile (Lane *et al.*, 1995).

The Rocky Mountain Research Station developed the Forest Service Water Erosion Prediction Project (FSWEPP: http://forest.moscowfsl.wsu.edu/fswepp/) interfaces (Elliot & Hall, 1997; Elliott, 2004), which provide the capability to evaluate erosion and sediment delivery from disturbed forest and rangelands. The application uses the Water Erosion Prediction Project (WEPP) model (Flanagan & Livingston, 1995) to estimate erosion rates and sediment delivered using input values developed at the Rocky Mountain Research Station (Elliot & Hall, 1997). The interface provides links to different applications capable of simulating sediment yield from burned areas, a road segment across a buffer, erosion from forest roads, erosion from rangeland, forestland, and forest skid trails. The applications are linked to the Rock:Clime climate generator with a database from more than 2600 weather stations. The different applications found on the US Forest Service WEPP Interfaces website are:
• *Cross Drain* – interface to the Water Erosion Prediction Project soil erosion model (WEPP) to determine optimum cross-drain spacing for existing or planned roads, and for developing and supporting recommendations concerning road construction, reconstruction, realignment, closure, obliteration, or mitigation efforts based on sediment yield.
• *WEPP: Road* – interface to the WEPP model that allows users easily to describe numerous road erosion conditions. The interface presents the results as a summary and extended WEPP output, and has an optional log to store the results from a series of runs.
• *Disturbed WEPP* – interface to the WEPP model to allow users easily to describe numerous disturbed forest and rangeland erosion conditions. The interface presents the results as a summary and extended WEPP outputs, and also presents the probability of a given level of erosion occurring the year following a disturbance.
• *WEPP FuME* – interface to the WEPP model (WEPP) to analyse soil erosion rates associated with fuel management activities. This interface estimates background erosion rates, and predicts erosion associated with mechanical thinning, prescribed fire, and the road network.
• *ERMiT* – the Erosion Risk Management Tool (ERMiT) is a web-based application that uses the WEPP model to estimate erosion, in probabilistic terms, on burned and recovering forest, range and chaparral lands, with and without the application of erosion mitigation treatments (Robichaud *et al.*, 2007). User inputs are processed by ERMiT to combine rain event variability with spatial and temporal variabilities of soil burn severity and soil properties, which are then used as WEPP input parameters. Based on 20 to 40 individual WEPP runs, ERMiT produces a distribution of rain event sediment delivery rates with a probability of occurrence for each of five post-fire years.

Examples of three of these applications are described in Chapter 16.

The WEPP (Water Erosion Prediction Project) web interface can be found at http://milford.nserl.purdue.edu/wepp/weppV1.html (Flanagan *et al.*, 2001, 2004). As noted in earlier chapters, the WEPP model is significantly more complex than the RUSLE model. Consequently, more extensive databases must be bundled with the model in the web interface to enable ready execution via the Internet. In this case, over 20,000 soil database records, climate described at over 2600 locations within the US, and an extensive set of land management examples comprising operation types and dates for cropland and rangeland, are bundled with the website. Four hillslope shapes can be represented (e.g. uniform, convex, concave, S-shape). Once simulation selections are made, and the simulations are completed, WEPP model input files in ASCII format are available on the website. In addition, a wide variety of detailed graphics can be displayed ranging from climatic inputs, plant and residue attributes as a function of simulation time, to time-varying hydraulic and erosion parameters and model outputs.

The well-known RUSLE erosion model (Renard *et al.*, 1997; and see Chapter 8) has been set up as a web-based application for the state of Michigan (Ouyang & Bartholic, 2001; http://www.iwr.msu.edu/rusle/). For this application, erosion can be calculated for agricultural and construction land uses. A simple graphical interface is displayed for the user to select an individual county within the state. Drop-down menus are then displayed for the user to enter run identification information, hillslope characteristics, and soil types from NRCS databases for the selected county. For the agricultural land-use case the user then selects from a list of cropping rotations and tillage practices by year for up to five years. From these selections the various factors of RUSLE equations are obtained from databases and lookup tables built into the system. The calculation for annual erosion is then completed. The *C*-factor (cover) can also be manually set.

The Water Erosion Prediction Project Climate Assessment Tool (WEPPCAT: http://typhoon.tucson.ars.ag.gov/weppcat/) is a further refinement of the hillslope version of WEPP in which simultaneous assessments of climate change and the effectiveness of end-of-field filter strips for mitigation of erosion can be evaluated. It is an easy-to-use, web-based system that allows users to adjust climate inputs for user-specified climate scenarios within the continental US. It allows the user to modify monthly mean maximum and minimum temperatures, the monthly mean number of wet days, monthly mean precipitation, and rainfall intensity in order to predict changes in surface water runoff and erosion rates. WEPPCAT allows the user to assess erosion changes under a large variety of land management

alternatives. It does not require specialized scientific expertise to run, and scenarios are quick and easy to set up.

The Sediment and Erosion Control Planning, Design and SPECification Information and Guidance Tool (SEDSPEC; Tang *et al.*, 2004) predicts small watershed peak runoff and will assist in the design of hydrological, sediment, and erosion control measures. The SEDSPEC system is composed of a model, database, and user interface. Two hydrological models (the Rational Method and TR–55) simulate short-term peak runoff based on site-specific hydrological soil groups and land uses. The hydrological models estimate peak runoff using design storm data stored in associated databases. The DSS integrates WebGIS technology to help users to estimate watershed boundaries and access a spatial database to obtain land use and hydrological soil group data for the watershed. As the final output, SEDSPEC calculates dimensions and costs of hydrological, sediment and erosion control structures based on users' specifications, and provides structure maintenance information. SEDSPEC will provide customized drawings of the structures, and there is a limited amount of interaction which allows users to determine what size structure fits their needs.

17.5.4 Watershed model applications

Unlike the hillslope erosion models, the development of watershed model applications is just beginning, and to date no application is currently available 'online' that specifically addresses erosion and sediment yield. One difficulty with watershed model applications is their need for geospatial information (e.g. digital elevation models, soil maps, land use/land cover). Making the geospatial information for a region locally available for an Internet application using a 2- or 3-tier architecture would be costly (i.e. creation costs, storage requirements, maintenance). The costs would be considerably greater if national or international applications are desired. Requiring users to provide their own data creates other problems related to data quality, storage requirements, greater user capability, and security. Importantly, watershed models are typically more complex then hillslope erosion models and take longer to execute, which requires more processing power or user patience. In the future distributed systems (n-tier architecture) would address some of these issues where datasets would be stored and accessed from different host servers with application datasets assembled as needed. Distributed systems will require high bandwidth access, which is not always present. Examples of Internet-based watershed model applications that address erosion and sediment yield include •AGWA and Web-based SWAT (Park *et al.*, 2009). •AGWA is discussed below.

One of the first watershed applications was •AGWA ('Dot AGWA'), the Internet version of AGWA that grew out of the PC-based Automated Geospatial Watershed Assessment (AGWA) tool developed by the USDA's Agricultural Research Service, in cooperation with the Environmental Protection Agency and the Universities of Arizona and Wyoming. Many of the initial concepts for •AGWA were conceived by Miller *et al.* (2004) and the system was fully developed and implemented as an alpha version (Cate *et al.* 2005, 2006; Cate, 2008). AGWA was developed as a multipurpose hydrological analysis system for use by watershed, water resource, land use, and biological resource managers and scientists developing watershed and basin-scale studies (Miller *et al.*, 2007; Semmens *et al.*, 2008). AGWA incorporates several spatial datasets, GIS mapping, analysis and visualization tools, and two watershed and erosion models into one package, providing easy access to these features. The two watershed models embedded within AGWA are SWAT (Arnold & Fohrer, 2005; http://www.brc.tamus.edu/swat/) for relatively large basin applications, and KINEROS2 (Goodrich *et al.*, 2006; Semmens *et al.*, 2008; http://www.tucson.ars.ag.gov/kineros) for small to medium watershed applications. This enables rapid multiscale watershed analysis.

•AGWA employs ESRI's ArcIMS and Spatial Data Engine (SDE) as well as Oracle's spatial database to provide the GIS data and interactions.

Java-based web server technology is used to connect •AGWA to the watershed models in the application. The web application is based on the Model-View-Controller (MVC) design pattern. This design pattern is useful in separating the presentation components of the system architecture from the data storage and processing components. In this architecture, the Model component allows the different system components to be represented as individual entities. The View is simply the user interface, and the Controller ties the Model and View components together. This separation is useful as it allows changes, replacements or alterations to one of the three components without major changes in other parts of the system. For example, if MySQL is the initial supporting database used in the system and it becomes inadequate when the user base expands greatly, then another database like PostgeSQL can be inserted into the existing system with minimal effort or interruption.

Users can define management scenarios, and like AGWA, have the application parameterize and run the models for the defined management plan. Both of the watershed models are spatially distributed so that simulation results for erosion and sediment transport, as well as hydrology, can be imported back into •AGWA and mapped back onto upland or channel elements within the internet view. Different output formats (i.e. XML, Word doc, HTML) for the resulting simulation output can also be specified by the user. As noted, the tool leverages client-server architecture so that changes and improvements in core components will not disrupt end-user interaction with the application. The alpha-based application is currently undergoing further development and is not supported for general Internet access at this time.

17.6 Example of an Internet-Based Application

Flanagan *et al.* (2004) described the Internet-based WEPP-GIS application (WEPP-GIS; http://milford.nserl.purdue.edu/wepp/gis2.php?IES=1). The application utilizes the core procedures developed by Cochrane and Flanagan (1999) and implemented in GeoWEPP (Renschler *et al.*, 2002; Renschler, 2003), but within an interface that only requires a web-browser and Internet connection on the user's computer (a 'thin-client' application). GIS Viewer software allows users to specify an area of interest to model with WEPP, then digital elevation model (DEM) data for the area are sent to topographic parameterization software to delineate watersheds, channels and hillslopes. The DEM data are processed on the server side, and then images of the delineated watershed and hillslopes are passed to the user's web-browser. Once the hillslopes and channels have been located, WEPP model simulations of representative hillslope profiles and channels, and/or all flowpaths in the watershed, are conducted. The simulated soil erosion results in graphical format are sent as images to the client computer. Subsequent model simulations using different land management practices can help to show the impact of conservation practices on hillslope runoff and erosion.

Plate 14 provides a schematic of the WEPP-GIS application. The application uses the open source MapServer environment from the University of Minnesota (http://mapserver.gis.umn.edu) as the basic Web GIS. The TOPAZ (Topographic PArameteriZation) (Garbrecht & Martz, 1997) digital landscape analysis tool is used for channel, watershed and sub-basin (hillslope) delineation. There are six major software components of the Internet-based WEPP GIS application. Users can select a US State of interest. They then can zoom in to find their specific area of interest. The data for display are obtained from the TerraServer site (http://terraservice.net) and from local spatial data on the National Soil Erosion Research Laboratory (NSERL) server. Image data are sent from the MapServer software (1 in Plate 14) to the client's web-browser, and MapServer also handles requests for zooming and panning in the display. After the location of interest has been identified, TopazPrep software (2 in Plate 14) extracts a region of the DEM to process with TOPAZ. TopazPrep is custom software coded in C++ and PHP. PHP is an open-source scripting language used for web development and it can be inserted into HTML (HyperText Markup Language).

The TOPAZ software (3 in Plate 14) is run at least twice. The first time is to delineate the entire network of channels within the displayed region of the DEM. Once the delineated channels are visible, the user can either accept them, or alter the critical source area and minimum channel length parameters and rerun TOPAZ until a satisfactory representation of the channel network is obtained. The user must then select the outlet point for the watershed of interest, after which TOPAZ is run a second time to delineate the watershed boundary and sub-basins (i.e. hillslope regions). The area the user can model is currently limited to 0.25 degrees latitude by 0.25 degrees longitude, in order to ensure that TOPAZ can handle the extracted DEM and have a reasonable response time. Once an acceptable watershed has been delineated, the WeppPrep (4 in Plate 14) program (custom software also written in C++ and PHP) generates WEPP inputs from the extracted DEM, land use, soils and TOPAZ watershed configuration. WeppPrep also executes the CLIGEN (5 in Plate 14) weather generator (Nicks *et al.*, 1995) to create a climate input file for WEPP. Finally, the WEPP model (6 in Plate 14) is run on the hillslopes/channels and/or flowpaths. Once the WEPP simulations are completed, WeppPrep prepares the output files, interprets the results and produces maps which are sent to the client using MapServer.

The WEPP-GIS application represents a client/server system (2- or 3-tier architecture) if only the NSERL server is used to provide spatial data. However, if users utilize topographic map images (digital raster graphs) and aerial photography to assist in locating the area where they wish to apply the WEPP watershed model, they are using a distributed system (n-tier architecture) since the data is retrieved on demand from TerraServer USA (http://terraservice.net) using a Web Mapping Services protocol. The WEPP-GIS is a 'thin-client', where the client (i.e. user) only needs a web browser to access the application. All other data processing is accomplished on the host server, in the above example the execution of TOPEZ, CLIGEN and WEPP, gathering input data for the map server and/or the models, and preparing the output.

17.7   Conclusion

The WEPP-GIS application provides a glimpse of how future Internet-based applications will be developed. With improvements in bandwidth and Internet access, applications will increasingly be based on distributed systems where input data will be accessed as needed and data will be stored temporarily. It is foreseeable that tools such as TOPEZ and CLIGEN will become application services in themselves, in which the WEPP-GIS client would request their services. What is known for certain is that Internet-based applications will evolve to become the primary mechanism by which most users apply erosion models.

Acknowledgements

The authors wish to acknowledge the Arizona Agricultural Experiment Station and the USDA Agricultural Research Service Southwest Watershed Research Center for their support. The authors also wish to acknowledge that the above review was not intended to be all-inclusive and we deeply apologize to all the worthy developers we did not cite.

References

Aggett, G. & McColl, C. (2006) Evaluating decision support systems for PPGIS applications. *Cartography and Geographic Information Science* **33**: 77–92.

Arnold, J.G. & Fohrer, N. (2005) SWAT2000: current capabilities and research opportunities in applied watershed modeling. *Hydrological Processes* **19**: 563–72.

Bruckman, A. (2002) The future of E-learning communities. *Communications of the ACM* **45**: 60–63.

Cate, A.J. (2008) *Data Mining and the Development of Internet-based Tools for Watershed Assessments.* PhD dissertation, University of Arizona.

Cate, A.J., Semmens, D.J., Burns, I.S., *et al.* (2005) *AGWA Design Documentation: Migrating to ArcGIS and the Internet.* EPA/600/R-05/056, ARS/181027.

Cate, A.J., Goodrich, D.C. & Guertin, D.P. (2006) *Integrating hydrologic models and spatial data in a distributed internet application.* Proceedings, 3rd Federal Interagency Hydrologic Modeling Conference, 2–6 April 2006, Reno, Nevada. CD-ROM.

Cochrane, T.A. & Flanagan, D.C. (1999) Assessing water erosion in small watersheds using WEPP with GIS and digital elevation models. *Journal of Soil and Water Conservation* **54**: 678–85.

Di Luzio, M., Srinivasan, R. & Arnold, J.G. (2009) Integration of watershed tools and SWAT Model into BASINS. *Journal of the American Water Resources Association* **38**: 1127–41.

Elliot, W.J. (2004) WEPP internet interfaces for forest erosion prediction. *Journal of the American Water Resources Association* **40**: 299–309.

Elliot, W.J. & Hall, D.E. (1997) Water Erosion Prediction Project (WEPP) Forest Applications. General Technical Report INTGTR-365, USDA Forest Service, Rocky Mountain Research Station, Ogden, Utah. Available at: http://forest.moscowfsl.wsu.edu/engr/library/Elliot/Elliot1997f/1997f.pdf [accessed July 2009].

EPA (1998) *Better assessment science integrating point and nonpoint sources, BASINS 2.0.* EPA-823-B-98-006.

Flanagan, D.C. & Livingston, S.J. (eds) (1995) *WEPP User Summary.* NSERL Report No. 11, National Soil Erosion Research Laboratory, West Lafayette, Indiana.

Flanagan, D.C., Renschler, C.S., Frankenberger, J.R., *et al.* (2001) Enhanced WEPP model applicability for improved erosion prediction. In Sponsor, M. (ed.), *MODSIM 2001 International Congress on Modelling and Simulation.* Modelling and Simulation Society of Australia and New Zealand, December 2001: 179–84. Available at: http://www.mssanz.org.au/MODSIM01/Vol%204/Flanagan.pdf

Flanagan, D.C., Frankenberger, J.R. & Engel, B.A. (2004) *Web-based GIS application of the WEPP Model.* ASAE/CSAE Meeting Paper No. 04-2024. ASAE, St. Joseph, Michigan.

Foote, K.E. & Kirvan, A.P. (1997) WebGIS. NCGIA Core Curriculum in Geographic Information Science. Available at: http://www.ncgia.ucsb.edu/giscc/units/u133/u133_f.html

Garbrecht, J. & Martz, L.W. (1997) *TOPAZ: an automated digital landscape analysis tool for topographic evaluation, drainage identification, watershed segmentation and subcatchment parameterization: overview.* ARS-NAWQL 95-1, US Department of

Agriculture, Agricultural Research Service, Durant, Oklahoma.

Goodrich, D.C., Unkrich, C.L., Smith, R.E. & Woolhiser, D.A. (2006) KINEROS2 – New features and capabilities. Proceedings, 3rd Federal Interagency Hydrologic Modeling Conference, 2–6 April 2006. Reno, Nevada. CD-ROM. Available at: http://acwi.gov/hydrology/mtsconfwkshops/conf_proceedings/3rdFIHMC/Poster_Goodrich.pdf

Grandison, T. & Sloman, M. (2000) *A survey of trust in internet applications.* IEEE Communications Surveys and Tutorials, Fourth Quarter. Available at: http://www.comsoc.org/pubs/surveys/

Hagen, L.J. (1991) A wind erosion prediction system to meet user needs. *Journal of Soil and Water Conservation* **46**: 106–11.

Hall, D.E. & Elliot, W.J. (2001) Interfacing soil erosion models for the World Wide Web. In Sponsor, M. (ed.), *MODSIM 2001 International Congress on Modelling and Simulation.* Modelling and Simulation Society of Australia and New Zealand, December 2001: 179–84. Available at http://www.mssanz.org.au/MODSIM01/Vol%201/Hall.pdf

Horrigan, J. (2009) *Home Broadband Adoption 2009.* Pew Internet & American Life Project, Pew Research Center. Available at: http://www.pewinternet.org/Reports/2009/10-Home-Broadband-Adoption-2009.aspx

Kingston, R., Carver, S., Evans, A. & Turton, I. (2000) Web-based public participation geographic information systems: an aid to local environmental decision-making. *Computers, Environment and Urban Systems* **24**: 109–25.

Kirkby, M.J., Jones, R.J.A., Irvine, B., *et al.* 2004. *Pan-European Soil Erosion Risk Assessment: The PESERA Map, Version 1, October 2003.* Explanation of Special Publication Ispra 2004 No. 73 (S.P.I.04.73). European Soil Bureau Research Report No. 16, EUR 21176, and 1 map in ISO B1 format. Office for Official Publications of the European Communities, Luxembourg.

Lane, L.J., Nichols, M.H. & Paige, G.B. (1995) Modeling erosion on hillslopes: concepts, theory, and data. In Binning, P., Bridgman, H. & Williams, B. (eds), *Proceedings of the International Congress on Modelling and Simulation (MODSIM'95), 27–30 November 1995.* University of Newcastle, Newcastle, NSW, Australia. Uniprint, Perth, Australia.

Miller, R.C., Guertin, D.P. & Heilman, P. (2004) Information technology in watershed management decision making. *Journal of the American Water Resources Association (JAWRA)* **40**: 347–57.

Miller, S.N., Semmens, D.J., Goodrich, D.C., *et al.* (2007) The Automated Geospatial Watershed Assessment Tool. *Journal of Environmental Modeling and Software* **22**: 365–77.

Montgomery, J. (1997) Distributing components. *BYTE* **22**: 93–8.

Morgan, R.P.C., Quinton, J.N., Smith, R.E., *et al.* (1998) The European Soil Erosion Model (EUROSEM): a dynamic approach for predicting sediment transport from fields and small catchments. *Earth Surface Processes and Landforms* **23**: 527–44.

National Research Council (1999) *New Strategies for America's Watersheds.* National Academy Press, Washington, DC.

Nicks, A.D., Lane, L.J. & Gander, G.A. (1995) Weather generator. In Flanagan, D.C. & Nearing, M.A. (eds), *USDA-Water Erosion Prediction Project (WEPP): Hillslope Profile and Watershed Model Documentation.* NSERL Report No. 10, USDA Agricultural Research Service, West Lafayette, Indiana, Chapter 2.

Ouyang, D. & Bartholic, J. (2001) Web-based GIS application for soil erosion prediction. In Ascough II, J.C. & Flanagan, D.C. (eds), *Soil Erosion Research for the 21st Century, International Symposium, Honolulu, HI, 3–5 January 2001.* ASAE.701P0007, St. Joseph, MI: 260–63.

Palmer, C. & Helen, M. (2001) Abstracts of recent articles and literature. *Computers & Security* **20**: 70–74.

Park, Y.S., Kim, J., Engel, B., *et al.* (2009) Development of web-based SWAT system. Presentation at the 2009 ASABE Annual International Meeting, Reno, NV, 21–24 June 2009. Available at: http://asae.frymulti.com/

Peng, Z.-R. (1997) *An assessment of the development of Internet GIS.* ESRI International User Conference, 8–11 July 1997.

Peng, Z.-R. & Tsou, M.-H. (2003) *Internet GIS.* John Wiley & Sons Inc., Hoboken, New Jersey.

Pickles, J. (ed.) (1995) *Ground Truth: the Social Implications of Geographic Information Systems.* Guilford Press, New York.

Plew, B. (1997) *GIS Online: Information Retrieval, Mapping, and the Internet.* OnWord Press, Albany, New York.

Renard, K.G., Foster, G.A., Weesies, G.A., *et al.* (1997) *Predicting Soil Erosion by Water: a Guide to Conservation Planning with the Revised Universal Soil Loss Equation (RUSLE).* USDA-ARS Agricultural Handbook No. 703, 384 p.

Renschler, C.S. (2003) Designing geo-spatial interfaces to scale process models: the GeoWEPP approach. *Hydrological Processes* **17**: 1005–17.

Renschler, C.S., Flanagan, D.C., Engel, B.A. & Frankenberger, J.R. (2002) *GeoWEPP – the Geospatial Interface to the Water Erosion Prediction Project.* ASAE Paper No. 02-2171. ASAE, St. Joseph, MI.

Robichaud, P.R., Elliot, W.J., Pierson, F.B., *et al.* (2007) *Erosion Risk Management Tool (ERMiT) User Manual.* USDA Forest Service, Rocky Mountain Research Station General Technical Report RMRS-FTR-188.

Semmens, D.J., Goodrich, D.C., Unkrich, C.L., *et al.* (2008) KINEROS2 and the AGWA modeling framework, 2008. In Wheater, H., Sorooshian, S. & Sharma, K.D. (eds), *Hydrological Modelling in Arid and Semi-Arid Areas.* Cambridge University Press, Cambridge: 49–69.

Tang, Z., Engel, B.A., Choi, J., *et al.* (2004) A web-based DSS for erosion control structure planning. *Applied Engineering in Agriculture* **20**: 707–14.

Tsou, M.-H. & Buttenfield, D.P. (1998) Client/server components and metadata objects for distributed geographic information services. In *Proceedings of GIS/LIS'98, Fort Worth, Texas, 10–12 November 1998.* American Society of Photogrammetry and Remote Sensing, Bethesda, Maryland: 590–599.

Wagner, L.E. (2001) WEPS 1.0: What it is and what it isn't. In Ascough II, J.C. & Flanagan, D.C. (eds), *Proceedings of International Symposium and Exhibition: Soil Erosion Research for the 21st century, Honolulu HI, 3–5 January 2001.* ASAE, St Joseph, MI: 372–5.

18   Applications of Long-Term Erosion and Landscape Evolution Models

G.R. WILLGOOSE[1] AND G.R. HANCOCK[2]

[1]*School of Engineering, Faculty of Engineering and the Built Environment,*
The University of Newcastle, Callaghan, New South Wales, Australia
[2]*School of Environment and Life Sciences, Faculty of Science, The University of Newcastle,*
Callaghan, New South Wales, Australia

18.1   A Short History of Landform Evolution Modelling

Landform evolution models were initially developed as a means of understanding the links between environmental processes acting on a landform (primarily runoff and erosion) and the form of the landscape resulting from the long-term action of those processes (i.e. the geomorphology). With their use we are able to (1) identify the geomorphological fingerprint of an individual process, so that aspects of the natural process acting on that landform can be identified and/or calibrated (or at least constrained) from the landform statistics, and (2) relate geomorphological scaling and organization principles to process, allowing simplified, but still physically-based, representations of catchment-scale processes. More recent developments have seen these models used for practical erosion modelling applications. It is these recent developments that will be the focus of this chapter. We first provide some context for the original models because their capabilities are underpinned by the science agenda justifying their original and continued development.

Gilbert (1909) was the first to note that hillslopes subjected to an erosion process with a rate that increased with slope, but which was inde-

pendent of distance downslope, resulted in slopes with convex profiles (hereafter a convex slope is one that has a downward curvature in its elevations in the direction of flow, while a concave slope has an upward curvature). Kirkby (1971) was the first to quantify the process interaction between distance down a hillslope and slope, defining the hillslope profile and its concavity in terms of causal processes. Kirkby did this by solving a differential equation for hillslope elevations and erosion processes, constructing in the process a simple landform evolution model.

In Kirkby's work, hillslopes were assumed to have parallel flow lines with a fixed downstream boundary condition (determined by the river bed elevation) so that neither flow convergence and divergence nor interaction with a dynamic river were considered in his solutions. Ahnert (1976) constructed the first landform evolution model where flow convergence was allowed, and applied it to understanding the evolution of mountain ranges (e.g. Ahnert, 1984).

The modern generation of landform evolution models is generally considered to have started with those presented by Willgoose *et al.* (1991a) and Howard (1994). These models simulated entire catchments rather than a single hillslope, the simultaneous evolution of the hillslopes and channels by different processes (and channel extension and retreat), and allowed for flow convergence on slopes. Moglen and Bras (1994) built on the work of Willgoose to allow

Handbook of Erosion Modelling, 1st edition. Edited by R.P.C. Morgan and M.A. Nearing. © 2011 Blackwell Publishing Ltd.

the modelling of flow divergence. The main difference in the approaches of Willgoose and Howard was in the dominant erosion process modelled. Willgoose used a transport-limited erosion process where the driver of erosion and deposition was whether the amount of sediment carried by the flow was above or below the transport capacity of the flow. Howard modelled a detachment-limited erosion process where the erosion rate was determined by the ability of the flow to detach particles from the land surface. Willgoose assumed that there was no limitation on the detachment rate. Howard implicitly assumed that sediment transport was always below the transport capacity, so his model was unable to model sediment deposition. Subsequent developments of both models have seen a convergence of their capabilities, with both now capable of modelling both transport- and detachment-limited erosion.

Subsequent developments in landform evolution models (there are now many tens of models worldwide) are detailed in Willgoose (2005), but highlights include: (1) quantitative testing of the models against experimental evolving landforms and field landforms (both natural and man-made); (2) inclusion of more physical processes for hillslopes (e.g. mass movements such as landslides and debris flows); (3) models for the dynamics of meandering rivers; (4) different numeric algorithms (although testing of their relative merits has been limited); and (5) extra-terrestrial geomorphology applications (mostly for Mars). For the purpose of this Handbook the main areas of active development are in the modelling of soil and vegetation dynamics and understanding how they co-evolve with the evolving landform. Vegetation and soil dynamics are at the research frontier, and we will return to them later in the chapter.

When we simultaneously model the evolution of environmental states (e.g. vegetation) as well as the landform elevations, the models should rightly be called landscape evolution models rather than landform evolution models. However, the original terminology has stuck, and in the literature the names are used interchangeably, commonly abbreviated in both cases to LEM.

18.2   Using LEMs as Erosion Models

At their most fundamental level, and at the risk of oversimplifying, LEMs are erosion models where the landform evolves with time in response to erosion and where the erosion changes in response to the changing landform. For instance, if a landform is allowed to evolve for one year, then the difference between the landform at the start and end of the year gives the erosion and/or deposition during that year. If the physics modelled is the same, and the amount of landform change in that year is small, then LEMs will (and do) give the same result as traditional erosion models (we use 'traditional' in this chapter to refer to models that do not simulate the evolution of the landform, models as discussed in the previous chapters of this book). It is from this starting point that we will discuss the use of LEMs.

All LEMs simulate some form of fluvial erosion for the hillslopes and channels. They may also model other processes that are crucial once the landform begins to evolve (e.g. soil creep, debris flows) and which may or may not be important for any specific application. If we focus for a moment just on the fluvial erosion models that are used in LEMs, the debate about the correctness and/or adequacy of the underlying physics is no different from traditional models. That said, there are some aspects of erosion physics that are central to how the landform will evolve, but which are typically not as critical in traditional models. We will highlight some of the important ones (e.g. the relative importance of discharge and slope in the erosion equation) below. Typically because of the computational effort in running an LEM simulation, the erosion physics is simpler relative to traditional models, though the core mechanism of shear-stress driven processes is correctly implemented.

Inter-rill erosion, typically rainsplash and rainflow, is also normally modelled in LEMs (usually with a Fickian diffusion term), although at the catchment scale for which most LEMs are applied, rainsplash is rarely a significant component. Inter-rill erosion is generally considered by practitioners to dominate for the first few metres of flow, which is normally below the resolution of most LEMs. Most LEMs discretize the catchment elevation onto a grid or triangulation with a resolution of tens of metres, so that rainsplash is only active at a sub-grid resolution. Applications with metre-scale resolution are rare because of computer limitations.

In addition to fluvial erosion, most LEMs include a variety of other processes including soil creep, landsliding and debris flows. These processes are not normally included in traditional models (but see Chapter 14), and represent one of the most significant process differences between LEMs and traditional models. One problem in discussing LEMs in general is that not all LEMs include all of these non-fluvial processes. Which of these processes are modelled is typically a reflection of the research heritage of the model. Furthermore, which of these processes is important for any particular application is also sometimes difficult to judge, even for researchers. However, we do know that to generate naturally occurring rolling soil-mantled hillslopes with convex hilltops, it is necessary to include at least soil creep and rainsplash (called 'diffusive' processes because they tend to smooth the landform) in addition to fluvial erosion (which generates the concave-up parts of the landscape). To state it more generally, we know that to generate natural landscapes with regions of convex and concave hillslopes we need at least two processes: fluvial and some other 'diffusive' process. Which diffusive process applies is site-specific, but for soil-mantled landscapes some form of creep or soil mass movement process normally needs to be modelled.

All LEM fluvial erosion models have a sediment transport process that increases in flux/unit width with increasing water discharge/unit width. This means that if flow converges as it moves downstream as, for instance, the result of the development of a rill, then the erosion within the rill will increase relative to the adjacent uniform sheetflow. Outside of the rill, erosion will decrease relative to uniform sheetflow because discharge/unit width is decreasing. This means

that, in the absence of other processes (e.g. rainsplash), rills and gullies develop as result of the positive reinforcement of the physics of fluvial erosion. This is the competition between rill and inter-rill erosion in traditional models. In principle, LEMs can directly model this process where the processes interact to change an initially smooth landform surface into a rilled surface, while traditional models, where the landform does not change in response to the erosion, cannot. Moreover, if the dependence of sediment transport is superlinear with discharge/unit width (i.e. *transport* ∝ *discharge^m* where $m > 1$), then the sediment transport from a given width of rilled/gullied catchment (i.e. one where there are series of rills separated by inter-rills along the contour) will be higher than for the same width of uniform depth sheetflow, so that the erosion rate is higher. Many traditional models use empirical rilling factors that allow for this increase in erosion as a result of rilling which, in principle, are unnecessary for LEMs because the rilling process can be modelled directly.

Gully erosion has a further important consequence. Many man-made containment structures for potentially hazardous and nuclear waste cover the waste material with a layer of benign low erodibility material – the capping layer. Since failure of the structure will occur if any of the waste is released, the capping layer must resist erosion. As we will show in the examples below, when overland flow convergence can occur as a result of landform evolution, the landscape becomes covered in a network of high erosion regions (the gullies) separated by the low erosion regions (the ridges). Failure occurs when the gullies penetrate through the capping. It is of little consequence to know that the average erosion across the whole structure is less than the thickness of the capping if the point of deepest erosion determines failure. The deepest erosion is much higher than the average because the average includes all the intervening ridges where erosion is very low.

In farmland and rehabilitated mine sites, contour banks are a major form of erosion control. They work by (1) reducing the sediment discharge on the hillslope by reducing hillslope lengths and (2) capturing sediment behind the bank. Both of these processes can be satisfactorily modelled with traditional models. However, as sediment accumulates behind the contour bank the hydraulic conveyance of the channel along the contour bank is reduced. Eventually, if the channel is not periodically maintained, water in a storm will overflow the contour bank as a result of channel blockage. At this stage the contour bank will be eroded and the flow behind the contour bank concentrated into a gully that develops below the bank. Modelling post-failure contour banks is well suited to LEM application. However, to date, post-failure analyses have not been common because they require being able to simulate the hydraulics of the flow behind the contour bank within a rainfall event. To date, within-event hydraulics has been seen as being too computationally intensive in the context of the multi-year erosion simulations carried out with a LEM.

Geomorphologically stable landforms have long been an objective of landform rehabilitation design. These landforms would degrade less rapidly (and presumably re-establish a natural ecosystem more quickly), have less off-site impact, and look more natural. LEMs allow us to model the evolution of these landforms, and quantitatively assess their stability and how far they are from equilibrium. For instance, we know that fluvial erosion leads to concave hillslopes, and the link between the erosion process and its natural concavity is known. Providing quantitative predictions of the stability of a concave landform is important because waste disposal structures are normally designed with a convex (not concave) profile, because this profile minimizes the landform footprint for a given volume of waste. This reduced footprint is by default one of the main design criteria for waste disposal rehabilitation, and LEMs have assisted in making the case that the increased footprint of a structure with concave-up components is a compromise worth making. We will talk about some of the real world challenges of building a geomorphic design in the examples below.

Finally, by their nature LEMS calculate the sediment transport balance on a DEM (either gridded elevations or a triangulated mesh), simulating erosion and deposition at a point from the sediment transport balance at that point. Thus they are inherently spatially distributed. Many modern traditional models are also spatially distributed, even if their landform does not evolve. Either way, because they are spatially distributed they are excellent candidates for inclusion as components of GIS or CAD design tools. Most GIS, however, are not good at handling an evolving landform (more typically GIS are used to display some evolving property on a fixed landform), so LEM capabilities are not fully exploited by GIS interfaces. Accordingly most have been integrated into scientific visualization tools (where an evolving landform is not a constraint) rather than GIS, or stand-alone custom interfaces. Unfortunately this has compromised their ability to integrate into existing land management tools, which are mostly GIS-based, although there are some initiatives on the horizon that might (at least partially) address this problem (e.g. CSDMS: Syvitski *et al.*, 2004; TelluSim: Willgoose, 2009).

This section has summarized some of the similarities and differences between LEMs and traditional erosion models, and highlighted how those differences influence the scope of applications for LEMs. To make these general statements more concrete, we now look at some examples of applications of LEMs over the last decade. These examples will highlight how LEM capability has elucidated aspects of the erosion assessment that are either difficult or impossible to do using traditional approaches.

18.3   Application Case Studies

18.3.1   Example 1: Encapsulation structures

In many industries there is a need to build earth- or rock-covered structures to encapsulate waste that may otherwise be a potential danger to the environment or life. These wastes include mining (tailings, below-grade ore, and waste rock), chemical, nuclear, and household solid waste.

In some industries the timescales required of containment are long. For instance, worldwide it is generally required that even the most benign waste from the nuclear industry must be contained for a minimum of 1000 years (e.g. Willgoose & Riley, 1998a,b; Crowell *et al.*, 2005). Long design lifetimes are also being increasingly required worldwide for the rehabilitation of mine sites. This is particularly true when: (1) it is anticipated there will be no active management of the site post-closure; (2) governments perceive that they may be required to fund clean-up in the event of failure; or (3) governments may be held legally responsible in the event of failure because they signed off the containment strategy.

Often this waste has to be contained above the ground surface. In this case, erosion of the encapsulation structure is a primary failure mode. Often above-ground structures are difficult to avoid. For instance, in the mining industry when rock is mined it is fragmented as part of the mining operation. The fragmentation of the rock increases its volume by 30–40%, primarily because of the newly created air voids between the rock fragments relative to what was previously solid rock. This means that for mining operations where the metal being extracted is only a small percentage of the volume (typically all mines except coal, iron ore, and aluminium) it is not possible to place all of the mine waste back in the hole from which it was mined, so some of the waste needs to be stored above the ground. Even for the exceptions there can be operational reasons why placing the waste back in the hole is unfeasible or very difficult.

The first application of a landform evolution model to the assessment of the long-term stability of an encapsulation structure was by Willgoose and Riley (1993, 1998a,b) who examined an above-ground structure (proposed as part of the rehabilitation design) for the Ranger Uranium Mine, Northern Territory, Australia. The discussion below follows Willgoose and Riley (1998a,b) with updates reflecting follow-on work.

Field runoff and erosion plots (with a range of areas and slopes) were used to calibrate an event hydrology model and an event erosion model.

The event hydrology model was then used to generate a 20-year, 15-minute resolution runoff series using recorded pluviograph data from a nearby meteorological station. This runoff series was then used with the event erosion model to generate a 15-minute resolution erosion series. The long-term average erosion model in the SIBERIA LEM, which relates sediment transport rate to catchment area and slope, was then calibrated to this erosion series to yield the transport law to be used in the long-term simulations. This calibration process meant that the erosion law used for the landform evolution simulations yielded the result of the average sediment transport, and did not model the effect of individual runoff and erosion events. Rainsplash erosion effects were calibrated by comparing paired plots that were covered with shadecloth and plots without rainsplash protection. Soil creep and other mass movement processes were not modelled because they were not believed to be important at this site.

The initial landform and the landform after 1000 years of erosion are shown in Fig. 18.1. The first thing to note is that the erosion that has occurred is not uniform in space. There are regions of localized high erosion (i.e. the valleys) separated by regions of low erosion (i.e. the ridges). The high-erosion regions have total erosion depths of up to 8 m, while some of the ridges have barely eroded. Note that due to the vertical exaggeration in the figures, the high-erosion areas look like gullies but are in fact more like valleys (8 m deep by 120 m wide). The valleys are initiated at the transition from the low slope top of the structure to the higher slope batters. They subsequently propagate upstream from the transition, while simultaneously cutting down. On the flatter natural areas surrounding the landform (and downstream of the valleys in the structure), alluvial fans are created. Over time they increase in depth and extent from the structure. These fans are barely visible in Fig. 18.1. The rapid valley downcutting stops when the valley floor cuts down to the height of the depositing alluvial fan (Fig. 18.2). Subsequently, depending on the geometry of the landform upstream of the initiation

(a)

(b)

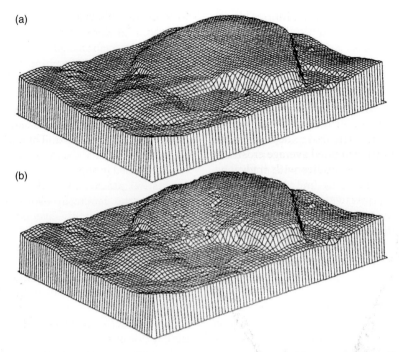

Fig. 18.1 Landform evolution simulations for the Ranger Uranium Mine, Northern Territory, Australia: (a) the initial design landform at 0 years, and (b) the design landform after 1000 years of erosion. Some features of note are: (1) the above-ground structure is the large doughnut-shaped structure with steep sides and flatter top in the centre of the picture; (2) the rounded hill in the centre left foreground, that appears from this viewing angle to be an extension of the structure, is a natural hill; (3) the large flat depression in the centre front is the rehabilitated mine pit; and (4) the flat triangular depression in the right foreground to the immediate right of the above-ground structure is a sedimentation basin. The approximate horizontal dimensions of the domain are 1.5 km by 1 km, the above ground structure is approximately 25 m high, and the grid spacing is 60 m. The simulations were carried out with a 30 m grid, but for ease of visualization only every second grid point is shown. The figure is vertically exaggerated for clarity. From Willgoose and Riley (1998a,b).

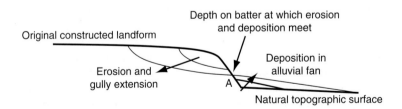

Fig. 18.2 A schematic of the balance between gully erosion on the above-ground landform and alluvial fan deposition on the surrounding natural terrain. Erosion and deposition proceed rapidly until the elevation of the outlet of the gully reaches the elevation of the top of the fan (point A in the figure). After this time the rate of downcutting of the gully and deposition of the fan depend on the balance between erosion and deposition, in a complex balance involving the upstream shape of the landform into which the gully is downcutting, and the downstream topography onto which the fan is being deposited.

point, we have observed either (1) infilling of the valley which is driven by continuing deposition on the alluvial fan, or (2) continued valley down-cutting and incision into the deposited alluvial fan and remobilization of the fan sediments. In the example in Fig. 18.1 the formation of the deep valleys was important because it was planned to encapsulate the waste with 2 m of environmentally benign material. The design depth of material was based on an estimated average erosion of 0.3 m found by previous studies with traditional models (and also found by Willgoose & Riley using SIBERIA). However, while the average erosion of 0.3 m suggested that the encapsulation layer of 2 m depth would not fail, the existence of 8 m valleys would clearly result in the failure of the structure. This was the major novel finding of Willgoose and Riley over previous work using traditional models at the site.

The exact location of the valleys depends on the initial roughness and drainage pattern on the landform. Willgoose and Riley (1998a,b) found that small changes in the exact position of flow concentration on the initial landform changed the exact location of the valleys. Thus small changes in the initial landform elevations (e.g. errors in construction, dozer tracks, mine waste compaction, etc.) would significantly change the location of the valleys 1000 years later. Morgan (1994) (see also Willgoose & Gyasi-Agyei, 1995) extended this work by doing Monte Carlo simulations where he randomly varied the initial topography with noise consistent with random settlement of up to 1.0 m, and then statistically analysed the resulting erosion simulations. Morgan used a slightly different landform design proposal from that in Willgoose and Riley because of limitations in computer speed at the time of the work (these days, multiple Monte Carlo simulations are easily done on a desktop computer). Figure 18.3 shows a plot of the landform, the average and maximum erosion depth, and a plot of the probability of the erosion being more than 2 m (i.e. the depth of the encapsulation layer) across all the erosion simulations. Firstly, the maximum erosion at the transition from the flat cap to the steep batter is about three times the

average, and this region of high erosion is concentrated at the slope transition as was previously observed by Willgoose and Riley. Secondly, the probability of failure of any point of the landform (i.e. erosion greater than 2 m, typically caused by a valley at that point) is uniformly distributed around the edge of the landform. No one point at the edge of the landform looked notably more at risk than adjacent points. Thus while an individual erosion history for the structure might have had a valley at a particular point, the possible range of locations of the high-erosion valleys is more or less uniformly distributed around the edge of the landform. One hope was that the Monte Carlo simulations would identify regions where the risk of valley erosion was high and areas where the risk was low. This would have then allowed us to concentrate erosion protection measures in the areas of high risk. Fig. 18.3 shows that the right-hand and left-hand corners of the landform have a higher risk of failure with a probability peaking near 0.8, while the top corner has a batter where the risk of failure is almost zero. This reduced risk is also mirrored by the distribution of the mean and maximum erosion depth. Hancock (2005) performed a similar analysis using a natural catchment DEM and calibrated erosion parameters. The work of Morgan (1994), Willgoose and Gyasi-Agyei (1995) and Hancock (2005) provides a framework for erosion risk assessment using LEMs.

Evans *et al.* (1995) and Evans and Loch (1996) compared the erosion loss predictions for RUSLE and the SIBERIA LEM when each was calibrated to the same erosion plot data, and found that they gave similar rates for areal average erosion. They confirmed the compatibility of traditional and LEM erosion loss predictions when the same data were used for model calibration and when the landform is not allowed to evolve during the simulations. Hancock *et al.* (2008b), at the Nabarlek uranium mine, compared erosion predictions using the SIBERIA LEM with those derived from a previous study using the RUSLE (Hancock *et al.*, 2006), and found good agreement in average erosion rates when calibrated to similar materials. This compatibility between traditional and

(a)

(b)

(c)

(d)

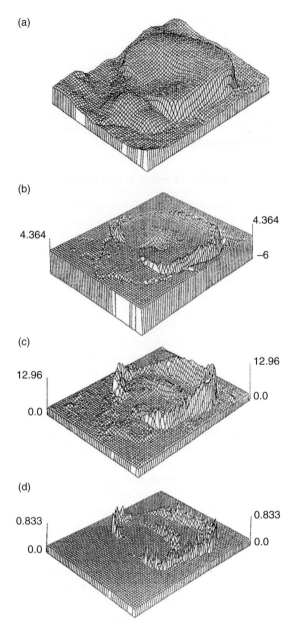

Fig. 18.3 Erosion risk plots for a rehabilitation design for Ranger Uranium Mine, Northern Territory. This design is an alternative to that illustrated in Fig. 18.1 but is shown from a similar viewpoint (note, for instance the hill in the centre-left foreground common to both figures). The figures are: (a) the original design landform showing the kidney-shaped 25 m high above-ground structure and a natural hill in the centre left foreground; (b) the mean erosion depth after 1000 years of Monte Carlo simulations of erosion (vertical units are metres); (c) the maximum erosion depth after 1000 years of Monte Carlo simulations (vertical units are metres); and (d) the probability that the erosion will be greater than 2 m at that point in the domain (from Morgan, 1994; Willgoose & Gyasi-Agyei, 1995).

LEM erosion estimates is important when we discuss calibration below.

A final component of the work at Ranger was validation of the predictions of the LEM. Numerous laboratory model studies have been carried out to validate the SIBERIA LEM (e.g. Hancock & Willgoose, 2001a,b, 2002), but only one model study has looked specifically at the cap-batter transition type erosion pattern behaviour discussed above (Hancock & Willgoose,

2004). These cap-batter gullies are a common ero-sional feature of mines sites. Plate 15 shows an example of such a gully at a mine in Northern Australia (Willgoose & Loch, 1996).

Willgoose and Loch (1996) identified a 50-year old mine site, Scinto, that had an eroding mine waste structure with the same type of cap-batter geometry as the proposed Ranger structure (Hancock *et al.*, 2000). The runoff and erosion at the site were monitored for three months with a series of plots and small catchments and SIBERIA calibrated in an identical way to that done for Ranger. The original landform was not known, but given that erosion had occurred in a series of valleys at the cap-batter transition it was decided simply to infill the valleys in the landform used for the initial condition and then run SIBERIA forward from that initial condi-tion. The match between SIBERIA and the observed landform was very good, with the volume and depth of eroded and deposited material and the location of the valleys being well modelled.

Bell and Willgoose (1998) carried out a three-month field trial over a single wet season at Ranger to simulate the evolution of a gully cre-ated by the transition from the low slope on the cap onto the high slope of the batter. As noted above, these gullies are a common feature of the LEM simulations. The study indicated: (1) that a key feature of the erosional development was the creation of a coarse armour layer at the base of the gully which significantly reduced erosion; and (2) the beginnings of the development of a depositional fan at the base of the slope where it transitioned onto the flatter surrounding land-scape. The LEM simulations of the gully erosion modelled the three dominant erosion events dur-ing the wet season of the trial and found that the gully was developed mostly during the largest event (also observed in the field), and that if the erosion rate of the original material was used, the LEM overpredicted the gully depth. In subsequent unpublished work we found that if the erosion rate of the armour material developed on the gully bottom was incorporated into a model where the erodibility of the material was a linear function of the cumulative erosion depth, then, in the LEM simulations, the gully depth was cor-rectly simulated. The conclusion of this work is that to model correctly the initial rate of gully development on rocky material, a model of the evolution of the soil erodibility is necessary. It should be noted that because of the short dura-tion of the trial, no conclusion can be inferred on the long-term equilibrium depth of the gullies, which are a function of the fully-developed height of the downstream alluvial fan.

18.3.2   *Example 2: Farm scale*

Gyasi-Agyei and Willgoose (1996) examined the impact of graded contour banks on long-term ero-sion. Contour banks are a common means of reducing erosion in landscapes. They do this in two ways.

• Firstly, they trap sediment behind the contour bank so that it is captured on the hillslope. This does not lower erosion from the hillslope, but does at least ensure that the eroded material does not leave the slope.

• Secondly, they reduce erosion by breaking up the downhill slope length into a series of short seg-ments. Almost all traditional erosion models indi-cate that as slope length increases, the amount of material eroded per unit area (in t ha^{-1}) from the slope also increases. This means that a series of shorter slopes erodes less than a single slope of the same total length. Contour banks break up the slope so that less material is eroded from the slope. Gyasi-Agyei and Willgoose used an LEM to simu-late the evolution of the landform over 100 years and found that the reduction in erosion is main-tained even as the landform evolves (Fig. 18.4). The contour banks were not along-contour but had a longitudinal slope of 1%. Technically the 1% slope means that they were graded rather than contour banks, but the differences for our purposes here are small. The graded bank allowed the hill-slope to shed water but keep water velocities behind the contour bank to a low level. Figure 18.5 shows that reduction in erosion due to contour banks actually improves with time, with a maxi-mum reduction in erosion in their case study of 95% after 100 years, even though the initial imp-rovement was only about 20% after 1 year.

(a)

(b)

(c)

(d)

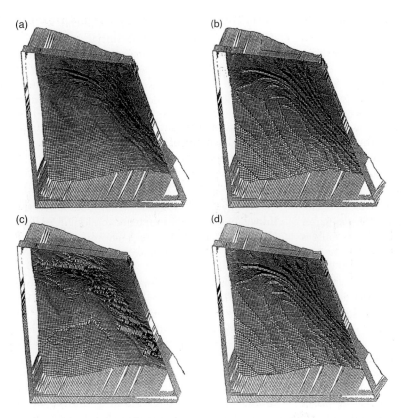

Fig. 18.4 Simulations of 100 years of erosion with and without contour banks. (a) Original landform without contour banks; (b) original landform with contour banks; (c) 100-year erosion without contour banks; (d) 100-year erosion with contour banks. Grid spacing is 1 m and the left-to-right dimension of the domain is approximately 100 m (from Gyasi-Agyei & Willgoose, 1996).

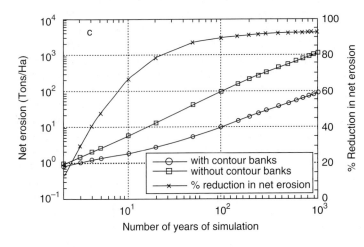

Fig. 18.5 The cumulative net erosion (i.e. net erosion = erosion − deposition) over 1000 years for the landform in Fig. 18.4. Note that these simulations assume that at no stage do the contour banks overtop, so that the contour banks are perfectly efficient throughout the 1000 years (from Gyasi-Agyei & Willgoose, 1996).

The mechanism for this improvement seemed to be that localized deposition occurred behind the contour banks as a result of slight concentration of flow down the slope, so that some points generated localized deposition when the flow reached the contour bank. These areas of localized deposition behind the banks then created pools in which flow down the grade along the contour bank could then deposit sediment. Thus as the landform evolved, it naturally created sediment-capture regions along the grade of the contour bank that did not exist in the original landform, where water flowed freely down the drain behind the contour bank. Clearly, a model that does not allow the landform to evolve in response to erosion and deposition cannot model this sediment capture mechanism; only an LEM can capture this behaviour.

The only assumption that was made in the modelling was that the contour banks were not overtopped at any stage during the simulation. To verify this assumption, post-processing of the simulation data indicated that deposition behind the banks did not fill up the storage capacity behind the contour bank. However, the simulations did not examine the effects of extreme events. Thus the simulations did not capture large events where the conveyance capacity of the channel behind the contour bank would be exceeded. This is not a fundamental limitation of LEMs, but reflected the limitations of the SIBERIA LEM used for the study, which could not model event-scale hydraulics. The authors are unaware of any LEM that can realistically model event-scale dynamics while also modelling landform evolution over many years.

18.3.3 Example 3: Geomorphic design

In applications where a landform is to be designed, the erosion assessment task is not only to minimize erosion (or alternatively make it the same as occurs naturally) by the use of surface treatments (e.g. vegetation or rock cover), but also to design a landform that minimizes erosion. For instance, we know that natural catchment long profiles are concave, so a concave slope is naturally more geomorphologically stable than a planar or convex slope. This principle has been long understood, at least qualitatively (e.g. Toy & Hadley, 1987). LEMs allow us to quantify this concavity in terms of the hydrology and erosion processes on the slope. LEMs can then be used to assess the long-term fate of those slopes relative to other design options (Loch & Willgoose, 2000a,b; Vasey *et al.*, 2000) so that an objective judgement can be made of the costs and benefits of the various alternatives.

One important point to note here is that in landform design the absolute erosion rate is not always important, because we are comparing the relative efficacy of design alternatives. Generally, a good landform design alternative is good no matter what the erodibility and absolute erosion rate of the materials used to construct the slope. Of course, if the objective is simply to meet a tonnes per hectare threshold then the erodibility will be important. More critical is the area-slope relationship for the materials (Willgoose *et al.*, 1991b; Willgoose, 1994).

As noted above, a common regulatory requirement of an above-ground containment structure is that it covers as little land area as possible to reduce the amount of impacted area to a minimum. The maximum volume of stored waste per unit area covered is provided by a structure that is convex. Yet natural landscapes are only convex near the hilltops and concave downstream. As we have noted, this convex–concave hillslope profile (commonly referred to as a catena profile) results from the balance between soil creep and fluvial erosion processes. A structure that is convex everywhere is therefore not a good match with natural landscapes, and is not a good approximation of an equilibrium hillslope profile. Figure 18.6 shows the longitudinal profile of three hillslopes covering the range of concavities typically observed in the field. The convex slope has 25% more volume stored in it compared with the planar slope, while the concave has 30% less material stored in it than the planar. The three slopes each have the same average slope, so if the profile is averaged as, for example, in USLE, then they would deliver the same sediment load. However, at the base of the slope the convex profile will yield the highest

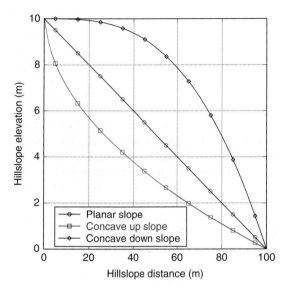

Fig. 18.6 Three hillslope longitudinal profiles with concavities ranging over that observed in nature. The concavities of these slopes (Equation (18.2)) are: (a) concave down slope, $\alpha = -1$; (b) planar slope, $\alpha = 0.0$; and (c) concave up slope, $\alpha = 0.5$.

sediment load off-site because it has the greatest slope at its base.

Moreover, LEM simulations show that if the initial slope is concave or planar, the hillslope will evolve to develop rills and gullies that further evolve toward a longitudinal concave profile. The authors have also seen many degraded mine sites where planar slopes have rapidly degraded through the development of gullies with a concave longitudinal profile, reflecting the area and slope dependence of fluvial erosion.

If the sediment transport equation is of the form

$$Q_s = KQ^m S^n \qquad (18.1)$$

where Q_s is the sediment transport capacity of the flow, K is the erodibility of the sediment, Q is the discharge per unit width, S is the slope (in units of m m^{-1}) and m and n are parameters of the erosion process, then the concavity of the equilibrium hillslope that this erosion process will generate is:

$$S \propto L^{\frac{1-m}{n}} \qquad (18.2)$$

where L is the distance down the slope from interfluve at the top of the slope, so that for a unit width slope $A = L$. For natural slopes $\alpha = \frac{m-1}{n}$ is normally in the range 0.4 to 0.8, and Equation (18.2) shows how α is dependent on the physics of the erosion process. Equation (18.2) provides a relationship that allows design of rehabilitated slopes that are geomorphologically stable.

To explore the implications of Equation (18.2) using LEM simulations, Hancock *et al.* (2003) explored the relative merits of a range of concave slopes as batters to several above-ground constructed landforms at a mine site. They fitted Equation (18.2) to natural landforms around the mine site and found a value of $\alpha = 0.35$. They also determined m and n in Equation (18.1) by calibrating the equation to rainfall runoff erosion simulator data to yield respective values of 2.53 and 2.67 for site 1, and 1.11 and 1.5 for site 2. The data from site 1 imply $\alpha = 0.57$ and from site 2 $\alpha = 0.08$. The main difference between the two sites was in the type of waste material. Hancock *et al.* then carried out a series of LEM simulations using the SIBERIA LEM to compare the predicted erosion rates in t ha^{-1}. Two sets of comparison profiles were created. The first had the concavity implied by the erosion data for site 1 (i.e. $\alpha = 0.57$), while the second set were constructed as per the erosion data for site 2 (i.e. $\alpha = 0.08$). These two sets of slopes were then compared with planar hillslopes with the same average slope and the same total height. For site 1, concave slopes had an erosion rate of 20% of that of the planar slopes, while site 2 had an erosion rate about 60% of that of the planar slope. There were several important conclusions from this work:

- The concave slopes had significantly lower erosion rates for the same average slope.
- The more concave slopes of site 1 ($\alpha = 0.57$) had a greater reduction in the erosion rate than the less concave slopes of site 2 ($\alpha = 0.08$). A plot of the reduction in erosion with concavity for site 1 is shown in Fig. 18.7.
- The reduction in erosion rate with concave slopes was very high. If the height of the slope was

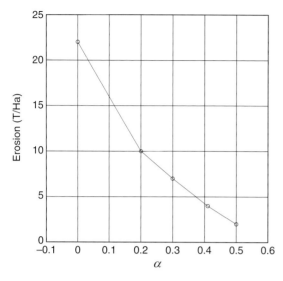

Fig. 18.7 The net erosion at the bottom of a concave slope (site 1 in the text) for a range of concavities α (based on data in Hancock *et al.*, 2003).

held constant (i.e. the height of the above-ground structure was held constant) it was possible to double the average slope of the hillslope, but it was still possible for a steeper concave hillslope to have less erosion than a flatter planar slope. The reason for this was partly the better performance of the concave hillslope and partly because the steeper hillslope was of shorter length because the height is constant.

18.4 Calibration of Landform Evolution Models

The calibration of an LEM for a specific site involves many of the same problems as faced when calibrating a traditional erosion model (see Chapter 3). For the fluvial erosion component of the model there are the soil cover and treatment effects, the grading of the soil material and the erosivity of the rainfall. The similarity between LEMs and physically-based erosion models is that the underlying physics is generally the same (although the details will depend on the LEM), so

a considerable amount of knowledge can be transferred from our work with traditional models (e.g. cover factors).

However, if the user wants to examine other non-fluvial processes then there are associated parameters that need to be calibrated. For instance, if the user wishes to model soil creep, then there are soil creep rates and other parameters (e.g. the dependence of the creep rate on the hillslope gradient) to be calibrated. Each new physical process involves a new set of parameters. However, a problem is that for these latter, non-fluvial, processes the LEMs are ahead of the available field data, and applicable parameter databases that may be used or adapted do not, in general, exist. Sometimes the models can be fitted to natural landforms to infer parameters, although there may be questions about the reliability of using natural materials as analogues of manmade materials. It is to these calibration issues that this section is (briefly) devoted.

In many of the applications that the authors have been involved in by the time there was an interest in doing LEM simulations, there had already been a considerable amount of erosion modelling done with traditional models. This meant that there was a set of parameters for a traditional erosion model that had already been fitted to the site. Accordingly the LEM could be fitted to the simulations of these pre-existing models.

Briefly, the calibration process is as follows. The LEM is run for a year. The LEM parameters are adjusted so that the erosion from LEM matches the pre-existing model. This use of the pre-existing model is advantageous for two reasons. Firstly, it makes use of any pre-existing knowledge base at the site. Secondly, it demonstrates that the LEM generates results that are consistent with a pre-existing model under the same experimental conditions (i.e. when the landform is not significantly evolving, assuming that a year is a short time in the evolution of the landform). This second result cannot be underestimated because it highlights that any inconsistency between failure mechanisms in the LEM and traditional simulations is a result of the landform evolution, not differences in the physics or the parameters. This may be important in the event of political and/or managerial resistance to acceptance of the LEM results.

To do the calibration above we need to do the simulations on some kind of landform. The simplest way to calibrate LEMs to an existing model is to create a series of synthetic hillslopes of different length/area and slope and run the existing model on them, outputting the erosion rate for each slope. The fluvial transport equation is normally expressed in the form

$$Q_s = KQ^mS^n = \hat{K}A^{\hat{m}}S^n \qquad (18.3)$$

where it is normally assumed that discharge is proportional to the catchment area so that $m = \hat{m}$.

The key components of the calibration of Equation (18.3) are (a) the area and slope dependence of the fluvial erosion process, (b) any minimum shear stress threshold on transport, and (c) any upper bound threshold on slope stability. For simplicity, Equation (18.3) reflects only the first of these three sets of parameters. Either by multiple regressions on area and slope, or by hand-fitting the LEM directly to erosion plot data (either field-based or computer-generated), these parameters of the erosion model in the LEM can be calibrated.

The main difficulty with this approach is that the ratio α of the area and slope exponents in Equation (18.3) (i.e. $\alpha = (m - 1)/n$; see Willgoose et al. (1991b) and Willgoose (1994) for further explanation) that is calibrated will be a function of the traditional model used. For natural catchments α is about 0.5, while for traditional models α ranges between 0 and 1 (Willgoose & Gyasi-Agyei, 1995). To place the α range of 0–1 implied by traditional models in context with respect to the evolution of natural slopes:
• a value of $\alpha = 0$ means that rilling cannot occur and that the long profile of the final equilibrium slope is a flat plane (see Fig. 18.6 and Example 18.3.3 for an explanation of how α controls the concavity of the slope that the landform will evolve to in the LEM); while
• $\alpha = 1$ indicates a slope that will rill quite strongly and will generate a slope with a strong

upward concavity (Fig. 18.6). If a key part of the landform design process is to design the concavity of the slope, then some caution is needed in calibration of the LEM to an existing traditional erosion model.

The user must therefore feel confident that the area and slope dependencies of the traditional erosion model are correctly calibrated, otherwise any inaccuracy in the traditional model will be simply transferred to the LEM through the calibration. A subtle difference in the area-slope dependency can make non-trivial differences in landform evolution. In a comparison study of two LEMs, SIBERIA and CAESAR, Hancock et al. (2010) showed that slight differences in this aspect of the two models resulted in subtle but noticeable differences in their long-term sediment transport predictions.

Given the concavity/area-slope dependence of the physics of landform evolution, the best data source for calibrating an LEM is field erosion plot data collected at the site for the soils of interest, where the areas and slopes of the experimental plots encompass the range of areas and slopes expected in the final landform to be assessed by the LEM. It is important to have a range of areas and slopes so that m and n can be calibrated from the data. Given the simplicity of Equation (18.3) it is generally sufficient to do a multiple regression with area and slope as the independent variables and sediment load as the dependent variable.

As part of a project to develop a general LEM for the Queensland Coal Industry (Bell et al., 1993), an easy graphical interface for the first author's SIBERIA model and associated engineering tools were developed, called EAMS-SIBERIA, and a database of parameters for unvegetated spoils and soils found in the Bowen Basin Coal Province in Queensland was derived. While generality of the parameters in the Bowen Basin database for other areas has never been fully tested, a number of studies has shown that the database is quite robust if soil texture properties are known (Hancock et al., 2008a,b). The location and rate of erosion is well modelled with EAMS-SIBERIA when using the Bowen Basin database. The only deviation from satisfactory performance in these studies was the inability to model the evolution of a significant gully on the mining landform. This was due to an inability to resolve a mine road, and consequent impacts on overland flow concentration, in the digital elevation model, rather than any inherent deficiency in the LEM or the database of parameters.

Using the Bowen Basin database, Martinez *et al.* (2009) showed that both EAMS-SIBERIA and RUSLE gave good matches to the observed erosion rate (since 1950) estimated by using caesium-137 for a natural catchment. The catchment where the comparison was done was a grazing paddock of 7 hectares and slopes ranging from 5–30%. Both models allowed for the vegetation cover by the use of USLE cover factors. In this application there was little change in the landform, so the comparison is more a test of the validity of the Bowen Basin database than of the LEM.

18.5   Landform Evolution Model Limitations

The discussion above highlights the new insights that can be made using LEMs. However, they are not without their limitations at the current time. At the science level, an evolving landform imposes limitations, which may or may not be important for any particular application. Many things that can be assumed to be constant in a traditional model will evolve in concert with the landform. Soils change as the soil armours in the bottoms of gullies. This gully and armour development changes the spatial distribution of the soil erodibility in a way that is intimately linked with the evolution of the landform. Similarly, the soil moisture distribution may also change with the evolving landform. This will change the distribution of vegetation density (particularly in arid regions), and therefore the cover factor across the landform.

At a practical level, an LEM must solve the erosion equations many millions of times as the landform evolves over time, whereas a traditional model only has to solve the erosion equations once. For an LEM this increases the

computing load significantly. Some of the LEMs are quite efficient, but they will always use more computer time than a traditional model. Small areas (a few square kms at a resolution of tens of metres, or a million computational nodes in space) can be easily run on a desktop computer with modest run times. Long simulations for large areas can still require significant compute times (CPU days).

The CPU time issue leads to one of the most common criticisms of LEMs by erosion modellers from a traditional background. This criticism is that the physics of LEMs is simpler than in their current model. This is generally in reference to the fluvial erosion model. Much of this criticism is, in our view, misguided. As noted above, the fundamental physics in the best LEMS and the most recent process-based traditional models are very similar. Many details such as spatial distribution of soil cover, practice and soil properties are commonly ignored by LEMs, but this is a practical issue reflecting their research heritage rather than a fundamental flaw. For instance, the EAMS erosion assessment package built around the first author's SIBERIA and TelluSim LEMs allows the input of these management factors, although at the current time it is more an exception to the rule in that regard (see Hancock *et al.* (2008b) for an example of using spatially variable soil erodibility in a LEM).

A limitation of the current fluvial erosion models in many LEMs that is only now being addressed (Coulthard *et al.*, 2000) is that they generally do not track the full soil grading of the eroded material. In the original science applications this was not seen as a critical failing, and ignoring the grading simplifies the computations considerably. Several LEMs now include sediment grading as part of their calculations. Some motivations for current efforts to include eroded grading are: (1) to model multilayer caps and the long-term behaviour of constructed rock armours on waste containment structures; (2) to simulate better any depositional behaviour; and (3) to allow the modelling of water quality parameters such as sorbed pollutants and radionuclides. This area is evolving rapidly.

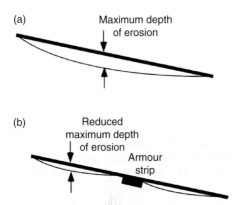

(a)   Maximum depth
                          of erosion

(b)   Reduced
                    maximum depth
                       of erosion
                                        Armour
                                         strip

Fig. 18.8 Using an armour strip to reduce gully and/or rill erosion in the upstream part of a catchment or hillslope. The heavy line is the original hillslope, while the lighter line is the erosion after some time, showing how the armour strip reduces the maximum depth of erosion: (a) hillslope without the armour strip protection; (b) hillslope with armour strip protection.

One emerging application of LEMs is in the design of erosion protection of landforms using rock armours. It can be very expensive to cover an entire landform with a rock armour layer to protect it from erosion. Yet in arid zones where it can be difficult to establish vegetation cover, this may be the only protection measure possible. It is possible to protect an area from erosion incision by creating a low erosion zone downstream of a sensitive area that effectively stops the upward propagation of gullies/rills, and constrains the depth of gullies/rills upstream of the low erosion zone (Fig. 18.8). A key feature is that the evolution of the landform is constrained by the low erosion of the armour strip. This armour strip pins the change in elevation at the mid-slope, and thus constrains the change in elevation upslope and downslope of the armour strip. LEM simulations confirm this behaviour. Notably, LEM simulations show that both protected and unprotected slopes initially have the same erosion rate, and it is only after some time that erosion is reduced on the hillslope with protection because the erosion protection relies on the evolution (or lack of it) on the armour strip relative to the rest of the slope.

Generalized rules for the design of these strips will no doubt emerge as we develop more experience in field applications.

At the managerial and decision-maker level, the main limitation to LEM application is that LEMs are still an unfamiliar tool. The LEM user may sometimes address this through education of the manager. In other cases this unfamiliarity can be important for highly political applications. LEMs have not yet been proven in court and this can be important for projects that will need to be legally defended. In these latter applications the LEMs unquestionably provide insight that cannot be obtained by any other means. The problem is that if millions of dollars are to be spent and justified using LEM simulations, some assurances are needed that the model's predictions (sometimes hundreds and thousands of years into the future) are, if not correct, at least not potentially misleading. An important related issue is the need to identify indicators based on short-term performance (e.g. 10–20 years of monitoring) that can be used by regulators, as part of rehabilitation sign-off, to provide confidence in long-term (e.g. 100–1000 years) LEM predictions. Progress has been slow in this area, but will be driven by practical applications. These practical applications will elucidate the types of validation required. These validation test cases are qualitatively different from the type of validation needed for traditional erosion models. For instance, we believe more work needs to be done to understand the area-slope dependence of erosion models, particularly with respect to case studies where concavity plays a large role in the performance of the structure.

We will now concentrate on the science issues. The main issue with LEMs is that many hillslope properties that can be directly measured for an existing hillslope actually evolve in concert with an evolving hillslope. This complicates the modelling of an evolving hillslope because this indicates that we need submodels for the evolution of these hillslope properties, and these submodels must respond dynamically to the evolution of the hillslope.

The best example of this need for submodels is the evolution of the grading of soil on the surface

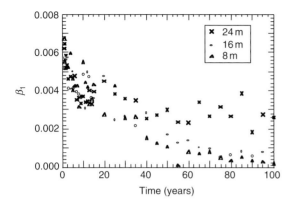

Fig. 18.9 The evolution of the erodibility (K in the erosion Equation (18.3)) of a hillslope under the action of armouring due to fluvial erosion using the ARMOUR fluvial erosion and armouring model. The distances are the surface erodibility at three distances along the slope measured from the slope divide. From Sharmeen & Willgoose (2007).

of the hillslope. As a slope evolves the erosion process strips out the finest fraction of the soil on the surface. This process leaves behind a coarser armour layer which is relatively more resistant to erosion. As the surface erodes further, even more fines are eroded and the surface continues to coarsen. Thus the erodibility of the surface is intimately tied to the cumulative erosion from the surface (Evans, 1998). It is common, after landform construction, to observe a period of high sediment transport off-site as fines are winnowed from the landform surface. This process can be modelled (Willgoose & Sharmeen, 2006; Sharmeen & Willgoose, 2007), although until recently it was believed to be too computationally intensive to model this armouring process in conjunction with landform evolution (Fig. 18.9). Recent advances (Cohen *et al.*, 2009) have overcome this problem, and combined landform evolution and soil pedogenesis models will soon start to emerge.

Coupled with the evolution of the soil surface is the evolution through weathering of material below the surface. The key importance of the evolution of the subsurface materials is that this grading drives the soil water-holding

capacity and the ability of the soil to support plant life. If the soils are actively evolving then the vegetation is also likely to be evolving. There are a number of ways in which vegetation can feed back to landform evolution, but the most obvious is through the modification of the vegetation cover factors in the erosion equation (Evans & Willgoose, 2000). Several researchers have used LEMS to model empirically the interaction between landforms and vegetation (e.g. Collins *et al.*, 2004; Saco *et al.*, 2007), but a fully-coupled physically-based soils and vegetation model that is suitable for use in LEMs is yet to emerge.

18.6   Future Trends in Landform Evolution Modelling

No discussion of a new approach to erosion modelling would be complete without some discussion of the future directions of LEM development. We have discussed some of the important science issues in the previous section. Here we will focus on issues related to their application to problems of a more applied nature.

It is still early days in the development of LEMs, particularly with respect to their application to real problems. At an exploratory level it is relatively easy for a researcher to write an LEM. Accordingly there has been an explosion of research-focused models in the last five years. Many of them have strong underlying similarities in the physics, but they typically reflect the interests of the researcher so that there is much that is unique in each particular model. Generally, these research-focused LEMs do not have the complete set of physics, support and analysis tools, and documentation that are required by the erosion practitioner. Nor, critically, do they have the set of validation tests needed by the practitioner to be able to defend the results of their models. In some cases, like the Monte Carlo aspects of valley development on landforms (where comparing a landform in the field with a single computer simulation is not possible even in principle), we are only at the early stages of developing testing methodologies (Willgoose et al., 2003; Hancock, 2003).

In particular there is a need for a tool with an easy-to-use interface, which is well documented, computationally robust, widely validated on the types of problems of interest, contains all the physics that might be needed for typical applications, and with a database of erosion parameters (e.g. using pedo-transfer functions). No LEM currently available meets all of these needs. Coulthard (2001) in a recent review, albeit aimed at a research audience, indicated that the first author's EAMS-SIBERIA rehabilitation design package most closely met these criteria. This reflects the wide range of (mainly) mine rehabilitation case studies on which it has been applied. It is anticipated that a new model, TelluSim, will supersede EAMS-SIBERIA in the near future (Willgoose, 2009). This new model will facilitate new applications in new areas and will make it easier to customize for specific applications and user needs.

18.7   Conclusions

Landform evolution models (LEMs) are emerging as a practical tool for simulating erosion for a range of problems that are not possible for traditional erosion models to address. Broadly speaking these problems are ones where features that take some time to develop (e.g. gullies), which develop only in response to erosion, and where the features are critical to the success or otherwise of the project. The authors' experience is mostly in the area of mine rehabilitation, waste containment and nuclear waste repositories. In these cases failure occurs when the maximum depth of erosion incision reaches the waste. In these cases it is the maximum depth of incision of gullies and valleys that develop on the landform rather than the average erosion that is critical. LEMs are ideally suited to address this type of problem because of their ability to model the incision process over time. This localized incision is a key part of the development of hills and valleys in landform evolution.

However, as we have noted there are other examples of erosion where the very act of modelling the landform evolution draws out features of the erosion process that are not explicated by traditional approaches. These revolve around the design of optimal landform – landforms that are in equilibrium with the erosion processes occurring on them. LEMs have quantified the rules for the shape of the landform that is required. This has eliminated the risky step of simply assuming that the best shape for a constructed slope is the same as the adjacent natural slope. LEMs have shown that if the soil materials are different, then the shape of the hillslope required is likely to be different.

Finally we have discussed how the evolution of the landform itself can be used as a way of protecting a site by the use of selective armouring of the surface. By using selective armouring on hillslopes we can reduce long-term hillslope erosion markedly, while avoiding the costly step of armouring the entire slope to provide erosion protection.

Acknowledgements

The first author is currently funded by an Australian Research Council Australian Professorial Fellowship (APF). Work described in this chapter has been funded by the Queensland Coal Association, Environmental Research Institute of the Supervising Scientist (eriss), Energy Resources of Australia, Rio Tinto and the Australian Research Council.

References

Ahnert, F. (1976) Brief description of a comprehensive three-dimensional process-response model for landform development. *Zeitschrift für Geomorphologie N.F. Supplement* **25**: 29–49.

Ahnert, F. (1984) Local relief and the height limits of mountain ranges. *American Journal of Science* **204**: 1035–55.

Bell, J.R.W. & Willgoose, G.R. (1998) *Monitoring of gully erosion at ERA Ranger Uranium Mine, Northern Territory, Australia.* Internal Report 274, Environmental Research Institute of the Supervising Scientist, Jabiru, NT, Australia.

Bell, L.C., Loch, R.J., Haneman, D. & Willgoose, G.R. (1993) *A post-mining landform research program for open-cut mines.* In Australian Minerals Industry Council Environmental Workshop, Australian Institute of Mining and Metallurgy, Burnie, Tasmania, Australia.

Cohen, S., Willgoose, G.R. & Hancock, G.R. (2009) The mARM spatially distributed soil evolution model: A computationally efficient modeling framework and analysis of hillslope soil surface organization. *Journal of Geophysical Research (Surface Processes)* **114**, F03001.

Collins, D.B.G., Bras, R.L. & Tucker, G.E. (2004) Modeling the effects of vegetation-erosion coupling on landscape evolution. *Journal of Geophysical Research* **109**: F03004.

Coulthard, T.J. (2001) Landscape evolution models: a software review. *Hydrological Processes* **15**: 165–73.

Coulthard, T.J., Kirkby, M.J. & Macklin, M.G. (2000) Modelling geomorphic response to environmental change in an upland catchment. *Hydrological Processes* **14**: 2031–45.

Crowell, K.J., Wilson, C.J., Lane, L.J., *et al.* (2005) *Impact of extreme events and soil hydraulic conductivity on the evolution of a mesa-top waste repository cover.* In Fall Meeting of the American Geophysical Union, San Francisco: H43D-0523.

Evans, K.G. (1998) *Runoff and erosion characteristics of a post-mining rehabilitated landform at Ranger Uranium Mine, Northern Territory, Australia and the implications for topographic evolution.* PhD thesis, University of Newcastle, Callaghan, Australia.

Evans, K.G. & Loch, R.J. (1996) Using the RUSLE to identify factors controlling erosion rates of mine spoils. *Land Degradation and Development* **7**: 267–77.

Evans, K.G. & Willgoose, G.R. (2000) Post-mining landform evolution modelling. II. Effects of vegetation and surface ripping. *Earth Surface Processes and Landforms* **25**: 803–23.

Evans, K.G., Loch, R.J., Hall, R.N. & Bagnall, C.P. (1995) *USLE parameter values for waste rock material at the Ranger uranium mine derived using PSA and the effect of these parameters on erosion rates.* Internal Report 177, Supervising Scientist for the Alligator Rivers Region, Canberra.

Gilbert, G. (1909) The convexity of hillslopes. *Journal of Geology* **17**: 344–50.

Gyasi-Agyei, Y. & Willgoose, G.R. (1996) Evaluation of the use of contour banks as post-mining rehabilitation control option using a digital terrain based

rainfall-runoff-erosion models. In Holzmann, H. & Nachtnebel, H.P. (eds), *Application of Geographic Information Systems in Hydrology and Water Resources Management*, HydroGIS '96. Institut für Wasserwirtschaft, Universität für Bodenkultur, Vienna, Austria: 143–50.

Hancock, G.R. (2003) Effect of catchment aspect ratio on geomorphological descriptors. In Wilcock, P.R. & Iverson, R.M. (eds), *Prediction in Geomorphology*. American Geophysical Union, Washington, DC: 217–30.

Hancock, G.R. (2005) Digital elevation model error and its effect on modelling soil erosion and catchment geomorphology sediment budgets II (Proceedings of symposium S1 held during the Seventh IAHS Scientific Assembly at Foz do Iguaçu, Brazil, April 2005). *International Association of Hydrological Sciences Publication* **292**: 119–26.

Hancock, G.R. & Willgoose, G.R. (2001a) The use of a landscape simulator in the validation of the SIBERIA catchment evolution model: declining equilibrium landforms. *Water Resources Research* **37**: 1981–92.

Hancock, G.R. & Willgoose, G.R. (2001b) The interaction between hydrology and geomorphology in a landscape simulator experiment. *Hydrological Processes* **15**: 115–33.

Hancock, G.R. & Willgoose G.R. (2002) The use of a landscape simulator in the validation of the SIBERIA landscape evolution model: transient landforms. *Earth Surface Processes and Landforms* **27**: 1321–34.

Hancock, G.R. & Willgoose, G.R. (2004) An experimental and computer simulation study of erosion on a mine tailings dam wall. *Earth Surface Processes and Landforms* **29**: 457–75.

Hancock, G.R., Evans, K.G., Willgoose, G.R., *et al.* (2000) Medium term erosion simulation of an abandoned mine site using the SIBERIA landscape evolution model. *Australian Journal of Soil Research* **38**: 249–63.

Hancock, G.R., Loch, R.J. & Willgoose, G.R. (2003) The design of post-mining landscapes using geomorphic principles. *Earth Surface Processes and Landforms* **28**: 1097–1110.

Hancock, G.R., Grabham, M.K., Martin, P., *et al.* (2006) A methodology for the assessment of rehabilitation success of post mining landscapes – sediment and radionuclide transport at the former Nabarlek uranium mine, Northern Territory, Australia. *Science of the Total Environment* **354**: 103–19.

Hancock, G.R., Crawter, D., Fityus, S.G., *et al.* (2008a) The measurement and modelling of rill erosion at angle of repose slopes in mine spoil. *Earth Surface Processes and Landforms* **33**: 1006–20.

Hancock, G.R., Lowry, J.B.C., Moliere, D.R. & Evans, K.G. (2008b) An evaluation of an enhanced soil erosion and landscape evolution model: a case study assessment of the former Nabarlek uranium mine, Northern Territory, Australia. *Earth Surface Processes and Landforms* **33**: 2045–63. DOI: 10.1002/esp.1653

Hancock, G.R., Lowry, J.B.C., Coulthard, T.J., *et al.* (2010) A catchment scale evaluation of the SIBERIA and CAESAR landscape evolution models. *Earth Surface Processes and Landforms* **35**: 863–75.

Howard, A.D. (1994) A detachment-limited model of drainage-basin evolution. *Water Resources Research* **30**: 2261–85.

Kirkby, M.J. (1971) Hillslope process-response models based on the continuity equation. In *Slopes: Form and Process*. Institute of British Geographers, London: 15–30.

Loch, R.J. & Willgoose, G.R. (2000a) *Design of stable rehabilitated landforms*. In AMEEF Innovation Conference, Brisbane: 13–20.

Loch, R.J. & Willgoose, G.R. (2000b) *Rehabilitated landforms: designing for stability*. In Workshop on environmental management in arid and semi-arid areas, Kalgoorlie: 24–6.

Martinez, C., Hancock, G.R. & Kalma, J.D. (2009) Comparison of fallout radionuclide (caesium-137) and modelling approaches for the assessment of soil erosion rates for an uncultivated site in south-eastern Australia. *Geoderma* **151**: 128–40.

Moglen, G.E. & Bras, R.L. (1994) *Simulation of observed topography using a physically-based evolution model*. TR 340, Ralph M. Parsons Laboratory, Dept. of Civil Engineering, MIT, Cambridge, MA.

Morgan, A. (1994) *Erosional stability of mine rehabilitation*. ERISS Internal Report 135, Supervising Scientist for the Alligator Rivers Region.

Saco, P.M., Willgoose, G.R. & Hancock, G.R. (2007) Eco-geomorphology and vegetation patterns in arid and semi-arid regions. *Hydrology and Earth System Sciences* **11**: 1717–30.

Sharmeen, S. & Willgoose, G.R. (2007) A one-dimensional model for simulating armouring and erosion on hillslopes. 2. Long-term erosion and armouring predictions for two contrasting mine spoils. *Earth Surface Processes and Landforms* **32**: 1437–53.

Syvitski, J., Paola, C., Slingerland, R., *et al.* (2004) *Building a community surface dynamics modeling system: rationale and strategy*. Report to National Science Foundation, Penn State University. Available at http://csdms.colorado.edu/wiki/CSDMS_docs

Toy, T.J. & Hadley, R.F. (1987) *Geomorphology and Reclamation of Disturbed Lands*. Academic Press, Orlando.

Vasey, A., Loch, R.J. & Willgoose, G.R. (2000) *Rough and rocky landforms*. In 4th International and 25th National Minerals Council of Australia Environmental Workshop, Perth.

Willgoose, G.R. (1994) A physical explanation for an observed area-slope-elevation relationship for declining catchments. *Water Resources Research* **30**: 151–9.

Willgoose, G.R. (2005) Mathematical modeling of whole-landscape evolution. *Annual Review of Earth and Planetary Sciences* **33**: 443–59.

Willgoose, G.R. (2009) *TELLUSIM: A Python Plug-in based computational framework for spatially distributed environmental and earth sciences modelling*. In 18th World IMACS/MODSIM Congress, Cairns, Australia.

Willgoose, G., Bras, R.L. & Rodriguez-Iturbe, I. (1991) A coupled channel network growth and hillslope evolution model .1. Theory. *Water Resources Research* **27**: 1671–84.

Willgoose, G., Bras, R.L. & Rodriguez-Iturbe, I. (1991b) A physical explanation of an observed link area-slope relationship. *Water Resources Research* **27**: 1697–1702.

Willgoose, G.R. & Riley, S.J. (1993) The assessment of the long-term erosional stability of engineered structures of a proposed mine rehabilitation. In Chowdhury, R.N. & Sivakumar, M. (eds), *Environmental Management: Geowater and Engineering Aspects*. Balkema, Wollongong University: 667–76.

Willgoose, G.R. & Gyasi-Agyei, Y. (1995) New technology in hydrology and erosion modeling for mine rehabilitation. In *APCOM XXV Application of Computers and Operations Research in the Mineral Industries*, Australian Institute of Mining and Metallurgy, Brisbane: 555–62.

Willgoose, G.R. & Loch, R.J. (1996) *An assessment of the Nabalek rehabilitation, Tin Camp Creek and other mine sites in the Alligator Rivers Region as test sites for examining long-term erosion processes and the validation of the SIBERIA model*. Internal Report No. 229, Environmental Research Institute of the Supervising Scientist, Jabiru, Australia.

Willgoose, G.R. & Riley, S.J. (1998a) *Application of a catchment evolution model to the prediction of long term erosion on the spoil heap at Ranger Uranium*

Mines: Initial analysis. Australian Government Publishing Service, Canberra.

Willgoose, G.R. & Riley, S.J. (1998b) An assessment of the long-term erosional stability of a proposed mine rehabilitation. *Earth Surface Processes and Landforms* **23**: 237–59.

Willgoose, G.R., Hancock, G.R. & Kuczera, G.A. (2003) A framework for the quantitative testing of landform evolution models. In Wilcock, P.R. & Iverson R.M. (eds), *Predictions in Geomorphology*. American Geophysical Union, Washington, DC: 195–216.

Willgoose, G.R. & Sharmeen, S. (2006) A one-dimensional model for simulating armouring and erosion on hillslopes. 1. Model development and event-scale dynamics. *Earth Surface Processes and Landforms* **31**: 970–91.

19   Gully Erosion: Procedures to Adopt When Modelling Soil Erosion in Landscapes Affected by Gullying

J.W.A. POESEN[1], D.B. TORRI[2]
AND T. VANWALLEGHEM[3]

[1]Department of Earth and Environmental Sciences, Katholieke Universiteit Leuven,
GEO-Institute, Celestijnenlaan, Heverlee, Belgium
[2]IRPI CNR, Perugia, Italy
[3]Department of Agronomy, Institute for Sustainable Agriculture – CSIC,
Finca Alameda del Obispo, Córdoba, Spain

19.1   Why Model Gully Erosion?

Most research dealing with soil erosion by water has focused on sheet (inter-rill) and rill erosion processes operating at the runoff plot scale. This can be concluded from two facts: (1) the many run-off plot studies reporting soil loss rates caused by sheet and rill erosion for various climatic and land use conditions; and (2) the use of both empirical and process-based field-scale and catchment-scale erosion models addressing mainly sheet and rill erosion for predicting soil erosion rates under environmental change or for establishing soil erosion risk maps at various scales (see other chapters in this book). In Europe, for instance, more than 2200 plot-years of data on annual soil loss by sheet and rill erosion have been published over the last decades (Cerdan et al., 2006), whereas during the same period less then 50 gully-years of data on annual soil loss by gully erosion have been reported (Poesen et al., 2006). However, in many landscapes and for a range of climatic and land-use conditions, one can observe the presence and dynamics of various types of gullies: ephemeral gullies, permanent or classical gullies, and bank or edge-of-field gullies.

Field-based evidence suggests that soil losses from sheet and rill erosion as measured on runoff plots or as predicted using most water erosion models are not realistic indicators of total catchment erosion rates or sediment yield, nor do they indicate satisfactorily the sources and redistribution of eroded soil within a catchment (de Vente & Poesen, 2005). Yet soil losses caused by gully erosion have rarely been accounted for in soil loss assessment programmes (Poesen et al., 2003).

The development of gully channels on-site causes a significant decrease in soil quality through very high soil losses (i.e. 10–100 t ha^{-1} y^{-1}; Poesen et al., 2002) and through the enhanced drainage and desiccation of the inter-gully areas which in dry environments may lead to limited soil water availability and significant crop yield reductions. In addition, the erosion channels lower the trafficability of the land, inducing an extra economic cost for farmers and residents.

Gully erosion often represents a major source of sediment. Area-specific sediment yield in Mediterranean environments differs on average one order of magnitude when comparing non-gullied with intensively gullied catchments (Poesen et al., 2002). Typically, gully channels occupy less than 5% of a catchment area, but their contribution to total catchment sediment yield is well above this percentage: i.e. from 10%

Handbook of Erosion Modelling, 1st edition. Edited by R.P.C. Morgan and M.A. Nearing. © 2011 Blackwell Publishing Ltd.

to 90% depending upon environmental controls (climate and weather, soil type, topography, land use, gully type) as well as on the spatial and temporal scales considered (Poesen *et al.*, 2003). Moreover, once gullies develop, the eroded channels increase the connectivity for runoff and sediment within a catchment significantly, leading to a rapid transfer of eroded soil from the uplands to the lowlands, hence contributing significantly to muddy floods and to pond and reservoir siltation (Verstraeten *et al.*, 2006). It is through (ephemeral) gully channels that a large fraction of soil eroded within a field or catchment is redistributed and delivered to watercourses.

From this discussion it becomes clear that gully erosion cannot be neglected when assessing the impacts of climatic and, in particular, land-use changes on the rates of soil erosion by water and on soil quality. However, as indicated above, relatively few data on gully erosion rates are available from the literature. Hence, modelling gully erosion remains an alternative approach. Therefore, this chapter provides background on gully types and factors controlling gully erosion, reviews strategies to model gully erosion and discusses the interactions between gully erosion, hydrological and other erosion processes.

19.2   Gully Erosion and Gully Types

Gully erosion is defined as the erosion process whereby runoff water accumulates and often recurs in narrow channels and, over short periods, removes the soil from this narrow area to considerable depths. *Permanent gullies* (Plate 16) are often defined for agricultural land in terms of channels too deep to ameliorate easily with ordinary farm tillage equipment; typically they range from 0.5 m to as much as 25 to 30 m depth (Soil Science Society of America, 2001).

In the 1980s, the term *ephemeral gully erosion* was introduced to include concentrated flow erosion greater than rill erosion but less than classical gully erosion, as a consequence of the growing concern that this sediment source used to be overlooked in traditional soil erosion assessments

(Foster, 1986; Grissinger, 1996a,b). Even though in the literature ephemeral gullies are recorded on many photographs of erosion, it is only during the last three decades that these erosion phenomena have been recognised as being a major part of the erosional systems on cropland (Evans, 1993). According to the Soil Science Society of America (2001), *ephemeral gullies* (Plate 17) are small channels eroded by concentrated overland flow that can be easily filled by normal tillage, only to reform again in the same location by additional runoff events. Poesen (1993) observed ephemeral gullies to form in concentrated flow zones, located not only in natural drainage lines (thalwegs or hollows) but also along (or in) linear landscape elements, such as drill lines, dead furrows, headlands, parcel borders and access roads. Channel incisions in linear landscape elements are usually classified as rills according to the traditional definitions that associate rill formation with the micro-relief generated by tillage or land forming operations (Haan *et al.*, 1994). However, such incisions may also become very large, so this classification seems unsuitable. In order to account for any type of concentrated flow channels that would never develop in a conventional runoff plot used to measure rates of soil loss by inter-rill and rill erosion, Poesen (1993) distinguished rills from (ephemeral) gullies by a critical cross-sectional area of 929 cm² (i.e. the 1 square foot criterion first proposed by Hauge (1977)). Other criteria include a minimum width of 0.3 m and a minimum depth of about 0.6 m (Brice, 1966), or a minimum depth of 0.5 m (Imeson & Kwaad, 1980). As to the upper limit of gullies, no clear-cut definition exists. For instance, Derose *et al.* (1998) studied sediment production by a large gully (i.e. 500 m wide and 300 m deep). In other words, the boundary between a large gully and a(n) (ephemeral) river channel is very vague. Nevertheless, it must be acknowledged that the transition from rill erosion to ephemeral gully erosion (Plate 17) to classical gully erosion (Plate 18) and to river channel erosion represents a continuum, and that any classification of hydraulically related erosion forms into separate classes, such as microrills, rills, megarills, ephemeral gullies and gullies, is, to some extent, subjective (Grissinger 1996a,b). In fact, Nachtergaele *et al.* (2002a) demonstrated that (ephemeral) gullies can be considered as channels characterized by a mean width (*W*) between that of rills and (small) rivers. For all these channels, *W* seems to be essentially controlled by peak flow discharge (*Q*), and the relation between both parameters can be expressed by the equation $W = a\,Q^b$, with *a* being a coefficient and the exponent *b* varying from 0.3 for rills, over 0.4 for (ephemeral) gullies, to 0.5 for (small) rivers. For gullies, the proposed *W-Q* relation only holds for concentrated flow incising relatively homogeneous soil material in terms of erodibility (i.e. soil erodibility remains constant with depth). If a resistant soil horizon is present at shallow depth (e.g. frozen layer, plough pan, Bt-horizon, fragipan, petrocalcic horizon, ironstone hardpan or bedrock), *W* will be much larger than the value predicted by this equation. Also, if a more erodible layer is present at shallow depth, this relation will no longer hold (Nachtergaele *et al.*, 2002a).

By definition, *bank gullies* or *edge-of-field gullies* (Plate 18) develop wherever concentrated runoff crosses an earth bank. Given that the local slope gradient of the soil surface at the bank riser is very steep (i.e. subvertical to vertical), bank gullies can rapidly develop at or below the soil surface by hydraulic erosion, piping, tunnelling and eventually mass movement processes, even though catchment areas are rather small (Poesen & Govers, 1990). Once initiated, bank gullies retreat by headcut migration into the more gentle sloping soil surface of the bank shoulder and further into low-angled pediments, river or agricultural terraces (Poesen *et al.*, 2002).

So far, no systematic compilation of morphological characteristics (e.g. length, width, depth) of the different types of gullies and their controlling factors (e.g. topography, soil type, land use, hydrology) in a wide range of environments has been made. Such quantitative data would be needed so as to allow land managers to foresee the type of gullies they might expect when land-use changes are taking place. These data could also be used to develop gully erosion models for the different gully types.

For more detailed information on gully erosion processes and controlling factors, the reader is referred to review papers by Bocco (1991), Dietrich and Dunne (1993), Bull and Kirkby (1997, 2002), Poesen *et al.* (2002, 2003) and Valentin *et al.* (2005).

19.3   Prediction of Gully Erosion

Questions of major interest to those wanting to model soil erosion in landscapes affected by gullying are:
(1) whether and where gullies may form in a given landscape;
(2) how fast gullies will erode (in terms of channel length, channel cross-section, gully headcut retreat); and
(3) how gully development will interact with hydrological and other soil erosion processes.

19.3.1   When and where do gullies develop?

Gully development is a threshold phenomenon. It occurs only when a threshold in terms of flow hydraulics, rainfall (or snowmelt), topography, pedology (or lithology) and land use has been exceeded. Here we discuss and illustrate the type and magnitude of these thresholds.

(i) **Hydraulic thresholds**   Gully channels form only if concentrated (overland) flow intensity during a rain event exceeds a threshold value. Horton (1945) first proposed the concept of a threshold force required for channel initiation. This force of flow is often expressed in terms of the boundary flow shear stress ($\tau_b = \rho g d s$ where ρ = density of runoff water, g = acceleration due to gravity, d = depth of flow and s = sine of the soil surface slope angle). The threshold force required to cause channel incision into the soil surface of the concentrated flow zone is termed the critical flow shear stress (τ_c). A key question is: how large should τ_c be for (ephemeral) gullies to initiate?

Critical flow shear stress values for incipient motion of individual soil particles have been well studied. Entrainment of loose silt and fine to medium sand grains occurs at τ_c values of less than 1 Pa (as deduced from the Shields curve;

Vanoni & Brooks, 1975: 99). For bare, cohesive top soils with soil shear strength values at saturation up to 10 kPa, laboratory experiments indicate that τ_c values can go up to 4 Pa (e.g. Rauws & Govers, 1988; Brunori *et al.*, 1989; Crouch & Novruzi, 1989). These τ_c values are of the same order as those reported for rill incision in bare topsoils in the field under drainage conditions, that is, 1.8–10.6 Pa depending on soil properties (texture, soil water content, content of calcium, iron, organic carbon and potassium; Gilley *et al.*, 1993). Soil shear strength values at saturation appear to be a good indicator for the value of τ_c (Poesen *et al.*, 1998; Knapen & Poesen, 2010). Experimental data collected by Huang and Laflen (1996) indicate that critical flow conditions for rilling under seepage conditions may be significantly less than those for drainage conditions. Land management practices may affect the critical flow shear stress values for concentrated flow erosion, as Franti *et al.* (1999) reported that τ_c values for no-till were about twice that for tilled soil. Along the same lines, Laflen and Beasly (1960) clearly demonstrated that compaction of the topsoil increased τ_c values. Living plant roots may increase critical flow conditions for rill channel development (Li, 1995; Sidorchuk & Grigorév, 1998; De Baets *et al.*, 2006). Knapen *et al.* (2007) recently reviewed published τ_c data and discussed the soil and environmental properties affecting the soil resistance to concentrated flow erosion.

In contrast to the number of publications on critical flow conditions for incipient rilling, very few studies report critical flow conditions for incipient gullying. During a rain event, many rills may develop, but only a few may grow into a gully provided that local flow intensities exceed those needed for the erosion of a gully channel. For cropland, Poesen *et al.* (2003) reported critical flow shear stresses during peak flow ranging between 3.3 and 32.2 Pa (mean = 14 Pa) for ephemeral gullies eroded in silt loam (loess-derived) topsoils in Belgium, whereas τ_c ranged between 16.8 and 74.4 Pa (mean = 44 Pa) for ephemeral gullies formed in stony sandy loams in Portugal. In general, an inverse relationship between concentrated flow

width and τ_c for ephemeral gully development in these study areas is observed (Poesen *et al.*, 2002). The significant difference in τ_c between both study areas cannot be explained by differences in land use, as in both cases ephemeral gullies developed in tilled cropland, but are attributed to different soil types. Whereas no rock fragments are present in the Belgian loess-derived soils, the rock fragment content of topsoils in southern Portugal amounts to 30% by mass on average. Poesen *et al.* (1999) and Rieke-Zapp *et al.* (2007) demonstrated experimentally that rock fragment content in topsoils significantly reduces their susceptibility to concentrated flow erosion. For non-cultivated land in Australian valley floors, Prosser (1996) reported τ_c values for gully initiation of 21 Pa for a bare clay soil, but for vegetated soils the values were 70 Pa for heavily degraded aquatic plants or tussock and sedge, >105 Pa for undisturbed aquatic plants, >180 Pa for lightly degraded tussock and sedge and, >240 Pa for undisturbed tussock and sedge. Grassed irrigation canals have also been found to resist flow shear stresses of up to 260 Pa before showing signs of scour (Reid, 1989, cited by Prosser 1996).

For prediction purposes, more data are needed on critical hydraulic conditions leading to gully initiation, development and infilling in a range of environments, as well as for different land management practices. Very few studies have attempted to measure critical hydraulic conditions for incipient gullying in field conditions, mainly because of logistical constraints. Therefore, several studies have instead attempted to assess critical environmental conditions for gullying that were more easily quantifiable in field conditions, such as rainfall, topography, soils (or lithology) and land use, as these factors control either the runoff hydraulics, the resistance of the soil surface to incision, or both.

(ii) Rainfall thresholds Threshold rain depths (P, mm) needed to initiate ephemeral gullies in cropland (i.e. 14.5 mm < P < 22 mm) are only slightly larger compared with those needed to initiate rills (i.e. 7.6 mm < P < 25 mm) (Poesen *et al.*, 2003). The range of observed threshold values for ephemeral gully development in cropland is attributed to different states of the soil surface (roughness, degree of sealing) as affected by tillage operations and antecedent rains. Nachtergaele (2001) analysed 38 ephemeral gully erosion events that occurred over a 15-year period in central Belgium and found critical P values of 15 mm in (late) winter ($n = 21$) and of 18 mm in (early) summer ($n = 17$), which is attributed to a difference in soil moisture content between the two seasons. Threshold rains for gully development in land under forestry operations in Australia are significantly larger ($P = 80–100$ mm) than those for ephemeral gully development on seedbeds. Sudden snowmelt on frozen / thawing soil presents a special case of a meteorological threshold condition at higher latitudes, higher altitudes or areas with a continental climate, which can lead to the rapid development of ephemeral gullies. Øygarden (2003) documented how the combination of frozen subsoils, saturated topsoils with low strength and intense rainfall led to the development of ephemeral gullies in Norway, even in areas with gentle slope gradients. These observations point to the fact that a gradual climate change to more unstable winter conditions (i.e. freezing and thawing combined with intense rain) is likely to increase the risk of (ephemeral) gully erosion. One of the difficulties encountered when assessing critical rain depths for gully initiation is the lack of representative rain data for the sites where gully erosion processes have been observed (Vandekerckhove et al., 2000).

(iii) Topographic thresholds Where do gullies develop in the landscape? Most models predicting soil erosion by water do not predict the location of gullies. Yet this issue is important for land managers and for predicting possible impacts of climatic or land-use changes on the spatial distribution and density of gullies. This question can be reformulated as follows: where do gully channels start and where do they end in the landscape?

Where do (ephemeral) gully channels start? An approach to predict locations where gully heads

might develop is presented by the threshold concept, first applied to geomorphological systems by Patton and Schumm (1975). This concept is based on the assumption that in a landscape with a given climate and land use, there exists for a given slope gradient of the soil surface (S) a critical drainage area (A) necessary to produce sufficient runoff to cause gully incision. As slope steepens, this critical drainage area decreases and vice versa. For different environmental conditions and different gully initiating processes (hydraulic erosion by Hortonian overland flow, saturation overland flow and seepage erosion, landslides), different topographic thresholds apply (Montgomery & Dietrich, 1994). Threshold lines for gully development by hydraulic erosion can be represented by a power-type equation (Begin & Schumm, 1979; Vandaele et al., 1996): $S = aA^b$ with a and b coefficients depending on the environmental characteristics. Table 19.1 presents an overview of studies reporting topographic thresholds for incipient gullying in a range of environments and reveals that b ranges in value from –0.10 to –0.80. Kirkby et al. (2003) have shown that power law equations describing sediment transport for water erosion occurring on runoff plots are consistent with S-A relations describing the location of ephemeral and permanent gully channel heads in the landscape.

The topographic threshold concept for gully initiation permits one to predict, for a given land use, the location in the landscape where gully channels may develop by providing a physical basis for the initiation of gullies. Poesen et al. (1998) compared ten published critical S-A datasets for ephemeral gullies and permanent gullies in different environments and found that not only the environmental characteristics, but also the methodology used to assess critical S and A, affect the reported topographic threshold for incipient gullying. Poesen et al. (2003) demonstrated that topographic threshold conditions for gully initiation in non-cultivated land (i.e. sagebrush and scattered trees, open oak woodland and grasslands, coastal prairie, logged forest and swampy, reed-covered valley floors) plot well above those needed to initiate ephemeral gullies in

Table 19.1 Overview of studies reporting topographic thresholds for incipient gullying in a range of environments. The topographic threshold is the value of b in the relationship $S = aA^b$ where S is the slope gradient of the soil surface and A is the drainage area.

Country	Land use	Value of b	Source
Europe			
Belgium	Cropland	−0.30 to −0.40	Vandaele *et al.* (1996)
Belgium: shallow (depth<0.8 m) ephemeral gullies	Cropland	−0.14	Vanwalleghem *et al.* (2005b)
Belgium: deep (depth>0.8 m) ephemeral gullies	Cropland	−0.15	Vanwalleghem *et al.* (2005b)
UK (South Downs)	Cropland	−0.25	Boardman (1992)
France	Cropland	−0.40	IGN (1983)
Portugal (Alentejo)	Cropland	−0.35	Vandaele *et al.* (1997)
Portugal (Alentejo)	Cropland	−0.30	Vandekerckhove *et al.* (2000)
Portugal (Alentejo)	Rangeland	−0.41	Vandekerckhove *et al.* (2000)
Portugal (Bragança)	Cropland	−0.23	Vandekerckhove *et al.* (2000)
Spain (Rambla Chortal)	Cropland	−0.13	Vandekerckhove *et al.* (2000)
Spain (Cerro Tonosa)	Cropland	−0.10	Vandekerckhove *et al.* (2000)
Spain (Rambla Chortal)	Cropland	−0.14	Vandekerckhove *et al.* (2000)
Spain (Sierra de Gata)	Rangeland	−0.27	Vandekerckhove *et al.* (2000)
Spain (Almeria)	Rangeland	−0.49	Poesen *et al.* (2002)
Greece (Lesvos)	Rangeland	−0.14	Vandekerckhove *et al.* (2000)
Africa			
Ethiopia (Tigray)	Rangeland	−0.50	Nyssen *et al.* (2002)
Tanzania (Makonde Plateau, SE Tanzania)	Rangeland	−0.80	Achten *et al.* (2008)
Tanzania (inland plains, SE Tanzania)	Rangeland	−0.36	Achten *et al.* (2008)
Rwanda	Cropland	−0.60	Moeyersons (2003)
Swaziland	Rangeland	−0.11 to −0.26	Morgan & Mngomezulu (2003)
America			
US (Colorado)	Sagebrush and scattered trees	−0.26	Patton & Schumm (1975)
US (Oregon)	Logged forest	−0.40	Montgomery & Dietrich (1988)
US (Southern Sierra Nevada)	Open oak woodland and grasslands	−0.60	Montgomery & Dietrich (1988)
US (California)	Coastal prairie	−0.40	Montgomery & Dietrich (1988)
Australia			
Australia (New South Wales)	Pasture with sparse vegetation	−0.63	Prosser and Abernethy (1996)
Australia	Pasture	−0.54	Prosser & Winchester (1996)
Australia (Victoria and New South Wales)	Forest with gullies at road drain outlets	No topographic threshold observed	Takken *et al.* (2008)
Asia			
China (Shaanxi)	Cropland	−0.24	Wu & Cheng (2005)
China (Inner Mongolia)	Cropland	−0.38	Cheng *et al.* (2006)
China (Shaanxi)	Cropland	−0.30	Cheng *et al.* (2007)
China (Heilongjiang)	Cropland	−0.15	Zhang *et al.* (2007)

cropland – the coefficient a for non-cultivated land is significantly larger than that for cropland. From a comparison of six critical S-A datasets corresponding to various Mediterranean study areas in Europe and collected using the same methodology, Vandekerckhove et al. (2000) found that vegetation type and cover were far more important than climatic conditions in explaining differences in topographic thresholds for different areas. In cultivated fields, topsoil structure and soil moisture conditions, as controlled by the antecedent rainfall distribution, are crucial factors affecting the S-A relationships, rather than daily rain for the gully initiating events. For rangelands, vegetation cover and type (annuals and perennials) at the time of gully head development appears to be the most important factor differentiating between topographic thresholds. The importance of vegetation biomass in concentrated flow zones for reducing gully initiation risk in semi-arid environments was also stressed by Graf (1979) and Nogueras et al. (2000). Graf (1979) observed critical flow shear force (F_c, dynes) for gully cutting into valley floors in Colorado to depend on the valley-floor biomass (B_v, kg m^{-2}) following the relation $F_c = 0.07\,(B_v)^2$. Along the same lines, Prosser and Slade (1994) demonstrated through flume experiments on an unincised valley floor near Canberra, Australia, the crucial role that vegetation cover plays in decreasing the susceptibility of valley floors to gully formation. In addition to the above-ground biomass, plant roots significantly contribute to the increase of the site resistance to concentrated flow erosion. De Baets et al. (2006, 2007) demonstrated that fibrous (grass) roots provide a larger increase in soil cohesion and hence in resistance to concentrated flow erosion compared with tap-rooted plant species.

From these observations it becomes clear that any land-use change implying a vegetation biomass decrease (either above- or below-ground), as well as a lowering of the erosion resistance of the topsoil by tillage operations in concentrated flow zones, will decrease the topographic threshold for incipient gullying. This implies that for a given slope gradient (S), the critical drainage area (A) for gully head development will decrease,

and therefore gully density will increase, as pointed out by Kirkby (1987).

Several studies have applied the topographical threshold concept in combination with a hydraulic threshold to predict areas at risk of gullying (e.g. Dietrich et al. 1993; Prosser & Abernethy, 1996; Jetten et al., 2006). Desmet and Govers (1997) and Desmet et al. (1999) investigated the relative importance of slope gradient (S) and drainage area (A) for the optimal prediction of the initiation and trajectory of ephemeral gullies. In the latter study, a striking discrepancy was found between the high A-exponent (i.e. 0.7–1.5) required to predict optimally the trajectory of the gullies, and the low A-exponent (i.e. 0.2) required to identify spots in the landscape where ephemeral gullies begin.

Where do (ephemeral) gullies end? Gullies usually end where the transporting capacity of the concentrated runoff drops and/or where the erosion resistance of the topsoil increases sharply. A sudden change from one land use to another might trigger sediment deposition instead of channel entrenchment (vegetation-controlled sediment deposition, e.g. Takken et al., 1999; Beuselinck et al., 2000; Steegen et al., 2000). In many field conditions, a lowering of the slope gradient with increasing drainage area causes a drop in transporting capacity, and hence a decrease in gully channel depth (slope-controlled sediment deposition). In contrast with critical S-A relations established for the location of gully heads, few S-A relations have been established for the location of sites where (ephemeral) gullies end (e.g. Poesen et al., 1998; Vandekerckhove et al., 2000; Nachtergaele et al., 2001a,b). Field measurements in different cropland areas of northern Europe reveal that topographically-induced sediment deposition at the lower end of ephemeral gully channels, which developed in loamy to loamy sand soils, usually occurs in a narrow range of local slope gradient under cropland of 2–4%. However, when the rock fragment content of the topsoil increases, slope-controlled sediment deposition occurs on steeper slopes of up to 25–30% (Poesen et al., 2002).

From this review we conclude that detailed information on the impact of various land uses on topographic thresholds needed to initiate gullies under a range of climatic conditions is rather scarce. Yet, such information is crucial for predicting where in the landscape gully development might be expected under different environmental conditions.

(iv) Pedological and lithological controls

Soil type Many studies have investigated the susceptibility of soils (soil erodibility) to inter-rill and rill erosion (see Bryan, 2000, for a review). Knapen *et al.* (2007) recently reviewed published data on the effects of soil properties on the resistance of soils to concentrated flow erosion (expressed in terms of critical flow shear stress and channel erodibility) and proposed a general soil erosion resistance ranking according to soil texture and tillage practices. Comparatively few studies have investigated the susceptibility of soils to gully erosion. Soil type and in particular the vertical distribution of the erosion resistance of various soil horizons largely controls the size and more specifically the depth and cross-sectional morphology of gullies. Ireland *et al.* (1939) were the first to point to the important role of the resistant Bt-horizons in controlling gully depth and gully head shape in the southeastern United States. Other studies conducted on gully development in duplex soils in Australia (e.g. Sneddon *et al.*, 1988) and on loess-derived soils in Europe (Poesen, 1993) also came to the same conclusion. Poesen (1993) found that soil shear strength at saturation of the various loess-derived soil horizons is a good indicator of their resistance to concentrated flow erosion. For loess-derived soils, Nachtergaele and Poesen (2002) showed that: (1) τ_c and channel erodibility (related to concentrated flow erosion) for a Bt-horizon were significantly larger compared with τ_c and erodibility for an Ap or a C horizon; and (2) an increasing antecedent moisture content of each horizon had a negative effect on their erodibility. In landscape positions where Bt-horizons are still present, ephemeral gully depth is limited to a maximum of 0.50m. However, for landscape positions where no Bt-horizon is present, concentrated flow may erode ephemeral gullies several metres deep (Poesen, 1993; Vanwalleghem *et al.*, 2005b). Erosion of Bt-horizons caused by various processes (i.e. water erosion, tillage erosion, removal of soil during root and tuber crop harvesting, land levelling) therefore largely increases the risk of deep gully development. Other soil horizons observed to be resistant to gully erosion are plough pans, fragipans, petrocalcic horizons, ironstone hardpan (petroplinthite) or unweathered bedrock. On the other hand, less permeable soil horizons can induce positive pore-water pressures in the overlying soil layers, which in turn lowers the erosion resistance of these soil horizons, particularly when seepage conditions (return flow) occur (e.g. Moore *et al.*, 1988; Huang & Laflen, 1996; Poesen *et al.*, 2003). This in turn may alter the topographic threshold for gully head initiation (e.g. Montgomery & Dietrich, 1994; Vandekerckhove *et al.*, 2000; Poesen *et al.*, 2002).

Lithology Many field observations clearly reveal that lithology significantly controls the size and density of gullies that can develop in a given landscape. For instance, the maximum depth of the permanent gully shown in Plate 16 (b) is controlled by the presence of hard unweathered bedrock at a depth of ca. 8m. The occurrence of landscapes heavily dissected by gullies in the Mediterranean (i.e. badlands) is strongly controlled by the presence of particular lithological conditions: unconsolidated or poorly sorted materials such as shales, gypsiferous and salty-silt marls and silt-clay deposits of Tertiary and Quaternary age (Poesen & Hooke, 1997; Gallart *et al.*, 2002). Faulkner *et al.* (2003) reported on the role of site geochemistry in morphological development of badlands. In tropical environments, the development of large gully systems typically occurs on unconsolidated loose sandy deposits (Plate 19). In contrast with sheet and rill erosion, relatively little is known about the properties of soils or parent materials and the associated processes that control the dynamics of their resistance to gully erosion.

(v) Land use thresholds

Gully development in prehistoric and historical times triggered by a combination of human-induced land-cover changes and extreme rainfalls has been documented for various parts of the world, for example: induced by intensive land use in central Belgium during Roman times (Vanwalleghem et al., 2006); caused by a change in catchment hydrology in response to human-induced vegetation change in the UK in the 9th and 10th centuries AD (Harvey, 1996); due to high land-use pressure and extreme rains in central Europe in the 14th century in Germany (Bork et al., 1998); during the Little Ice Age in Slovakia (Stankoviansky, 2003); caused by the introduction of cattle (leading to overgrazing) and a climatic shift in the southwestern US (Webb & Hereford, 2001); and in eastern Australia since European settlement 200 years ago (Prosser & Winchester, 1996).

Several recent case studies have documented the significant impacts of a gradual or sudden shift in land use on the triggering of gullying or the increase in gully erosion rates. For instance, field observations in central Belgium indicate that the increase in area under maize over the last two decades has resulted in an increased ephemeral gully erosion risk (Nachtergaele, 2001). Faulkner (1995) reported on the triggering of gully erosion associated with the expansion of unterraced almond cultivation after hasty clearance of native Mediterranean matorral in southern Spain. This land-use change also caused the development or reactivation of bank gullies along ephemeral streams in southeastern Spain (Oostwoud Wijdenes et al., 2000). Bork et al. (2001) documented the effect of agricultural intensification in the second half of the 20th century in the Upper Yangtze river basin (southwest China) on rapid gully development and the subsequent gully stabilization as a consequence of reforestation by air-seeding. Several studies conducted in a range of environments have documented the impact of road construction on the increased gully erosion risk on steep slopes (e.g. Moeyersons, 1991; Montgomery, 1994; Wemple et al., 1996; Croke & Mockler, 2001; Nyssen et al., 2002; Takken et al., 2008). Gully incision is significantly more likely below

culverts on steep slopes with longer than average contributing ditch length (Wemple et al., 1996; Nyssen et al., 2002). Montgomery (1994) showed that for a given slope gradient, the drainage area required to support a gully head is smaller for road-related runoff than for undisturbed slopes. Contributing road length and the gradient of the discharge hillslope have been successfully used to separate gullied and non-gullied flow pathways within catchments (Croke & Mockler, 2001). Burkard and Kostaschuk (1997) attributed the increased growth rates of bank gullies along the shoreline of Lake Huron to increased snowfall and extreme flow events, but also to the extension of municipal drains and the use of subsurface drainage. Vanacker et al. (2003) have documented the impact of collapsing irrigation canals and the mismanagement of excess irrigation water on the extension of the rill and gully network in a semi-arid region of Ecuador. Several studies have reported the strong impact of urbanization through vegetation removal, the drastic increase in runoff response and the concentration of runoff on gully erosion risk, particularly in Third World countries (e.g. Plate 19).

Many more detailed case studies are needed if we want full understanding of the impact of various types of land-use change and its interaction with extreme weather conditions on gully development. Also, more research is needed on the (socio-economic) drivers of land-use changes causing increased or decreased gully erosion risk.

19.3.2 Gully erosion models

This section reviews models to predict gully location in landscapes, the soil losses caused by ephemeral and permanent gully erosion, and the rate of retreat of (bank) gully headcuts.

(i) Modelling the location of ephemeral gullies and permanent gullies As discussed earlier, gully location and morphology depend upon topographical, geological, land cover and hydrological characteristics. The exact position of gullies in the landscape depends on the specific processes that are responsible for gully formation, but basically, their location is controlled by the way water is concentrated and where the capacity of the water is sufficiently high to cut a channel. Three processes have been identified that are important in gully formation: overland flow, seepage and shallow landsliding. Any overland flow, whether it results from infiltration or saturation excess, will exert a shear stress on the underlying soil surface which, if it exceeds a critical or threshold value, will result in channel initiation and development. Seepage erosion involves the entrainment of material as a result of water flowing through and emerging from the soil (Dietrich & Dunne, 1993), lowering its erodibility through seepage forces (Gabbard *et al.*, 1998). This erosion process is responsible for the initiation of subsurface pipes and tunnels, which have been found to be important for gully formation especially in semi-arid areas (Faulkner *et al.*, 2004). Finally, shallow landslides can concentrate water in the landscape and act as headcuts for the development of gullies. Montgomery and Dietrich (1994) defined a theoretical framework for delineating the range of topographic conditions (local slope gradient and drainage area) where each of these processes acts. In practice, however, interactions are likely to occur and gullies are formed by simultaneous or subsequent action of various processes.

Although the topographic threshold concept, as explained above, has some limitations (Chaplot *et al.*, 2005), it is of great practical significance to predict locations in the landscape where gully heads might develop. For each pixel in the landscape, the upslope drainage area (*A*) and local slope gradient (*S*) must be calculated and, using an appropriate critical *S-A* relation for that environment (see Table 19.1), one can then assess the risk of having a gully head developing in that pixel. Using such an approach, Prosser and Abernethy (1996) predicted the extent of a stable gully network successfully. Along the same lines Desmet *et al.* (1999), Jetten *et al.* (2006) and Knapen and Poesen (2010) successfully predicted the location of ephemeral gullies in cropland.

Ephemeral gully channels end in a downslope direction where massive sediment deposition and fan-building occurs. This is where either surface roughness increases suddenly (e.g. where a different land use begins – land use-induced sediment deposition) or where local slope gradient decreases (i.e. slope-induced sediment deposition: Beuselinck *et al.*, 2000). Here transport capacity of the concentrated flow will drop sharply leading to massive (coarse) sediment deposition. Very few *S-A* relationships for sediment deposition exist. For several European cropland conditions, Nachtergaele *et al.* (2001a,b) reported data indicating that the topographic threshold (*S-A* relationship) for sediment deposition at the bottom end of ephemeral gullies was smaller than the corresponding *S-A* relationship for incipient ephemeral gullying. The difference between the critical topographical conditions for ephemeral gully initiation and those for sedimentation vary among different environments, and depend, among other factors, on the texture and rock fragment content of the topsoils (Vandekerckhove *et al.*, 2000; Poesen *et al.*, 2002). These few datasets allow one to locate the initiation point and the sediment deposition point of an ephemeral gully, based on topographic attributes (*S* and *A*) and on rock fragment content. Consequently, ephemeral gully length can be derived by routing concentrated flow from the gully head towards the fan at the gully end. For other environments, more data are needed to predict ephemeral gully length.

Desmet *et al.* (1999) investigated the possibility of predicting the location of ephemeral gullies using an inverse relationship between local slope gradient (*S*) and upslope contributing area per unit length of contour (*As*). Predicted locations of ephemeral gullies were confronted with the locations recorded in three intensively cultivated catchments over a 5-year observation period. The optimal relative area (*As*) exponent (relative to the slope exponent) ranged from 0.7 to 1.5. A striking discrepancy was found between the high relative area exponent required to predict optimally the entire trajectory of the ephemeral gullies, and the low relative area exponent (0.2) required to identify the spots in the landscape where ephemeral gullies begin. This indicates that zones in the landscape where ephemeral gullies start are more controlled by slope gradient, while the presence of concavities control the trajectory of the gullies until the slope gradient is too low and (coarse) sediment deposition dominates. Such an approach can be improved by incorporating the presence of linear landscape elements, soil surface state, vegetation cover and root density, and possibly rain, to the input parameters. Souchère *et al.* (2003) presented an expert-based model for predicting the location and the volumes of ephemeral gullies.

Even with detailed topographic data, it is sometimes hard to predict where exactly the water will concentrate. Anthropogenically-induced linear roughness elements (e.g. drill and plough furrows, headlands) in landscapes with gentle slopes might make this even more difficult by altering the true catchment area (e.g. Souchère *et al.*, 1998; Takken *et al.*, 2001). Water often concentrates in furrows created by tillage or along rural roads, which are difficult – if not impossible – to include in topographical maps. Even with topographic maps of high resolution, a simple pass of agricultural equipment could disturb the existing drainage structure of a field.

Many studies investigating critical *S*-*A* thresholds for incipient gullying (see Table 19.1) have noted the large scatter of the observation data points. This scatter has been attributed to spatial variations in erosional or hydrological processes (Dietrich *et al.*, 1993; Prosser & Dietrich, 1995; Prosser & Abernethy, 1996) or land-use pattern changes (Prosser & Soufi, 1998; Desmet *et al.*, 1999; Vandekerckhove *et al.*, 2000). These studies typically classify about 70–80% of the channel heads correctly. Istanbulluoglu *et al.* (2002) formalized the description of this variability by interpreting the topographic threshold C $(AS^\alpha = C)$ as a random variable. Its probability distribution is then derived physically from the random variability of quantities involved in the erosion process. They considered median grain size, roughness, and excess rainfall. This resulted in a probabilistic channel incision zone that shifted according to the variability in these input factors. The resulting probability distribution of the threshold C was shown to follow a gamma distribution. The

advantage of the probabilistic model of Istanbulluoglu *et al.* (2002) over a single initiation threshold is that the latter will predict significant erosion only in locations where channelization is predicted on a long-term basis. The probabilistic model then provides a way to account for the less frequent contribution to erosion due to channelization even at locations that do not meet the single channelization threshold. Comparing their resulting stochastic model with field data from gullies in the Idaho Batholith, they concluded that a large part of the observed variation could be attributed to grain size differences.

While of great practical use, estimating gully location using topographical conditions alone has its limitations, given the variety of other environmental variables that control gully initiation and development. Therefore, this simple approach might result in high prediction errors in some cases. Vandekerckhove *et al.* (1998), for example, applied this topographic threshold concept to different landscapes in Spain and Portugal, and their results revealed that *S* and *A* were weakly correlated with gullying. The prediction of gully erosion was considerably strengthened by including additional information on land use, soil stoniness, and soil horizon hydraulic conductivity. Knapen and Poesen (2010) demonstrated that ephemeral gully initiation points and dimensions are not only topographically controlled but also depend upon the erosion resistance of the topsoil. Various erosion models use this information in their water flow routing routines. Therefore, Chaplot *et al.* (2005) tested to what extent the direct flow velocity estimations, as obtained from an existing surface water routing algorithm, could be used to predict the location of what they called 'linear erosion elements', which included rills and small ephemeral gullies. By defining a critical velocity threshold of 0.062 m s^{-1}, the location and extent of linear erosion elements corresponded relatively well with observations in a 0.62 km^2 watershed in Laos.

Statistical techniques have been applied with respect to the mapping of gully location. Hughes *et al.* (2001), for example, applied a regression tree model to the prediction of gully density on a national scale, for the relatively densely populated areas in Australia, covering some 1.7 million km². This methodology was used because of the complexity of the relationship between gully density and environmental factors. Their final model included land use, geology, texture, rainfall and other climate indices and topographical attributes.

(ii) Modelling soil losses by (ephemeral) gully erosion One of the first models for simulating soil loss by ephemeral gully erosion that was accessible to field practitioners was the Ephemeral Gully Erosion Model (EGEM), developed by scientists from USDA-NCRS (Merkel *et al.*, 1988; Woodward, 1999). The main limitation of EGEM is that the length and topographic location of the gullies need to be known exactly. Gully evolution in the model is therefore limited to incision and widening, neglecting its lengthwise growth. The channel erosion routines are a slightly modified version of those used in the CREAMS model, and also used in WEPP (Foster, 1986). Soil detachment rate is dependent upon the difference between actual sediment load and maximum sediment load in the flow according to the following equation:

$$D = KC(\tau - \tau_c) \qquad (19.1)$$

where D is the detachment rate (g m⁻² s⁻¹), KC the channel erodibility, τ the average flow shear stress (N m⁻²) and τ_c the critical shear stress for particle entrainment (N m⁻²).

In order to satisfy Equation (19.1), gullies are first deepened (maintaining the initial width) until they reach a less erodible soil layer, typically the tillage depth or topsoil depth, and then widened until they reach the estimated maximum width. Since the maximum depth is also user-defined, this maximum width is a critical variable. This width (W_{max}) was calculated using the following regression equation:

$$W_{max} = 179 Q^{0.552}\, n^{0.556}\, S^{0.199}\, \tau_c^{-0.476} \qquad (19.2)$$

where Q is the peak runoff rate (m³ s⁻¹), n is Manning's roughness coefficient (m⁻¹/³ s), S is

critical runoff slope (m m⁻¹) and τ_c is the critical flow shear stress (N m⁻²).

Nachtergacle *et al.* (2001a,b) were among the first to thoroughly test the model's performance against field data on ephemeral gullies formed in stony soils (Mediterranean environment) and in loess-derived soils (temperate maritime climate). In spite of the fact that both gully length and maximum depth were input parameters, they showed that EGEM was not capable of predicting measured ephemeral gully volumes satisfactorily. Pooling all data from 112 ephemeral gullies, they observed a strong relationship between gully volume (V, m³) and gully length (L, m): $V = 0.048\, L^{1.29}$ with an R² of 0.91. Capra *et al.* (2005) confirmed the conclusions of Nachtergaele *et al.* and also observed a significant relationship between V and L for 92 ephemeral gullies formed in clayey soils (Sicily): $V = 0.0082\, L^{1.42}$ with an R² of 0.64. Recently, Zhang *et al.* (2007) reported the following relation for 21 ephemeral gullies formed in clay loam soils in northeast China: $V = 0.015\, L^{1.43}$ with an R² of 0.67. These findings imply that predicting ephemeral gully length is a valuable alternative for the prediction of ephemeral gully volume. Nachtergaele *et al.* (2001a, b) also presented a simple procedure based on topographic thresholds to predict ephemeral gully length. The empirical *V-L* relations can also be used to convert ephemeral gully length data extracted from, say, aerial photographs into ephemeral gully volumes.

Apart from the static gully length that was needed *a priori*, several other drawbacks of EGEM were reported. Nachtergaele *et al.* (2001a,b) pointed to the practical problems associated with field determination of parameters for Equation (19.2), such as n and τ_c. Furthermore, runoff discharge and channel width vary distinctly in time and space over the gully's length, which are not taken into account. Instead, concentrated flow discharge was assumed constant in time over the entire length of the runoff event. Calculations according to equations (19.1) and (19.2) were performed at the gully mouth. In space, some constants were then used to scale gully width and transport capacity, calculated at the gully mouth, over the rest of the gully channel. One of the other problem issues identified by Nachtergaele et al. (2001a,b), although not specifically related to EGEM but typical of many erosion models, is the faulty calculation of channel erodibility (Torri et al. 1987) and critical flow shear stress. The latter is calculated according to Smerdon and Beasly (1961):

$$\tau_c = 0.311 \cdot 10^{0.0128Pc} \qquad (19.3)$$

where Pc is clay content (%) of the soil.

Channel erodibility (KC) is then assumed to be inversely linearly related to τ_c. However, an extensive literature review has recently shown the limitations of this simple approach (Knapen et al., 2007). Furthermore, the flow's transport capacity is calculated according to Yalin's method. This implies that sediment deposition cannot be simulated and that particle-size characteristics are simplified to some representative grain diameter, so sediment sorting cannot be represented.

With the objective of overcoming some of the limitations of EGEM, the physical basis and predictive capabilities of EGEM were further elaborated. This resulted in an improved version of EGEM, called the Revised Ephemeral Gully Erosion Model (REGEM) (Gordon et al., 2007). Four major changes can be emphasized: (1) dynamic gully length by explicitly modelling headcut migration processes; (2) spatially and temporally varied runoff discharge during events; (3) improved gully width estimates, based on discharge alone; (4) five different particle sizes accounted for in the sediment transport calculations.

The computational framework has been designed in order to incorporate the model in the Annualized Agricultural Nonpoint Source Pollution Model (AnnAGNPS). AnnAGNPS uses amorphous areas or subwatersheds, which implies that some of the sub-pixel gully erosion processes need to be simplified. One disadvantage of this approach is that local topography is simplified to some average subwatershed characteristics. Therefore, gully initiation points need to be defined by the user, hereby introducing some degree of subjectivity. However, usually this is not problematic, since ephemeral gullies

are recurring phenomena and their position can be derived from earlier observations. If this information is not available, for example for a modelling study of an unknown area, a separate analysis using local slope and contributing area thresholds could identify potential initiation points in a more objective manner. This analysis then has to be done outside the REGEM framework.

A first important change in the hydrological module is that it has been adapted to simulate unsteady flow. REGEM still remains based on TR-55 methods, using the empirical curve number method to estimate runoff processes (USDA Soil Conservation Service, 1986), but now the calculated runoff volume and peak runoff rate are used to construct a triangular hydrograph for each event. Since the model now considers unsteady and spatially varied flow, these are able to influence the headcut migration rate, channel width and depth, and the rates of sediment entrainment, transport and deposition.

Two sources of sediment are considered in REGEM: headcut erosion and channel widening. The gully incision processes start with a localized incision, which is considered of little relevance since its eroded volume is not taken into account. Once the headcut is formed, this is one of the main processes of sediment production. The processes of plunge-pool erosion and headcut migration are represented by the fully analytical model of Alonso et al. (2002), discussed further on in this chapter.

These headcut migration processes dynamically interact with the hydrological model component. After every time step, during which the gully head can migrate, the contributing area and resulting runoff discharge are modified accordingly. As the gully grows, the actual drainage area is reduced, so that the discharge at the gully head diminishes accordingly. Since the real shape of the catchment in which the modelled gully is located has been approximated by one single amorphous cell in AnnAGNPS, some approximations need to be introduced. The drainage area contributing to the runoff at each timestep j is expressed as a function of the actual gully length, L_j, and the maximum gully length possible, L_{max}:

$$A_i = 1 - \left[\left(\frac{L_i}{L_{\max}} \right)^{5/3} \right] \qquad (19.4)$$

Equation (19.4) results from the simplified representation of the drainage area as a right-angled triangle. For the same reason, L_{max} is not directly available from the real path of the concentrated flow. L_{max} is defined as a function of the given drainage area A_d (ha) of each gully. In the present model, it is calculated by an empirical function fitted by Leopold *et al.* (1964).

$$L_{\max} = 80.3 A_d^{0.6} \qquad (19.5)$$

This equation was originally fitted for rivers, with considerably larger dimensions and drainage areas than ephemeral gullies.

The consideration of gully headcut migration is one of the main changes with respect to EGEM. The fact that gully length is no longer a static variable is a big step forward, although it still has to be assessed how the model performs in heterogeneous material (e.g. in the presence of a resisting soil layer). The underlying plunge-pool erosion and headcut migration model of Alonso *et al.* (2002) has been validated by Bennett (1999), Bennett *et al.* (2000) and Bennett and Casalí (2001), who conducted their experimental measurements in homogeneous material. Although Gordon *et al.* (2007) reported that gullies did not reach maximum length in their simulations, for longer simulations, where the gullies might reach maximum length, this length fully depends on the gully initiation point defined by the user and the maximum gully length as derived from Equation (19.5). Since Nachtergaele *et al.* (2001a,b) identified gully length as the single most important parameter controlling ephemeral gully soil loss, it seems critical to evaluate the performance of this empirical equation in different environments. Available datasets for ephemeral gullies in the Belgian and Chinese loess belt do not seem to corroborate this simplified approach. Gully length (*L*)–drainage area (*A*) relations from Nachtergaele *et al.* (2001a,b) and Cheng *et al.* (2006) did not yield any statistical

relationship between *L* and *A*. Respectively, *P* values of 0.2 (*n* = 67) and 0.6 (*n* = 49) were obtained.

In order to determine shear stresses exerted by the flow on the gully walls, gully width is a critical parameter to calculate flow velocity from discharge. Gully width (*W*) is now determined directly from flow discharge (*Q*) through the regression equations in the form of:

$$W = aQ^b \qquad (19.6)$$

where the values of the coefficients reported by Nachtergaele *et al.* (2002a) and Torri *et al.* (2006) are used: $a = 2.51$ and $b = 0.412$. Torri *et al.* (2006) observed that the exponent *b* varies continuously with channel width, indicating that more research is needed to improve the predictive capacity of this equation. The Manning equation is then applied, with Manning's *n* as a user-defined input parameter, to derive flow depth and velocity iteratively so that flow shear stresses can be calculated.

Sediment detachment rate is calculated according to Equation (19.1). The critical flow shear stress τ_c in Equation (19.3) is further adapted according to tillage practices. Soil erodibility *KC* is determined from τ_c according to the relation:

$$KC = 100\tau_c^{-0.5} \qquad (19.7)$$

Although Gordon *et al.* (2007) recognized that at present, no comprehensive field dataset is available that would allow a full validation of REGEM, they tested the model's performance against data from four field sites in central Mississippi (Smith, 1992). Detailed measurements across multiple cross-sections were available for four gullies. They reported that simulated gully lengths and channel widths approximated reasonably well to the observations (RMSE of 31% and 52% respectively).

In the light of these results it is clear that although REGEM conceptually tackles some of the limitations of EGEM, at present, no extensive validation of REGEM has been performed. Nevertheless, REGEM is conceptually a very useful tool to assess the potential contribution and dynamics of ephemeral gully erosion.

Another physically-based approach that could be used to predict sediment evacuated by gullies was developed by Istanbulluoglu et al. (2003). They measured cross-sections at several points within gullies that had formed during a single intense rainstorm after a forest fire, in the Idaho Batholith. Using the planimetric map of the gullies they predicted the eroded volume between the different cross-sections with standard sediment transport equations as a function of dimensionless shear stress. The latter variable was calculated from field topographical data (i.e. runoff contributing area and local slope) and the characteristics of the rainfall event (i.e. intensity and duration). Vanwalleghem et al. (2009) further tested this approach in a wider range of conditions and concluded that good results could be obtained with one single sediment transport equation for incision that ranged in size from small rills (several cm wide) to large gullies (several metres wide). One important condition, however, is that the incisions formed during a single event with well-known hydrological conditions.

Gully sidewall failure is known to contribute significantly to channel widening. Statistical techniques have been applied to analyse and model gully sidewall erosion. Martínez-Casasnovas et al. (2004) used logistic regression to model the presence–absence of sidewall failure. Although their model did not include a measure of soil loss by sidewall failures, they successfully predicted wall collapse (overall accuracy = 87%) as a function of variables related to topography and soil hydrological conditions.

Istanbulluoglu et al. (2005) developed a physically-based approach for the initiation of slab failures in gullies. Based on a force balance equation of an assumed failure geometry, they implemented their numerical model both in a simple one-dimensional model and in a more complex three-dimensional landscape evolution model (CHILD). They showed that their model could explain 60% of the variability in observed bank heights in a study area in Colorado. With

the more complex landscape evolution model, they concluded that slab failures have profound effects on the tempo of topographic evolution.

Sidorchuk (1999) and others (reviewed by Vanwalleghem et al. 2005c) observed that gully channel formation is very rapid during the period of gully initiation, when morphological characteristics of a gully (i.e. length, depth, width, area and volume) are far from stable. This period typically takes about 5% of a gully's lifetime. For the remaining part of a gully's lifetime, its size is near a stable, maximum value. Sidorchuk (1999) proposed two gully erosion models corresponding to these two stages of gully development: (1) a dynamic model predicting rapid changes of gully morphology during the first short period of gully development; and (2) a static model to calculate the final morphometric parameters of stable gullies. The dynamic gully model is based on the solution of the equations of mass conservation and gully bed deformation. The static gully model is based on the assumption of a final morphological equilibrium of a gully. Both model types were tested using data on gully morphology and dynamics from Yamal peninsula (Russia) and New South Wales (Australia) (Sidorchuk & Grigorév, 1998; Sidorchuk, 1999). These models have been used at the landscape level to evaluate the effects of gully development at a larger scale. Sidorchuk et al. (2001, 2003) applied the stable (or static) gully model to the catchment of the Mbuluzi river in Swaziland, after having subdivided it into erosion response units (ERU). An ERU is a 3D terrain unit, where an erosion process is scarcely spatially varying with respect to other ERUs (i.e. variation of process type and intensity increases while further increasing the ERU size). The application of the modelling required the definition of a stable channel width, which resulted for the Mbuluzi river to be $W_b = 0.5A_c^{0.3}$ where W_b is the gully bed width and A_c is the contributing area (m^2).

(iii) Modelling gully headcut retreat Once initiated, (bank) gullies essentially expand by gully headcut retreat, and to a lesser extent by gully wall retreat (e.g. Plate 18 and Plate 19). Whether a bank gully retreats by a single headcut or by multiple headcuts is controlled by factors such as topography, material type and land use. Oostwoud Wijdenes *et al.* (1999) classified gully headcuts into four types (i.e. gradual, transitional, abrupt and rilled-abrupt) and observed in southeast Spain that gully head morphology could be used as an indicator for gully development stage, and hence for sediment production.

Several studies report that, besides topography (drainage area), land use (change) has a significant impact on bank (gully) head erosion (e.g. Burkard & Kostaschuk, 1997; Oostwoud Wijdenes *et al.*, 2000; Wang *et al.*, 2008). For instance, Burkard and Kostaschuk (1997) attributed the increased growth rates of bank gullies along the shoreline of Lake Huron partly to the increased extension of municipal drains and the use of subsurface drainage. Land use changes involving the removal of semi-natural vegetation (*matorral*) and the extension of almond cultivation intensified the bank gully head activity in southeast Spain (Oostwoud Wijdenes *et al.*, 2000). Wang *et al.* (2008) observed in Yunnan (China) mean gully headcut retreat rates (over 4 years) to range between 0.2 and 0.4 m y^{-1} for 15–20% vegetation cover (grass, shrubs and forest) in the gully catchment, 3.73 m y^{-1} for cropland, and 4.69 m y^{-1} for bare land without any vegetation. Several studies show that lithology also has a clear impact on (bank) gully headcut activity: for the same land use type in Spain, headcuts in marls, sandy loams and loams were significantly more active compared with headcuts that developed in gravels and conglomerates (Oostwoud Wijdenes *et al.*, 2000; de Luna Armenteros *et al.*, 2004). Similar observations were reported for Romania by Radoane *et al.* (1995). These authors reported that the mean rate of gully headcutting was over 1.5 m y^{-1} for gullies developing in sandy deposits, and under 1 m y^{-1} for gullies cut in marls and clays.

Several studies have attempted to quantify and predict gully area increase or gully headcut retreat (R) in a range of environments, including linear measurements (e.g. Thompson, 1964; Seginer, 1966; Soil Conservation Service, 1966; De Ploey, 1989; Burkard & Kostaschuk, 1995, 1997;

Radoane *et al.*, 1995; Oostwoud Wijdenes & Bryan, 2001; Vandekerckhove *et al.*, 2001, 2003), area measures (e.g. Beer & Johnson, 1963; Burkard & Kostaschuk, 1995, 1997), volumetric measures (e.g. Stocking, 1980; Sneddon *et al.*, 1988, Vandekerckhove *et al.*, 2001, 2003) and weight measures (e.g. Piest & Spomer, 1968). According to Stocking (1980), volumetric measures are the best compromise, avoiding difficult considerations of bulk density of soils no longer *in situ*. The resulting equations typically link R (obtained from detailed cross-sectional surveys of gully channel cross-sections taken at periodic intervals throughout the study period, or from aerial photographs to estimate changes in channel dimensions over time) with parameters such as drainage area (A) above the gully head (an index for surface or subsurface runoff volume), rainfall depth, erodibility, height of the headcut, relief energy of drainage basin, and runoff response of the drainage area. Here we list these equations, based on field measurements, as they help in understanding the effect of environmental factors on R.

For the deep loess area of western Iowa (US), Beer and Johnson (1963) expressed gully growth as:

$$R = 81.41\, R_t^{0.0982}\, A_t^{-0.0440}\, L_g^{0.7954}\, L_w^{-0.2473}\, e^{-0.0014dp} \quad (19.8)$$

where R is the gully surface growth over the observation period (m^2), R_t is the index of surface runoff (mm), A_t is the terraced area of the catchment (m^2), L_g is the length of the gully at the beginning of period (m), L_w is the length from end of gully to catchment divide (m), e is the natural logarithm, and dp is the deviation of precipitation from normal (mm).

In the US, Thompson (1964) studied gully head advancement in Minnesota, Iowa, Alabama, Texas, Oklahoma and Colorado, and developed an empirical equation:

$$R = (7.13 \times 10^{-5})\, A^{0.49}\, S^{0.14}\, P^{0.74}\, E \quad (19.9)$$

where R is the gully head advancement for the time period of interest (m), A is the drainage area above the gully head (m²), S is the slope (%) of approach channel above the gully head; P is the summation of rainfall from 24-hour rains equal to or larger than 12.7 mm for the time period of interest (mm), and E is the clay content (%) of the eroding soil profile.

Seginer (1966) studied the advancement of gully headcuts in southern Israel and proposed the following equation:

$$R = a\,A^{0.50} \qquad (19.10)$$

where R is the mean medium-term (15 years) annual gully head retreat rate (m y⁻¹), A is the area of the drainage basin (km²), and a is the coefficient ranging between 2.1 and 6.0, depending on the studied catchment.

The US Soil Conservation Service (1966) analysed headward advancement for 210 gullies in six widely scattered land resource areas east of the Rocky Mountains, and proposed the following relationship:

$$R = 0.36\,A^{0.46}\,P^{0.20} \qquad (19.11)$$

where R is the mean annual gully head advance (m y⁻¹), A is the drainage area above gully head (ha); and P is the annual precipitation (mm) on days with precipitation in excess of 12.7 mm day.

Stocking (1980, 1981) examined the gully headcut retreat rate of 66 valley-bottom gullies in central Zimbabwe in an area of sodium-rich, fine, sandy soils over different time spans, producing a series of multiple-factor regression models. Here two equations are reported, the first describing waterfall-head gullies and the second piped-head gullies:

$$R = 0.00687 P^{1.34} A_c H^{0.52} \qquad (19.12)$$

$$R_p = 0.00793\,A_p^{0.57} I^{1.72} \qquad (19.13)$$

where R and R_p are the soil volume (m³) loss at headcut per rain event, P is the event rain depth (mm), A_c is the catchment area at headcut (km²), H is the headcut height (m) of the waterfall head,

A_p is an antecedent moisture index (mm); and I is a piping index.

These relationships, established in the same study area during the same period, show that gullies retreat at very different rates, reacting at minimal perturbation (e.g. piped gully headcut – Stocking (1981) commented that piping strongly increases the rate of retreat) or only during intense storms because the causative factors can be quite different. Several studies of this type produced equations similar to those produced by Stocking (1981). These equations are all of local values because the role of soil and soil characteristics is not parameterized. For example, the exponents of Equation (19.12) vary between 1.02 and 1.88 for P, 0.38 and 1.00 for A_c, and 0.42 and 1.27 for H.

The two gully headcut retreat equations developed by Stocking (1981) indicate that when soil dispersion occurs, seepage and piping can significantly modify the type of processes and erosion rates and can even change the dominant factors involved. At present, for gully development in dispersive soils no mathematical model is available but a series of observations allows one to understand better gully generation and expansion. Dispersivity characteristics are well known to affect dramatically the erodibility of the material (Fitzpatrick et al., 1992). When dispersivity is coupled with seepage, then subsurface erosion, pipes and gullies are certain to occur. Faulkner et al. (2004) observed that variations in sodium absorption ratio (SAR) in the Mocatán catchment (Almeria, Spain) follow hydraulic gradients, with peaks corresponding to the most incised part of the studied gully, locally increasing the soil erodibility. They also observed that SAR increases with clay content. When clay swells, rain infiltration is reduced and even completely impeded, forcing water to move laterally instead of vertically in the soil profile, hence enhancing the development of pipes running subparallel to the soil surface. This process can also prevent pipes from deepening and widening. In some cases pipes are confined under a crust rich in calcium, which replaces sodium during leaching. Most of the rills develop from these pipes after pipe roof collapse. The depth-to-width ratio of linear incisions in dispersive material can be quite high. Part of this description also applies to other areas where piping and gullied badlands develop on geological substrata of different ages and materials (e.g. in Tuscany, Italy, on Plio-Pleistocene marine sediments: Torri & Bryan, 1997)). The presence of vegetation, which favours the formation of large connected pores due to root growth, also favours leaching with translocation of sodium. These processes strongly undermine soil resistance to concentrated flow erosion. Faulkner *et al.* (2008) used these observations to explain the coupling of pipes to a rejuvenating channel in Mocatán. A series of incisions reconnects slopes and channels, causing an increase in the hydraulic gradient along the slopes. This in turn favours piping and pipe enlargement, with subsequent collapse of pipe roofs. Hence lateral gullies can develop quickly. The new (lateral) gullies have walls along which similar processes can develop, first piping and then further gullying. This feedback process will only end after a long period of slope-base stability. Examples of these types of processes can be observed in many places with badlands such as the *biancane* badlands of Tuscany (Torri & Bryan, 1997).

Radoane *et al.* (1995) reported on an extensive study of gully head advancement in Moldavia (Romania) and proposed the following equations:

$$R = aA^b\, L^c\, E^d\, P^e \quad \text{(for 22 gullies cut in marls and clays)} \tag{19.14}$$

$$R = a + bA + cE + dL + eP \tag{19.15}$$
(for 16 gullies cut in sandy rocks)

where R is the medium-term (14 years) mean retreat rate of the gully head (m y^{-1}), A is the drainage basin area upstream of the gully head (ha), L is the gully length (m), E is the relief energy of the drainage basin (m), P is the drainage basin inclination (m (100m)$^{-1}$), and a, b, c, d and e are empirical coefficients or exponents.

By far the most important independent variable controlling gully recession rate was drainage-basin area, explaining 54% of the variance in the case of marls and clays, and 68% in the case of sandy rocks.

Burkard and Kostaschuk (1997) examined the behaviour of 44 gullies cut into clays of glaciolacustrine and glacial origin along the eastern shoreline of Lake Huron, Canada, over a 62-year period (using aerial photographs) and proposed the following equation:

$$R = 0.3996\, A^{0.59} \tag{19.16}$$

where R is the mean annual gully area growth (m^2 y^{-1}) and A is the drainage area (m^2)

Vandekerckhove *et al.* (2001) monitored 46 active bank gully heads in southeast Spain and found that the present drainage basin area (A, m^2) was the most important factor explaining mean short-term (2 years) gully headcut retreat rate (R, m^3 y^{-1}), following the equation:

$$R = 0.04\, A^{0.38} \tag{19.17}$$

Vandekerckhove *et al.* (2003) showed that when predicting R in southeast Spain, the weight given to drainage basin area (A) increases from the short term (i.e. a few years) to the long term (decades, centuries), and attribute this to several reasons:
(1) So far, most studies have focused on the medium-term retreat rate of gullies. Little is known about the processes and factors controlling the short-term gully head erosion of gullies. Predicting long-term gully head retreat rates seems to be easier than short-term retreat rates because of the stochastic nature of some gully wall subprocesses such as tension crack development, soil toppling and soil fall, piping and fluting.
(2) Spatial rainfall variability may be held responsible for important variations in short-term headcut retreat rates within the study areas.
(3) In the long term, one may expect an increased contribution of extreme rainfall events to gully headcut retreat whereby the role of the drainage basin area becomes more pronounced, as contributing runoff is produced from a larger fraction of the gully catchment during such events and runoff transmission losses are smaller compared with low-intensity rain events.
Several attempts to model gully growth, based on theoretical–physical considerations, have been

reported by Faulkner (1974), De Ploey (1989), Kemp (1990), and Robinson and Hanson (1994). Merz and Bryan (1993) and Stein and Julien (1993) described more refined approaches for nickpoints at incipient rilling and headcut migration in gullies and rills. Hanson et al. (2001) proposed a detailed model based on failure at headcuts. They used the geometry of the plunge pool and waterfall combined with the hypothesis that erosion rate is proportional to the excess of effective stress on the critical stress needed to cause failure. Alonso et al. (2002) described the headcut migration as a self-similar propagation process. Central to their model is the description of the erosive action of the jet that enters and scours the gully plunge pool. Flores-Cervantes et al. (2006) extended their approach and implemented it in the three-dimensional landscape evolution model CHILD (Tucker et al., 2001). Prasad and Römkens (2003) presented a holistic and energy-based conceptual framework for modelling headcut dynamics.

Although several attempts have been made to develop empirical and process-based models for predicting either gully subprocesses or gully erosion rates in a range of environments, there are still no reliable (i.e. validated) models available allowing one to predict impacts of environmental change on gully erosion rates at various temporal and spatial scales, and their impacts on sediment yield, hydrological processes and landscape evolution.

19.4 Interaction Between Gully Erosion, Hydrological and Other Erosion Processes

What is the impact of gully erosion on hillslope hydrological processes such as infiltration and drainage? Once gullies develop, water infiltration rates through the gully bottom may be significantly greater compared with that of the soil surface in the intergully areas, if the gully channel develops into more permeable horizons. Through the gully bed and banks, significant runoff water transmission losses can then take place, particularly in semi-arid and arid environments as shown by Esteves and Lapetite (2003) in Niger. Such water transmission losses have also been reported to occur in smaller erosion channels (i.e. rills, see Poesen & Bryan, 1989; Parsons et al., 1999) as well as in larger (ephemeral) river channels (for a review, see Beven, 2002). Recent studies (e.g. Leduc et al., 2001; Avni, 2004) indicate that gully development in semi-arid areas may therefore lead to significant groundwater recharge. On the other hand, if gullies develop into hillslopes with temporary water tables, they will cause an enhanced drainage and a rapid water table lowering, which results in a significant drying-out of the soil profiles in the intergully areas, as observed by Moeyersons (2000) in Africa. In addition, Okagbue and Uma (1987) reported that gullies located at the seepage areas of groundwater systems in southeastern Nigeria may become very active during the peak recharge times of the rainy season, because high porewater pressures reduce the effective strength of the unconsolidated materials along the seepage faces. The seepage forces caused by exit hydraulic gradients at the levels of seepage on the gully walls produce boiling conditions, piping and tunnelling that undermine the gully walls and activate their retreat. Most erosion models are driven by hydrological models (runoff). The previous discussion clearly indicates that there are also important feedback mechanisms – gully erosion may in turn also control the intensity of some hydrological processes (water transmission losses or groundwater depletion) at the hillslope scale. These interactions deserve more attention.

How does gully erosion interact with other soil erosion processes? Once gullies develop they often trigger other soil degradation processes such as piping, soil fall or soil topple (driven by gravity) after tension crack development and undercutting. Furthermore, gully channels enhance the export of sediment produced at the intergully areas (sheet and rill erosion) by increasing the connectivity in the landscape (e.g. Stall, 1985; Poesen et al., 2002, 2006), which leads to an increased risk of sediment deposition in the lower parts of the landscape. If no gully control measures are taken, gully growth rates usually decline exponentially (e.g. Graf, 1977; Rutherford et al.,

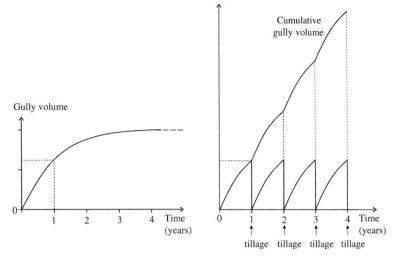

Fig. 19.1 Illustration of the interaction between ephemeral gully development and tillage erosion/deposition. Left: evolution of (ephemeral) gully volume over time if no tillage operations take place. Note the digressive increase of gully volume (based on review by Vanwalleghem *et al.*, 2005c). Right: evolution of (cumulative) ephemeral gully volume over time, if gully channels are filled in annually by tillage operations.

1997; Nachtergaele *et al.*, 2002b; Poesen *et al.*, 2006). However, in cropland areas, ephemeral gullies are usually filled in by tillage (tillage erosion and tillage deposition) within less than a year of their initiation. During subsequent storms (years), the infilled soil material is usually eroded again by concentrated flow, thereby increasing the plan-form concavity of the site. The newly created plan-form concavity increases the probability of concentrated flow erosion. So ephemeral gully erosion and tillage erosion reinforce each other (Poesen *et al.*, 2003). Gordon *et al.* (2008) recently demonstrated through modelling that total soil erosion rates in several geographical regions (Belgium, US) could be 250% to 450% greater when ephemeral gullies are tilled and, hence, reactivated annually as opposed to no-till conditions. These results demonstrate that routine filling of ephemeral gully channels during tillage practices may result in markedly higher rates of soil loss compared with allowing these gullies to persist on the landscape, demonstrating a further advantage of adopting no-till management practices (Fig. 19.1).

In various parts of Europe, landscapes heavily dissected by gullying (badlands) have been levelled, thereby causing strong soil profile truncation in the intergully areas and infilling of gullies

with this material (e.g. Revel & Guiresse, 1995; Poesen & Hooke, 1997; Borselli *et al.*, 2006). Such land levelling operations have often resulted in renewed gully incision of the levelled land, as well as in shallow landsliding causing large soil losses (Clarke & Rendell, 2000; Borselli *et al.*, 2006). In other words, important interactions exist between concentrated flow erosion and tillage erosion (Poesen, 1993) as well as with erosion caused by land levelling.

The significant interactions between gully erosion on the one hand and hydrological (i.e. infiltration, drainage) as well as other soil erosion processes (piping, mass wasting, tillage erosion and erosion by land levelling) on the other, need to be better understood for improving our predictions of hydrological response and land degradation rates under different environmental conditions. This improved understanding is the basis for taking appropriate and effective measures to control soil erosion.

Acknowledgements

The authors thank the many colleagues for fruitful discussions on gully erosion and for stimulating visits to their gully sites. Financial support from the European Commission through the Integrated Project DESIRE (Desertification mitigation and remediation of land – a global approach for local solutions; FP6, sub-priority 1.1.6.3. Global Change and Ecosystems, Contract no 037046) is acknowledged.

References

Achten, W., Dondeyne, S., Mugogo, S., *et al.* (2008) Gully erosion in South Eastern Tanzania: spatial distribution and topographic thresholds. *Zeitschrift für Geomorphologie* **52**: 225–35. DOI: 10.1127/0372-8854/2008/0052-0225

Alonso, C.V., Bennett, S.J. & Stein, O.R. (2002) Predicting head cut erosion and migration in concentrated flows typical of upland areas. *Water Resources Research* **38**: 1–15.

Avni, Y. (2004) Gully incision inducing ongoing desertification in the arid regions of the Middle East: Example from the Negev Highlands, southern Israel. In Li, Y., Poesen, J. & Valentin, C. (eds), *Gully Erosion under Global Change*. Sichuan Science and Technology Press, Chengdu, China.

Beer, C.E. & Johnson, H.P. (1963) Factors in gully growth in the deep loess area of western Iowa. *Transactions of the American Society of Agricultural Engineers* **6**: 237–40.

Begin, Z.B. & Schumm, S.A. (1979) Instability of alluvial valley floors: a method for its assessment. *Transactions of the American Society of Agricultural Engineers* **22**: 347–50.

Bennett, S.J. (1999) Effect of slope on headcut growth and migration in upland concentrated flows. *Geomorphology* **30**: 273–90.

Bennett, S.J., Casalí, J., Robinson, K.M. & Kadavy, K.C. (2000) Characteristics of actively eroding ephemeral gullies in an experimental channel. *Transactions of the American Society of Agricultural Engineers* **43**: 641–9.

Bennett, S. & Casalí, J. (2001) Effect of initial step height on headcut development in upland concentrated flows. *Water Resources Research* **37**: 1475–84.

Beuselinck, L., Steegen, A., Govers, G., *et al.* (2000) Characteristics of sediment deposits formed by intense rainfall events in small catchments in the Belgian Loam Belt. *Geomorphology* **32**: 69–82.

Beven, K. (2002) Runoff generation in semi-arid areas. In Bull, L.J. & Kirkby, M.J. (eds), *Dryland Rivers: Hydrology and Geomorphology of Semi-Arid Channels*. John Wiley & Sons, Chichester: 57–105.

Boardman, J. (1992) Current erosion on the South Downs: implications for the past. In Bell, M. & Boardman, J. (eds), *Past and Present Soil Erosion. Archaeological and Geographical Perspectives.* Oxbow Monograph 22, Oxbow Books, Oxford: 9–20.

Bocco, G. (1991) Gully erosion: processes and models. *Progress in Physical Geography* **15**: 392–406.

Bork, H.R., Bork, H., Dalchow, C., *et al.* (1998) *Landschaftsentwicklung in Mitteleuropa.* Klett-Perthes, Gotha, Germany.

Bork, H.R., Li, Y., Zhao Y., *et al.* (2001) Land use changes and gully development in the upper Yangtze River Basin, SW-China. *Journal of Mountain Science* **19**: 97–103.

Borselli, L., Torri, D., Øygarden, L., *et al.* (2006) Land levelling. In Boardman, J. & Poesen, J. (eds), *Soil Erosion in Europe*. John Wiley & Sons, Chichester, UK: 644–58.

Brice, J.B. (1966) Erosion and deposition in the loess-mantled Great Plains, Medicine Creek drainage basin, Nebraska. *US Geological Survey Professional Paper* 352H: 235–339.

Brunori, F., Penzo, M. & Torri, D. (1989) Soil shear strength: its measurement and soil detachability. *Catena* **16**: 59–71.

Bryan, R.B. (2000) Soil erodibility and processes of water erosion on hillslope. *Geomorphology* **32**: 385–415.

Bull, L.J. & Kirkby, M.J. (1997) Gully processes and modelling. *Progress in Physical Geography* **21**: 354–74.

Bull, L.J. & Kirkby, M. (2002) Channel heads and channel extension. In Bull, L.J. & Kirkby, M.J. (eds), *Dryland Rivers: Hydrology and Geomorphology of Semi-Arid Channels*. John Wiley & Sons, Chichester, UK: 263–98.

Burkard, M.B. & Kostaschuk, R.A. (1995) Initiation and evolution of gullies along the shoreline of Lake Huron. *Geomorphology* **14**: 211–19.

Burkard, M.B. & Kostaschuk, R.A. (1997) Patterns and controls of gully growth along the shoreline of Lake Huron. *Earth Surface Processes and Landforms* **22**: 901–11.

Capra, A., Mazzara, L.M. & Scicolone, B. (2005) Application of the EGEM model to predict ephemeral gully erosion in Sicily, Italy. *Catena* **59**: 133–46.

Cerdan, O., Poesen, J., Govers, G., *et al.* (2006) Sheet and rill erosion. In Boardman, J. & Poesen, J. (eds), *Soil Erosion in Europe*. John Wiley & Sons, Chichester, UK: 501–13.

Chaplot, V., Giboire, G., Marchand, P. & Valentin, C. (2005) Dynamic modelling for gully initiation and development under climate and land-use changes in northern Laos. *Catena* **63**: 318–28.

Cheng, H., Wu, Y., Zou, X., *et al.* (2006) Study of ephemeral gully erosion in a small upland catchment on the Inner-Mongolian Plateau. *Soil & Tillage Research* **90**: 184–93.

Cheng, H., Zou, X., Wu, Y., *et al.* (2007) Morphology parameters of ephemeral gully in characteristics hillslopes on the Loess Plateau of China. *Soil & Tillage Research* **94**: 4–14.

Clarke, M. & Rendell, H. (2000) The impact of the farming practice of remodelling hillslope topography on badland morphology and soil erosion processes. *Catena* **40**: 229–50.

Croke, J. & Mockler, S. (2001) Gully initiation and road-to-stream linkage in a forested catchment, southeastern Australia. *Earth Surface Processes and Landforms* **26**: 205–17.

Crouch, R. & Novruzzi, T. (1989) Threshold conditions for rill initiation on a vertisol, Gunnedah, NSW, Australia. *Catena* **16**: 101–10.

De Baets, S., Poesen, J., Gyssels, G. & Knapen, A. (2006) Effects of grass roots on the erodibility of topsoils during concentrated flow. *Geomorphology* **76**: 54–67.

De Baets, S., Poesen, J., Knapen, A. & Galindo, P. (2007) Impact of root architecture on the erosion-reducing potential of roots during concentrated flow. *Earth Surface Processes and Landforms* **32**: 1323–45. DOI: 10.1002/esp.1470

De Luna Armenteros, E., Laguna Luna, A., Poesen, J. & Giráldez Ververa, J.V. (2004) Evolución de un sistema de cárcavas activas en el sureste español. *Ingeniería del Agua* **11**: 5–73.

De Ploey, J. (1989) A model for headcut retreat in rills and gullies. *Catena Suppl.* **14**: 81–6.

Derose, R.C., Gomez, B., Marden, M. & Trustrum, N. (1998) Gully erosion in Mangatu Forest, New Zealand, estimated from digital elevation models. *Earth Surface Processes and Landforms* **23**: 1045–53.

Desmet, P.J.J. & Govers, G. (1997) Two-dimensional modelling of the within-field variation in rill and gully geometry and location related to topography. *Catena* **29**: 283–306.

Desmet, P.J.J., Poesen, J., Govers, G. & Vandaele, K. (1999) Importance of slope gradient and contributing area for optimal prediction of the initiation and trajectory of ephemeral gullies. *Catena* **37**: 377–92.

De Vente, J. & Poesen, J. (2005) Predicting soil erosion and sediment yield at the basin scale: Scale issues and semi-quantitative models. *Earth Science Reviews* **71**: 95–125.

Dietrich, W.E. & Dunne, T. (1993) The channel head. In Beven, K. & Kirkby, M.J. (eds), *Channel Network Hydrology*. John Wiley & Sons, Chichester, UK: 175–219.

Dietrich, W.E., Wilson, C.J., Montgomery, D.R. & McKean, J. (1993) Analysis of erosion thresholds, channel networks, and landscape morphology using a digital terrain model. *Journal of Geology* **101**: 259–78.

Esteves, M. & Lapetite, J.M. (2003) A multi-scale approach of runoff generation in a Sahelian gully catchment: a case study in Niger. *Catena* **50**: 255–71.

Evans, R. (1993) On assessing accelerated erosion of arable land by water. *Soils and Fertilizers* **56**: 1285–93.

Faulkner, H. (1974) An allometric growth model for competitive gullies. *Zeitschrift für Geomorfologie Supplement Band* **21**: 76–87.

Faulkner, H. (1995) Gully erosion associated with the expansion of unterraced almond cultivation in the coastal Sierra de Lujar, S. Spain. *Land Degradation & Rehabilitation* **9**: 179–200.

Faulkner, H., Alexander, R. & Wilson, B.R. (2003) Changes to the dispersive characteristics of soils along an evolutionary slope sequence in the Vera badlands, southeast Spain: implications for site stabilisation. *Catena* **50**: 243–54.

Faulkner, H., Alexander, R., Teeuw, R. & Zukowskyj, P. (2004) Variations in soil dispersivity across a gully head displaying shallow sub-surface pipes, and the role of shallow pipes in rill initiation. *Earth Surface Processes and Landforms* **29**: 1143–60.

Faulkner, H., Alexander, R. & Zukovskyj, P. (2008) Slope–channel coupling between pipes, gullies and tributary channels in the Mocatán catchment badlands, Southeast Spain. *Earth Surface Processes and Landforms* **33**: 1242–60.

Fitzpatrick, R.W., Boucher, S.C., Naidu, R. & Fritsch, E. (1992) Environmental consequences of soil sodicity. In Naidu, R., Sumner, M.E. & Rengasamy, P. (eds), *Australian Sodic Soils*, CSIRO, Australia, Chapter 6.1.

Flores-Cervantes, J.H., Istanbulluoglu, E. & Bras, R.L. (2006) Development of gullies on the landscape: a model of headcut retreat resulting from plunge pool erosion. *Journal of Geophysical Research* **111**: F01010. DOI: 10.1029/2004JF000226

Foster, G.R. (1986) Understanding ephemeral gully erosion. In *Soil Conservation: Assessing the Natural Resource Inventory*, Volume 2. National Academy of Science, Washington, DC: 90–128.

Franti, T.G., Laflen, J.M. & Watson, D.A. (1999) Predicting soil detachment from high-discharge concentrated flow. *Transactions of the American Society of Agricultural Engineers* **42**: 329–35.

Gabbard, D.S., Huang, C., Norton, L.D. & Steinhardt, G.C. (1998) Landscape position, surface hydraulic gradients and erosion processes. *Earth Surface Processes and Landforms* **23**: 83–93.

Gallart, F., Solé, A., Puigdefàbregas, J. & Lázaro, R. (2002) Badland systems in the Mediterranean. In Bull, L.J. & Kirkby, M.J. (eds), *Dryland Rivers: Hydrology and Geomorphology of Semi-Arid Channels*. John Wiley & Sons, Chichester: 300–326.

Gilley, J.E., Elliot, W.J., Laflen, J.M. & Simanton, J.R. (1993) Critical shear stress and critical flow rates for initiation of rilling. *Journal of Hydrology* **142**: 251–71.

Gordon, L.M., Bennett, S.J., Bingner, R.L., *et al.* (2007) Simulating ephemeral gully erosion in AnnAGNPS. *Transactions of the American Society of Agricultural and Biological Engineers* **50**: 857–66.

Gordon, L.M., Bennett, S.J., Alonso, C.V. & Bingner, R.L. (2008) Modeling long-term soil losses on agricultural fields due to ephemeral gully erosion. *Journal of Soil and Water Conservation* **63**: 173–81.

Graf, W.L. (1977) The rate law in fluvial geomorphology. *American Journal of Science* **277**: 178–91.

Graf, W.L. (1979) The development of montane arroyos and gullies. *Earth Surface Processes and Landforms* **4**: 1–14.

Grissinger, E. (1996a) Rill and gullies erosion. In Agassi, M. (ed.), *Soil Erosion, Conservation, and rehabilitation*. Marcel Dekker, New York: 153–67.

Grissinger, E. (1996b) Reclamation of gullies and channel erosion. In Agassi, M. (ed.), *Soil Erosion, Conservation and Rehabilitation*. Marcel Dekker, New York: 301–13.

Guerra, A.J.T., Bezerra, J.F.R., Fullen, M.A., *et al.* (2007) Urban gullies in São Luis city, Maranhão state, Brazil. In Casali, J. & Giménez, R. (eds), *Progress in Gully Erosion Research*, Universidad Pública de Navarra: 58–9.

Haan, C.T., Barfield, B.J. & Hayes, J.C. (1994) *Design Hydrology and Sedimentology for Small Catchments*. Academic Press, London.

Hanson G.J., Robinson K.M. & Cook K.R. (2001) Prediction of headcut migration using a deterministic approach. *Transactions of the American Society of Agricultural Engineers* **44**: 525–31.

Harvey, A.M. (1996) Holocene hillslope gully systems in the Howgill Fells, Cumbria. In Anderson, M.G. & Brooks, S.M. (eds), *Advances in Hillslope Processes*. John Wiley & Sons, Chichester, UK: 731–52.

Hauge, C. (1977) Soil erosion definitions. *California Geology* **30**: 202–3.

Horton, R.E. (1945) Erosional development of streams and their drainage basins: hydrophysical approach to quantitative morphology. *Geological Society of America Bulletin* **56**: 275–370.

Huang, C.H. & Laflen, J.M. (1996) Seepage and soil erosion for a clay loam soil. *Soil Science Society of America Journal* **60**: 408–16.

Hughes, A.O., Prosser, I.P., Stevenson, J., *et al.* (2001) *Gully erosion mapping for the National Land and Water Resources Audit*. Technical Report 26/01, CSIRO, Canberra.

IGN (1983) *Erosion des terres agricoles d'après photographies aériennes*. IGN, Ligescourt-Somme, 23 pp.

Imeson, A.C. & Kwaad, F.J.P.M. (1980) Gully types and gully prediction. *Geografisch Tijdschrift* **14**: 430–41.

Ireland, H.A., Sharpe, C.F.S. & Eargle, D.H. (1939) *Principles of Gully Erosion in the Piedmont of South Carolina*. USDA Technical Bulletin 633, 142 pp.

Istanbulluoglu, E., Tarboton, D.G., Pack, R.T. & Luce, C.A. (2002) A probabilistic approach for channel initiation. *Water Resources Research* **38**: 1325.

Istanbulluoglu, E., Tarboton, D.G., Pack, R.T. & Luce, C.H. (2003) A sediment transport model for incision of gullies on steep topography. *Water Resources Research* **39**: 1103–17.

Istanbulluoglu, E., Bras, R.L., Flores-Cervantes, H. & Tucker, G.E. (2005) Implications of bank failure and fluvil erosion for gully development: field observations and modelling. *Journal of Geophysical Research – Earth Surface* **110**: F01014.

Jetten, V., Poesen, J., Nachtergaele, J. & van de Vlag, D. (2006) Spatial modelling of ephemeral gully incision: a combined empirical and physical approach. In Owens P.N. & Collins A.J. (eds), *Soil Erosion and Sediment Redistribution in River Catchments*. CAB International, Wallingford, UK: 195–206.

Kemp, A.C. (1990) Towards a dynamic model of gully growth. *International Association of Hydrological Sciences Publication* **189**: 121–34.

Kirkby, M.J. (1987) Modelling some influences of soil erosion, landslides and valley gradient on drainage density and hollow development. *Catena Suppl.* **10**: 1–11.

Kirkby, M.J., Bull, L.J., Poesen, J., *et al.* (2003) Observed and modelled distributions of channel and gully heads – with examples from SE Spain and Belgium. *Catena* **50**: 415–34.

Knapen, A. & Poesen, J. (2010) Soil erosion resistance effects on rill and gully initiation points and dimensions. *Earth Surface Processes and Landforms* **35**: 217–28.

Knapen, A., Poesen, J., Govers, G., *et al.* (2007) Resistance of soils to concentrated flow erosion: A review. *Earth-Science Reviews* **80**: 75–109.

Laflen, J.M. & Beasly, R.P. (1960) Effects of compaction on critical tractive forces in cohesive soils. *University of Missouri, Agricultural Experiment Station, Research Bulletin* 749.

Leduc, C., Favreau, G. & Schroeter, P. (2001) Long-term rise in a Sahelian water-table: the Continental Terminal in South-West Niger. *Journal of Hydrology* **243**: 43–54.

Leopold, L.B., Wolman, M.G. & Miller, J.P. (1964) *Fluvial Processes in Geomorphology*. Freeman and Co., San Francisco.

Li, Y. (1995) *Plant Roots and Soil Anti-scouribility on the Loess Plateau*. National Natural Science Foundation of China, Science Press, Bejing, 133 pp.

Martínez-Casasnovas, J.A., Ramos, M.C. & Poesen, J. (2004) Assessment of sidewall erosion in large gullies using multi-temporal DEMs and logistic regression analysis. *Geomorphology* **58**: 305–21.

Merkel, W.H., Woodward, D.E. & Clarke, C.D. (1988) Ephemeral gully erosion model (EGEM). In *Modelling Agricultural, Forest and Rangeland Hydrology*. American Society of Agricultural Engineers Publication 07-88, St Joseph, MI: 315–23.

Merz, W. & Bryan, R.B. (1993) Critical conditions for rill initiation on sandy loam Brunisols: laboratory and field experiments in southern Ontario, Canada. *Geoderma* **57**: 357–85.

Moeyersons, J. (1991) Ravine formation on steep slopes – forward versus regressive erosion – some case-studies from Rwanda. *Catena* **18**: 309–24.

Moeyersons, J. (2000) Desertification and man in Africa. *Bulletin of the Royal Academy of Overseas Science, Brussels*, **46**: 151–70.

Moeyersons, J. (2003) The topographic thresholds of hillslope incisions in southwestern Rwanda. *Catena* **50**: 381–400.

Montgomery, D.R. (1994) Road surface drainage, channel initiation, and slope instability. *Water Resources Research* **30**: 1925–32.

Montgomery, D. & Dietrich, W. (1988) Where do channels begin? *Nature* **336**: 232–4.

Montgomery, D.R. & Dietrich, W.E. (1994) Landscape dissection and drainage area-slope thresholds. In Kirkby, M.J. (ed.), *Process Models and Theoretical Geomorphology*. John Wiley & Sons, Chichester: 221–46.

Moore, I.D., Burch, G.J. & Mackenzie, D.H. (1988) Topographic effects on the distribution of surface soil water and the location of ephemeral gullies. *Transactions of the American Society of Agricultural Engineers* **32**: 1098–1107.

Morgan, R.P.C. & Mngomezulu, D. (2003) Threshold conditions for initiation of valley-side gullies in the Middle Veld of Swaziland. *Catena* **50**: 401–14.

Nachtergaele, J. (2001) *A spatial and temporal analysis of the characteristics, importance and prediction of ephemeral gully erosion*. PhD thesis, Department of Geography-Geology, K.U. Leuven.

Nachtergaele, J. & Poesen, J. (2002) Spatial and temporal variations in resistance of loess-derived soils to ephemeral gully erosion. *European Journal of Soil Science* **53**: 449–63.

Nachtergaele, J., Poesen, J., Steegen, A., *et al.* (2001a) The value of a physically based model versus an empirical approach in the prediction of ephemeral gully erosion for loess-derived soils. *Geomorphology* **40**: 237–52.

Nachtergaele, J., Poesen, J., Vandekerckove, L., *et al.* (2001b) Testing the ephemeral gully erosion model (EGEM) for two Mediterranean environments. *Earth Surface Processes and Landforms* **26**: 17–30.

Nachtergaele, J., Poesen, J., Sidorchuk, A. & Torri, D. (2002a) Prediction of concentrated flow width in ephemeral gully channels. *Hydrological Processes* **16**: 1935–53.

Nachtergaele, J., Poesen, J., Oostwoud Wijdenes, D. & Vandekerckhove, L. (2002b) Medium-term evolution of a gully developed in a loess-derived soil. *Geomorphology* **46**: 223–39.

Nogueras, P., Burjachs, F., Gallart, F. & Puigdefabregas, J. (2000) Recent gully erosion in El Cautivo badlands (Tabernas, SE Spain). *Catena* **40**: 203–15.

Nyssen, J., Poesen, J., Moeyersons, J., *et al.* (2002) Impact of road building on gully erosion risk: a case study from the northern Ethiopian highlands. *Earth Surface Processes and Landforms* **27**: 1267–83.

Okagbue, C.O. & Uma, K.O. (1987) Performance of gully erosion control measures in southeastern Nigeria. *International Association of Hydrological Sciences Publication* **167**: 163–72.

Oostwoud Wijdenes, D.J. & Bryan, R.B. (2001) Gully-head erosion processes on a semi-arid valley floor in Kenya: a case study into temporal variation and sediment budgeting. *Earth Surface Processes and Landforms* **26**: 911–33.

Oostwoud Wijdenes, D., Poesen, J., Vandekerckhove, L., *et al.* (1999) Gully-head morphology and implications for gully development on abandoned fields in a semi-arid environment, Sierra de Gata, Southeast Spain. *Earth Surface Processes and Landforms* **24**: 585–603.

Oostwoud Wijdenes, D.J., Poesen, J., Vandekerckhove, L. & Ghesquiere, M. (2000) Spatial distribution of gully head activity and sediment supply along an ephemeral channel in a Mediterranean environment. *Catena* **39**: 147–67.

Øygarden, L. (2003) Rill and gully development during an extreme winter runoff event in Norway. *Catena* **50**: 217–42.

Parsons, A.J., Wainwright, J., Stone, P.M. & Abrahams, A.D. (1999) Transmission losses in rills on dryland hillslopes. *Hydrological Processes* **13**: 2897–2905.

Patton, P.C. & Schumm, S.A. (1975) Gully erosion, Northwestern Colorado: a threshold phenomenon. *Geology* **3**: 88–90.

Piest, R.F. & Spomer, R.G. (1968) Sheet and gully erosion in the Missouri Valley Loessial Region. *Transactions of the American Society of Agricultural Engineers* **11**: 850–53.

Poesen, J. (1993) Gully typology and gully control measures in the European loess belt. In Wicherek, S. (ed.), *Farm Land Erosion in Temperate Plains Environment and Hills*. Elsevier Science Publishers, Amsterdam: 221–39.

Poesen, J. & Bryan, R. (1989) Influence de la longueur de pente sur le ruissellement: rôle de la formation de rigoles et de croûtes de sédimentation. *Cahiers ORSTOM, Série Pédologie* **25**: 71–80.

Poesen, J. & Govers, G. (1990) Gully erosion in the loam belt of Belgium: typology and control measures. In Boardman, J., Foster, D.L. & Dearing, J.A. (eds), *Soil Erosion on Agricultural Land*. John Wiley & Sons, Chichester, UK: 513–30.

Poesen, J. & Hooke, J.M. (1997) Erosion, flooding and channel management in Mediterranean Environments of southern Europe. *Progress in Physical Geography* **21**: 157–99.

Poesen, J., Vandaele, K. & van Wesemael, B. (1998) Gully erosion: importance and model implications. In Boardman, J. & Favis-Mortlock, D.T. (eds), *Modelling Soil Erosion by Water*. Springer-Verlag, Berlin Heidelberg, NATO-ASI Series, I-55: 285–311.

Poesen, J., de Luna, E., Franca, A., *et al.* (1999) Concentrated flow erosion rates as affected by rock fragment cover and initial soil moisture content. *Catena* **36**: 315–29.

Poesen, J., Vandekerckhove, L., Nachtergaele, J., *et al.* (2002) Gully erosion in dryland environments. In Bull, L.J. & Kirkby, M.J. (eds) *Dryland Rivers: Hydrology and Geomorphology of Semi-arid Channels*. John Wiley & Sons, Chichester, UK: 229–62.

Poesen, J., Nachtergaele, J., Verstraeten, G. & Valentin, C. (2003) Gully erosion and environmental change: importance and research needs. *Catena* **50**: 91–133.

Poesen, J., Vanwalleghem, T., de Vente, J., *et al.* (2006) Gully erosion in Europe. In Boardman, J. & Poesen, J. (eds) *Soil Erosion in Europe*. John Wiley & Sons, Chichester, UK: 515–36.

Prasad, S.N. & Römkens, M.J.M. (2003) Energy formulation of head cut dynamics. *Catena* **50**: 469–88.

Prosser, I.P. (1996) Thresholds of channel initiation in historical and Holocene times, southeastern Australia. In Anderson, M.G. & Brooks, S.M. (eds), *Advances in Hillslope Processes*. John Wiley & Sons, Chichester, UK: 687–708.

Prosser, I.P. & Abernethy, B. (1996) Predicting the topographic limits to a gully network using a digital terrain model and process thresholds. *Water Resources Research* **32**: 2289–98.

Prosser, I.P. & Dietrich, W.E. (1995) Field experiments on erosion by overland flow and their implication for a digital terrain model of channel initiation. *Water Resources Research* **31**: 2867–76.

Prosser, I.P. & Slade, C.J. (1994) Gully formation and the role of valley-floor vegetation, southeastern Australia. *Geology* **22**: 1127–30.

Prosser, I.P. & Soufi, M. (1998) Controls on gully formation following forest clearing in a humid temperature environment. *Water Resources Research* **34**: 3661–71.

Prosser, I.P. & Winchester, S.J. (1996) History and processes of gully initiation and development in eastern Australia. *Zeitschrift für Geomorphologie Supplement Band* **105**: 91–109.

Radoane, M., Ichim, I. & Radoane, N. (1995) Gully distribution and development in Moldavia, Romania. *Catena* **24**: 127–46.

Rauws, G. & Govers, G. (1988) Hydraulic and soil mechanical aspects of rill generation on agricultural soils. *Journal of Soil Science* **39**: 111–24.

Reid, L.M. (1989) *Channel initiation by surface runoff in grassland catchments*. PhD thesis, University of Washington, Seattle.

Revel, J.C. & Guiresse, M. (1995) Erosion due to cultivation of calcareous clay soils on the hillslides of south west France. I. Effect of former farming practices. *Soil & Tillage Research* **35**: 147–55.

Rieke-Zapp, D., Poesen, J. & Nearing, M. (2007) Effects of rock fragments incorporated in the soil matrix on concentrated flow hydraulics and erosion. *Earth Surface Processes and Landforms* **32**: 1063–76. DOI: 10.1002/esp.1469.

Robinson, K.M. & Hanson, G.J. (1994) A deterministic headcut advance model. *Transactions of the America Society of Agricultural Engineers* **37**: 1437–43.

Rutherford, I.D., Prosser, I.P. & Davis, J. (1997) Simple approaches to predicting rates and extent of gully development. In Wang, S.S.Y., Langendoen, E.J. & Shields, F.D. (eds), *Management of Landscapes Disturbed by Channel Incision*. Proceedings, International Conference, University of Mississippi, Oxford, Mississippi, 19–23 May 1997. Center for Computational Hydroscience and Engineering, University of Mississippi: 1125–30.

Seginer, I. (1966) Gully development and sediment yield. *Journal of Hydrology* **4**: 236–53.

Sidorchuk, A. (1999) Dynamic and static models of gully erosion. *Catena* **37**: 401–14.

Sidorchuk, A. & Grigorév, V. (1998) Soil erosion on the Yamal Peninsula (Russian Arctic) due to gas field exploitation. *Advances in GeoEcology* **31**: 805–11.

Sidorchuk, A., Marker, M., Moretti, S., & Rodolfi, G. (2001) Soil erosion modelling in the Mbuluzi river catchment (Swaziland, South Africa) – part I: modelling the dynamic evolution of gullies. *Geogr. Fis. Dinam. Quat.* **24**: 177–87.

Sidorchuk, A., Marker, M., Moretti, S. & Rodolfi, G. (2003) Gully erosion modelling and landscape response in the Mbuluzi River catchment of Swaziland. *Catena* **50**: 507–26.

Smerdon, E.T. & Beasly, R.P. (1961) Critical tractive forces in cohesive soils. *Agricultural Engineering* **42**: 26–9.

Smith, L. (1992) *Investigation of ephemeral gullies in loessial soils in Mississippi*. Technical Report GL-93-11. US Army Corps of Engineers, Waterways Experiment Station, Vicksburg, MS.

Sneddon, J., Williams, B.G., Savage, J.V. & Newman, C.T. (1988) Erosion of a gully in duplex soils. Results of a long-term photogrammetric monitoring program. *Australian Journal of Soil Research* **26**: 401–8.

Soil Conservation Service (1966) Procedure for determining rates of land damage, land depreciation and volume of sediment produced by gully erosion. Technical Release 32. In FAO (ed.), *Guidelines for Watershed Management*. FAO Conservation Guide 1, Rome: 125–41.

Soil Science Society of America (2001) *Glossary of Soil Science Terms*. Soil Science Society of America, Madison, WI. Available at http://www.soils.org/sssagloss/

Souchère, V., King, D., Darroussin, J., et al. (1998) Effects of tillage on runoff directions: consequences on run-off contributing area within agricultural catchments. *Journal of Hydrology* **206**: 25–67.

Souchère, V., Cerdan, O., Ludwig, B., et al. (2003) Modeling ephemeral gully erosion in small cultivated catchments. *Catena* **50**: 489–505.

Stall, J.B. (1985) Upland erosion and downstream sediment delivery. In El-Swaify, S.A., Moldenhauer, W.C. & Lo, A. (eds), *Soil Erosion and Conservation*. Soil Conservation Society of America, Ankeny, IA: 200–205.

Stankoviansky, M. (2003) Historical evolution of permanent gullies in the Myjava Hill Land, Slovakia. *Catena* **51**: 223–39.

Steegen, A., Govers, G., Nachtergaele, J., et al. (2000) Sediment export by water from an agricultural catchment in the Loam Belt of Central Belgium. *Geomorphology* **33**: 25–36.

Stein, O.R. & Julien, P.Y. (1993) Criterion delineating the mode of headcut migration. *Journal of Hydraulic Engineering* **119**: 37–49.

Stocking, M.A. (1980) Examination of factors controlling gully growth. In De Boodt, M. & Gabriels, D. (eds), *Assessment of Erosion*. John Wiley & Sons, Chichester: 505–20.

Stocking, M.A. (1981) Causes and prediction of the advance of gullies. In *Proceedings of the South-East Asian Regional Symposium on 'Problems of Soil Erosion and Sedimentation'*, Bangkok, Thailand, 27–29 January 1981: 37–47.

Takken, I., Beuselinck, L., Nachtergaele, J., et al. (1999) Spatial evaluation of a physically-based distributed erosion model (LISEM). *Catena* **37**: 431–47.

Takken, I., Govers, G., Steegen, A., Nachtergaele, N. & Guérif, J. (2001) The prediction of runoff flow directions on tilled fields. *Journal of Hydrology* **248**: 1–13.

Takken, I., Croke, J. & Lane, P. (2008) Thresholds for channel initiation at road drain outlets. *Catena* **75**: 257–67.

Thompson, J.R. (1964) Quantitative effect of watershed variables on rate of gully-head advancement. *Transactions of the American Society of Agricultural Engineers* **7**: 54–5.

Torri, D. & Bryan, R. (1997) Micropiping processes and biancana evolution in southeast Tuscany, Italy. *Geomorphology* **20**: 219–35.

Torri, D., Poesen, J., Borselli, L. & Knapen, A. (2006) Channel width-flow discharge relationships for rills and gullies. *Geomorphology* **76**: 273–9.

Torri, D., Sfalanga, M. & Chisci, G. (1987) Threshold conditions for incipient rilling. *Catena* **8**: 97–105.

Tucker, G.E., Lancaster, S.T., Gasparini, N.M. & Bras R.L. (2001) The channel-hillslope integrated landscape development (CHILD) model. In Harmon, R.S. & Doe W.W. (eds), *Landscape Erosion and Evolution Modeling*. Springer, New York: 349–88.

US Soil Conservation Service (1966) *Procedures for determining rates of land damage, land depreciation and volume of sediment produced by gully erosion*. Technical Release 32, USDA, Washington, DC, 18 pp.

USDA Soil Conservation Service (1986) *Urban Hydrology for Small Watersheds*. Technical Release 55, 2nd edn, NTIS PB87-101580. Springfield, Virginia.

Valentin, C., Poesen, J. & Li, Y. (2005) Gully erosion: impacts, factors and control. *Catena* **63**: 132–53.

Vanacker, V., Govers, G., Poesen, J., *et al.* (2003) The impact of environmental change on the intensity and spatial pattern of water erosion in a semi-arid mountainous environment. *Catena* **51**: 329–47.

Vandaele, K., Poesen, J., Govers, G. & van Wesemael, B. (1996) Geomorphic threshold conditions for ephemeral gully incision. *Geomorphology* **16**: 161–73.

Vandaele, K., Poesen, J., Marques da Silva, J.R., *et al.* (1997) Assessment of factors controlling ephemeral gully erosion in southern Portugal and central Belgium using aerial photographs. *Zeitschrift für Geomorphologie* **41**: 273–87.

Vandekerckhove, L., Poesen, J., Oostwoud Wijdenes, D. & de Figueiredo, T. (1998) Topographical thresholds for ephemeral gully initiation in intensively cultivated areas of the Mediterranean. *Catena* **33**: 271–92.

Vandekerckhove, L., Poesen, J., Oostwoud Wijdenes, D., *et al.* (2000) Thresholds for gully initiation and sedimentation in Mediterranean Europe. *Earth Surface Processes and Landforms* **25**: 1201–20.

Vandekerckhove, L., Poesen, J., Oostwoud Wijdenes, D. & Gyssels, G. (2001) Short-term bank gully retreat rates in Mediterranean environments. *Catena* **44**: 133–61.

Vandekerckhove, L., Poesen, J. & Govers, G. (2003) Medium-term gully headcut retreat rates in Southeast Spain determined from aerial photographs and ground measurements. *Catena* **50**: 329–52.

Vanoni, V.A. & Brooks, N.H. (1975) *Sedimentation Engineering*. Manuals and Reports on Engineering Practice 54, ASCE, New York, 745 pp.

Vanwalleghem, T., Bork, H.R., Poesen, J., *et al.* (2005a) Rapid development and infilling of a historical gully under cropland, central Belgium. *Catena* **63**: 221–43.

Vanwalleghem, T., Poesen, J., Nachtergaele, J. & Verstraeten, G. (2005b) Characteristics, controlling factors and importance of deep gullies under cropland on loess-derived soils. *Geomorphology* **69**: 76–91.

Vanwalleghem, T., Poesen, J., Van Den Eeckhaut, M., *et al.* (2005c) Reconstructing rainfall and land use conditions leading to the development of old gullies. *Holocene* **15**: 378–86.

Vanwalleghem, T., Bork, H.R., Poesen, J., *et al.* (2006) Prehistoric and Roman gullying in the European loess belt: a case study from central Belgium. *Holocene* **16**: 393–401.

Vanwalleghem, T., Giráldez, J.V., Jiménez-Hornero, F.J. & Laguna, A. (2009) Evaluating a general sediment transport model for linear incisions under field conditions. *Earth Surface Processes and Landforms* **34**: 1852–7.

Verstraeten, G., Bazoffi, P., Lajczak, A., *et al.* (2006) Reservoir and pond sedimentation. In Boardman, J. & Poesen, J. (eds), *Soil Erosion in Europe*. Wiley, Chichester, UK: 759–74.

Wang, X., Zhong, X., Liu, S. & Li, M. (2008) A non-linear technique based on fractal method for describing gully-head changes associated with land-use in an arid environment in China. *Catena* **72**: 106–12.

Webb, R.H. & Hereford, R. (2001) Floods and geomorphic change in the southwestern United States: an historical perspective. *Proceedings, Seventh Federal Interagency Sedimentation Conference, 25–29 March, Reno, Nevada*: IV30–IV37.

Wemple, B.C., Jones, J.A. & Grant, G.E. (1996) Channel network extension by logging roads in two basins, western Cascades. *Water Resources Bulletin* **32**: 1195–1207.

Woodward, D.E. (1999) Method to predict cropland ephemeral gully erosion. *Catena* **37**: 393–9.

Wu, Y. & Cheng, H. (2005) Monitoring of gully erosion on the Loess Plateau of China using a global positioning system. *Catena* **63**: 154–66.

Zhang, Y., Wu, Y., Liu, B., *et al.* (2007) Characteristics and factors controlling the development of ephemeral gullies in cultivated catchments of black soil region, Northeast China. *Soil & Tillage Research* **96**: 28–41.

Part 3

Future Developments

20 The Future of Soil Erosion Modelling

M.A. NEARING[1] AND P.B. HAIRSINE[2]

[1]USDA-ARS, Southwest Watershed Research Center, Tucson, AZ, USA
[2]CSIRO Land and Water Division, Canberra, Australian Capital Territory, Australia

Niels Bohr reputedly commented that 'Prediction can be very difficult, especially about the future' (Rosovsky, 1991). Certainly that statement is true for prediction of processes in nature, including soil erosion. It also may hold true when attempting to predict the future direction of a field of research. In 1990 Nearing et al. published a paper in the *Soil Science Society of America Journal* entitled 'Soil Erosion Prediction Research Needs'. A retrospective assessment of that paper indicates that it was largely unsuccessful in outlining the important advances and changes in this field of science since it was published. One might want to keep that in mind when reading this chapter.

The Nearing et al. (1990) review of research needs was written during a time when the development of process-based soil erosion models was at the forefront of the science. This was a line of research that began sometime in the late 1960s and early 1970s (Meyer & Wischmeier, 1969), and was near its peak of effort at the time. A team of scientists from the USDA was developing the Water Erosion Prediction Project (WEPP) model (Nearing et al., 1989; Laflen et al., 1997), which relied on a numerical, steady-state solution of the sediment continuity equation, and which focused heavily on modelling inter-storm variations in the determinant system properties such as soil

erodibility, soil moisture, soil surface conditions, plant canopy and ground cover. Another team from Europe, in a project funded by the European Union, developed a model called EUROSEM (Morgan et al. 1998), which used a dynamic solution to the sediment continuity equation driven in part by a hydrological model based on the kinematic wave equation (Woolhiser et al., 1990). EUROSEM is a single storm model that focuses on infiltration, runoff and erosion from individual storms, and allows the user to define initial system conditions for storms. In Australia, Hairsine and Rose (1992a,b) developed a dynamic solution to the sediment continuity equation that encompassed what were at that time novel and important descriptions of fundamental erosion mechanics not explicitly included in other models. These were based on the concept of balancing simultaneous entrainment, deposition and re-entrainment of particles rather than relying on an independent sediment transport equation. A simplified product from this line of research was later introduced as the GUEST model (Misra & Rose, 1996; Yu et al., 1997). In the Soviet Union, Larionov (1993) had been working on yet another process-based description of soil erosion for the purposes of prediction.

Certainly there has been a great deal of value derived from the development of process-based soil erosion models, both in terms of practical application and advancement of the science. The engineers who worked on the clean-up of the Rocky Flats Superfund site in Colorado claimed

Handbook of Erosion Modelling, 1st edition. Edited by R.P.C. Morgan and M.A. Nearing. This chapter © 2011 M.A. Nearing. Published 2011 by Blackwell Publishing Ltd.

to have saved 600 million dollars by using the WEPP model for remediation design (Clark *et al.*, 2006), which was undoubtedly by any estimation a greater sum than had been spent in the development of the model itself. Hundreds of scientific studies have been conducted over the last two decades related to applying and improving process-based erosion models, including EUROSEM, WEPP, LISEM (DeRoo *et al.*, 1996), the Hairsine-Rose models, GUEST, and so on, the sum of which has resulted in a greatly improved understanding and quantification of soil erosion and sediment yields.

The four major conclusions of the Nearing *et al.* (1990) research-needs review paper were that the future of erosion modelling research would follow the paths of advancing: '...(i) fundamental erosion relationships, (ii) soil and plant parameters and their effects on erosion, (iii) databases, user interfaces, and conservation system design, and (iv) model development and analysis.' They further stated that 'Development of process-based erosion prediction technology has required the delineation and description of fundamental erosion processes and their interactions. Further improvement in prediction technology will require further delineation and mathematical descriptions.' These statements generally represent reductionist science, which was the norm at least in this area of science at the time. Both the advantages and limitations of this approach were discussed by Govers (1996), who concluded in part stating:

'The selection of priority subjects for process studies should be driven by the deficiencies between model predictions and field observations: the construction of an alternative, more sophisticated model to include an additional effect is only meaningful if a strategy can be devised which allows a validation of the model so that its presumed superiority can be proved.'

Many of the realised advances in soil erosion science and modelling over the last two decades have arisen coincidentally with our increased understanding of the limitations of the process-based soil erosion models, and many of those limitations centre around variability and uncertainties of many different types. These include issues related to, but not limited to: (a) natural variability in rates of soil loss from replicated soil erosion plots; (b) temporal variability, the importance of extreme erosion events and vulnerable site conditions, and the associated problems of interpreting short-term data records; (c) extreme spatial variability of erosion on the landscape, our lack of measured data to quantify that variability, and our limited ability to model the variability correctly; and (d) the effects of input data variability on model projections, particularly relative to cumulative model output uncertainty. The reader will find that variability and uncertainty have been common themes throughout this book, which represents a significant shift in thinking from two decades ago.

An eminent European scientist once stated that if there were to be a Nobel Prize for Soil Erosion Science, it would have to go to the study published by Wendt *et al.* (1986). In that study the authors reported soil erosion rates for 40 cultivated, fallow, experimental plots located in Kingdom City, MO, in 1981. All of the 40 plots were cultivated and in other ways treated identically. The coefficients of variation between plots for the erosion rates measured for each of 25 storms ranged from 18% to 91%. Based on the data from that study, they calculated that the 95% confidence interval for quantifying the mean erosion rate of two replicated plots for a given storm was plus or minus 175% of the mean value. In other words, the confidence interval for the mean erosion from the two plots would range from essentially zero to nearly twice the mean measured value. Also, the results of the study indicated that 'only minor amounts of observed variability could be attributed to any of several measured plot properties, and plot differences expressed by the 25 events did not persist in prior or subsequent runoff and soil loss observations at the site.' The study was suggesting that replicated plots may give greatly different measures, and variable trends, of soil erosion for all conditions being equal.

Nonetheless, a survey of the literature will show that when scientists attempt to test a new soil erosion model or the application of a model in a new environment, they almost invariably rely upon measured data for comparison with model output results. Also, invariably, the documented fact that the data have enormous natural variability is ignored: the measured data are assumed to be correct. If one were to model, for example, the 40 replicated plots from the Wendt *et al.* (1981) study, with essentially the same soil, cover and rainfall conditions, the input parameters for the model would be nearly or exactly the same for all the plots. Given that the models are essentially all deterministic in nature, the output of the model would be a single value. In that case one could see where the modelled value falls within the distribution of the 40 measured values. However, if one only has a single (or two at best) measured erosion values with which to compare, one has no idea where that measured value lies within the distribution associated with the natural measurement variability. In most cases that variability would be much larger than recognized. Nearing *et al.* (1999) provided a more universal scheme for characterizing replicated plot variability, and Nearing (2000) attempted to develop a procedure for using that information in model validation studies, but those concepts have been neither widely recognized nor implemented.

A major limitation that erosion modellers face in quantifying and comparing soil erosion rates is the lack of long-term data. The paradigm the world over for funding scientific research is the two- to five-year grant, which is a serious problem in terms of collecting long-term data. A study by Edwards and Owens (1991) found that soil erosion measurements on nine small watersheds in Ohio over 28 years were dominated by a few large storms. The five largest erosion-producing events out of more than 4000 accounted for 66% of the total erosion. On one watershed, one storm caused more than half of the 28-year total. Nearing *et al.* (2007) looked at 11 years of data from six small watersheds in the Walnut Gulch Experimental Watershed in southeastern Arizona. In each case the single largest storm on the record contributed between 9% and 11% of the total sediment yield for the 11-year period of record, and approximately 50% of the sediment yield came from between six and ten events during the 11 years. Lane and Kidwell (2003) looked at data from four small watersheds in the Santa Rita Experimental Range in southern Arizona measured over 16 years. They found that the year with the largest erosion event accounted for between 18% and 26% of the total measured sediment. This temporal variability is one reason why we need soil erosion models. Appropriately constructed, a process-based model may have the ability to extrapolate a short record of measured erosion to a longer time frame. Nonetheless, the problem is that models developed and parameterized from short records that do not contain the extreme event probably will not effectively represent the extreme event. The most likely scenario will be that the impact of the extreme event will be under-predicted. This is one area obviously ripe for further research.

Jetten *et al.* (2003) published a review on the application of models in terms of spatial distributions of erosion rates within watersheds. Not surprisingly they found that the models were able to characterize sediment yields from watershed outlets only moderately well, a result they attributed to 'the high spatial and temporal variation of erosion and sediment transport and our inability to assess and/or describe this variability in terms of the input parameters normally used in erosion models.' The models performed even more poorly in terms of characterizing the spatial erosional patterns within the watersheds: 'The application of the LISEM tested here shows that accurate predictions at the grid-cell resolution at which the model is run are impossible.' They found that the finer and more detailed the resolution for the model inputs and grid, the worse were the spatial predictions. Obtaining good spatial predictions of measured erosion requires extensive and detailed spatial datasets (van Oost *et al.*, 2004). The number of scientific papers in the literature related to modelling spatial distributions of soil erosion is relatively large, particularly with the increasing use of GIS in modelling, but studies that make any attempt to evaluate the spatial predictive capability of the models using measured data are very few.

If we have learned anything over the past two decades it is that increased model complexity does not correspond to improved capability to predict soil erosion rates and sediment yields. It is important to keep in mind, however, that improved prediction capability, in terms of improved ability to quantify erosion rates and amounts as a function of system properties and inputs, is not the only goal for the models. Models also form a structure for integrating our understanding of soil erosion processes. Complex models may also play a role in addressing some problems that simple models cannot – climate change, for example. Govers (1996) noted this in his review paper on soil erosion models, as did Williams et al. (1996) in discussions of modelling climate change impacts on soil erosion. Also, complex models do not necessarily need to remain complex in the application phase. A good example of this was the evolution of the Hairsine-Rose model framework (Hairsine & Rose, 1992a,b) to that of the GUEST model described in Chapter 11. Another example was the use of the framework of the WEPP model (Laflen et al., 1997) to the simpler, more targeted-use, and less data-intensive Rangeland Hydrology and Erosion Model (Wei et al., submitted), as well as the web-based WEPP Climate Assessment Tool (Bayley et al., in preparation).

Model complexity can lead to increased prediction uncertainty. Chapter 4 addressed the issue of model uncertainty in a great deal of detail. The most mathematically accurate, and hence common, manner to assess the propagation of input errors is with the use of Monte Carlo simulations using distributions of input parameter variation (e.g. Wei et al., 2008). Conceptually, however, the first-order error (FOE) framework (Wu et al., 2006) allows one easily to visualize error summation in the models as a function of complexity. Every input parameter for a model carries with it some degree of uncertainty, which can be expressed using FOE by using a coefficient of variation (CV). Prediction uncertainty associated with parameter definition will propagate through all models to generate some level of uncertainty in the model response, which within the FOE analysis is expressed as a CV of the model response. The degree to which the error propagates is directly proportional to the sensitivity of the model output to the model input parameter and to the input uncertainty (CV). First-order errors sum with each additional input parameter, so the decision on whether to add an additional input parameter to a model is whether or not the new process described by the equations that use the parameter adds more to prediction capability than is lost through the additional error propagated due to the uncertainty in the value of the input parameter. This is more or less equivalent to the statement attributed to Govers (1996) above.

Hairsine and Sander (2009) recently provided a further description of the trade-offs in the development of models of soil erosion by water. Figure 20.1 shows the conceptual trade off between data availability, model complexity and model performance as proposed by Grayson and Bloschl (2000) for hydrological prediction. For any application with a given level of data available, there will be an optimum level of model complexity that will allow one to reach optimum predictive performance (see bold solid line in Fig. 20.1). In order to move forward with increasing model predictive performance, model complexity must move forward hand-in-hand with data availability. To make progress, we should be constantly moving in the direction of the solid arrow in Fig. 20.1. When one is using a model that is parameter-rich and informed by a relatively small amount of data, predictive performance deteriorates due to parameter identifiability problems because the model is too complex. We contend that this is the current situation for most existing models of soil erosion and related sediment transport in most predictive environments. Thus these models plot in the bottom right-hand corner of Fig. 20.1, in which case the path to greater predictive capability is along the bold

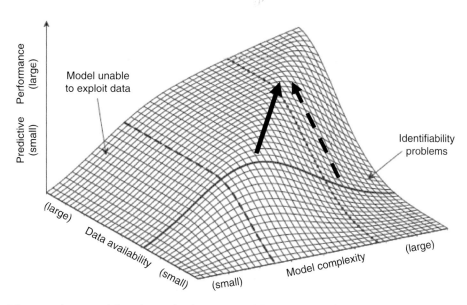

Fig. 20.1 Schematic diagram of the relationship between model complexity, data availability and predictive performance (reproduced with permission from Grayson & Bloschl, 2000). The added arrows are: the full line which is the ideal path of model development where added complexity is supported by new data; and the dashed line which is the more common scenario where new data acts to better constrain existing models.

dashed arrow: more often than not we need more and better data rather than more complex models to improve erosion predictions.

And so we come back to the subject of the chapter and ask: 'In what direction will the future of soil erosion modelling go?' The 1950s to the 1970s was the period of empirically-based erosion equation development, culminating in the second Universal Soil Loss Equation release in 1978 (Wischmeier & Smith, 1978). The 1970s to the 1990s was a period of development of process-based erosion models facilitated by computer simulation modelling. As discussed above, much of the progress during the 1990s and 2000s has focused on understanding and representing uncertainty associated with model applications. In most recent years we have seen, and (we predict) into the future we will continue to see, advances in spatial modelling and up-scaling, interfaces that utilize GIS, the increased use of remotely sensed data, and web-based delivery systems tied

to large databases for soils, topography, land use and weather. This does not mean that we will not also see advances in other aspects of the modelling. Empirical modelling has continued, for example, as evidenced by the publication of the Revised Universal Soil Loss Equation in 1997 (Renard *et al.*, 1997), and there is much work yet to be done on solving the problems associated with model and data uncertainty, as discussed above. Uncertainty will continue to play a large role in our thinking on erosion, and ideally will lead to new ways of both modelling soil erosion and thinking about how we manage land for purposes of conservation. Nonetheless, practical goals associated with management decisions will require a set of model requirements that stress data reliability and availability, ease of use, capability for routing water and sediment through watersheds, and ability to delineate primary trends in erosion rates as a function of management practices and changing climate.

The material in this book goes a long way to pointing to the probable future of soil erosion modelling. A perusal of the applications chapter of this book indicates that watershed- and basin-scale assessments have become a dominant interest to modellers (see chapters 9, 13, 14, 16, 18 and 19). As we look to larger areas and up-scaling soil erosion models (chapters 6 and 19), spatial interfaces complete with GIS will continue to play a big role in new model development. The use of the Internet as a delivery mechanism for erosion models will certainly also be important in the future (chapters 16 and 17).

Continued data collection for supporting erosion modelling will be critical. Paucity of data remains a major limitation to the development of reliable models. The types of data needed will correspond to the new emphases in the model applications. Whereas in the past, plot studies have been the basic data of the erosion modeller, in the future modellers will increasingly need spatial data on erosion patterns and sources (Walling, 2005). Fortunately, there is a wide range of sediment sourcing tools now available to test models that accumulate sediment across complex environments. Motha et al. (2004) applied a combination of minor and major element chemistry and sediment magnetic properties to assess the proportion of sediment reaching a river from different sources, including gravelled and ungravelled roads and hillslope erosion from different soil types. Rhoton et al. (2008) performed similar work in the Walnut Gulch Experimental Watershed in southeastern Arizona. On a smaller spatial scale, Polyakov et al. (2004) used rare earth element oxides to measure the erosion, redistribution and deposition of sediment in a small agricultural watershed in Ohio. Sediment tracers that differentiate between surface and subsurface soil have been used widely to assess the proportion of river sediment coming from hillslope, stream bank and gully erosion (see review by Mabit et al., 2008). In some instances these techniques have been extended to assess the contributions of rill and sheet to hillslope erosion (Wallbrink & Murray, 1993). Sediment tracers that differentiate between soil and sediment that was sourced from

land uses with specific vegetation types are under active research (Gibbs, 2008).

Sediment that is deposited can be used as an accumulated record of the transport history. Where this history is well defined, it can serve as a testing ground for predictive models run retrospectively. A key element in this approach is the association of the deposited sediment with events in the rainfall record. A wide range of evidence has been used to establish the age of recent (within the past 100 years) layers of sediment deposition. These include charcoal layers from known fires, and labelling by atmospheric events including the atmospheric testing of nuclear weapons (e.g. ^{137}Cs as reviewed by Walling and He (1999) and Walling et al. (1999), and plutonium as introduced by Everett et al. (2008)). Until recently, many of the sediment dating techniques used in geochronology and anthropology were not applicable to the last 100 years. The advent of optically stimulated luminescence and other techniques has enabled the assessment of the age of modern deposited sediment and residence times of sediment in fluvial systems (Gale, 2009). Sediment dating data are now available for recent agricultural history (Olley et al., 1998). The application of single-grained, optically stimulated luminescence has now extended this development to small samples and ages greater than 2 years (Gale, 2009; Pietsch, 2009).

Each of these forms of sediment dating and tracing techniques serve to provide further data to our models of sediment transport. Specifically, we can test the question of 'Are we getting the right answer for the right reason?' in terms of when the sediment was transported, from what soil type it was sourced, what were the eroding processes (sheet, rill or gully), and what land use was in place at the source of the sediment.

Integrating the use of remote-sensing data into the erosion modelling process has the potential to offer an opportunity to verify independently and provide initial conditions to models, but also to change the way we conceptualize modelling of erosion. Two key methodologies include the determination of effective vegetative soil cover and the direct sensing of sediment concentration in surface waters. Remote sensing has long been used to assess vegetative cover at regional to global scales. There is increasing use of libraries of time series of images to assess time variations in cover (Lu *et al.*, 2003). These time series permit assessment of spatial and temporal variations of cover inputs to models and the evaluation of hindcast predictions from crop and pasture models. Limitations in the classifications of some forms of cover, specifically bleached dead vegetation, have been resolved by Guerschman *et al.* (2009). Remotely sensed interpretations of soil moisture and soil surface roughness could also be important mechanisms for informing erosion models (Rahman *et al.*, 2008). At the catchment scale there are some prospects for the use of remote sensing in estimating the concentration of surface water sediment concentrations directly. A further form of remote sensing with ramifications for erosion modelling is the development and availability of rainfall radar data (Steiner *et al.*, 1995). These data could be made useful in providing more spatially and temporally realistic inputs of rainfall rates than do conventional and often sparse networks of rain gauges. Where all of these techniques could have the advantage of providing a spatial-rich source of data for evaluation and identifying the initial conditions of models, near real-time remote sensing combined with models also opens up the prospects of data–model fusion where initial conditions for models, rainfall rate inputs and parameter estimation are continuously and automatically updated. These techniques have been extensively developed in terrestrial and ocean biogeochemical models (e.g. Barrett *et al.*, 2005).

Ultimately, the model builder must combine lines of evidence to assess and improve models. To do this well the product should track the ridge indicated by the solid arrow in Fig. 20.1. The use of multiple lines of evidence will result in more robust models that engender more confidence for use in predictive environments. A case study of adaptive changes to a model and the consequent prediction of spatial erosion processes was provided by Rustomji *et al.* (2008). This study showed that default parameters from a national assess-

ment could be significantly improved by local information so as to give a progressive refinement of sediment source maps used to target management actions.

Underlying the discussion above is also the increasing understanding that 'Stationarity is Dead' (Milly *et al.*, 2008). We live in a rapidly changing world with respect to both land use and climate (Chapter 15). As population increases, stresses on land resources will continue to increase, often in those areas of the world that are already severely stressed. When we add to that trends of increasing rainfall amounts (Karl & Knight, 1998), rainfall intensities (Groisman *et al.*, 2005), and rainfall erosivity (Nearing, 2001), it becomes evident that our erosion models must be able to represent a world of changing land use and a changing climate.

References

Barrett, D.J., Hill, M.J., Hutley, L.B., *et al.* (2005) Prospects for improving savanna biophysical models by using multiple-constraints model-data assimilation methods. *Australian Journal of Botany* **53**: 689–714.

Bayley, T.W., Nearing, M.A., Guertin, D.P., *et al.* (In preparation) The Water Erosion Prediction Project Climate Assessment Tool: predicting soil erosion in the face of changing precipitation regimes.

Clark, D.L., Janecky, D.R. & Lane, L.J. (2006) Science-based cleanup of Rocky Flats. *Physics Today* **59**(9): 34–40.

De Roo, A.P.J., Wesseling, C.G. & Ritsema, C.J. (1996) LISEM: A single-event physically based hydrological and soil erosion model for drainage basins. 1: Theory, input and output. *Hydrological Processes* **10**: 1107–17.

Edwards, W.M. & Owens, L.B. (1991) Large storm effects on total soil erosion. *Journal of Soil and Water Conservation* **46**: 75–8.

Everett, S.E., Timsa, S.G., Hancock, G.J., *et al.* (2008) Comparison of Pu and ^{137}Cs as tracers of soil and sediment transport in a terrestrial environment. *Journal of Environmental Radioactivity* **99**: 383–93.

Gale, S.J. (2009) Quaternary dating the recent past. *Geochronology* **4**: 374–7.

Gibbs, M.M. (2008) Identifying source soils in contemporary estuarine sediments: A new compound-specific isotope method. *Estuaries and Coasts* **31**: 344–359. DOI 10.1007/s12237-007-9012-9

Govers, G. (1996) Soil erosion process research: a state of the art. *Academiae Analecta* 58(1). Mededelingen van de Koniklijke Academie voor Wetenschappen, Letteren en Schone Konsten van België, Brussels.

Grayson, R. & Bloschl, G. (2000) Spatial processes, organisation and patterns. In Grayson, R. & Bloschl, G. (eds), *Spatial Patterns in Catchment Hydrology: Observations and Modelling*. Cambridge University Press, Cambridge.

Groisman, P.Y., Knight, R.W., Easterling, D.R., *et al.* (2005) Trends in intense precipitation in the climate record. *Journal of Climatology* **18**: 1326–50.

Guerschman, J.P., Hill, M.J., Renzullo, L.J., *et al.* (2009) Estimating fractional cover of photosynthetic vegetation, non-photosynthetic vegetation and bare soil in the Australian tropical savanna region upscaling the EO-1 Hyperion and MODIS sensors. *Remote Sensing of the Environment* **113**: 928–45.

Hairsine, P.B. & Rose, C.W. (1992a) Modeling water erosion due to overland flow using physical principles: I. Sheet Flow. *Water Resources Research* **28**: 237–43.

Hairsine, P.B. & Rose, C.W. (1992b) Modeling water erosion due to overland flow using physical principles: 2. Rill flow. *Water Resources Research* **28**: 245–50.

Hairsine, P.B & Sander, G.C. (2009) Comment on "A transport-distance based approach to scaling erosion rates": Parts 1, 2 and 3 by Wainwright *et al. Earth Surface Processes and Landforms* **34**: 882–5. DOI: 10.1002/esp.1782

Jetten, V., Govers, G. & Hessel, R. (2003) Erosion models: quality of spatial predictions. *Hydrological Processes* **17**: 887–900.

Karl, T.R. & Knight, R.W. (1998) Secular trend of precipitation amount, frequency, and intensity in the United States. *Bulletin of the American Meteorological Society* **79**: 231–42.

Laflen, J.M., Elliot, W.J., Flanagan, D.C., *et al.* (1997) WEPP-Predicting water erosion using a process-based model. *Journal of Soil and Water Conservation* **52**: 96–102.

Lane, L.J. & Kidwell, M.R. (2003) Hydrology and soil erosion. *USDA Forest Service Proceedings* RMRS-P-30: 92–100.

Larionov, G.A. (1993) *Erosion and Wind Blown Soil*. Moscow State University Press, Moscow, 200 pp.

Lu, H., Raupach, M.R., McVicar, T.R. & Barrett, D.J. (2003) Decomposition of vegetation cover into woody and herbaceous components using AVHRR NDVI time series. *Remote Sensing of the Environment* **86**: 1–18.

Mabit, L., Benmansour, M. & Walling, D.E. (2008) Comparative advantages and limitations of the fall-out radionuclides Cs-137, Pb-210(ex) and Be-7 for assessing soil erosion and sedimentation. *Journal of Environmental Radioactivity* **99**: 1799–1807.

Meyer, L.D. & Wischmeier, W.H. (1969) Mathematical simulation of the process of soil erosion by water. *Transactions of the American Society of Agricultural Engineers* 12: 754–8, 762.

Milly, P.C.D., Betancourt, J., Falkenmark, M., *et al.* (2008) Stationarity is dead: whither water management. *Science* 319: 573–4.

Misra, R.K. & Rose, C.W. (1996) Application and sensitivity analysis of process-based erosion model GUEST. *European Journal of Soil Science* **47**: 593–604.

Morgan, R.P.C., Quinton, J.N., Smith, R.E., *et al.* (1998) The European Soil Erosion Model (EUROSEM): a dynamic approach for predicting sediment transport from fields and small catchments. *Earth Surface Processes and Landforms* **23**: 527–44.

Motha, J.A., Wallbrink, P.J., Hairsine, P.B. & Grayson, R.B. (2004) Agricultural lands and unsealed roads as sources of suspended sediment in a predominantly agricultural water supply catchment in south-eastern Australia. *Journal of Hydrology* **286**: 1.

Nearing, M.A. (2000) Evaluating soil erosion models using measured plot data: accounting for variability in the data. *Earth Surface Processes and Landforms* **25**: 103–43.

Nearing, M.A. (2001) Potential changes in rainfall erosivity in the United States with climate change during the 21st century. *Journal of Soil and Water Conservation* **56**: 229–32.

Nearing, M.A., Foster, G.R., Lane, L.J. & Finkner, S.C. (1989) A process-based soil erosion model for USDA-Water Erosion Prediction Project technology. *Transactions of the American Society of Agricultural Engineers* **32**: 1587–93.

Nearing, M.A., Lane, L.J., Alberts, E.E. & Laflen, J.M. (1990) Prediction technology for soil erosion by water: status and research needs. *Soil Science Society of America Journal* **54**: 1702–11.

Nearing, M.A., Govers, G. & Norton, L.D. (1999) Variability in soil erosion data from replicated plots. *Soil Science Society of America Journal* **63**: 1829–35.

Nearing, M.A., Nichols, M.H., Stone, J.J., *et al.* (2007) Sediment yields from unit-source semi-arid watersheds at Walnut Gulch. *Water Resources Research* **43**: W06426. DOI: 10.1029/2006WR005692

Olley, J.M., Caitcheon, G. & Murray, A.S. (1998) The distribution of apparent dose as determined by

Optically Stimulated Luminescence in small aliquots of fluvial quartz: implications for dating young sediments. *Quaternary Geochronology* **17**: 1033–40.

Pietsch, T.J. (2009) Optically stimulated luminescence dating of young (<500 years old) sediments: testing estimates of burial dose. *Quaternary Geochronology* **4**: 406–22.

Polyakov, V.O., Nearing, M.A. & Shipitalo, M. (2004) Tracking sediment redistribution in a small watershed: implications for agro-landscape evolution. *Earth Surface Processes and Landforms* **29**: 1275–91.

Rahman, M.M., Moran, M.S., Thomas, D.P., *et al.* (2008) Mapping surface roughness and soil moisture using multi-angle radar imagery without ancillary data. *Remote Sensing of the Environment* **112**: 391–402.

Renard, K.G., Foster, G.R., Weesies, G.A., *et al.* (1997) *Predicting soil erosion by water – a guide to conservation planning with the revised universal soil loss equation (RUSLE).* Agricultural Handbook No. 703, US Government Printing Office, Washington, DC.

Rhoton, F.E., Emmerich, W., DiCarlo, D.A., *et al.* (2008) Identification of suspended sources using soil characteristics in a semiarid watershed. *Soil Science Society of America Journal* **72**: 1102–12.

Rosovsky, H. (1991) *The University: an Owners Manual.* Norton & Co., New York, 312 pp.

Rustomji, P., Caitcheon, G., & Hairsine, P.B. (2008) Combining a spatial model with geochemical tracers and river station data to construct a catchment sediment budget. *Water Resources Research* **44**: W01422. DOI: 10.1029/2007WR006112

Steiner, M., Houze, R.A. & Yuter, S.E. (1995) Climatological characterization of 3-dimensional storm structure from operational radar and rain-guage data. *Journal of Applied Meteorology* **34**: 1978–2007.

Van Oost, K., Beuselinck, L., Hairsine, P.B. & Govers, G. (2004) Spatial evaluation of a multi-class sediment transport and deposition model. *Earth Surface Processes and Landforms* **29**: 1027–44.

Wallbrink, P.J. & Murray, A.S. (1993) Use of fallout radionuclides as indicators of erosion processes. *Hydrological Processes* **7**: 297–304.

Walling, D.E. (2005) Tracing suspended sediment sources in catchments and river systems. *Science of the Total Environment* **344**: 159–84.

Walling, D.E. & He, Q. (1999) Improved models for estimating soil erosion rates from cesium-137 measurements. *Journal of Environmental Quality* **28**: 611–22.

Walling, D.E., He, Q. & Blake, W. (1999) Use of Be-7 and Cs-137 measurements to document short- and medium-term rates of water-induced soil erosion on agricultural land. *Water Resources Research* **35**: 3865–74.

Wei, H., Nearing, M.A., Stone, J.J. & Breshears, D.D. (2008) A dual Monte Carlo approach to estimate model uncertainty and its application to the Rangeland Hydrology and Erosion Model. *Transactions of the American Society of Agricultural and Biological Engineering* **51**: 515–20.

Wei, H., Nearing, M.A., Stone, J.J., *et al.* (Submitted) A Rangeland Hydrology and Erosion Model. *Rangeland Ecology and Management.*

Wendt, R.C., Alberts, E.E. & Hjelmfelt, A.T. Jr. (1986) Variability of runoff and soil loss from fallow experimental plots. *Soil Science Society of America Journal* **50**: 730–36.

Williams, J., Nearing, M.A., Nicks, A., *et al.* (1996) Using soil erosion models for global change studies. *Journal of Soil and Water Conservation* **51**: 381–5.

Wischmeier, W.H. & Smith, D.D. (1978) *Predicting rainfall erosion losses: a guide to conservation planning.* USDA Agriculture Handbook No. 537. USDA-SEA, US Government Printing Office, Washington, DC.

Woolhiser, D.A., Smith, R.E. & Goodrich, D.C. (1990) *KINEROS: a Kinematic Runoff and Erosion Model: Documentation and User Manual.* USDA Agricultural Research Service ARS-77.

Wu, J.R., Zou, Z. Rui & Yu, S.L. (2006) Uncertainty analysis for coupled watershed and water quality modeling systems. *Journal of Water Resources Planning and Management ASCE* **132**: 351–61.

Yu, B., Rose, C.W., Ciesiolka, C.A.A., *et al.* (1997) Toward a framework for runoff and soil loss prediction using GUEST technology. *Australian Journal of Soil Research* **35**: 1191–1212.

Index